D0419406

CONTENTS

Preface vii

SECTION 1. GENERAL INTRODUCTION

1. Introduction 3
2. Physiological Anatomy of Plant Groups in Relation to Disease . . 7

SECTION 2. DISEASES OF CEREALS AND GRASSES

3. Barley Diseases 23
4. Corn Diseases 74
5. Millet Diseases 115
6. Oat Diseases 122
7. Rice Diseases 153
8. Rye Diseases 173
9. Sorghums, Sudan Grass, and Johnson Grass Diseases 188
10. Sugarcane Diseases 206
11. Wheat Diseases 225
12. Grass Diseases 293

SECTION 3. DISEASES OF LEGUMES

13. Alfalfa and Sweetclover Diseases 335
14. Clover Diseases 366
15. Soybean Diseases 383
16. Peanut Diseases 396
17. Diseases of Other Legume Crops 401

SECTION 4. DISEASES OF FIBER AND OTHER FIELD CROPS

18. Cotton Diseases 409
19. Flax Diseases 428
20. Tobacco Diseases 443

Appendix A. Diseases of Field Crops Arranged by Causal Factor (Suggested for a Three-hour Course) 469
Appendix B. Bacteria and Fungi Parasitic on Field Crops Arranged under Orders and Families with More Common Suscepts 472

Index 491

Appreciation is expressed for the cooperation and assistance of the many investigators who have supplied information and who have assisted in its interpretation.   Much of this has been supplied during personal conferences with individuals and groups during the past several years in which this revision was in preparation.

JAMES G. DICKSON

tendency toward the specialization of agricultural investigations of crop plants argues further for the presentation of the diseases on this basis as a convenient reference volume for the investigator. The student using the volume as a text in diseases of field crops will acquire, perhaps, a better comprehension of the disease and more stimulation toward individual thinking by reference to several chapters for the information on a given disease. For the convenience of the teacher and as a guide to the student, the diseases are regrouped in the appendix on the basis of (1) the primary causal factor, a suggested list for class presentation on this basis, and (2) a list of the bacteria and fungi arranged by order and family. The diseases are arranged under each crop plant on the basis of the primary causal factor as follows: (1) nonparasitic, (2) virus, (3) bacterial, (4) phycomycetous, (5) ascomycetous, (6) hypomycetous (fungi-imperfecti), and (7) basidiomycetous. The detailed discussion of a disease occurring on several crop plants is given under one crop only, and cross reference is made to this discussion in the other chapters.

The international Botanical Congress, at the Stockholm meeting in 1953, approved the proposal to use the binomials for both the perfect and imperfect stages of fungi. To comply with this change in rules, the binomial for the imperfect stage of the pathogens has been removed from synonymy and indicated as the conidial stage of the pathogen.

The taxonomic status of the numerous bacterial and fungal pathogens as expressed in the binomials used for their identification is natural only in accordance with the basic information available at the time they are described and named. Therefore, new binomials are appearing in the literature as the scope of basic morphological and developmental information advances. The resultant changes in binomials appear chaotic to the initiate; however, they express the natural, dynamic status of classification. The recording of these several binomials for the same pathogen in logical sequence offers a challenge to the teacher and the author, especially during the transition period when several names are in common use for a given species. The dogmatic attitude argues for the arbitrary selection of a binomial on the best evidence available and on long usage. The dynamic approach suggests the presentation of the several binomials in use in order to enable the reader to comprehend fully this changing status in biology. In the present compilation, the several binomials are given even at the risk of creating some confusion. This appears essential in an accurate compilation of information for reference use.

Illustrations contributed by the various cooperators are indicated in the legends. The remainder are from original photographs taken largely by Eugene Herrling of the department of plant pathology of the University of Wisconsin.

PREFACE

The purpose of this volume is to present in as brief a space as feasible the important information pertaining to diseases of the field crops. Investigations on the diseases of this group of economic plants and upon the crop plants themselves are continually contributing new facts and modifying former points of view; therefore, a volume of this type can never be complete for the relatively large number of diseases included. During the past decade more basic knowledge and specific information has accumulated on diseases of field crops than in any like period in the 30 years since the initiation of this compilation. The study of virus diseases has evolved into a major branch of plant pathology and a number of new diseases of field-crop plants have been described. The pathology and distribution of nematodes and their relation to the expression of diseases incited by soil-inhabiting bacteria and fungi have been defined. The genetical investigations of bacteria and fungi have advanced to the factorial interpretation of developmental morphology, chemical and physiological processes of nutrition, and variability and mutation. Virulence and avirulence of physiological races of pathogens are expressed on a factor basis and compared with factors conditioning susceptibility and resistance in the hosts. In the transition interim, the classification of physiological races of many of the pathogens of field crops is in a state of change; therefore, tables and keys concerned with physiological races have been omitted for many of the pathogens. The number of genetic factors discovered that condition resistance to disease has increased greatly. The use of interspecific hybrids in exploiting these factors in practical disease control has passed from theoretical consideration to common use. Knowledge of the inheritance of these factors in practical disease control has passed from theoretical consideration to common use. Knowledge of the inheritance of these factors, their isolation, and their recombination in genotypes readily usable for breeding and basic investigations is in progress.

The diseases are listed on the basis of the primary cause of the disease under the crop plants included. While many of the diseases occur on more than one crop plant, the economic importance and varietal reactions within the various crop species differ considerably. The rather general

To My Son

DISEASES OF FIELD CROPS

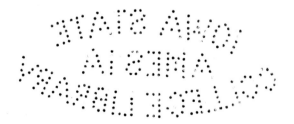

THE MAPLE PRESS COMPANY, YORK, PA.

DISEASES OF FIELD CROPS

JAMES G. DICKSON, Ph.D.

Professor of Plant Pathology, University of Wisconsin
Agent, Section of Cereal Crops, Division of Field Crops
Agricultural Research Service, U.S. Department of Agriculture

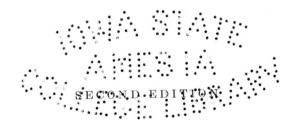
SECOND EDITION

McGRAW-HILL BOOK COMPANY, INC.

New York   Toronto   London

1956

McGRAW-HILL PUBLICATIONS IN THE
AGRICULTURAL SCIENCES
R. A. BRINK, *Consulting Editor*

DISEASES OF FIELD CROPS

McGRAW-HILL PUBLICATIONS IN THE AGRICULTURAL SCIENCES

R. A. Brink, *Consulting Editor*

GENERAL INTRODUCTION

CHAPTER 1

INTRODUCTION

The history and development of field crops have been associated with plant diseases. Throughout the years, one or another plant malady has threatened the economy of crop production, with the result that the malady was brought under control or the crop shifted to other environments. Economic pressures frequently have been involved in these adjustments. The history of these battles has been an interesting one, and it has illustrated the ability of man and the adaptability of plants. Undoubtedly the same battle existed in the past when subsistence farming was the problem of the family, although generally the conflict was neither so spectacular nor so broad in its implications. With the advent of commercial farming and the subsequent production of large continuous acreages of the same variety of a given crop and the product's entry into commerce, these adjustments in crop production assumed regional and frequently national or international significance. The history of plant pathology also has been shaped by such conflicts.

Diseases of the cereal crops were among the first studied. The early studies of plant diseases were dominated by dogma and tradition. Knowledge was vague regarding the nature of disease, and the role of microorganisms was unknown. Many of the diseases of crop plants were attributed to weather conditions, frequently a conspicuous factor associated with the unhealthy condition of the crop. Later, when fungi were observed to be associated with diseased plants, they were interpreted, at first, as the result rather than the cause of the unhealthy conditions. For a period emphasis was placed upon the collection and classification of the fungi. In the latter part of the eighteenth century and continuing into the early part of the nineteenth century, increasing emphasis was placed on systematic mycology. Bulliard, Persoon, Nees von Esenbeck, Schweinitz, Léveille, Fries, and Berkeley contributed much to the early classification of the fungi. During the latter half of the nineteenth century Fuckel, Karsten, the Tulasnes, Corda, Saccardo, Thaxter, and many others advanced the mycological study.

Economic pressure for more food in the heavily populated European countries was indirectly the force that stimulated the investigation of

3

diseases and their control. Prévost and later De Bary proved by experimentation the nature of parasitism, the epidemic character of fungus diseases, the importance of methods of inoculation and infection, and the practical application of disease control. These two investigators founded the science of phytopathology. Kühn was among the first to organize a general, practical attack on the diseases of cultivated crops with special emphasis on disease control. During the latter half of the nineteenth century, publications were numerous. Among many others who contributed special service to plant pathology or indirectly through mycological publications were Brefeld, Frank, Hartig, Schroeter, Sorauer, and Winter in Germany; Oudemans in Holland; Cornu, Millardet, and Prillieux in France; Comes in Italy; Woronin in Russia; Eriksson, Henning, and Jensen in Sweden; Massie, Plowright, and Ward in England; Burrill, Clinton, Farlow, Halstead, Hitchcock, Kellerman, Selby, Swingle, and many others in the United States.

The investigations on diseases of field crops in the United States started early with the development of the U.S. Department of Agriculture and the state agricultural experiment stations. Much of this early work was directed more especially toward immediately practical problems of disease control. Seed treatments, sanitation, disease reaction of varieties of the different cereals, and study of the various fungi constituted the earlier investigations. The rediscovery of Mendel's laws of heredity and their application to genetics soon offered new techniques for disease control. The difference in disease reaction of specific varieties and specialization of the fungus parasites first recognized by Eriksson and Ward and later studied in detail by Salmon, Freeman, Stakman, Reed, and others directed attention to disease resistance as a control measure. Orton and Bolley working with the wilt diseases developed the early techniques and proved the practicability of disease resistance as a control measure. Nilsson-Ehle and Biffin soon after proved that the characters for rust resistance were inherited on a definite factorial basis. The advances in this field during the past three decades have placed disease resistance as of first importance in the control of diseases and the improvement of field crops.

The study of the life cycle of the fungi, especially in relation to the fusion of gametes and the genetical implications of these fusions to variation in pathogenicity, has become increasingly important in the development of basic plant pathology. The recent use of the factorial interpretation of developmental morphology, nutritional physiology, and physiological specialization in microorganisms has supplied new methods for experimental investigations in plant pathology. Demonstrations of the inheritance of these factors on a Mendelian basis have resulted in the use of genetic techniques in determining the degree of heterozygosity of a given isolate or inbreeding lines homozygous for specific factors before employing them in experimentation. Physiological specialization has been placed on the basis of factors conditioning virulence or avirulence in the pathogen, rather than on factors conditioning susceptibility or resistance in the host. The isolation of these factor pairs and their incorporation into specific genotypes, or a common genotype which differs only in a factor pair so closely linked that it functions with the factor, have been accomplished for some few pathogens and for several field-crop plants. The value of such material in investigations on the potentialities of the pathogen and its control by the use of resistant varieties has been accepted, although not fully comprehended.

The investigations on the influence of environmental conditions upon the development of the diseases of field crops have expanded largely since Sorauer's early work. The earlier advances in general plant physiology and the study of the normal physiology of cultivated plants led naturally to the study of the abnormal physiology of diseased plants. The environmental complex that predisposes the plant to attack in the case of parasitic diseases or induces abnormal functioning or development in nonparasitic diseases was studied especially by Ward and Blackman in England, by L. R. Jones and associates in the United States, and more recently by many others. The application of special environments to the production of epiphytotics has materially increased the accuracy of the tests for disease resistance and the factorial analysis of resistance. With the more intensive cultivation of field crops the so-called "deficiency diseases" have become of increasing importance.

The virus diseases first studied by Mayer, Iwanowski, and Beijerinck and demonstrated as transmissible by insects by Takami and E. D. Ball are becoming of increasing importance in the pathology of field crops. Improved techniques in transmission and identification of viruses on various plants have demonstrated their presence in most of the cultivated crop plants. The rapidly increasing list of virus diseases and their effect upon plant yield has demonstrated fully their importance in economical field-crop production.

In the plant virus disease the investigator is concerned apparently with a less highly organized pathogen, perhaps a chemical structure of high molecular order that increases in the proper host protoplasm. The virus then spreads systemically or remains localized. However, disease development appears to be dependent upon factors similar to those influencing bacterial and fungal diseases. Frequently a third biological entity, the insect vector of the virus, is involved in the complex.

The science of plant pathology is young. New diseases of cultivated plants appear. Diseases of little economic importance shift to serious maladies when a susceptible crop variety is grown or a new physiologic race of the pathogen develops. Disease epiphytotics are local or distributed widely; they increase to cause alarming losses, then subside as control measures are applied or seasonal environments change. Disease control helps to prevent these fluctuations in crop yield, but rarely eliminates the pathogen. Disease control through disease resistance depends largely upon a better understanding of the genetics of the pathogen and host. Factors conditioning resistance are found, isolated, and recombined into agronomically adapted genotypes of the crop plants. Experience suggests a series of these economic genotypes, each with a factor for resistance to a specific factor for virulence in the pathogen. Resistance to other pathogens added to these genotypes in the same manner results in a reserve of varieties usable for planned rotation in production or special emergency control measures. The development and use of fungicides and insecticides represent an additional important means of disease control. New chemicals functioning as systemics and modern machinery for their application are possible economical control measures of the future. The two methods of control, the reserve of resistant genotypes and chemical control, combined, represent the practical objective of plant pathology.

CHAPTER 2

PHYSIOLOGICAL ANATOMY OF PLANT GROUPS IN RELATION TO DISEASE

Disease development and, conversely, disease resistance are determined in part by morphological structures and physiological processes in the plants involved. The morphology and physiology of the plant are governed by the genetic complement (genotype) operating within the plant structures (phenotype) in the external environment. In other words, the expression of the complex of genetic factors through structure (morphology) and function (physiology) are conditioned within limits by the external environment. The importance of a basic understanding and experienced familiarity with the physiological anatomy of the plants upon which the disease occurs is essential in the comprehension of disease development and in the application of disease control. In the nonparasitic diseases, the investigator is concerned with the plant and its reaction to excesses or deficiencies in the environment. In the diseases caused by bacteria and fungi, the interplay of two plants, the host and pathogen, and the conditioning environment are involved in disease development. Or stated in more detail, the potential capacity of the bacterium or fungus to incite disease is dependent upon genetic constitution expressed through morphological structures and physiological processes and the environmental complex, including the suscept, under which the potential parasite is developing. Likewise the susceptibility, tolerance, or resistance of the suscept is conditioned by a similar complex. The presence of natural avenues of entrance for the parasite, the existence of tissue barriers, the type of embryonic development, the rate of tissue maturation, and the presence or absence of suitable nutrient complexes or single compounds are illustrative of a few of the anatomical and physiological factors that determine the course of disease development, economic importance, and control. Therefore it is essential that the plant pathologist be familiar with the physiological anatomy of the suscepts as well as the morphology and physiology of the pathogen.

1. Physiological Anatomy of the Gramineae. The developmental anatomy of the cereals and grasses is associated with disease development and type of disease in several phases of growth and maturation. Space does not permit a comprehensive discussion of the subject.   However, reference should be made to Avery (1930), Haberlandt (1914), Hayward (1938), McCall (1934), Percival (1921), and others for the developmental anatomy of the Gramineae.

The mature grain is a caryopsis.   The kernel either threshes free from the floral bracts as in most wheats, or it is enclosed in the lemma and palea, as for example, in most barley and oat varieties.   The caryopsis consists of the adherent pericarp and remains of the integuments and nucellus, the protective tissues enclosing the endosperm, and the embryo embedded against the endosperm.   A semipermeable membrane of lipoid composition is deposited on the outer surface of the nucellar tissue and the inner epidermis of the inner integument.   At maturity the membrane is continuous on these compressed residual tissues and extends through the conductive tissue of the chalaza.   The membrane (cuticle) over the back of the scutellum is joined with this semipermeable membrane, sealing in the epithelial surface in contact with the endosperm. The role of this differentially permeable membrane system is discussed by Brown (1907), Dickson and Shands (1941), Tharp (1935), and others. The pericarp tissues when moist constitute a suitable medium for many of the fungi parasitic on the cereals and grasses.   The semipermeable membrane functions not only in holding soluble reserves within and preventing the entrance of most chemicals into the kernel, but also as a barrier to prevent fungi and bacteria in the pericarp entering the endosperm and embryo, as described by Johann (1935), Pugh et al. (1932), and others.   Mechanical injury of the membrane results in rapid invasion of the endosperm reserves and embryonic tissues by soil microorganisms during the early stages of germination.   Parasites invading through the ovary wall, as the fungi causing ergot and the loose smuts, must enter early after pollination, or this avenue is barred.   The erect position and compact nature of the spike in many of the Gramineae tend to increase fungus and bacterial invasion of the pericarp as free moisture is held in the floral bracts enclosing the caryopsis, resulting in conditions favorable for invasion of the basal portion of the kernel by blight and rot organisms.   In grain enclosed permanently in the floral bracts, the lemma and palea provide a protective covering for spores and mycelium held between the pericarp and bracts.   The inoculum in this position is in close proximity to the embryonic tissues of the young seedling, and infection occurs before the tissues become resistant.   The attached floral bracts protect the seed-borne inoculum from the direct action of fungicides, as trapped air prevents complete wetting of the inner surfaces. Consequently, the volatile compounds are more effective as seed fungicides in such grains than toxic substances acting by direct contact of the solutions with the fungi and bacteria.

The physiological anatomy and development of the seedling of the

Gramineae frequently are favorable for disease development. The anatomy of the seedling and differences in development in the various genera of the family are discussed fully by Avery (1930), McCall (1934), Sargent and Arber (1915), and others. Brown (1936), Jones *et al.* (1926), Kolk (1930), Pearson (1931), and others describe the avenues of entrance and tissues invaded as well as the physiology of disease development. The composition and maturation of seedling tissues, especially as influenced by environment, are important in relation to disease development and the expression of resistance to seedling blights, seedling smut infection, and other diseases. The emergence of the seminal roots by digestion and mechanical pressure through the coleorhiza and especially the cortex of the lower internode and the cotyledonary node offers avenues of entrance to cortical-invading fungi. This is especially significant in the seedling type illustrated by corn, in which internodal elongation occurs below the coleoptilar node or in tissues unprotected by the coleoptile (Fig. 1).

The role of the coleoptile as a protective sheath is important. Seedling infection by the smut fungi is largely through the coleoptile while it is very young. The tissues soon mature and prevent entrance into the enclosed growing point. The formation of crown and tiller primordia early in the seedling development is important in the establishment of the "systemic" type of infection common in the cereal and grass smuts.

The structure and development of the basal tissues of the culm, or crown, are important in the development of crown rot, foot rot, and stalk rot diseases. In winter cereals and perennial grasses the crown is the important structure associated with winterkilling. The crown consists of secondary culms arising from axillary buds at the basal nodes of the main axis and lateral culms. Under ordinary environments this complex of branches occurs in the upper inch of soil. In certain of the perennial grasses, new culm primordia that develop from the axillary buds require one or two seasons before culm elongation occurs (Bond, 1940). This delayed culm development apparently is associated with the time interval between infection and spore development in certain of the grass smuts. The crown roots arise from the pericyclic region of the first internode and in the intercalary meristems at the base of the lower internodes. The crown roots rupture the cortex by pressure and digestion. If wound response and deposition of suberin, lignin, or lipoid substances in the adjacent cortical cell walls are delayed, the root ruptures offer avenues of entrance for soil-infesting organisms. Insect injuries in these tissues also offer channels of entrance of crown rot parasites. Stuckey (1941) discusses seasonal growth of the grass roots, and Weaver (1926) published good descriptions of the root systems of the cereal crops.

The anatomy and development of the culm are such as to function in protecting the primordia and embryonic tissues. Newton and Brown

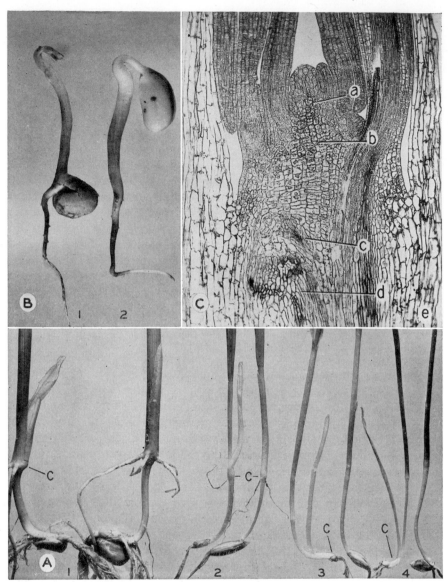

Fɪɢ. 1. Seedling development and internodal elongation in cereals and legumes. (A) Seedlings of corn, oats, barley, and wheat, illustrating the two common types of internodal elongation in relation to the coleoptilar node (c): corn (1) and oats (2) in which the internode between the cotyledonary and coleoptilar nodes elongates, leaving the internode unprotected by the coleoptile, and barley (3) and wheat (4) in which the internodal elongation is above the coleoptilar node or enclosed within the protective sheath.   The coleoptile is split and pulled aside in the left-hand seedling of each pair to show the base and the internode enclosed.   The development of seminal and crown roots is shown also.   (B) Pea (1) and soybean (2) seedlings, illustrating the two types of seedling development in the Leguminosae.   (C) Longisection of the oat seedling before internodal elongation showing (a) culm primordia, (b) differentiating crown nodes and internodes, (c) coleoptilar node, (d) scutellar node, and (e) the coleoptile.

(1934), Griffiths (1928), Zehner and Humphrey (1929), and others have shown that introduction of spores of rust fungi and of the corn smut fungus into the differentiating embryonic tissues results in a general heavy infection. Inoculum entering the funnel at the top of the leaf whorl frequently comes in contact with embryonic or susceptible tissues and results in infection. The leaf sheath with the ligule tightly encircling the culm at the juncture of the sheath and the base of the leaf blade prevents inoculum from reaching the younger tissues of the basal part of the culm internode and the axillary bud. The position of the leaf blade and the basal region of elongation frequently results in leaf blotch diseases, killing the entire blade.

The type of inflorescence and development of the reduced flowers varies in the family. Generally the compact spike is damaged more by blights and rots than the open panicle. The floral bracts protect the young ovary from flower-infecting organisms. The cross-pollinated species in which the floral bracts are open longer are damaged more by ergot than the self-pollinated species. Humid, wet weather prolongs the period of open floral bracts in both self- and cross-pollinated species, as well as providing conditions favorable for germination and infection by the parasite and, therefore, usually results in higher infections with ergot, loose smut, and kernel blights. Dead anther, pollen, and stylar tissues all furnish excellent nutritive media for the initial development of many of the blight and rot fungi, and these organisms, once established as saprophytes, invade the adjacent living tissues by contact infection to produce disease.

The development of exterior and interior membranes in association with tissue maturity or wound response are important in water economy, gaseous exchange, and the invasion of parasitic organisms. The development of such membranes in the Gramineae is not greatly different from that in the other families of crop plants. Therefore the discussion of this important phase of physiological anatomy and that of natural openings in relation to disease development will be presented later in the chapter.

2. Physiological Anatomy of the Leguminosae. The leguminous crops included in the discussion of field-crop diseases are largely perennial or biennial in growth habit. The annual crop plants of this family differ somewhat in root and stem development from the perennial or biennial types. However, anatomically, they are sufficiently similar to be included in the general discussion for this family. Compton (1912), Hayward (1938), Jones (1928), Lute (1928), Martin (1914), Pammel (1899), Wilson (1913), and Winton (1914) discuss various phases of the structure and morphology of this group.

The seeds are borne in pods, the latter varying in shape and structure.

The pod tissues frequently are invaded by fungi.   The infections in many instances extend through the wall, rotting the pod and enclosed seeds or resulting in seed infection.   The epidermis of the seed in most members of this family consists of a row of palisade cells varying in cell-wall thickness, permeability, and durability in different species and under different environments.   Apparently the walls of the palisade cells function in a manner somewhat similar to the semipermeable membrane in the caryopsis of the Gramineae.   Mechanical injury such as scarification overcomes the impermeability but frequently increases seedling loss by cracking the seed coat.   Breaking the seed coat in harvesting and handling the seed increases the losses from seed rots and seedling blights during the early stages of germination.   The seed at maturity consists of the two cotyledons in which the reserves are stored and the embryonic epicotyl and hypocotyl all enclosed within the integuments.   Parasites established in the seed coat or in the cotyledons or gaining entrance to the young seedling through cracks in the seed coat during the very early stages of germination cause severe damage.

In germination, the primary root emerges near the hilum or in the micropylar area.   In the subsequent development, two types of germination are represented in the family: (1) epigeal, in which the cotyledons, usually enclosed within the seed coat, are pushed above the soil surface by the rapid elongation and straightening of the hypocotyl; (2) hypogeal, in which the hypocotyl elongation is limited, the cotyledons remain in position within the seed coat in the soil, and the epicotyl, more fully differentiated prior to germination, pushes up through the soil (Fig. 1). In this latter type of development, seedling blight and lesions below the soil level are the common types of disease.   In the former, seedling blight during the very early stages of germination occurs, followed by a later lesioning or "damping-off" from invasion of the hypocotyl near the soil surface or by a blighting of the embryonic epicotyl.   Disease lesions on the cotyledons develop, and fungus sporulation frequently is abundant.

Root development consists of a primary taproot with numerous laterals.   Weaver (1926) illustrated the root system of several members of this family.   Elongation, branching, and secondary thickening of the root system is rapid and rather continuous during the first growing season.   As described by Fred et al. (1932), nodule formation is abundant during this period of active root growth.   Secondary cambial roots develop during the following season.   In both biennials and perennials, noncambial smaller root branches develop in the spring of the second and in the succeeding years in the case of perennials, as described by Jones (1943).   These apparently function in increasing absorbing area during the period of rapid spring growth and deteriorate during the summer to reappear in limited numbers in the fall.   Decay is severe in these transient roots during the summer period, at least in the vicinity of Madison, Wis.

Jones (1928), Jones and McCulloch (1926), Jones and Weimer (1928), Weimer (1927), and others have described winter injury in the older roots and crown and the relation of these injuries to disease development, especially in alfalfa. The capacity of the plant to recover from such injuries through the formation of phellogen and periderm or by wall thickening and the deposition of suberin-like substances into the walls is related to both crown rots and bacterial-wilt development.

The crown in the older plants consists of the basal stem branches and the axillary and secondary buds. The development of the crown is influenced by environmental conditions and age of the plant. Diseases also modify the number of buds developing as in crown wart and witches' broom of alfalfa. The annual stems developing from the crown buds are generally angular in shape with a fairly large outer layer of collenchyma tissue. Secondary thickening and changes in turgor in the collenchyma frequently result in longitudinal splitting of the epidermis. These openings unless rapidly closed by periderm serve as avenues of entrance for fungi and bacteria.

The stem elongation and development of the inflorescence in many legumes are affected by insects. Species of aphids and leaf hoppers cause severe damage through retarded apical bud development and yellowing of the foliage. The blighting of the flower buds and dropping of flowers are associated with these and other insects.

3. **Physiological Anatomy of Other Dicotyledonous Plants.** The physiological anatomy of cotton, flax, and hemp is discussed in detail by Hayward (1938). As in the legumes, seedling structure and development are associated closely with disease development. Both seed-borne and soil-borne organisms cause blighting, lesioning, and damping-off during early seedling development. In cotton and flax, the structure, composition, and development of the seed are associated with seed infection and damage by microorganisms. The industrial quality of the fibers is frequently damaged by disease in the more humid climates. Conant (1927), Jewett (1938), and Johnson (1924) in the discussion of tobacco diseases gave consideration to structure and development of the plant in relation to disease.

4. **General Discussion of Plant Defense Mechanisms.** The crop plants in economical production are, generally speaking, the result of long years of selection and adaption to environment. In the relatively short-lived herbaceous plants, the reaction to major diseases plays an increasingly important role in the selection and survival of a variety. The cost of protective fungicidal sprays and dusts is too great in field-crop economy to permit their general use; therefore crop-plant develop-

ment is based upon selecting disease-escaping, disease-tolerant, or disease-resistant varieties. These adaptations in plants generally are associated with gradual or abrupt changes in morphology and physiology. The nature of these changes, their dependence upon the general or more specific genetic composition of the plant for their stability, and the relation of composition and physiology of the plant as a whole or of specific tissues for their expression are important factors in the economy of the plant species or variety. The difference between susceptible and resistant plants may not be primarily dependent upon a particular structure or substance, but rather upon the more basic metabolic or maturation processes or upon some inherent quality or characteristic that functions even more directly. Certain of these morphological and physiological characteristics apply more or less generally to all plant species, and therefore they are presented briefly at the end of this chapter.

Natural openings are important avenues of entrance for certain organisms producing disease in plants. In field crops, the stomata are probably the most important, as hydathodes and lenticels are less extensively developed than in other plant groups. The rust fungi constitute the large group of parasites entering the tissues by way of the stomata. Although extent of infection is associated perhaps with stomatal size and the period the stomata are open, as shown by Hart (1929, 1931) and Peterson (1931), the more basic type of resistance to the rusts is associated with cellular physiology and composition.

The rate and character of cuticle formation on the epidermal surfaces of the plant are important in disease development. A relatively large number of fungi are capable of penetrating the young cuticle. In general, the resistance to mechanical penetration increases with the maturation and thickening of this surface membrane. Priestley (1921, 1943) summarized the literature and suggested the aggregate nature of cutin and suberin. Blackman and Wellsford (1916), Brown and Harvey (1927), Dey (1919), Leach (1923), Young (1926), and others have studied the penetration of the cuticular membrane by fungi. This surface membrane is important in the water economy of the plant, as discussed by Maximov (1929) and Priestley and coworkers (1922, 1923, 1930). The cuticle and surface structures, hairs, etc., prevent wetting and uniform distribution of water over the exposed tissue surfaces. This phenomenon is important in distribution of inoculum and entrance through natural openings as well as direct surface penetration by plant parasites. Apparently the quantity and quality of cuticle formation are associated with the physiological maturity not only of the epidermal cells, but also of those of the underlying tissues. The deposition of the similar aggregate of substances, suberin, on or in the walls of cells within certain tissues is associated with the physiology of the cells.

Suberized cell walls function not only in water economy, but also in retarding or checking the advance of parasitic organisms. Periderm, which serves as a protective tissue after the loss of epidermis and cortex, contains this suberin aggregate. The periderm, whether derived from phellogen of pericyclic origin, secondary phloem parenchyma, or from the subepidermal cortical parenchyma, is similar in its function and the general presence of suberin. The endodermis is impregnated with suberin as the roots and underground stems mature. The radial and end walls of the endodermal cells are suberized first, followed frequently by suberization of the tangential walls. Collenchyma and sclerenchyma cell walls frequently contain suberin as the tissues approach maturity. Lignification also is commonly associated with many of these thickened walls. The development of these protective and strengthening tissues occurs in the sequence of tissue differentiation, development, and maturation. Environment, however, influences the rate and frequently the extent of wall thickening.

The rate and type of thickening of cellular structures are activated frequently by wound response. The regular sequence in development of these anatomical structures apparently is hastened and expanded by mechanical injuries, invasion of parasites, etc. Such injuries usually stimulate cell-wall thickening and suberization, periderm formation, or ultimately both, in cells of cambium, parenchyma, pericycle, and less frequently collenchyma. An active type of defense mechanism is stimulated in such responses. The role of such a defense mechanism is described by Conant (1927), Dickson et al. (1923), Fellows (1928), Jones (1928), Pearson (1931), Weimer and Harter (1921), and others. The response in the corn root by the deposition of suberin-like substances in the cortical cells around secondary-root ruptures and the thickening of the endodermis and suberization of cells of the pericycle and cortex in advance of fungus invasion illustrate this type of reaction.

The composition of the cell walls and the middle lamellae affects the rate of advance of the parasite and tissue necrosis. The changes in composition associated with maturation of many tissues are correlated apparently with susceptibility or resistance to disease. Harlow (1932), Ritter (1925), and others suggested the change in composition of the middle lamella, in woody tissues at least, from pectin-like compounds in the embryonic state to lignin upon maturity. Brown (1915, 1916, 1917), DeBary (1886), Hawkins and Harvey (1919), Jones (1909), Pearson (1931), and Ward (1888) discussed the general physiology of intercellular advance of fungi and bacteria in both vegetative and storage tissues. Gäumann (1927), Ritter and Fleck (1926), Sponsler and Dore (1926), and others suggested that the degree of hydration of the cellulose molecular aggregates and the associated substances, lignin and suberin, is correlated with tissue maturation.   The physical and chemical changes in the cell membranes are a direct or indirect response of the cell protoplasm.   The rate of these changes is influenced by the genetic constitution and the influence of the environment upon the plant metabolism.

The physical and chemical properties of the cell protoplasm are associated directly with plant defense mechanisms.   Cold resistance and drought resistance are due partly to the composition and physical state of the cell protoplasts, as discussed by Akermann (1927), Harvey (1935), Levitt (1941), Maximov (1929), and Newton and Martin (1930).   Specific compounds existing in the plant cells or more commonly present in less active molecular combinations are important in the defense mechanism of the cell, as discussed by Walker (1924, 1929, 1941).   The cell protoplasts are concerned directly in the nutrition of the obligate parasite. In this group of parasites the hyphae or specialized branches, haustoria, enter the cells of the suscept and function in a balanced type of metabolism during the cycle of fungus development.   The living cells of the suscept, apparently influenced by the stimulus of the parasite, supply the essential nutrients for the balanced metabolism of both.   In the obligate parasite, as illustrated by the rusts and powdery mildew fungi, this is accomplished at the expense of the suscept.   In the symbionts, as in the nodule bacteria, there is a mutual beneficial relationship.   Parasitism of this type is usually very specialized.   Physiologic races of the fungi are restricted to species or even varieties in the plants capable of functioning in this compatible physiological or nutritional relationship.   Conversely, disease resistance is concerned directly with the incompatible relation of the two protoplasts.   The fungus hyphae coming in contact with the suscept cells or protoplasm either cease to develop further, or, more commonly, the protoplasts of the contacted cells of the suscept are disorganized, leaving the parasite isolated from living suscept tissue.   This response is described by Allen (1923, 1926, 1927), Gibson (1904), Humphrey and Dufrenoy (1944), Stakman (1914), Ward (1902, 1905), and others.   The development of disease-escaping, disease-tolerant, or disease-resistant plants is concerned with the morphology and physiology of the suscept plants and, in the case of parasitic diseases, the parasite.

The variation of the parasite in its pathogenic characteristics is equally important.   The basis of variation in fungi, bacteria, and viruses is still in the exploratory stage of development.   Some variation is due to segregation following nuclear fusion, some to mutation, and some to nuclear reassortment.   Variation is expressed in physiological changes and less commonly in alterations in morphology.   Flor (1954), Johnson (1953), and Stakman (1940) summarize the information on variability in this group of lower plants.

Insect damage and the relation of anatomy, physiology, and composition of field-crop plants to insect attack parallel somewhat that of plant diseases. Leach (1940) discusses the relation of insects to the transmission of diseases and gives a good summary of many types of insect damage. Metcalf and Flint (1951) present some of the more important insect problems of field crops. Resistance to insect attack and damage is important in field-crop economy as well as in disease control.

REFERENCES

AKERMANN, A. Studien über den Kaltetod und die Kalteresistanz der Pflanzen. *Birlingska Boktryckeriet.* Lund. 1927.

ALLEN, R. F. Cytological studies of infection of Baart, Kanred and Mindum wheats by *Puccinia graminis tritici* forms 3 and 19. *Jour. Agr. Research* **24**:571–604. 1923.

———. Cytological studies of forms 9, 21 and 27 of *Puccinia graminis tritici* on Khapli Emmer. *Jour. Agr. Research* **32**:701–725. 1926.

———. A cytological study of orange leaf rust, *Puccinia triticina* physiologic form 11 on Malakoff wheat. *Jour. Agr. Research* **34**:697–714. 1927.

AVERY, G. S., JR. Comparative anatomy and morphology of embryos and seedlings of maize, oats, and wheat. *Bot. Gaz.* **89**:1–39. 1930.

BLACKMAN, V. H., and E. J. WELSFORD. Studies in the physiology of parasitism. II. Infection of *Botrytis cinerea*. *Ann. Bot.* **30**:389–398. 1916.

BOND, T. E. F. Observations on the disease of sea lyme-grass (*Elymus arenarius* L.) caused by *Ustilago hypodytes* (Schlecht.) Fries. *Ann. Appl. Bot.* **27**:330–337. 1940.

BROWN, A. J. On the existence of a semipermeable membrane enclosing the seed of the Gramineae. *Ann. Bot.* **21**:79–87. 1907.

BROWN, W. Studies in the physiology of parasitism. I. The Action of *Botrytis cinerea*. *Ann. Bot.* **29**:313–348. 1915.

———. Studies in the physiology of parasitism. III. On the relation between the "infection drop" and the underlying host tissue. *Ann. Bot.* **30**:399–406. 1916.

———. Studies in the physiology of parasitism. IV. On the distribution of cytase in cultures of *Botrytis cinerea*. *Ann. Bot.* **31**:489–498. 1917.

———. The physiology of host-parasite relations. *Bot. Rev.* **2**:236–281. 1936.

——— and C. C. HARVEY. Studies in the physiology of parasitism. X. On the entrance of parasitic fungi into the host plant. *Ann. Bot.* **41**:643–662. 1927.

COMPTON, R. H. An investigation of the seedling structure in the Leguminosae. *Jour. Linn. Soc. (London) Bot.* **41**:1–122. 1912.

CONANT, G. H. Histological studies of resistance in tobacco to *Thielavia basicola*. *Am. Jour. Bot.* **14**:457–480. 1927.

DE BARY, A. Ueber einige Sclerotinien und Sclerotienkrankheiten. *Bot. Ztg.* **44**:377–474. 1886.

DEY, P. K. Studies in the physiology of parasitism. V. Infection by *Colletotrichum lindemuthianum*. *Ann. Bot.* **33**:305–312. 1919.

DICKSON, J. G., et al. The nature of resistance to seedling blight of cereals. *Proc. Nat. Acad. Sci.* **9**:434–439. 1923.

——— and H. L. SHANDS. Cellular modification of the barley kernel during malting. *Am. Soc. Brewing Chemists Proc.* 1–10. 1941.

FELLOWS, H. Some chemical and morphological phenomena attending infection of the wheat plant by *Ophiobolus graminis*. *Jour. Agr. Research* **37**:647–661. 1928.

FLOR, H. H.   Flax rust in seed flax improvement.   *Adv. Agron.* **6**:152–161.   1954.

FRED, E. B., *et al.*   Root nodule bacteria and leguminous plants.   *Univ. Wis. Studies in Science* No. 5.   1932.

GÄUMANN, E.   Der Einfluss der Keimungstemperatur auf die chemische Zusammensetzung der Getreidekeimlinge.   *Zeitschr. Botan.* **25**:385–461.   1932.

GIBSON, C. M.   Notes on infection experiments with various Uredineae.   *New Phytologist* **3**:184–191.   1904.

GRIFFITHS, M. A.   Smut susceptibility of naturally resistant corn when artificially inoculated.   *Jour. Agr. Research* **36**:77–89.   1928.

HABERLANDT, G.   Physiological plant anatomy.   Translation by M. Drummond. Macmillan & Company, Ltd.   London.   1914.

HARLOW, W. M.   The chemical nature of the middle lamella.   *N.Y. State Col. Forestry Tech. Pub.* 21.   1927.

HART, H.   Relation of stomatal behavior to stem rust resistance in wheat.   *Jour. Agr. Research* **39**:929–948.   1929.

————.   Morphologic and physiologic studies on stem rust resistance in cereals.   *U.S. Dept. Agr. Tech. Bul.* 266.   1931.

HARVEY, R. B.   An annotated bibliography of the low temperature relations of plants. Burgess Publishing Co.   Minneapolis.   1935.

HAWKINS, L. A., and L. B. HARVEY.   Physiological study of the parasitism of *Pythium debaryanum* on potato tuber.   *Jour. Agr. Research* **18**:275–297.   1919.

HAYWARD, H. E.   The structure of economic plants.   The Macmillan Company. New York.   1938.

HUMPHREY, H. B., and J. DUFRENOY.   Host-parasite relationship between the oat plant (Avena spp.) and crown rust (*Puccinia coronata*).   *Phytopath.* **34**:21–40. 1944.

JEWETT, F. L.   Relation of soil temperature and nutrition to the resistance of tobacco to *Thielavia basicola*.   *Bot. Gaz.* **100**:276–297.   1938.

JOHANN, H.   Histology of the caryopsis of yellow dent corn, with reference to resistance and susceptibility to kernel rots.   *Jour. Agr. Research* **51**:855–883.   1935.

JOHNSON, J.   Tobacco diseases and their control.   *U.S. Dept. Agr. Bul.* 1256.   1924.

JOHNSON, T.   Variation in the rusts of cereals.   *Biol. Rev.* **28**:105–157.   1953.

JONES, F. R.   Winter injury of alfalfa.   *Jour. Agr. Research* **37**:189–211.   1928.

————.   Development of the bacteria causing wilt in the alfalfa plant as influenced by growth and winter injury.   *Jour. Agr. Research* **37**:545–569.   1928.

————.   Growth and decay of the transient (noncambial) roots of alfalfa.   *Jour. Am. Soc. Agron.* **35**:625–634.   1943.

———— and L. McCULLOCH.   A bacterial wilt and root rot of alfalfa caused by *Aplanobacter insidiosum* L. Mc.   *Jour. Agr. Research* **33**:493–521.   1926.

———— and J. L. WEIMER.   Bacterial wilt and winter injury of alfalfa.   *U.S. Dept. Agr. Cir.* 39.   1928.

JONES, L. R.   Pectinase, the cytolytic enzyme produced by *Bacillus carotovorus* and certain other soft rot organisms.   *N.Y. (Geneva) Agr. Exp. Sta. Tech. Bul.* 11. 1909.

———— *et al.*   Wisconsin studies upon the relation of soil temperature to plant disease.   *Wis. Agr. Exp. Sta. Res. Bul.* 71.   1926.

KOLK, L. A.   The relation of host and pathogen in the oat smut *Ustilago avenae*.   *Bul. Torrey Bot. Club* **57**:443–499.   1930.

LEACH, J. G.   The parasitism of *Colletotrichum lindemuthianum*.   *Minn. Agr. Exp. Sta. Tech. Bul.* 14.   1923.

————.   Insect transmission of plant diseases.   McGraw-Hill Book Company, Inc. New York.   1940.

LEVITT, J. Frost killing and hardiness of plants. Burgess Publishing Co. Minneapolis. 1941.

LUTE, A. M. Impermeable seed of alfalfa. *Colo. Agr. Exp. Sta. Bul.* 326. 1928.

MARTIN, J. H. Comparative morphology of some Leguminosae. *Bot. Gaz.* **58**:154–167. 1914.

MAXIMOV, N. A. Internal factors of frost and drought resistance in plants. *Protoplasma* **7**:259–291. 1929.

———. The plant in relation to water. English translation by R. H. Yapp. George Allen & Unwin, Ltd. London. 1929.

McCALL, M. A. Developmental anatomy and homologies in wheat. *Jour. Agr. Research* **48**:283–321. 1934.

METCALF, C. L., and W. P. FLINT. Destructive and useful insects, their habits and control. McGraw-Hill Book Company, Inc. New York. 1951.

NEWTON, M., and A. M. BROWN. Studies on the nature of disease resistance in cereals. I. The reactions to rust of mature and immature tissues. *Can. Jour. Research* C **11**:564–588. 1934.

NEWTON, R., and W. M. MARTIN. Physico-chemical studies on the nature of drought resistance in crop plants. *Can. Jour. Research* C **3**:336–427. 1930.

PAMMEL, L. H. Anatomical characters of seed of Leguminosae, chiefly genera of Grey's Manual. *Trans. Acad. Sci. St. Louis* **9**:91–274. 1899.

PEARSON, N. L. Parasitism of *Gibberella saubinetii* on corn seedlings. *Jour. Agr. Research* **43**:569–596. 1931.

PERCIVAL, J. The wheat plant; a monograph. Gerald Duckworth & Co., Ltd. London. 1921.

PETERSON, R. F. Stomatal behavior in relation to the breeding of wheat for resistance to stem rust. *Sci. Agr.* **12**:155–173. 1931.

PRIESTLEY, J. H. Suberin and cutin. *New Phytologist* **20**:17–29. 1921.

——— et al. Physiological studies in plant anatomy. *New Phytologist* **21**:58–80, 113–139, 210–229, 252–268. 1922. **22**:30–44. 1923.

———. Studies in the physiology of cambial activity. *New Phytologist* **29**:56–73, 96–140, 316–354. 1930.

———. The cuticle in Angiosperms. *Bot. Rev.* **9**:593–616. 1943. (Good literature list.)

PUGH, G. W., et al. Relation of the semipermeable membranes of the wheat kernel to infection by *Gibberella saubinetii*. *Jour. Agr. Research* **45**:609–626. 1932.

RITTER, G. J. Distribution of lignin in wood. *Jour. Ind. Eng. Chem.* **17**:1194–1197. 1925.

——— and L. C. FLECK. Chemistry of wood. IX. Spring wood and summer wood. *Jour. Ind. Eng. Chem.* **18**:608–609. 1926.

SARGENT, E., and A. ARBER. The comparative morphology of the embryo and seedling in the Gramineae. *Ann. Bot.* **29**:161–222. 1915.

SPONSLER, O. L., and W. H. DORE. The structure of Ramies cellulose as derived from X-ray data. *Colloid Symp. Monogr.* **4**:174–262. 1926.

STAKMAN, E. C. A study in cereal rusts: Physiological races. *Minn. Agr. Exp. Sta. Bul.* 138. 1914.

———. The genetics of pathogenic organisms. *Am. Assoc. Adv. Science Pub.* 12. 1940.

STUCKEY, I. H. Seasonal growth in grass roots. *Am. Jour. Bot.* **28**:486–491. 1941.

THARP, W. H. Developmental anatomy and relative permeability of barley seed coats. *Bot. Gaz.* **47**:240–271. 1935.

WALKER, J. C. On the nature of disease resistance in plants. *Trans. Wis. Acad. Sciences, Arts, Letters* **21**:225–247. 1924.

————. Some remarks on the physiological aspects of parasitism. *Proc. Int. Cong. Plant Sci.* **2** :1263–1270. 1929.

————. Disease resistance in vegetable crops. *Bot. Rev.* **7** :458–506. 1941.

WARD, W. MARSHALL. A lily disease. *Ann. Bot.* **2** :319–382. 1888.

————. On the relations between host and parasite in the bromes and their brown rust, *Puccinia dispersa* Erikss. *Ann. Bot.* **16** :233–315. 1902.

————. Recent researches on the parasitism of fungi. *Ann. Bot.* **19** :1–54. 1905.

WEAVER, J. E. Root development of field crops. McGraw-Hill Book Company, Inc. New York. 1926.

WEIMER, J. L. Observations on some alfalfa root troubles. *U.S. Dept. Agr. Cir.* 425. 1927.

———— and L. L. HARTER. Wound-cork formation in the sweet potato. *Jour. Agr. Research* **21** :637–647. 1921.

WILSON, O. T. Studies on the anatomy of alfalfa (*Medicago sativa* L.). *Kans. Univ. Sci. Bul.* 7. 1913.

WINTON, K. B. Comparative histology of alfalfa and clovers. *Bot. Gaz.* **57** :53–63. 1914.

ZEHNER, M. G., and H. B. HUMPHREY. Smuts and rusts produced in cereals by hypodermic injection of inoculum. *Jour. Agr. Research* **38** :623–627. 1929.

DISEASES OF CEREALS AND GRASSES

CHAPTER 3

BARLEY DISEASES

The cultivated barleys comprise chiefly two species, *Hordeum vulgare* L., six-rowed, and *H. distichon* L., two-rowed barleys. Varieties of the former species are grown more extensively than the latter. In the genus *Hordeum*, as in *Triticum* and *Avena*, the basic chromosome number is seven pairs. Multiples of this basic number occur in the wild species. The cultivated barleys all have seven pairs of chromosomes. Barley has been used extensively in the study of linkage relations, including resistance to certain diseases, as summarized by Robertson, Wiebe, and Shands (1955).

The cultivated barleys are grouped into three classes based on the character of growth: winter, intermediate, and spring. Winter and spring barleys comprise the major economic groups (Åberg and Wiebe, 1946). The winter barleys are not winter-hardy; therefore they are grown in a belt extending east and west across the United States approximately south of 42° latitude. They occupy similar areas with mild winters throughout the world. In the extreme Southern United States and especially the Southwest, fall-sown spring barleys predominate. Spring varieties occupy the important barley acreage in the Northern United States and Canada and similar areas throughout the world.

Barley is grown under a wide range of environmental conditions, chiefly for grain, although it is used as a forage crop in limited sections. The crop is most productive in regions of cool climate during the growing season and in well-drained finer silt or clay soils. The wide range in growth characters, period required for development, and barley types largely account for its very extensive distribution throughout world agriculture (Carlton, 1920, Harlan, 1936, and Weaver, 1943).

The developmental anatomy of the barley plant is relatively similar to wheat (Chap. 2). The development of brown pigmentation in association with tissue injury and necrosis results in disease symptoms somewhat different from those in wheat and oats.

Barley diseases cause large losses in both yield and quality of grain. Estimated average annual losses in the United States for the 10-year period 1930 through 1939 amounted to 7.1 per cent of the crop, or over

14 million bushels (Plant Disease Survey). Two years in which scab occurred in epiphytotic form and one of stem rust are included in the 10-year period. Losses from barley stripe are much lower in the United States and Canada than prior to the general use of resistant varieties and organic mercury-seed treatments.

1. Nonparasitic Diseases. The nonparasitic disturbances in barley are manifested by leaf yellowing, spotting, striping, or irregular necrotic areas and frequently the incomplete emergence of the spike from the boot. The symptoms expressed in nonparasitic diseases frequently are similar to those incited by viruses, especially in the Gramineae.

Barley is moderately tolerant to high salt concentration although not especially drought-tolerant (Hayward and Wadleigh, 1949). Leaf spotting and general leaf necrosis are symptoms expressed in response to both excesses and deficiencies of certain mineral elements, notably boron and copper (Christensen, 1934, Eaton, 1944, Schropp, 1940). Plant culm elongation and spike emergence are influenced by unbalanced mineral nutrients and specific mineral deficiencies. Sterility in the basal or apical spikelets frequently is associated with drought or low temperatures during critical periods of flower development. Frost-damaged kernels with irregularly filled endosperm occur frequently in the colder climates.

2. Virus Diseases. Two virus diseases of the mosaic type are common in both winter and spring barleys and some other cereals and grasses. Apparently both diseases have occurred during the past thirty years, but their description, importance, and etiology have been defined only recently.

Stripe mosaic, first shown to cause losses up to 30 per cent in the barley varieties Glacier (C.I. 6976),[1] Plush (C.I. 6093), and Redman (C.I. 12638) wheat, was described earlier on barley as false stripe. The disease is distributed widely, and many of the barley and wheat varieties are susceptible. The barley stripe mosaic occurs on wheat, rye, and many grasses. Symptoms on barley include chlorotic—later brown—mottled stripes in the leaf blade, especially near the base, reduced internodal elongation of the culms, and increase in tiller development in some varieties (Hagborg, 1954) (Fig. 2). Chevron (C.I. 1111) barley inoculated in the seedling stage shows marked chlorosis and chlorotic to brown stripes, stunting, and excessive tillering. This variety is used for identifying and indexing the virus (McKinney, 1954). The virus is systemic and seed-borne in the hosts studied. The virus is transmitted mechanically by leaf rubbing, by pollen, and probably by insect vectors. Rod-shaped particles are present in all parts of infected plants (Gold *et al.*,

[1] Accession numbers of the Section of Cereal Crops, U.S. Department of Agriculture.

FIG. 2. Barley stripe mosaic, a virus-incited disease (*A*) contrasted with barley stripe (*B*) incited by *Helminthosporium gramineum*.

1954).   Indexing seed for freedom from virus, hot-water treatment of small lots of seed, sanitation, and resistant varieties (C.I. 3212, C.I. 3212-1) are the best means of control.

Yellow dwarf of the cereals and grasses was described and aphid transmission demonstrated in 1953 (Oswald and Houston, 1953).   The disease is distributed widely in both winter and spring barley areas.   Apparently the disease was present earlier in varying amounts on especially barley

Fig. 3. Symptoms of the yellow dwarf virus on barley showing plant dwarfing (A) and leaf symptoms (B).

and oats, but it was included in the complex of nonparasitic maladies common to these crop plants. Susceptible barley varieties show bright-yellow blotches, golden yellow developing from the leaf tip through the entire leaf blade, and dark-green stripes extending into the yellow in the color-transition portion of the blade (Fig. 3). The early infections result in extreme stunting and excessive tillering of the plants, limited or no spikes emerge, and reduced root development is common (Fig. 3). Later infections produce the leaf-yellowing symptoms, limited spike development, and reduced kernel formation and filling. The disease frequently appears first along the margins of the fields. The development of the symptoms of the disease on oats is similar to those described for barley, but the yellow-green blotches rapidly turn yellow or red to reddish brown (see Chap. 6). The symptoms on wheat are dark-green stripes extending into the chlorotic or yellow areas of the leaf blade, reduced tillering, and chlorotic, dwarfed plants when infection occurs early. Wheat infected after the tillering stage shows little reduction in culm elongation (see Chap. 11). No leaf mottle, characteristic of the other mosaic diseases, occurs on any of the cereals. Apparently the virus is not transmitted mechanically. Seed or soil transmission has not been demonstrated. Aphid transmission has been demonstrated with the corn aphid, *Rhopalosiphum* (*Aphis*) *maidis* (Fitch); apple-grain aphid, *R. fitchii* (Sand) [*R. prunifoliae* (Fitch)]; grain aphid, *Macrosiphum granarium* (Kirby); grass aphid, *M. dirhodum* (Walker), and greenbug, *Toxoptera graminum* (Rond.). Spherical particles have been shown to be associated with the virus-infected plants.[1] Nonviruliferous aphids, especially greenbugs, feeding on barley cause injury; therefore caution is necessary both in transmission studies and in determining the disease syndrome. Economical control of the yellow dwarf is difficult. Many commercial varieties of barley and oats are susceptible. Date of seeding timed to avoid aphid-population buildup during the seedling stage of plant development is important; early seeding in spring barley, late in fall-sown grain, and use of insecticides are advised. Resistant and tolerant varieties offer a possible means of control. The four barleys C.I. 1227, 1237, 2376, and Abate are rated resistant in California, but appear susceptible in Wisconsin.

3. Bacterial Blight, *Xanthomonas translucens* (L. R. Jones, A. G. Johnson, & Reddy) Dowson. The bacterial blights of the cereals and grasses are differentiated into two groups. (1) The bacterial colony is in a gelatinous matrix and advances between the cells of the tissues of the suspect. This group is relatively numerous. (2) The bacterial colony develops without the gelatinous matrix and is localized somewhat in tissue cavities with water soaking and chlorosis of cells surrounding the colony. The bacterial exudate is conspicuous on the lesions in the former and absent in the the latter. The bacterial blight of barley is typical of the first.

The bacterial blight occurs widely on many of the economic varieties and some wild *Hordeum* species. A similar disease occurs on the other

[1] Communication from Dr. A. H. Gold, University of California.

cereals and many grasses (Wallin, 1946).   The disease is distributed widely throughout North and South America (Fang *et al.*, 1950, Jones *et al.*, 1917) as well as northern Europe and Asia (Jaczewski, 1935). Bacterial blight is more prevalent through the spring barley areas where usually it is of minor importance, although local or general epiphytotics on very susceptible varieties are reported as reducing yields of foliage and grain.

SYMPTOMS AND EFFECTS.   Small linear water-soaked areas, frequently quite numerous on localized areas of the leaf blade and sheath, develop after several days of rainy, damp weather.   These lesions elongate and coalesce into irregular narrow glossy-surfaced stripes.   The stripe frequently shows water-soaked, light-yellow, light-brown, and dark-brown regions, depending upon the age of the coalesced lesions (Fig. 4).   The center of the lesion is translucent in the later stages of development. Minute drops of white resinous exudate or a thin film of the exudate are characteristic on the lesion surface.   Numerous lesions usually result in a slow yellowing and death of the leaf blade progressing from the apex downward.   Similar lesions develop on the leaf sheath and floral bracts. Severe late infections usually result in retarded spike elongation and in abundant exudate and blighting of the spike and adjacent tissues. Lesions on the kernels are small and inconspicuous.   The characteristic symptom is the narrow, linear, translucent, and glossy lesion.

THE ORGANISM

Xanthomonas translucens (L. R. Jones, A. G. Johnson, & Reddy) Dowson.[1]

[*Phytomonas translucens* (L. R. Jones, A. G. Johnson, & Reddy) Bergey *et al.*]

(*Bacterium translucens* L. R. Jones, A. G. Johnson, & Reddy)

[*Pseudomonas translucens* (L. R. Jones, A. G. Johnson, & Reddy) Stapp.]

The cylindrical rod-shaped bacteria with rounded ends are motile by means of polar flagella, and no spores are formed.   The bacterial colonies develop in a gelatinous matrix.   Specialized varieties and races occur on the cereals and grasses (Hagborg, 1942, Wallin, 1946).   Four serological types were differentiated by Fang *et al.* (1950).

ETIOLOGY.   The bacteria enter the young tissues through natural openings and wounds.   Advancement in the mesophyll and parenchymatous tissues is between the cells, especially when the tissues are water-soaked. First infections occur during the seedling stage early in the spring.   The secondary infections occur on the younger tissues throughout the growing

---

[1] Several systems of nomenclature are in use for the bacterial plant pathogens.   The writer has followed the classification presented by Bergey *et al.* (1939, 1946).   The reader is referred also to Elliott (1930, 1951) and Riker and Baldwin (1942).

FIG. 4. Leaf (A) and sheath (B) lesions of bacterial blight of barley incited by *Xantho-monas translucens*.

season whenever high moisture prevails. The exudate is splashed by meteoric water, transmitted by contact, and carried extensively by insects. Sucking and biting insects are important in dissemination and infection (Leach, 1940). The bacteria embedded in the exudate remain dormant under unfavorable conditions and resume active development when conditions become favorable. The bacteria remain viable over long periods when dehydrated and sealed in the gelatinous matrix in the lesions. The organism is carried from season to season in crop residue and on the seed.

CONTROL. Crop rotation and seed treatment with mercury compounds assist in the control of the disease. Sporadic outbreaks occur, however, regardless of the use of control measures. Varietal reaction to the disease varies greatly. Jones and Johnson (1917) reported Oderbrucker (C.I. 1272) and Chevalier as the most resistant of about 40 varieties. In the epiphytotic of 1944, Oderbrucker (C.I. 4666), a Chevron (C.I. 1111) × Bolivia (C.I. 1257) selection, and many selections of the Manchuria group showed resistance. Some of the smooth-awned varieties were low in infection.

4. Pythium Root Rot, *Pythium* spp. The root necrosis and browning incited by the soil-borne *Pythium* spp. frequently are severe on barley and other crops. The disease is widespread on barley and wheat in the Central and Northern plains area of North America. Severely infected plants are pale green and short. Dark-brown lesions are evident especially on the smaller roots and sporangia of the pathogen in host cells of the invaded tissues. *P. graminicola* Subrm. is the more common species on barley. See corn and wheat, Chaps. 5 and 11, for the detailed discussions.

5. Downy Mildew, *Sclerospora macrospora* Sacc. or *Sclerophthora macrospora* (Sacc.) Thir., Shaw & Naras. The downy mildew occurs infrequently on barley in the winter barley areas. Symptoms are similar to those described on wheat in Chap. 11. Oswald and Houston (1951) described a downy mildew on barley, first reported on barley in California in 1929, with smaller oospores than those of *S. macrospora*.

6. Powdery Mildew, *Erysiphe graminis* DC. The disease occurs generally on the cereals and grasses with the exception of those in the tribes Maydeae, Andropogoneae, Zoysieae, Paniceae, and Oryzeae (Mains and Dietz, 1930, Marchal, 1902, 1903, Reed, 1909, and Salmon, 1900, 1904). Economically, the disease is generally of more importance on barley than on the other cereal crops.

Many varieties of the cultivated barleys as well as wild species of *Hordeum* are susceptible to powdery mildew. The disease is general in distribution throughout the humid and semihumid areas of the world (Honecker, 1934, 1935, 1937, Marchal, 1903, and Salmon, 1904), and it is generally more severe on both winter and spring barleys in the areas where cool, humid, and cloudy weather persists during the growing period.

Heavy infection with powdery mildew increases respiration and reduces yield. Yarwood (1934) has shown an increase in rate of respiration with the clover powdery mildew. Honecker (1937) reported an increased protein content in barley from mildew-infected plants. Other investigators have studied the disease on wheat (see Chap. 11).

Yields are reduced when mildew infection is severe during the period of active plant growth and grain development. The effect upon yield has been demonstrated by using dusts and sprays as a control on a mildew-susceptible variety. Similar effects have been demonstrated by the use of resistant backcross material. Under a severe mildew epiphytotic in 1942,[1] Oderbrucker homozygous for waxy and mildew resistance yielded about 30 per cent more than Oderbrucker homozygous for a starchy endosperm and mildew susceptibility or than the susceptible Oderbrucker parent. The lines heterozygous for mildew resistance and waxy were intermediate in yield. Schaller (1951) found powdery mildew development prior to heading in barley, reduced the number of kernels, and after heading it reduced kernel weight. He recorded up to 27.4 per cent reduction in yield of grain.

SYMPTOMS AND EFFECTS. The powdery mildew develops on the epidermis of blades, leaf sheaths, and floral bracts. The superficial mycelium and conidia are first light gray in color, the mycelium darkens with age, and later numerous round dark perithecia develop on these areas (Fig. 5). The tissues of the suscepts beneath the mycelium vary in response to the fungus. In the more susceptible varieties, chlorosis and browning accompany the aging of the mycelium. In many varieties, light- to dark-brown pigmentation and frequently necrosis occur beneath or adjacent to the superficial mycelium (Fig. 5). The response is similar to the so-called "flecking" in the reaction to rust. The characteristic symptom is the gray powdery-surfaced lesions scattered or completely covering the leaf blade, with yellowing, browning, and gradual drying out of the leaf tissue. The symptom is also an indication of the damage to the suscept.

THE FUNGUS. *Erysiphe graminis* DC.

Specialized varieties of *Erysiphe graminis* occur on the cereals and related grasses as well as on certain groups of the grasses. These varieties of the fungus are similar morphologically, but they are restricted parasitically to certain genera of the Gramineae; trinomials are used for their designation. The variety occurring on barley is *E. graminis hordei* E. Marchal.

The mycelium is superficial, branched, white, and later turns gray to brown. Coni-

[1] Unpublished data from H. L. Shands, Department of Agronomy, University of Wisconsin, Madison, Wis.

Fig. 5. The response of barley varieties to *Erysiphe graminis hordei*. (A) Susceptible. (B) Resistant, sparse mycelial development with, left, abundant cell necrosis and, right, restricted necrosis without parasitic establishment of the fungus.

diophores form soon after the mycelium is established. They are medium in length with a terminal generative cell. The conidia are light gray, ovoid, measure 25–30 by 8–10 microns, and are borne in chains. Cleistothecia are dark brown, round to sub-spherical, about 220 microns in diameter, and scattered. The appendages are rudimentary, short, and pale brown. The asci are numerous, cylindric to ovate-oblong, with usually eight spores. In the North Central and Central area of the United States, many cleistothecia are empty or contain incompletely developed asci.

ETIOLOGY. The primary infection occurs from ascopores or conidia. Cleistothecia and ascospores are produced on mature tissues of the suscept under favorable weather conditions. Cleistothecial development occurs in late summer on the cereal crops and grasses in the North Central United States although, apparently, relatively few ascospores develop until spring. The mycelium persists from season to season in the areas where the winters are mild enough for infected leaves to survive; however, ascospores are the important source of primary inoculum in the spring-grain area. The spores are largely wind-borne. Conidial formation, dissemination, and germination are best in a humid, cool atmosphere, but in the absence of free water, according to Cherewick (1944) and Yarwood (1936).

Infection in *Erysiphe graminis* is by direct cuticular penetration of haustoria forming branches into the epidermal cells. The mycelium spreads on the epidermis of the leaves and floral bracts. Conidiophores are formed, and abundant diurnal development of conidia follows. The spread of the initial mycelium is rapid, and secondary infections occur in great abundance, especially during periods of cool, cloudy weather. As the suscepts approach maturity cleistothecia develop in the mycelial mats (Harper, 1905, Reed, 1909, Smith, 1900). Powdery mildew development is aggressive during the period of rapid growth and spike development of the cereals.

CONTROL MEASURES. The disease can be controlled by the use of sulfur dusts, and the inoculum is reduced by the use of potassium or sodium sulfide (1.0 per cent solution in water) or copper sulfate (1.05 per cent in water) sprays to which is added a suitable spreader, such as 0.03 per cent glyceryl alkyl resin (Yarwood, 1945). These methods of control are not economical except on an experimental basis; therefore the use of mildew-resistant varieties represents the most practical method of control in both the cereals and grasses.

DISEASE RESISTANCE AND PHYSIOLOGIC RACES. A relatively large number of barleys within the different species and groups are resistant to one or more of the physiologic races of *Erysiphe graminis hordei*. According to Mains and Martini (1932), Shands (1939), and Tidd (1937), Arlington (C.I. 702), Duplex (C.I. 2433), Chevron (C.I. 1111), unnamed selection (C.I. 2444), and others are resistant to at least five races of the parasite. Briggs (1935, 1937, 1938), Dietz (1930), Tidd (1937), and others have studied the inheritance of resistance. A number of mostly dominant factor pairs are involved in the expression of resistance in different barley varieties and to the various physiologic races, as summarized by Briggs and Stanford (1943). Recent investigations indicate that many factor pairs for resistance occur in the world collection of barleys. One allelic series of genes for resistance is located on chromosome II, and at least two other loci situated in chromosomes IV and VII are indicated.

The reaction in the seedling stage of 6 differential varieties of barley to 10 physiologic races of *Erysiphe graminis hordei* as reported by Cherewick (1944), Mains and Dietz (1930), and Tidd (1937) is shown in the following table.

TABLE 1

Physiologic race	Reaction on barley varieties					
	Nepal (C.I. 595)	Heils Hanna (C.I. 682)	Goldfoil (C.I. 928)	Peruvian (C.I. 935)	Black Hull-less (C.I. 666)	Chevron (C.I. 1111)
1	I	I	R	R	I	R
2	S	S	R	I	I	R
3	S	S	R	S	S	R
4	S	S	R	I	S	R
5	S	S	S	S	S	R
6	R	S	R	R	I	R
7	R	S	S	R	R	R
8	S	S	R	R	S	S
9	I	S	R	R	I	S
10	I	S	R	S	I	R

R—resistant; I—intermediate; S—susceptible.

7. Fusarium Blight or Scab, *Gibberella zeae* (Schw.) Petch or *G. roseum f. cerealis* (Cke.) Snyder & Hansen and *Fusarium* Spp. The *Fusarium* blight occurs on all the cereals and many of the grasses. The head blight is especially severe on rye, barley, and wheat. The complete discussion of the disease is included in Chap. 11.

DISTRIBUTION AND IMPORTANCE ON BARLEY. The head blight occurs in the Eastern and Central United States and adjacent Canada. The area includes the humid spring barley and the northern portion of the humid winter barley sections. Damage to both yields and quality frequently is high in this area, and losses become more sporadic and less frequent in the drier prairie and plains areas. The seedling blight and root rot phase of the disease extends into the drier areas. The disease is common on barley in the humid regions of Europe, Africa, and Asia.

The fungi causing the disease show a geographical distribution within the different regions. In the corn-belt section of the Eastern and Central United States, *Gibberella zeae* (Schw.) Petch is the common species. In the more northern area, *Fusarium culmorum* (W. G. Sm.) Sacc. and *F. avenaceum* (Fr.) Sacc. and some other species predominate. In northern Europe, *F. culmorum*, and in southeastern Europe, *G. zeae* are the chief organisms on barley. In northeastern Asia, *G. zeae* causes heavy annual losses in barley, as reported by Atanasoff (1923), Bennett (1932, 1933, 1935), Dickson (1930), Naumov (1914), and Palchevsky (1891).

According to Snyder and Hansen (1945), the three species listed above are varieties of *Fusarium roseum f. cerealis* (Lk.) Snyder & Hansen, Graminearum, Culmorum, and Avenaceum, respectively (see wheat, Chap. 11).

The disease causes reduction in yield of grain, and it damages quality. The blighted barley kernels are shriveled or lighter in weight, yet cannot be separated completely from the sound grain. The infected kernels contain substances causing acute emesis in pigs, dogs, and humans. Cattle, sheep, and mature chickens seemingly are not affected by the diseased grain, according to Christensen and Kernkamp (1936), Dickson (1942), Hoyman (1941), Mains et al. (1930), Mundkur (1934), Roche and Bohstedt (1931), and others. Barley containing 4 per cent of blight-damaged kernels, including blight caused by various fungi, is given the special designation "blighted" in the Federal grain-grading procedure. Blighted barley is discounted heavily on the commercial markets of the United States and cannot be exported to foreign markets.

SYMPTOMS. Spikes are dwarfed and compressed with infected spikelets closed rather than spreading. All or part of the spike is infected (Fig. 6). Hulls (lemma and palea) are light to dark brown with a dead, lusterless surface. Conidial or perithecial masses commonly develop on the surface, especially during moist weather (Fig. 7). The kernels are shrunken and light brown in color. The pericarp surface is rough or scabby in appearance. The starch mass is grayish in color and flour-like in texture. The scab is not distinguished easily from the *Helminthosporium* blight unless symptoms are very characteristic or spore masses are present on the kernel. Plating the kernels on acidified agar is a reliable method of differentiating the two.

Seedling infections in barley are primarily from seed-borne inoculum. Restricted reddish-brown cortical lesions occur when the infected seed is sown in cool, moist soil. Seedling blight before or after emergence occurs in a warm soil. Seedling infection of clean seed from mycelium in the soil is common when the soil temperatures are high. Crown and basal culm rot occurs commonly in the later stages of development of the barley plant.

CONTROL. Clean seed and seed treatment help control seedling infection. The mercury dusts are effective in controlling seedling damage from infected seed. Sanitation, rotation, and early planting help in reducing crown infection and head blight. Soil preparation to obtain complete coverage of barley, wheat, and corn-crop residues helps reduce inoculum for head blight infection.

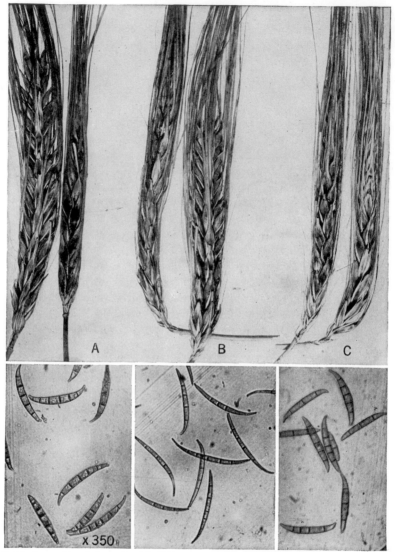

FIG. 6. Barley spikes infected with *Gibberella* and *Fusarium* species and the conidia of three species. *G. zeae*, *F. culmorum*, and *F. avenaceum* or *F. roseum f. cerealis.* (*A*) Graminearum, (*B*) Culmorum, and (*C*) Avenaceum. Conidia highly magnified.

Varieties resistant to head blight offer the best means of control. The commercial varieties commonly grown in the areas where scab is prevalent are susceptible to the disease. No variety highly resistant to scab has been found. The more resistant barleys are Svansota (C.I. 1907) (a two-rowed barley), Chevron (C.I. 1111), Korsbyg (C.I. 918), Cross

(C.I. 1613 and 2492), an unnamed selection (C.I. 1918), and Peatland (C.I. 5267) (six-rowed barleys). These barleys are being used in breeding for scab resistance (Shands, 1939). See *Fusarium* Blight, Chap. 11 for the complete discussion.

FIG. 7. Barley kernels blighted by *Gibberella zeae* or *Fusarium roseum f. cerealis* var. Graminearum showing discoloration and conidial and perithecial development.

8. Ergot, *Claviceps purpurea* (Fr.) Tul. Ergot is found frequently on barley through the North Central United States and Canada. The disease occurs in other similar areas where rye or common grass suscepts are infected. Damage is rarely as high as in rye, durum wheats, and many grasses. See Ergot on rye, Chap. 8, for the complete discussion.

9. Net Blotch, *Helminthosporium teres* Sacc., *Pyrenophora teres* (Died.) Drechsl. The disease is common on all the cultivated barleys and apparently is restricted to these species. Net blotch is distributed with barley culture throughout the temperate humid regions of the world. The disease occurs more abundantly in the cooler climates or where barley is grown in the cooler period of the year. It is very common throughout northern Europe. Like the stripe disease, it was introduced with barley into North America. Under most conditions the disease is of minor importance, although under favorable conditions it causes considerable reduction of foliage, especially as the crop approaches maturity.

SYMPTOMS. The first symptom on the seedling leaf is the development of brown reticulate blotches at or near the tip of the blade. The lesions rarely develop from the base of the seedling leaf as in spot blotch. Local lesions on the young leaves develop from the seedling stage until maturity. The young infections show the characteristic netted blotch. The darker brown necrotic areas of the blotch are distributed irregularly in narrow, indefinitely margined lines both parallel and perpendicular to the leaf axis. This makes a dark-brown reticulate pattern within the areas of lighter brown (Fig. 8). Later, as the necrosis of the mesophyll tissue expands and the blotches coalesce longitudinally, the characteristic appearance changes to dark-brown limited stripes with irregular margins. The net-like pattern is evident in these older lesions only along the margins. In the advanced stages of infection, a series of several to as many as 10 irregularly margined stripes extend parallel, frequently the full length of the blade. The stripes do not continue into the leaf sheath as in the stripe disease. Conidiophores and conidia develop sparsely on the lesions. Small linear brown lesions occur on the floral bracts. The light-brown discoloration of the lemma without the conspicuous netted appearance is characteristic of kernel infection. Seed-borne infection is determined accurately only by plating the kernels on acidified potato dextrose agar. Seed-borne infection is higher in the more northern areas; Machacek and Wallace (1942) reported 0 to 64 per cent in a Canadian survey.

THE FUNGUS

Pyrenophora teres (Died.) Drechsl.
Helminthosporium teres Sacc. Conidial stage
(*Pleospora teres* Died.)
(*Helminthosporium hordei* Eidam.)

The mycelium is white to olivaceous in the tissues and makes a very sparse tufted growth on media. Bodies resembling vegetative resting spores appear in culture. Conidial development on suscept and in culture usually is limited. The conidiophores are light brown to olivaceous, occurring singly or in groups of two to three, and the swollen basal cell is usually larger than in the other *Helminthosporium* spp. on barley. The average dimensions are 120–200 by 7–9 microns. Conidia are yellowish olivaceous, never dark olivaceous, thin-walled, constricted at the septa with much rounded apical cells. The basal cell is larger, resulting in a subcylindrical shape. Germ tubes develop from all cells of the conidium. The perithecia develop abundantly on barley stubble and straw, especially the following spring. They are superficial or partly submerged, elongated, irregular in shape, and about 0.5 mm. in diameter. Setae and

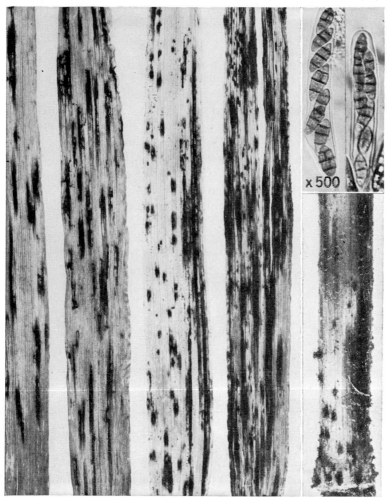

x 500

FIG. 8. Net blotch of barley caused by *Pyrenophora teres* showing the characteristic leaf lesions at different stages of development and the perithecia and ascospores of the fungus.

conidiophores are abundant on the surface. The ostiolar beak is not common. The numerous asci are subcylindrical with a ring-like thickening at the apical end, and each contains eight spores. The ascospores are light brown, 3-septate, with the center cell usually divided longitudinally in the mature spores. The spores are much constricted at the septa (Fig. 8). Germ tubes are formed from all cells (Drechsler, 1923)

Wehmeyer (1949) groups the species on the Gramineae under *Pleospora trichostoma* (Fr.) Ces. & de Not. The similar morphology in the perfect stage of several of the species with thin-walled, cylindrical conidia that develop germ tubes from both apical and intermediate cells of the conidium in the imperfect (*Helminthosporium*) stage perhaps argues for combination. However, *Helminthosporium teres* and *H. avenae*, similar in the imperfect stage, are dissimilar in the perfect stage; 2 transverse septa and 3 transverse septa, respectively, in the ascospores. A more systematic study of this group is necessary before combinations are justified.

ETIOLOGY. The early seedling infection of *Pyrenophora teres* is from seed-borne mycelium or ascospores produced on old straw and stubble. Ascospores or conidia are responsible for continued infection whenever conditions are favorable. The infection on fall or spring barley is abundant during cool, humid weather. The lesions enlarge and coalesce throughout the growing season. The mycelium grows into the sheath and culm tissues as the disease progresses and the plants mature. Local infections on the floral bracts occur from spike emergence to shortly after flowering. Perithecia develop on the barley stubble and straw in the late fall and again in the early spring. The disease is usually very abundant on volunteer barley late in the fall.

CONTROL. Seed treatment with the standard mercury dusts controls the seed-borne inoculum. Sanitation and crop rotation are important in reducing the ascosporic inoculum. Barley sown on or near fields with barley stubble on the surface is infected heavily the following spring and summer. Differences in varietal reaction to the disease are reported by Geschele (1928) and Ravn (1900). Most commercial varieties in North America are relatively susceptible to either eastern or western isolates of the pathogen. Shands[1] has shown that four North African barleys, C.I. 4975, Rabat (C.I. 4979), C.I. 6311, and Anoidium (C.I. 7269) are resistant to eastern and susceptible to western isolates of the pathogen. Manchurian and some similar varieties are resistant to the western, but varied widely in their response to the eastern isolates. Varieties resistant to both eastern and western isolates are Tifang (C.I. 4407-1), Manchu (C.I. 4795), Ming (C.I. 4797), Manchuria (C.I. 739), and Kindred (C.I. 6969). Resistance is conditioned by single, dominant factor pairs. Schaller (1955) also reported resistance in Tifang (C.I. 4407-1) to a western isolate as conditioned by a single dominant factor pair.

10. Stripe Disease, *Helminthosporium gramineum* Rabh. The barley stripe is today a relatively minor disease in spring barley culture in the United States and Canada. It is prevalent and causes considerable damage in the California barley area and in the South Central winter barley section of the United States. Damage from stripe is severe in limited sections of northern Europe and in northern and eastern Asia as well as in portions of Turkey, Iran, and Transcaucasia. The distribution

[1] R. G. Shands. Unpublished data in Annual Report 1951.

of the disease is widespread, as reported by Mitra (1931), Nishikado (1929), Ravn (1900), and Smith (1929, 1930).

SYMPTOMS AND EFFECTS. The symptoms of the disease are conspicuous from the late tillering stage until the crop is mature (Drechsler, 1923, Ravn, 1900). The first symptoms at the tillering stage are yellow striping of the older leaf blades and sheaths. Some seedling blight occurs in severely infected seed of susceptible varieties. The yellow stripes soon turn brown as tissue necrosis progresses, and finally the tissues dry out and fray as the leaves mature. During the period of culm elongation, the symptoms are distinctive: the young leaves unfolding show the yellow striping with the successive necrosis and browning conspicuous on the leaves below (Fig. 9). The elongation of the culms of the striped plants varies from rosette-like development to fully elongated plants. Varieties of barley, races of the pathogen, and environmental conditions influence culm elongation in diseased plants, as discussed by Arny and Shands (1945), Christensen and Graham (1934), Isenbeck (1937), Leukel et al. (1933), Mitra (1931), Shands (1934), and Shands and Dickson (1934). The spikes fail to emerge in many diseased plants. Those that emerge are blighted, twisted, compressed, and brown in color. On the Pacific Coast, the kernels are damaged less and the brown infected kernels fre-

FIG. 9. Barley plants showing the yellow to brown stripes of leaf blades and sheaths of plants infected with *Helminthosporium gramineum*. (A–B) The brown stripes, shredding of the leaves, and brown, blighted condition of the spikes of mature plants. (C) The brown stripes on the lower leaves and yellow stripes on the upper leaves of younger plants.

quently appear in the threshed grain. When the healthy plants are heading, the striped plants show a gray to olive-gray color due to the development of conidiophores and conidia over the mass of lesioned tissue. After conidial development of the fungus, the tissues of the infected plants split, fray out, and collapse. The florally infected kernels rarely show indications of infection, although the fungus can be plated from infected kernels on suitable media.

THE FUNGUS

Helminthosporium gramineum Rabh.
[*Brachysporium gracile* Wallr. var. *gramineum* (Rabh.) Sacc.]
(*Napicladium hordei* Rost.)
(*Heterosporium gramineum* Oud.)

Diedicke (1902, 1904) described the ascigerous stage under the binomial *Pleospora graminea* Died., Noack (1905) described it and used *P. trichostoma* Noack, and Paxton (1922) described what he considered to be the perfect stage from material collected in California. Ito and Kuribayashi (1931) discussed the taxonomy and described *Pyrenophora graminea* (Rabh.) Ito & Kuribay.; however, the connection with the conidial stage was not demonstrated and the morphology was similar to *P. teres*. Later investigations have not confirmed these reports.

The morphology of this and other species is given in detail by Drechsler (1923). The mycelium in the tissues is abundant, subhyaline to light yellow. In culture the mycelium varies greatly in color from gray through olive to black. The mycelium in culture does not produce conidia unless exposed to cyclic changes of light and darkness. The conidiophores and conidia develop on the plants only as the suscepts are heading. Conidiophores are borne in clusters, usually three to five. The basal segment is enlarged, the distal portion is slender, the color is gray to olivaceous. The conidia are subhyaline to yellowish brown, straight, subcylindrical to slightly tapering, have rounded ends, thin-walled, 1- to 7-septate without constrictions at the septa, and average 105 by 20 microns in size. Both end cells and less commonly the central cells form germ tubes. Secondary conidia form from the germ tubes, as described by Dreschler (1923), Sprague (1950). Christensen and Graham (1934) have shown morphological differences between some races.

ETIOLOGY. Natural infection from conidia of *Helminthosporium gramineum* occurs at or soon after flowering of the barley spikelet. The mycelium is established on or in the pericarp or in embryo tissues before maturity of the kernel. The final establishment of the mycelium in parasitic relationship with the young seedling tissues generally occurs during seed germination. Seedling infection can be induced in high percentages by placing the actively developing mycelium of the fungus in contact with the germinating barley kernels. This method of inoculation is used extensively in stripe-resistance studies, as reported by Arny (1945), Isenbeck (1930), and Shands (1934). After seedling infection, the parasite develops in the culm primordia and grows with the differentiating tissues during seedling development. The systemic distribution of the mycelium continues with the differentiation and development of the plant structures. This apparently is the only species in the genus *Helminthosporium* in which a systemic type of infection occurs. Conidial production is synchronized with the heading, blossoming, and early stages of kernel development of the suscept. The conidial formation is probably a direct response to the physiology of the suspect tissues, as the same response occurs under greenhouse culture during the winter months. The fungus is carried over from season to season on or in infected seed. The mycelium remains viable in dry seed for an indefinite length of time.

Environmental conditions influence floral infection and disease development in the seedling. Isenbeck (1937) and Ravn (1900) reported that sufficient moisture to wet the spores is necessary for floral infection. This limits the regions where the stripe disease occurs to areas where dew or precipitation occurs during the period of barley flowering. Christensen and Graham (1934), Isenbeck (1937), Kiessling (1916), Leukel *et al.* (1933), Ravn (1900), Shands (1934), and others reported on the influence of moisture, temperature, and fertility on seedling development of stripe. A cool, moist, fertile soil favors stripe development in the seedling and developing plant.

CONTROL MEASURES. Seed treatment with the organic mercury dusts, such as Ceresan, control the seed-borne infection. Resistant varieties offer the best means of control of the disease.

DISEASE RESISTANCE AND FUNGUS SPECIALIZATION. Physiologic specialization occurs, although investigations are not comprehensive enough to determine the number or stability of the races or arrange a key for their identification, as discussed by Christensen and Graham (1934), Shands and Dickson (1934), and others. Resistance in varying degrees is common in varieties of most of the barley types. Arny (1945), Isenbeck (1930), Shands and Arny (1944), Suneson (1950), and others reported varieties intermediate to highly resistant, and they have discussed the inheritance of resistance. The evidence indicates more than one factor pair determining resistance to stripe in many varieties and a single factor pair in certain varieties and cultures of the fungus. Lion (C.I. 923), Peatland (C.I. 5267), and Chevron (C.I. 1111) have been used extensively in breeding stripe-resistant varieties. Resistance has been incorporated into many commercial varieties, with the result that stripe is now a minor disease in the spring barley area of the United States. The following commercial varieties are resistant enough for practical stripe control; Wisconsin Barbless (C.I. 5105), Glabron (C.I. 4577), Trebi (C.I. 936), Regal (C.I. 5030), Newal (C.I. 6088), Velvon (C.I. 6109), and Mars (C.I. 7015), six-rowed; and Spartan (C.I. 5027), Vance (C.I. 4586), and Hannchen (C.I. 531) two-rowed.

11. Spot Blotch. *Helminthosporium sativum*, Pam., King, & Bakke. The spot blotch disease is different in several respects from many of the other diseases caused by species of *Helminthosporium*. The fungus attacks a wide range of grasses and is common on wheat and barley. The mycelium is very resistant to unfavorable conditions and is abundant in gramineous crop residues both in and on the soil surface. Local lesions occur on seedlings, plant crowns, culms, leaves, floral structures, and kernels. The seedling and crown damage occurs in relatively dry, hot areas as well as abundant infections of all tissues in humid, warm regions.

GEOGRAPHIC DISTRIBUTION AND IMPORTANCE. The disease is wide-spread in North America, and it is common although probably not so

FIG. 10. Seedling blight of barley caused by *Helminthosporium sativum*. Symptoms range from stunted seedlings with dark-green leaves and brown lesions on the basal leaf sheaths to blighted seedlings.

extensively distributed in South America, Europe, Asia, Australia, and Africa, as reported by Dickson (1930), Hynes (1935, 1937), Ito and Kuribayashi (1931), Kuribayashi (1917), Lindfors (1918), Smith (1930), Sorokin (1890), and others. Damage on both barley and wheat is severe where susceptible varieties are grown extensively.

SYMPTOMS AND EFFECTS. The seedling blight is characteristically a dry rot type of tissue necrosis. The dark-brown to black lesions usually occur first on the coleoptile and progress inward. The seedling is killed before emergence or more frequently after emergence. The seedling leaves of infected plants are dark green, erect with dark-brown lesions near the soil line that soon extend into the leaf blade. Development of

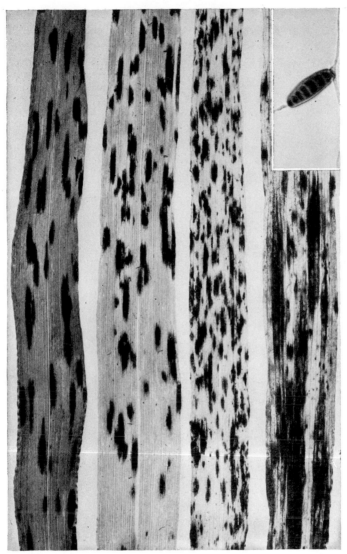

FIG. 11. Spot blotch of barley caused by *Helminthosporium sativum* and the germinating conidium of the fungus. The elongate brown spots with definite margin coalesce later to form brown irregular stripes. The conidia (insert) germinate by germ tubes from the apical cells only.

infected seedlings is retarded, and tillering frequently is excessive. The seedling symptoms are similar in barley, wheat, and some grasses. The crown rot develops at or below the soil surface. Cortical and leaf-sheath tissues are lesioned, tiller buds are blighted, and the crown-root system is invaded. The necrosis is accompanied by a dark-brown discoloration of the tissues (Fig. 10).

The leaf spot varies in size and shape. The individual lesions are round to oblong with definite margins. The color is a uniform dark brown without conspicuous water soaking accompanying necrosis. The spots coalesce to form the blotches that frequently cover large areas of

Fig. 12. Spike blight (*A*) and kernel blight (*B*) of barley caused by *Helminthosporium sativum.*

leaf blade. They are more restricted on the leaf sheath. The older lesions develop an olivaceous cast due to the abundant development of conidiophores and conidia. The heavily infected leaves dry out and mature early. The fungus continues to sporulate on the dead tissues (Fig. 11).

Lesions on the floral bracts and kernels range from small black spots to dark-brown discoloration of the surface. The characteristic "black point" or blackened embryo end of the kernel is one of the common symptoms on wheat and barley (Fig. 12). The extensive invasion of crown and culm tissues usually results in shorter culms, partial emergence of the spike, and sterility or poorly filled kernels. Head blight occurring early also causes sterility or killing of individual kernels soon after pollination (Figs. 11 and 12). The dark-gray to black mycelium with abundant conidial development in most cultures is characteristic when the diseased tissues are plated on acidified potato-dextrose agar.

THE FUNGUS

> *Cochliobolus sativus* (Ito & Kurib.) Drechsl. (*Ophiobolus sativus* Ito & Kurib.)
> *Helminthosporium sativum* Pam., King, & Bakke. Conidial stage
> (*Helminthosporium acrothecioides* Lindf.)
> (*Helminthosporium inconspicuum* Peck)
> (*Helminthosporium sorokinianum* Sacc.)

The mycelium is olivaceous to black when mature and develops abundantly, including conidial production on media. Conidiophores emerge from stomata or between epidermal cell walls, singly or two to three, rarely more. The basal cell is swollen, has a heavy wall, is dark olivaceous, and the conidial scars are conspicuous. Conidiophores on agar cultures are short modified branches of hyphae. The conidia are slightly to distinctly curved, thick-walled, reddish to dark olivaceous brown, 1- to 10-septate, widest near the middle, and the ends round off abruptly. The size, shape, and color vary greatly, depending upon the culture and the environment. The conidia germinate from the apical cells only (Drechsler, 1923, 1934) (Fig. 11). The ascigerous stage was described by Ito and Kuribayashi (1931) as follows.

Perithecia are black to brown, pseudoparenchymatous, flask-shaped with ostiolar beak, and measure 340–470 by 370–530 microns. Many hyphae and conidiophores are associated with the young perithecia, which disappear as the perithecia mature. Asci are numerous fusiform or cylindrical, straight or slightly curved, rounded at the apex, shortly stipitate at the base with hyaline wall, and contain usually four to eight ascospores coiled in a close helix. The spores are flagelliform or filiform, obtusely pointed at both ends, somewhat broader at the base, hyaline or light olive, and measure 160–360 by 6–9 microns. Drechsler (1934) transferred the group of species with coiled ascospores and bipolar germination of conidia to the new genus *Cochliobolus*.

Helminthosporium sativum was described by Pammel, King, and Bakke (1910). *H. sorokinianum* was reported by Sorokin (1890) and redescribed by Saccardo (1892, *Sylloge*, **10**:415–416, 615–616) as occurring on the spikes of wheat and rye in Ussuria, Russia. Drechsler (1923) discussed the possible synonomy of the two binomials, but without comparison of the types did not reduce *H. sativum* to synonymy. Mouranshkinski (1924) first placed *H. sativum* as a synonym for the Ussurian fungus. Luttrell (1955) examined the Saccardo-type specimen of *H. sorokinianum* and those of *H. sativum* and reported their synonymy. In contrast Palchevsky (1891), reporting on the same collection of fungi from Ussuria as Sorokin, contrasted the fungus as different from *H. gramineum* Rabh, but not a "special new form." The author in 1930 studied the several specimens of *Helminthosporium* collected by Voronin in 1889 and the type specimen of Sorokin filed in the Jaczewski Herbarium, Leningrad, U.S.S.R., and found they could not be included even in the wider range of conidial morphology currently recognized for *H. sativum*. With long usage of the binomial *H. sativum* and the uncertainty of the morphology of the early Ussurian specimens, it would appear unwise to relegate *H. sativum* to synonymy.

Tinline (1951) produced the perfect stage by mating compatible lines of the fungus on autoclaved barley kernels on nutrient media and studied the complete life cycle of the fungus in culture. He used the binomial *Cochliobolus sativus* (Ito & Kuribayashi) Drechsl. and confirmed the description given by Kuribayashi (1929) and Ito and Kuribayashi (1931). The fungus is heterothallic in so far as it is hermaphroditic, self-sterile, intergroup sterile, and intergroup fertile. At least two compatible groups are distributed at random in nature. Dastur (1942) described a similar fungus morphologically that produced perithecia on "black point" wheat kernels as *C. tritici* Dastur. Arny (1951) found that resistance in the seedling was conditioned by a single, dominant factor pair with specific lines of the pathogen. The fungus in nature comprises a highly variable complex maintained by the polynucleate condition of the mycelium and frequent anastomoses between mycelial cells. Pure lining of the fungus results in clones stable in cultural characters and specific for pathogenicity. Clones of the fungus have been isolated that incite both crown rot and leaf spot on most of the barley varieties. Christensen and Schneider (1948) demonstrated that a monosporous line passed through Marquis wheat for ten successive generations was relatively stable with a mutation frequency of 1:2,900.

ETIOLOGY. Seedling and crown infections occur from seed-borne mycelium or from crop residues in the soil. The organism develops aggressively as a saprophyte on crop residue or mature tissues of the cereals and grasses. Disease development is usually more severe in late-sown grain or in a warm soil. Infection of embryonic tissues is by direct penetration, natural openings, or injuries. Frequently insect injury of crown tissues is followed by the invasion of *Helminthosporium sativum*. Leaf infections develop under warm, moist conditions, and they spread rapidly from secondary conidial infections. The abundant conidial inoculum results in severe infection of young tissues whenever environmental conditions are favorable. Plants that are retarded in development by injuries or unfavorable growing conditions are usually more susceptible to attack. Seed infection is frequently quite high. Extensive platings of both barley and wheat kernels from the North Central area during the past 5 years show an average of over 5 per cent infection. In some seasons samples show as high as 75 per cent infection. Similar surveys in Canada indicate high infections in years favorable for the disease, as reported by Greaney and Machacek (1942).

CONTROL. The control of the disease is difficult and should be given more attention than in the past. Sanitation and crop rotation are important, but in the spring-grain area, where grains and grasses comprise such a large percentage of acreage, suitable rotations offer difficulties. Seed treatments with the mercury compounds have been effective in increasing stands of vigorous seedlings. Resistant varieties offer the best means of control. Hayes *et al.* (1923), Christensen (1945), and others have shown differences in susceptibility, although most commercial barleys and wheats are diseased under favorable conditions. The most resistant six-rowed barleys found to date are in the Manchurian group. Peatland (C.I. 5267), Chevron (C.I. 1111), and Kindred (C.I. 6969) are resistant to most races of the pathogen. Hannchen (C.I. 531) and Svansota (C.I. 1907) are moderately resistant two-rowed varieties. Resistance apparently is conditioned by several single factor pairs inherited independently (Arny, 1951). Christensen (1922, 1925, 1937) reported specialization and variation in the fungus.

Mackie and Paxton (1923) have described a species of *Helminthosporium* on barley in California differing slightly from *H. sativum* which they have described as *H. californicum* Mackie and Paxton. Mackie (1928) reported Chevalier as resistant to this rusty blotch.

12. Rhynchosporium Scald, *Rhynchosporium secalis* (Oud.) J. J. Davis. The scald disease occurs commonly on barley, rye, and *Bromus inermis* Leyss. Many other grasses are infected by this species and *Rhynchosporium orthosporum* Cald. Specialized races of the parasites occur which limit the distribution of inoculum between suscepts. The disease is distributed in the cooler humid and semihumid sections of North and South America, Europe, and Asia (Caldwell, 1937, Dickson, 1930). The disease is of minor importance in the North Central North American spring barley area, as the commercial barleys are moderately resistant. Scald causes considerable defoliation in the barley section of the Pacific Coast and in areas in Europe and Asia where susceptible varieties are grown.

DESCRIPTION. The blotches are conspicuous on the leaf blades and sheaths. The lesions are first water-soaked ovate to irregular scald-like blotches. The color of the lesions changes rapidly from bluish green to zonated scald and brown pigmented rings and finally to a bleached straw color with a brown margin on barley and straw color on rye (Fig. 13).

THE FUNGUS

Rhynchosporium secalis (Oud.) J. J. Davis
(*Marsonia secalis* Oud.)
(*Rhynchosporium graminicola* Hein.)

The mycelium is hyaline to light gray, developing sparsely as a compact stroma under the cuticle of the suscept. The conidia, borne sessilely on cells of the fertile stroma, are hyaline, 1-septate, cylindrical to ovate, with a short oblique apical beak on most spores, and measure 12–20 by 2–4 microns. *Rhynchosporium orthosporum* Cald. occurring on a number of grasses is differentiated by the longer, more cylindrical conidia without the apical beak.

ETIOLOGY. The leaf spots develop abundantly during cool weather. The mycelial stroma on the leaves apparently persists on living or dead leaf tissues to furnish the primary conidial inoculum. Leaf scald on winter rye from stromata formed the previous autumn frequently occurs during the first frost-free days of spring. Lesions develop on barley throughout the winter-growing season in the Pacific Coast barley area.

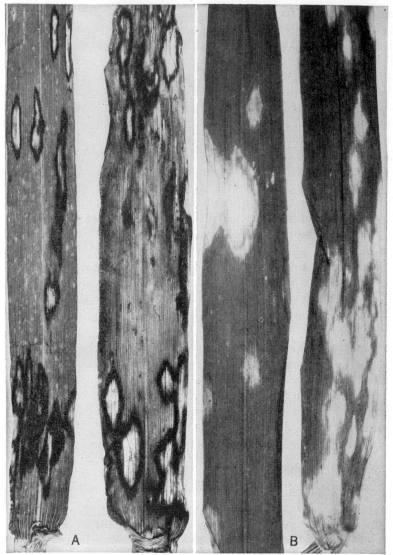

FIG. 13. Leaf scald caused by *Rhynchosporium secalis* on barley (*A*) and rye (*B*).

Infection is usually abundant in late autumn on volunteer spring barley. The fungus overwinters on the infected dead leaves and probably on the crop residue to produce new infections on barley seedlings the following spring. Secondary conidial inoculum is abundant during the cool, humid spring-growing period and again in late autumn. The conidia are airborne, frequently from considerable distances, and are distributed by rain as well. Infection is by direct penetration of the cuticle on the young leaf. The subcuticular mycelial stroma develops which soon ruptures the cuticle, and conidial production ensues. In the winter barley sections, lesions occur infrequently on the floral bracts, although seed-borne inoculum has not been demonstrated.

CONTROL. Crop rotation, elimination of crop residue, and resistant varieties offer the best means of control. Sporadic local occurrence of the disease is common on susceptible varieties even though local crop rotation and sanitation measures are used. Spring burning of residue of perennial grasses helps reduce the overwintered inoculum of the grasses.

RESISTANT VARIETIES AND SPECIALIZATION. Most of the commercial barley varieties grown in the North Central United States are moderately resistant to the disease. Otrada Beardless (C.I. 5631) and a Hanna selection are resistant at Davis, Calif. Resistance apparently is conditioned by a single factor pair (Mackie, 1929, Riddle and Briggs, 1950). Reports in the literature are conflicting regarding specialization of the fungus. The evidence in the United States and southern Russia indicates specialized races in both species of the genus. Six races of *Rhynchosporium secalis* are reported, each restricted to a single or closely related species, according to Caldwell (1937).

13. Anthracnose, *Colletotrichum graminicolum* (Ces.) Wils. The disease is common on the cereals and grasses. It is of little importance on barley unless the crop is grown on sandy dry soils low in fertility. See Anthracnose, Chap. 8, for the complete discussion.

14. Septoria Leaf Blotch, *Septoria passerinii* Sacc. The *Septoria* leaf blotches are common on the cereals and grasses. A large number of species have been described with the differentiation based largely on spore morphology (Sprague, 1944, 1950). The etiology of this group of parasites is similar; therefore the complete discussion of the diseases is given in Chap. 11.

The blotch caused by *Septoria passerinii* occurs on cultivated barleys and several wild *Hordeum* spp. The disease is distributed widely in North and South America, and it is common in both Europe and Asia. It causes damage on barley, as defoliation occurs.

DESCRIPTION. The lesions are linear with indefinite margins as the yellowish-brown area blends into the green of the leaf blade and sheath. Numerous small dark-brown pycnidia, embedded in the tissue, develop on the straw-colored older portion of the blotch.

THE FUNGUS

Septoria passerinii Sacc.
(*Septoria murina* Pass.)

The mycelium in culture is scant and hyaline to buff-colored. Conidia formed in culture are cylindrical or slightly curved, 3-septate, and hyaline. Pycnidia are subepidermal, dark brown, smooth, globose or mostly subglobose, 80 to 150 microns in diameter, with a smooth ostiole formed under the stomata of the suscept. Pycnospores are 3-septate, usually measure 1.7–3 by 23–46 microns, are hyaline, straight or slightly curved, and rounded at the ends. Microconidia develop in the pycnidia under some conditions.

Septoria nodorum Berk. is reported on barley kernels (Machacek, 1945), (Chap. 11).

15. Loose Smut, *Ustilago nuda* (Jens.) Rostr. INTRODUCTORY DISCUSSION. The smuts are relatively numerous on the cereals and grasses; for example, three species occur on the cultivated barleys. Mycologists and plant pathologists have named the *Ustilago* spp., in many instances, on the basis of the parasitic potentialities of the fungus rather than on differentiating morphological characters. This tendency makes it impossible to identify the species in certain of the cereal smuts without knowing the suscept upon which the fungus occurs. The consolidation of these species on sound morphological criteria might be desirable, especially with the information now available on life cycle, cytology, and genetics of this group of plant pathogens. This is in accord with the concept of the early mycologists before information on specialization was more complete and knowledge on the type of germination was available or when the classification was based largely upon gross morphology. Persoon's (1801) grouping of subspecies under the binomial *Uredo segetum* is an example. More recently, Cunningham (1924), Fisher (1943), Rodenhiser (1926), Tapke (1943), and others have discussed such recombinations based on morphology. Specialization could be designated then within the species by the use of trinomials indicating the suscept specialization. This method is in common use in the case of other plant pathogens of the cereals and grasses, notably in *Puccinia graminis* Pers., *P. rubigo-vera* (DC.) Wint., *Erysiphe graminis* DC., and others. Long usage of many of the accepted binomials for these smut fungi, however, argues against this combination of species and the resultant confusion. Stevenson and Johnson (1944) revised certain of the binomials to comply with the International Rules of Botanical Nomenclature. Fischer (1953) revised the nomenclature of the North American smut fungi.

SUSCEPTS AND DISTRIBUTION OF THE LOOSE SMUT. The true loose smut, or "deep-borne" loose smut, occurs on barley and a limited number of grasses. Many of the cultivated barleys are relatively susceptible, as well as the selections of *Hordeum spontaneum* Koch tested. The loose smut is distributed widely in regions where humid, cool weather occurs during the period barley is heading. The disease causes reductions in yield approximately equivalent to the percentage of loose smut (Semeniuk and Ross, 1942). The smutted plants develop sufficiently well to compete for moisture and soil nutrients and yet produce no grain.

Losses are higher, therefore, than occur in diseases in which the plants are killed before the competition for moisture becomes acute.

DESCRIPTION. The smutted spikes emerge from the boot slightly earlier than the spikes on healthy plants. The sori are enclosed in a fragile membrane which soon ruptures, releasing the brown to dark-brown dusty spore mass (Fig. 14). The sori frequently develop in the leaves when the plants make a rank vegetative growth. The smutted spikes are conspicuous when the barley is heading, as the erect smutted spikes stand above the healthy plants during this period of spore spread. The brown spore mass is wind-borne over the field, while the healthy plants are pollinating and extruding the dehisced anthers. At maturity of the crop, the bare rachises of the smutted plants frequently stand erect above the level of the filled heads.

THE FUNGUS

Ustilago nuda (Jens.) Rostr.
(*Ustilago segetum* var. *hordei* f. *nuda* Jens.)
(*Ustilago segetum* var. *nuda* Jens.)
(*Ustilago hordei* var. *nuda* Jens.)
[*Ustilago nuda* (Jens.) K. & S.]
See Fischer (1953) for synonymy.

The mycelium in culture is hyaline changing to buff, sparse, and predominantly binucleate. Chlamydospore-like cells develop in the old cultures on malt or barley-seedling agar. The mycelium in suscept is hyaline changing to brown, irregularly lobed, and predominantly binucleate. The fertile mycelial cells become subspherical to spherical and develop finely echinulate brown epispore walls in the formation of the chlamydospores. The mature spores are lighter-colored on one side, finely echinulate, and 5 to 9 microns in diameter (Fig. 14, insert *A*). The fresh spores germinate readily to form a one- to four-celled promycelium (basidium). No sporidia are produced (Fig. 14, insert *B*). Fusions occur between compatible cells of the promycelium by means of short or long conjugation tubes. Branches or hyphae develop from the fused cells or the conjugation tubes. Limited branching also occurs from unfused (haploid) cells, according to Christensen (1935), and Lang (1909, 1917). The earlier literature is summarized by Roemer *et al.* (1938) and Thren (1941). The sporidial development at low temperatures originally described by Huttig is questionable.

ETIOLOGY. The wind- or air-borne chlamydospores of *Ustilago nuda* which lodge in the susceptible barley flowers germinate, conjugation occurs between the compatible haploid cells of the promycelium, and the binucleate parasitic hyphae penetrate through the stigma or the young ovary wall. The mycelium becomes established in the pericarp, remains of the integuments, parenchymatous tissue of the chalazal bundle, and embryo tissues before the infected kernels are mature, and it remains dormant chiefly in the scutellum of the infected kernel until germination of the grain, as described by Brefeld (1903), Brioli (1913), Freeman and

FIG. 14. Spikes of barley (A) and wheat (B) with the spore masses of *Ustilago nuda* replacing the floral tissues. Spore morphology and type of germination of the spores, greatly magnified, are shown in the inserts.

Johnson (1909), Hecke (1904), Hori (1907), Lang (1917), Simmonds (1946), and others. The chlamydospores usually lose their germinative power after a period of a few months, although spores stored at low temperatures frequently remain viable for longer periods, as reported by Stakman (1913) and others. Maddox (1896) was probably the first to demonstrate floral infection in studies with the loose smut of wheat. Environmental conditions, during the period the flower tissues of the barley plant are susceptible, are important in relation to infection. Humid, cool weather with light showers and dews is favorable for dissemination of spores and infection. Inoculation by introducing the spores, either dry or in suspension, into the flowers before pollination

TABLE 2

Days before or after pollination	1	Pol.	1	2	3	4	5	6
Number of spikes inoculated	10	1237	419	482	472	92	33	4
Average loose smut, per cent	43.0	3412	25.1	20.8	20.1	9.0	9.6	3.3

to 3 to 4 days after pollination results in high percentages of infected kernels. Table 2, representing the averages of a large number of inoculations on barley, indicates the relation of stage of kernel development to infection.[1]

The development of *Ustilago nuda* in the infected seed is resumed with the germination and growth of the seedlings. Environmental conditions show only a limited influence on the development of the fungus during the seedling stage, although survival of infected seedlings is influenced greatly by environment during the period of seedling development in all of the smuts. Spores are formed in the leaves when vegetative development is rank. The spores are distributed for some days before and after pollination of the healthy flowers. Floral infection again establishes the fungus in the kernel. Seedling infection from dormant spores carried over on the seed or in the soil, as in the other smuts of the cereals, does not occur naturally.

Tisdale and Tapke (1924) described what they considered a seedling infection with loose smut. Tapke (1932, 1943) later showed this to be a loose type of smut with dark echinulate spores that germinate to form basidia and sporidia, which he described as *Ustilago nigra* Tapke. This fungus is capable of producing seedling infection. Biedenkopf (1894) first described, inadequately, this type of spore germination and considered the fungus an intermediate type of specific rank, *U. medians* Bied.

[1] From unpublished data of H. L. Shands, University of Wisconsin, Madison, Wis.

Specialized races of *U. nuda* occur. On the basis of Fischer's (1953) classification, specialized varieties are differentiated for barley and wheat, and within these varieties physiologic races are separated on the basis of varietal reaction of barley, wheat, and grasses. Some barley varieties such as Odessa (C.I. 934) are homozygous susceptible to all races of the barley pathogen and apparently also to *U. hordei* and *U. avenae* (*U. nigra*). Other varieties as discussed earlier are resistant to specific races of the pathogen; for example Trebi is resistant to the common race of *U. nuda hordei* and susceptible to the Trebi race; while some eight races of the pathogen have been reported from various parts of the world, accurate race differentiation is dependent on further investigation of the genetics of both host and pathogen. The expression of pathogenicity in the smut fungi is associated with the dicaryotic stage of the fungus. Limited investigations have shown most dicaryon clones relatively stable for pathogenicity; however, detailed determinations of homozygosity for pathogenicity are relatively limited. Practically, smut resistance is an economical smut-control measure, although new races of the pathogens are screened out that attack the resistant varieties with continued use and wider distribution of the resistant varieties. The use of seed treatment to prevent the increase of new races of the pathogen and the incorporation of additional factors for smut resistance as they are found reduce smut damage in both cereal and grass production.

Specialization in the smut fungi was first demonstrated by Kniep in 1919. The difference in the use of the term in relation to the smut fungi as compared with those causing the rust and powdery mildew diseases was discussed by Christensen and Rodenhiser (1940). The rust and powdery mildew fungi are propagated by urediospores or conidia, respectively, and thereby true biotypes are maintained. In the smut fungi the mature chlamydospores represent the diploid phase in the life cycle of the fungus which cannot be propagated independently. The chlamydospores germinate to form basidia or promycelia, which represent the beginning of the haploid phase as the reduction divisions occur during this so-called "germination" process. Therefore the resultant cells of the promycelium and sporidia are comparable to the sporidia and ascospores of the rust and powdery mildew fungi, respectively. Before infection occurs in the smut fungi, there must be fusion between compatible haploid lines. Consequently, each new generation of chlamydospores may consist of a new group of related biotypes. For practical purposes, however, the terms "physiologic races" or "specialized races" are used in the smut fungi for collections of chlamydospores having the same general virulence on certain differential suscepts.

CONTROL. Steeping the infected barley in water apparently is the best means of control of the internal, dormant mycelium of the pathogen. The specific use of hot water for loose smut control probably started with

Jensen (1888) and Kellerman and Swingle (1890). More recently the 24- to 30-hour steep in water at 20 to 24°C. or a 6-hour steep and storage in an airtight container under conditions favorable for carbon dioxide accumulation and the formation of intermediate respiratory compounds in the kernels has been used. Loose-smut control in both treatments is dependent somewhat upon the variety, size of kernel, and season and area where the grain was grown; therefore preliminary treatments to determine the proper conditions are advisable. Both treatments reduce germination of the barley approximately 10 per cent in obtaining complete control of the loose smut. The grain must be dried quickly at a temperature not exceeding 38°C. (100°F.). As the smut spores are wind-borne, the treated grain must be grown in an isolated plot to obtain smut-free seed. (See *Wis. Agr. Exp. Sta. Cir.* 416, 1955, and Leukel *et al.*, 1938.)

The early investigations on the inheritance of resistance to loose smut in both barley and wheat are incomplete largely because of low smut infections and incomplete understanding of pathological histology of the smuts. Improvement in methods of floral inoculation by means of the partial-vacuum method (Moore, 1936) and by the introduction of the dry spores into the flowers, using a hypodermic needle with a rubber bulb from a dropper pipette attached to the base of the needle, or by other methods of blowing the spores into all flowers, give more reliable infection percentages (Shands and Schaller, 1946). Inoculation techniques that result in the infection of all susceptible plants frequently demonstrate intermediate smut reaction of varieties based on smutted spikes. The pathological histology in this group of diseases, including the seedling-infecting smuts, involves penetration and early development of the pathogen in maternal tissues in the florally infecting smuts; penetration and systemic development in the seedling and developing plant; and finally, spore formation in the inflorescences in both the floral- and seedling-infecting smuts. Inheritance studies based on smutted spikes, therefore, involve penetration of young plant tissues and the nutritional physiology and disease defense mechanism of the host plant from seedling stage of development to maturity. Histological studies indicate that the expression of resistance involves several host tissues. In the loose smuts initial infection is through the maternal tissues of the ovary; therefore tissues of two generations of the host (ovary and embryo) frequently are involved in the expression of resistance. Reciprocal hybrids frequently demonstrate the influence of the maternal tissues in the expression of resistance.

RESISTANCE AND SPECIALIZATION. Many of the commercial barleys are susceptible to loose smut. Trebi (C.I. 936) and some varieties derived from it are resistant to the common race of the pathogen. Some of the barleys resistant to loose smut used in breeding are Abyssinia

(C.I. 668), Anoidium (C.I. 7269), Dorsett (C.I. 4821), Jet (C.I. 967), Valki (C.I. 5478), Trebi, and several winter barleys (Poehlman 1949, Shands and Schaller, 1946). Resistance to loose smut in both barley and wheat is conditioned by several single factor pairs, mostly dominant, occurring in the different varieties singly or, more frequently, in combination. Apparently one or more of the genes for resistance in the multiple gene relationship function as modifiers or weak genes for resistance with some races of the pathogen. This is the present genetic basis for the intermediate percentages of smutted plants that occur consistently in inheritance studies with the smut diseases (Konzak, 1953, Mohajir et al., 1952, Smith, 1951). Preliminary linkage data indicate that these factor pairs for resistance occur on several chromosomes (Robinson et al., 1955). Shands (1946) has shown a close linkage between stem rust and loose smut resistance in the cross Chevron (C.I. 1111) × Trebi (C.I. 963) with resistance to each disease conditioned by a single factor pair.

16. Black or Semiloose Smut, *Ustilago avenae* (Pers.) Rostr. or *Ustilago nigra* Tapke. Two species, *viz.*, *U. avenae* (Pers.) Rostr. on oats and certain grasses and *U. nigra* Tapke on barley are similar in morphology, life cycle, and the symptoms they produce on the respective suscepts. The two differ only in the plants they are capable of parasitizing. Fischer (1953) has combined these two species under *U. avenae* and listed the synonymy.

The distribution of the black semiloose smut on barley probably is extensive in Europe, Asia, and North and South America. As this smut can be differentiated from the true loose smut only by means of spore germination, the data on distribution are not extensive.

DESCRIPTION. The characteristic symptoms of the black semiloose smut on barley are the relatively dark-brown to black spore mass and the variation in the range of looseness of the spore mass. This smut is not distinguishable by symptoms alone; microscopic examination of the chlamydospores and germination of the spores of the fungus are necessary. The membranes enclosing the sori vary from relatively fragile to persisting or semicovered types. There is apparently considerable variation in the membrane texture due to the influence of the suscept (Fig. 15). The smutted spikes generally appear later than in the case of the loose smut or more nearly comparable to the time of appearance of the covered smut. The spores are shed later and over a longer period than in the case of the loose smut.

THE FUNGUS

Ustilago avenae (Pers.) Rostr. or
Ustilago nigra Tapke

Ustilago medians Bied. is used also as the binomial for this fungus, as discussed by Tapke (1943).

Mycelium is not formed abundantly in culture, although in certain isolates mycelium is produced sparsely. The fungus can be carried for relatively long periods in culture by the continuous budding of the sporidia. This is especially characteristic on culture media high in nutrients. Mycelium in the tissues is hyaline at first and gradually darkens to black at maturity. Chlamydospores are formed in the floral structures, and the spore mass is covered by a membrane varying in its persistence. The chlamydospores are spherical to subspherical, 6.5 by 7 microns, dark brown to black, with echinulations varying from slight to pronounced (Fig. 15, insert A). The spores germinate to form a promycelium (basidium) and oblong to elongate sporidia (Fig. 15, insert A). The sporidia increase by yeast-like budding or function as gametes to initiate the binucleate stage of the fungus. Allison (1937) and Bever (1945) reported hybrids between this species and *Ustilago hordei* (Pers.) Lagerh.

ETIOLOGY. The chlamydospores are carried over on the seed. The inoculum consists of spores carried on the floral bracts or frequently enclosed within the lemma and palea. The spores may be distributed

TABLE 3

Days before or after pollination	1	Pol.	1	2	3	4
Number kernels inoculated....	629	812	1012	1039	963	902
Average black loose smut, per cent.....................	5.8	17.9	18.4	15.3	16.3	14.7

during the flowering period or later during harvesting procedures. The dry chlamydospores remain viable for long periods. Spores introduced into the flowers by inoculation do not show the influence of young flowers on infection as in the case of the loose smut, as indicated in Table 3.[1]

Treatment of a similar number of inoculated kernels with formaldehyde (1 part commercial formalin to 250 parts water) gave complete control of this smut. Environmental conditions during and following the heading of the barley plant have little influence on smut infection by *Ustilago avenae*. Dormant spores may be carried over in the soil under dry conditions, notably in winter barley areas.

The chlamydospores germinate under conditions favorable for the germination of the barley kernels. Sporidial formation and gametic fusions occur under the same conditions. Penetration is chiefly through the coleoptile into the growing point of the very young seedling. The seedling infection and systemic establishment of the parasite in the culm primordia are influenced by environment, as reported by Josephson (1946). The fungus mycelium develops with the growth and differentiation of the plant tissues, and finally spores are produced replacing the ovaries and the floral bracts.

[1] From unpublished data of L. M. Josephson, formerly University of Wisconsin, Madison, Wis.

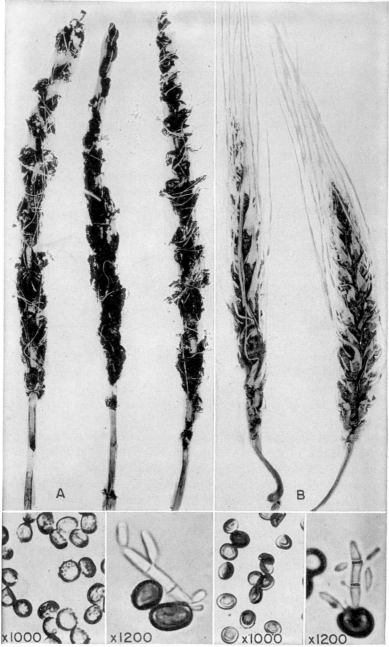

Fig. 15. The black or semiloose smut (A) and the covered smut (B) of barley caused by *Ustilago avenae* and *U. hordei*, respectively. The chlamydospores and type of germination of each are shown in the inserts.

CONTROL. This smut, as in the case of the covered smut, is controlled by the ordinary seed treatments. The organic mercury ducts, such as Ceresan, have been used extensively for the control of the seed-borne spores of this fungus. Formaldehyde, in liquid or dust form, is also fairly effective as a treatment. Inasmuch as the spores are frequently within the floral bracts (hulls), the seed-treatment compound must be relatively volatile to be effective in the control of the seed-borne spores.

RESISTANCE AND SPECIALIZATION. The reaction of barley varieties to this fungus is different from that in the loose smut. The varietal reaction is in general more similar to that of the covered smut. The commercial varieties of barley showing considerable resistance to this smut

TABLE 4

Physiologic race	Reaction on barley varieties					
	Excelsior (C.I. 1248)	Hannchen (C.I. 531)	Himalaya (C.I. 1312)	Lion (C.I. 923)	Nepal (C.I. 595)	Odessa (C.I. 934)
1	R	R	R	R	R	S
2	R	S	R	R	R	S
3	R	R	R	I	R	S
4	R	S	R	S	R	S
5	R	S	S	I	S	S
6	S	I	I	I	S	S
7	R	S	I	R	S	S

R—resistant: I—intermediate: S—susceptible.

are Manchurian varieties as a group and certain of the smooth-awned varieties, notably Newal (C.I. 6088), Wisconsin Barbless (C.I. 5105), and Glabron (C.I. 4577). Certain varieties have been resistant to most races of the fungus, viz., Persicum (C.I. 2448), Pannier (C.I. 1330), and Lyallpur (C.I. 3403). Tapke (1943) reported seven distinct physiological races of *Ustilago nigra* on five varieties of barley, as shown in the table above. Race 4 occurred more frequently in its distribution in the United States than all others combined.

Josephson (1946), using three additional varieties as differentials— Lompoc (C.I. 1213), Manchuria O.A.C. 21 (C.I. 1470), and Wisconsin Barbless (C.I. 5105)—differentiated nine races.

17. Covered Smut, *Ustilago hordei* (Pers.) Lagerh. The covered smuts occurring on barley and oats are listed under *Ustilago hordei* (Pers.) Lagerh. and *U. kolleri* Willie or formerly *U. levis* (K. & S.) Magn., respectively. The covered smuts on these two cereals are similar in symptoms as well as in life cycle and morphology of the fungi. Fischer (1953) combines these under *U. hordei* (Pers.) Lagerh. and lists the extensive host range on the cereals and grasses.   See also Fischer (1938, 1945).

The covered smut is world-wide in its distribution and is perhaps more extensively distributed than either of the other two species on barley. The disease causes losses in the barley crop in the United States averaging approximately as much as the other two smuts combined.

DESCRIPTION.   The characteristic symptom of the covered smut on barley is a rather persistent membrane enclosing the sorus until the plants are fully mature.   Quite frequently the grayish-white sori are enclosed within the partly modified floral bracts of the spikelet.   The awn on the awned varieties also develops in most instances (Fig. 15). The smutted heads emerge at about the same time as those of the healthy plants and are conspicuous especially as the crop reaches maturity. The smut masses and dark-brown to black smooth-walled spores on the kernels are conspicuous in the threshed grain from fields where smut infection is more than a few per cent.   The Federal grain grades designate such barley "smutty," and the barley grade carries the designation "smutty."

THE FUNGUS

Ustilago hordei (Pers.) Lagerh.
(*Uredo segetum* subsp. *hordei* Pers.)
(*Ustilago carbo* var. *hordei* D.C.)
(*Ustilago segetum* var. *hordei* Rab.)
(*Ustilago segetum* var. *hordei* f. *tecta* Jens.)
(*Ustilago avenae* var. *levis* Kell. & Swing.)
(*Ustilago levis* Kell. & Swing.)
(*Ustilago segetum* var. *tecta* Jens.)
(*Ustilago hordei* var. *tecta* Jens.)
(*Ustilago jensenii* Rostr.)
(*Ustilago hordei* Kell. & Swing.)
(*Ustilago kolleri* Wille)

Mycelium in culture is sparse, most of the development being by yeast-like budding of the numerous sporidia.   In the tissues the mycelium is at first hyaline, later turning to dark brown or black.   The chlamydospores are lighter-colored on one side, sub-spherical to angular with a smooth outer membrane, and range from 5 to 9 microns in diameter (Fig. 15, insert *B*).   The spores germinate to form characteristically a four-celled promycelium (basidium) and four ovate to oblong sporidia with the abundant development of secondary sporidia (Fig. 15, insert *B*).

ETIOLOGY.   Infection of the very young seedlings occurs from the seed-borne chlamydospores.   In some drier areas the infection may occur from spores in the surface soil.   Penetration is through the young coleoptile into the embryonic growing point, and the further development of the parasite occurs in association with the differentiating tissues of the crown buds and floral structures especially. Spores are produced, replacing the kernel and, less commonly, the floral bracts. The seedling infection is influenced by soil environmental conditions, especially soil temperature, moisture, and, to a lesser extent, soil fertility, as reported by Faris (1924), Schaffnit (1926), and others.

CONTROL. The seed-borne spores are controlled by the use of the organic mercury dusts as well as by formaldehyde dusts and solutions.

RESISTANT VARIETIES AND SPECIALIZATION. The varietal reaction to this smut is similar to that of the black semiloose smut. There are, however, other varieties that show some differential response. The following varieties are relatively resistant to this fungus: Persicum (C.I. 2448), Pannier (C.I. 1330), Hillsa (C.I. 1604), Lyallpur (C.I. 3403), and a hull-less barley (C.I. 2448). Shands (1956) has shown that resistance in Brachitic (C.I. 6572) is conditioned by a single dominant factor pair; minor factors influencing the expression of resistance are suggested in another hybrid studied.

Specialization in pathogenicity of the barley races of this parasite has been shown by Aamodt (1935), Faris (1924), Rodenhiser (1928), Tapke (1937, 1943), and others. The reaction of 8 barley varieties to the 13 physiologic races as reported by Tapke (1937, 1945) is shown in Table 5.

TABLE 5

Physio-logic race	Reaction of barley varieties							
	Excel-sior (C.I. 1248)	Hima-laya (C.I. 1312)	Hann-chen (C.I. 531)	Lion (C.I. 923)	Nepal (C.I. 595)	Odessa (C.I. 934)	Pannier (C.I. 1330)	Trebi (C.I. 936)
1	R	R	I	R	R	S	R	I
2	R	R	R	I	S	S	R	R
3	I	R	R	I	S	S	R	R
4	R	I	I	R	S	S	I	I
5	R	R	R	I	R	S	R	I
6	R	R	I–S	I	R	S	R	S
7	R	R	R	R	S	S	R	I
8	R	R	R	R	R	S	R	R
9	I	R	I	I	S	S	R	R
10	R	I	I	I	S	S	I	I
11	R	I	I–S	I	I	S	R	S
12	R	I	I	R	S	S	R	R
13	I	R	I	I	S	S	R	I

R—resistant: I—intermediate: S—susceptible.

The physiologic race 6 has been reported by Tapke (1937, 1945) as the most prevalent through the spring barley area of the United States.

18. Stem Rust, *Puccinia graminis* Pers. Stem rust on *Hordeum* spp. is caused by one or more of three of the specialized varieties, *viz.*, *Puccinia graminis tritici* Eriks. & Henn., *P. graminis secalis* Eriks., and certain specialized races that are confined more particularly to barley. Stem rust is distributed extensively on both cultivated and wild species of *Hordeum* throughout the humid and semihumid temperate regions of the world. Losses generally are not so heavy on barley over a period of years as those occurring on wheat. Barley, however, rates second in stem rust losses in the United States, and in certain years, as in 1937, losses are very severe over relatively wide areas. The disease is more prevalent in the North Central spring barley area of Canada and the United States than in any of the other barley sections of North America. Stem rust is also severe on barley across northern Europe and Asia. The important stem rust–resistant varieties used for breeding purposes are Chevron (C.I. 1111), Peatland (C.I. 5267), and several plant selections similar to these two varieties. Kindred (C.I. 6969) and Mars (C.I. 7015) are stem rust–resistant commercial varieties. Stem rust is discussed more fully in Chap. 11.

19. Stripe Rust, *Puccinia glumarum* (Schm.) Eriks. & Henn. The stripe rust occurs on cultivated and wild *Hordeum* spp. as well as being distributed extensively on *Triticum* spp. and many grasses. In the United States and Canada this disease is of relatively minor importance on barley, as it is restricted to the Pacific Coast and Intermountain area where most of the commercial varieties grown are moderately resistant. Considerable damage is reported on barley in northern Europe, parts of northern Asia, and in Argentina, Chile, and Peru in South America.

DESCRIPTION. The stripe rust on barley is different in appearance from either the leaf rust or stem rust. The linear citron-yellow uredia are conspicuous on the leaf blades and sheaths as well as on the floral bracts when conditions are favorable for infection. Frequently the uredia unite end to end to form narrow stripes extending considerable distances on the leaf blades and sheaths of susceptible varieties. The telia form narrow, fine lines, dark brown in color and covered by the epidermis. The uredia are conspicuous from very early spring until mid-summer; the telia develop sparsely in late summer to maturity of the crop. A relatively large number of barleys are resistant to this rust, as reported by Bever (1938), Newton and Johnson (1936), Straib (1935), and others. For a complete discussion of the stripe rust see Chap. 11.

20. Leaf Rust, *Puccinia anomala* Rostr., or preferably *P. hordei* Otth. The leaf rust of barley occurs extensively in both the winter and spring barley areas of the Eastern and Central United States. In this area the rust is found almost every season, and in some seasons it develops in epiphytotic form, especially in the southern spring barley area. This rust is distributed generally in most of the countries where barley is grown. *Puccinia hordei* is probably confined to cultivated barley species and a few closely related wild grasses. The aecial stage of this rust fungus is found on *Ornithogalum umbellatum* L. (the common star-of-Bethlehem) and some other species, as reported by Mains and Jackson (1924) and Tranzschel (1914). Although *O. umbellatum* is distributed generally, the aecial infection is uncommon in North America. The main distribution of the rust apparently is a northward spread of uredial inoculum from the southern winter barley area. Certain races of the wheat leaf rust fungus, *P. rubigo-vera tritici* (Eriks.) Carleton or *P. recondita* Rob. *et* Desm., also infect barley in the Mississippi Valley area and westward (Johnston, 1936).

DESCRIPTION. The uredia are small, round, and light yellowish brown in color. This rust is relatively inconspicuous until uredial development is quite abundant. The telia are round to oblong, brown, and covered by the epidermis. The telial stage is less abundant, especially in the more northern sections of the barley area. Both uredia and telia develop on the leaf blades and leaf sheaths and rarely occur on the floral structures.

THE FUNGUS

Puccinia hordei Otth. or *P. anomala* Rostr.
(*Puccinia rubigo-vera simplex* Koern.)
(*Puccinia simplex* Eriks. & Henn.)
(*Aecidium ornithagaleum* Bubak.)

Buchwald (1943) has shown that *Puccinia hordei* Otth. was used in 1871, and *P. anomala* Rostr. in 1878.

The pycnia and aecia occur as elevated light orange-yellow areas on the leaves of *Ornithogalum*. The aeciospores are globoid, hyaline, and minutely verrucose. The uredia are round, small, yellowish brown in color, and they show little or no rupturing of the epidermis. The urediospores are broadly ellipsoid, light yellow in color, finely echinulate, with germ pores distributed on all faces of the spore. The telia are round to oblong, are covered by the epidermis, and produce mainly one-celled spores. The teliospores germinate to form characteristically the four-celled basidium and sporidia.

ETIOLOGY. In the more southern winter barley regions, the leaf rust develops in the uredial stage throughout the winter growing period. The urediospores do not overwinter in the spring barley area. The northward spread of the uredial infection is prevalent as the crop develops. Infections in the more northern sections frequently are not evident until relatively late in the spring or early summer. Secondary spread from urediospores is abundant during warm, humid, summer weather. Telial development is limited in the spring barley area. The aecial stage does not develop abundantly, naturally in the United States, although

Mains and Jackson (1924) reported it in Indiana, and it is produced at Madison, Wis., when barley straw containing telia is placed near *Ornithogalum umbellatum*.

CONTROL AND RESISTANCE. The economical control of this rust is through the use of resistant varieties. Many of the cultivated barleys grown in the United States are moderately resistant. Barleys resistant to the leaf rust are Callas (C.I. 2440), Mecknos Moroc (C.I. 1379), Peruvian (C.I. 935), Quinn (C.I. 1204), Bolivia (C.I. 1257), Juliaca (C.I. 1114, C.I. 2329), and many others in the barley collection of the U.S. Department of Agriculture. Two selections of Orge—B100 and B101—were resistant to the two physiologic races in the United States and to the Australian race. Two physiologic races of *P. anomala* have been reported by Mains and Martini (1932) in the United States, and Waterhouse (1928) reported one additional race in Australia.

REFERENCES

AAMODT, O. S., and W. H. JOHNSTON. Reaction of barley varieties to infection with covered smut (*Ustilago hordei*). *Can. Jour. Research* **12**:590–613. 1935.

ÅBERG, E. A., and G. A. WIEBE. Classification of barley varieties grown in the United States and Canada in 1945. *U.S. Dept. Agr. Tech. Bul.* 907. 1946.

ALLISON, C. C. Studies on the genetics of smuts of barley and oats in relation to pathogenicity. *Minn. Agr. Exp. Sta. Tech. Bul.* 119. 1937.

ARNY, D. C. Inheritance of resistance to barley stripe. *Phytopath.* **35**:781–804. 1945.

———. Inheritance of resistance to spot blotch in barley seedlings. *Phytopath.* **41**:691–698. 1951.

ATANSOFF, D. Fusarium blight of the cereal crops. *Mededeel. Landbouwhsc.* **D.27**:1–132. 1923.

BENNETT, F. T. *Gibberella saubinetii* on British cereals. *Ann. Appl. Biol.* **17**:43–58. 1930. **18**:158–177. 1931. **20**:377–380. 1933.

———. Fusarium species on British cereals. *Ann. Appl. Biol.* **19**:21–34. 1932. **20**:272–290. 1933. **22**:479–507. 1935.

BERGEY, D. H., *et al.* Bergey's manual of determinative bacteriology. The Williams & Wilkins Company, 6th ed. Baltimore. 1946.

BEVER, W. M. Reaction of wheat, barley and rye varieties to stripe rust in the Pacific Northwest. *U.S. Dept. Agr. Cir.* 501. 1938.

———. Hybridization and genetics in *Ustilago hordei* and *U. nigra*. *Jour. Agr. Research* **71**:41–59. 1945.

BIEDENKOPF, H. *Ustilago medians*, ein neuer Brand auf Gerste. *Zeitschr. Pflanzenkrank.* **4**:321–322. 1894.

BREFELD, O. Neue Untersuchungen und Ergebnisse über die natürliche Infektion und Verbreitung der Brandkrankheiten des Getreides. *Nachr. Klub. Landw. Berlin,* **466**:4224–4234. 1903.

BRIGGS, F. N. The inheritance of resistance to mildew. *Am. Nat.* **72**:34–41. 1938.

——— and E. H. STANFORD. Linkage of factors for resistance to mildew in barley. *Jour. Gen.* **37**:107–117. 1938.

——— and ———. Linkage relations of the Goldfoil factor for resistance to mildew in barley. *Jour. Agr. Research* **66**:1–5. 1943.

BROILI, J. Versuche mit Brandinfektion zur Erzielung brandfreier Gerstenstämme. *Naturw. Zeitschr. Forst. Landw.* **8**:335–344. 1910.

BUCHWALD, N. F. VON. Über *Puccinia hordei* Otth. (syn. *P. simplex* (Kcke.) Erikss. und Henn), und *P. hordei-murini* n.n. (syn. *P. hordei* Fckl.). *Annales Mycologici* **41**:306–316. 1943.

CALDWELL, R. M. Rhynchosporium scald of barley, rye and other grasses. *Jour. Agr. Research* **55**:175–198. 1937.

CARLTON, M. A. The small grains. The Macmillan Company. New York. 1920.

CHEREWICK, W. J. Studies on the biology of *Erysiphe graminis* DC. *Can. Jour. Research* C **22**:52–86. 1944.

CHRISTENSEN, C. Haploide Linien von *Ustilago tritici*. *Der Züchter* **7**:37–39. 1935.

CHRISTENSEN, J. J. Studies on the parasitism of *Helminthosporium sativum*. *Minn. Agr. Exp. Sta. Tech. Bul.* 11. 1922.

———. Physiologic specialization and mutation in *Helminthosporium sativum*. *Phytopath.* **15**:785–796. 1925.

———. Non-parasitic leaf spots of barley. *Phytopath.* **24**:726–742. 1934.

——— and F. R. DAVIES. Nature of variation in *Helminthosporium sativum*. *Mycologia* **29**:85–99. 1937.

——— and T. W. GRAHAM. Physiologic specialization and variation in *Helminthosporium gramineum* Rabh. *Minn. Agr. Exp. Sta. Bul.* 95. 1934.

——— and H. C. H. KERNKAMP. Studies on the toxicity of blighted barley to swine. *Minn. Agr. Exp. Sta. Tech. Bul.* 113. 1936.

——— and H. A. RODENHISER. Physiologic specialization and genetics of the smut fungi. *Bot. Rev.* **6**:389–425. 1940.

——— and C. L. SCHNEIDER. The effect of repeated passage of *Helminthosporium sativum* through the host on genetic variation and pathogenicity. *Phytopath.* **38**:5. 1948.

——— and E. C. STAKMAN. Relation of Fusarium and Helminthosporium in barley seed to seedling blight and yield. *Phytopath.* **25**:309–327. 1935.

CUNNINGHAM, G. H. The Ustilagineae, or smuts of New Zealand. *Trans. New Zealand Inst.* **55**:397–433. 1924.

DASTUR, J. F. Notes on some fungi isolated from "black-point" affected wheat kernels in the Central Provinces. *Indian Jour. Agr. Sci.* **12**:731–742. 1942.

DICKSON, J. G. Cereal disease studies in Europe and Asia. Madison, Wis. 1930. (Mimeographed.)

———. Scab of wheat and barley and its control. *U.S. Dept. Agr. Farmers' Bul.* 1599. 1942.

DIEDICKE, H. Über den Zusammenhang zwischen *Pleospora* und *Helminthosporium* Arten I and II. *Centralbl. für Bakt.* **9**:317–329. 1902. **11**:52–59. 1904.

DIETZ, S. M. The varietal response and inheritance of resistance in barley to *Erysiphe graminis hordei*. *Iowa State Col. Jour. Sci.* **5**:25–33. 1930.

DRECHSLER, C. Some graminicolous species of *Helminthosporium*. *Jour. Agr. Research* **24**:641–739. 1923.

———. Phytopathological and taxonomic aspects of Ophiobolus, Pyrenophora, Helminthosporium, and a new genus, Cochliobolus. *Phytopath.* **24**:953–983. 1934.

EATON, F. M. Deficiency, toxicity and accumulation of boron in plants. *Jour. Agr. Research* **69**:237–277. 1944.

ELLIOTT, C. Manual of bacterial plant pathogenes. 2d ed. Chronica Botanica Co. Waltham, Mass. 1951.

68   DISEASES OF FIELD CROPS

———. Recent developments in the classification of bacterial plant pathogenes. *Bot. Rev.* **9**:655–666. 1943.

FANG, C. T., O. N. ALLEN, A. J. RIKER, and J. G. DICKSON. The pathogenic, physiological, and serological reactions of the form species of *Xanthomonas translucens*. *Phytopath.* **40**:44–64. 1950.

FARIS, J. A. Factors influencing infection of *Hordeum sativum* by *Ustilago hordei*. *Am. Jour. Bot.* **11**:189–214. 1924.

FISCHER, G. W. Some new grass smut records from the Pacific Northwest. *Mycologia* **30**:385–395. 1938.

———. Studies of the susceptibility of forage grasses to cereal smut fungi. *Phytopath.* **29**:490–494. 1939.

———. Some evident synonymous relationships in certain graminicolous smut fungi. *Mycologia* **35**:610–619. 1943.

———. Manual of the North American smut fungi. The Ronald Press Company. New York. 1953.

FREEMAN, E. M., and E. C. JOHNSON. The loose smuts of barley and wheat. *U.S. Dept. Agr. Bur. Plant Ind. Bul.* 152. 1909.

GESCHELE, E. The response of barleys to the parasitic fungus, *Helminthosporium teres*. Sacc. *Bul. Appl. Bot., Gen. and Plant Breeding* **19**:371–384. 1928. (In Russian.)

GOLD, A. H., C. A. SUNSEON, B. R. HOUSTON, and J. W. OSWALD. Electron microscopy and seed and pollen transmission of rod-shaped particles associated with the false stripe virus disease of barley. *Phytopath.* **44**:115–117. 1954.

GREANEY, F. J., and J. E. MACHACEK. Prevalence of seed-borne fungi in cereals in certain seed-inspection districts of Canada. *Sci. Agr.* **22**:419–437. 1942.

HARBORG, W. A. F. Classification revision in *Xanthomonas translucens*. *Can. Jour. Research* **C 20**:312–326. 1942.

———. Dwarfing of wheat and barley by the barley stripe-mosaic (false stripe) virus. *Can. Jour. Bot.* **32**:24–37. 1954.

HARLAN, H. V., and M. L. MARTINI. Problems and results in barley breeding. *U.S. Dept. Agr. Yearbook*, pp. 303–346. 1936.

HARPER, R. A. Sexual reproduction and organization of the nucleus in certain mildews. *Carnegie Inst. Wash. Pub.* **37**:1–104. 1905.

HAYES, H. K., *et al.* Reaction of barley varieties to *Helminthosporium sativum*. *Minn. Agr. Exp. Sta. Tech. Bul.* 21. 1923.

HAYWARD, H. E., and C. H. WADLEIGH. Plant growth on saline and alkali soils. *Adv. Agron.* **1**:1–38. 1949.

HECKE, L. Ein innerer Krankheitskeim des Flugbrandes im Getreidekorn. *Zeitschr. Landw. Versuchsw. Österr.* **7**:59–64. 1904.

HONECKER, L. Über die Modifizierbarkeit des Befalles und das Auftreten verschiedener physiologischer Formen beim Mehltau der Gerste, *Erysiphe graminis hordei* Marchal. *Zeitschr. Züchtung* **A 19**:577–602. 1934. *Züchter* **7**:113–119. 1935.

———. Die Stellung der Gerste in der Erzeugungsschlacht mit besonderer Berücksichtigung der Braugerste. *Prakt. Blat. Pflanzenb. und Pflanzensch.* **14**:325–342. 1937.

HORI, S. Seed infection by smut fungi of cereals. *Bul. Imp. Central. Agr. Exp. Sta. Japan* **1**:163–176. 1907. (In Japanese.)

HOYMAN, W. G. Concentration and characterization of the emetic principal present in barley infected with *Gibberella saubinetii*. *Phytopath.* **31**:871–885. 1941.

HYNES, H. J. Species of Helminthosporium and Curvularia associated with root-rot of wheat and other graminaceous plants. *Jour. Proc. Roy. Soc. N.S. Wales* **70**:378–391. 1937.

———. Studies on Helminthosporium root-rot of wheat and other cereals. *N.S. Wales. Dept. Agr. Sci. Bul.* 47. 1935. *Bul.* 61. 1938.

ISENBECK, K. Untersuchungen über *Helminthosporium gramineum* Rabh. im Rahmen der Immunitätszüchtung. *Phytopath. Zeitschr.* **2**:503–555. 1930.

———. Die Bedeutung der Faktoren Temperatur und Licht für die Frage der Resistenzverschiebung bei verschiedenen Sommergersten gegenüber *Helminthosporium gramineum. Kühn-Archiv.* **44**:1–54. 1937.

ITO, S., and K. KURIBAYASHI. The ascigerous forms of some graminicolous species of Helminthosporium in Japan. *Jour. Fac. Agr. Hokkaido Imp. Univ., Sapporo* **29**:85–125. 1931.

JACZEWSKI, A. A. Bacterial plant diseases. State Printing Office. Moscow-Leningrad. 1935. (In Russian.)

JENSEN, J. L. The propagation and prevention of smut in oats and barley. *Jour. Roy. Soc. England* **24**:397–415. 1888.

JOHNSTON, C. O. Reaction of certain varieties and species of the genus Hordeum to leaf rust of wheat, *Puccinia triticina. Phytopath.* **26**:235–245. 1936.

JONES, L. R., A. G. JOHNSON, and C. S. REDDY. Bacterial blight of barley. *Jour. Agr. Research* **11**:625–643. 1917.

Josephson, L. M. Studies on the intermediate loose smut of barley. Ph.D. thesis. University of Wisconsin. Madison, Wis. 1946.

KELLERMAN, W. A., and W. T. SWINGLE. Report on the loose smuts of cereals. *Kans. Agr. Exp. Sta. Rept.* **2**:213–288. 1890.

KIESSLING, L. Über die spezifische Empfindlichkeit der Gerste gegenüber der Streifenkrankheit. *Zeitschr. Pflanzenzücht.* **5**:31–40. 1917.

KONZAK, C. F. Inheritance of resistance in barley to physiologic races of *Ustilago nuda. Phytopath.* **43**:369–375. 1953.

KURIBAYASHI, K. The ascigerous stage of *Helminthosporium sativum. Trans. Sapporo Nat. Hist. Soc.* **10**:138–145. 1929.

LANG, W. Zur Ansteckung der Gerste durch *Ustilago nuda. Ber. Deutsch. bot. Ges.* **35**:4–20. 1917.

LEACH, J. G. Insect transmission of plant diseases. McGraw-Hill Book Company, Inc. New York. 1940.

LEUKEL, R. W., J. G. DICKSON, and A. G. JOHNSON. Effects of certain environmental factors on stripe disease of barley and the control of the disease by seed treatment. *U.S. Dept. Agr. Tech. Bul.* 341. 1933.

——— et al. Wheat smuts and their control. *U.S. Dept. Agr. Farmers' Bul.* 1711. 1938.

LINDFORS, T. Mykologische Notizen. *Svensk. Bot. Tidsskr.* **12**:221–227. 1918.

LUTTRELL, E. S. A Taxonomic revision of *Helminthosporium sativum* and related species. *Am. Jour. Bot.* **42**:57–68. 1955.

MACHACEK, J. E. The prevalence of Septoria on cereal seed in Canada. *Phytopath.* **35**:51–53. 1945.

——— and H. A. H. WALLACE. Non-sterile soil as a medium for tests of seed germination and seed-borne disease in cereals. *Can. Jour. Research* C **20**:539–557. 1942.

MACKIE, W. W. Inheritance of resistance to rusty blotch in barley. *Jour. Agr. Research* **36**:965–975. 1928.

———. Inheritance of resistance to barley scald. *Phytopath.* **19**:1141. 1929.

———— and G. E. PAXTON. A new disease of cultivated barley in California caused by *Helminthosporium Californicum* N. Sp. *Phytopath.* **13**:562. 1923.

MADDOX, F. Smut and bunt. *Agr. Gaz. Tasmania* **4**:92–95. 1896. *Agr. Gaz. N.S. Wales* **21**:58–59. 1910.

MAINS, E. B., *et al.* Scab of small grains and feeding trouble in Indiana in 1928. *Proc. Ind. Acad. Sci.* **39**:101–110. 1930.

———— and S. M. DIETZ. Physiologic forms of barley mildew, *Erysiphe graminis hordei* Marchal. *Phytopath.* **20**:229–239. 1930.

———— and H. S. JACKSON. Aecial stages of the leaf rusts of rye, *Puccinia dispersa* Eriks. and Henn., and of barley, *P. anomala* Rostr. in the United States. *Jour. Agr. Research* **28**:1119–1126. 1924.

———— and M. L. MARTINI. Susceptibility of barley to leaf rust (*Puccinia anomala*) and to powdery mildew (*Erysiphe graminis hordei*). *U.S. Dept. Agr. Tech. Bul.* 295. 1932.

MARCHAL, E. De la specialisation du parasitisme chez l'*Erysiphe graminis*. *Compt. rend. Acad. Sci. Paris* **135**:210–212. 1902. **136**:1280–1281. 1903.

McKINNEY, H. H. Culture methods for detecting seed-borne virus in Glacier barley seedlings. *Plant Dis. Reptr.* **38**:152–162. 1954.

MITRA, M. A comparative study of species and strains of Helminthosporium on certain Indian cultivated crops. *Trans. British Myc. Soc.* **15**:254–293. 1931.

MOHAJIR, A. R., D. C. ARNY, and H. L. SHANDS. Studies on the inheritance of loose smut resistance in spring barleys. *Phytopath.* **42**:367–373. 1952.

MOORE, M. B. A method for inoculating wheat and barley with loose smuts. *Phytopath.* **26**:397–400. 1936.

MOURASHKINSKI, K. E. Materials for the study of fusariose of cereals. I. Species of the genus Fusarium on cereals in Siberia. *Trans. Siberian Agr. Acad.* **3**:87–120. 1924.

MUNDKUR, B. B. Some preliminary feeding experiments with scabby barley. *Phytopath.* **24**:1237–1243. 1934.

NAUMOV, N. A. Quelques observations sur une espèce du genre Fusarium rottachée au *Gibberella saubinetii* Sacc. *Bul. Soc. Mycology, France* **30**:54–63. 1914.

NEWTON, M., and T. JOHNSON. Stripe rust, *Puccinia glumarum*, in Canada. *Can. Jour. Research* C **14**:89–108. 1936.

NISHIKADO, Y. Studies on the Helminthosporium diseases of Gramineae in Japan. *Ber. Ohara Inst. Landw. Forsch.* **4**:111–126. 1929.

NOACK, F. *Helminthosporium gramineum* und *Pleospora trichostoma*. *Zeitschr. Pflanzenkrank.* **15**:193–205. 1905.

OSWALD, J. W., and B. R. HOUSTON. A downy mildew of barley in California. *Phytopath.* **41**:942. 1951.

———— and ————. The yellow-dwarf virus disease of cereal crops. *Phytopath.* **43**:128–136. 1953.

PALCHEVSKY, N. A. Diseases of cereal crops of South Ussuaria. *Acad. Sci. St. Petersburg*, pp. 1–43. 1891. (In Russian. Translation in Department of Plant Pathology, University of Wisconsin.)

PAMMEL, L. H., C. M. KING, and A. L. BAKKE. Two barley blights with comparison of species of *Helminthosporium* upon cereals. *Iowa Agr. Exp. Sta. Bul.* 116, pp. 178–190. 1910.

PAXTON, G. E. Studies on *Helminthosporium* species found on cultivated barley in California. *Phytopath.* **12**:446–447. 1922.

PERSOON, D. C. H. Synopsis methodica fungorum. Dieterich, p. 224. Göttingae. 1901.

POEHLMAN, J. M.   Sources of resistance to loose smut, *Ustilago nuda*, in winter barleys of foreign origin.   *Agron. Jour.* **41**:191–194.   1949.

RAVN, F. K.   Nogle *Helminthosporium*.   Arter og de af dem fren.kaltde Sygdomme hos Byg og Havre.   *Bot. Tidsskr.* **23**:101–321.   1900.   (Translation in U.S. Department of Agriculture Library and Department of Plant Pathology, University of Wisconsin.)

REED, G. M.   The mildews of the cereals.   *Bul. Torrey Bot. Club* **36**:353–388.   1909.

RIDDLE, O. C., and F. N. BRIGGS.   Inheritance of resistance to scald in barley.   *Hilgardia* **20**:19–27.   1950.

RIKER, A. J., and I. L. BALDWIN.   Names for the bacterial plant pathogenes.   *Chronica Botanica* **7**:250–252.   1942.

ROBERTSON, D. W., G. A. WIEBE, and R. G. SHANDS.   A summary of linkage studies in barley: Supplement II, 1947–1953.   *Agron. Jour.* **47**:418–425.   1955.

ROCHE, B. H., and G. BOHSTEDT.   Scabbed barley and oats and their effects on various classes of livestock.   *Am. Soc. Animal Prod. Res. Ann. Meeting* **23**:219–222.   1931.

RODENHISER, H. A.   Physiologic specialization of *Ustilago nuda* and *Ustilago tritici*.   *Phytopath.* **16**:1001–1007.   1926.   **18**:955–1003.   1928.

ROEMER, T., *et al*.   Die Züchtung resistenter Rassen der Kulturpflanzen.   Paul Parey.   Berlin.   1938.

SALMON, E. S.   A monograph of the Erysiphaceae.   *Torrey Bot. Club Mem.* **9**:1–292.   1900.

————.   Cultural experiments with the barley mildew, *Erysiphe graminis* DC.   *Am. Mycology* **2**:70–99.   1904.

————.   Specialization of parasitism in the Erysiphaceae.   *New Phytologist* **3**:55–60, 109–121.   1904.

SAMPSON, K.   Life cycles of smut fungi.   *Brit. Myc. Soc. Trans.* **23**:1–23.   1939.

SCHAFFNIT, E.   Zur physiologie von *Ustilago hordei* K. and S.   *Ber. deutsch. bot. Gesell.* **44**:151–156.   1926.

SCHALLER, C. W.   Inheritance of resistance to loose smut *Ustilago nuda* in barley.   *Phytopath.* **39**:959–979.   1949.

————.   The effect of mildew and scald infection on yield and quality of barley.   *Agron. Jour.* **43**:183–188.   1951.

————.   Inheritance of resistance to net blotch of barley.   *Phytopath.* **45**:174–176.   1955.

SCHROPP, W.   Bor und Gramineen.   *Forschungs dienst* **10**:138–160.   1940.

SEMENIUK, W., and J. G. ROSS.   Relation of loose smut to yield of barley.   *Can. Jour. Research* C **20**:491–500.   1942.

SHANDS, H. L.   Temperature studies on stripe of barley.   *Phytopath.* **24**:362–383.   1934.

———— and D. C. ARNY.   Stripe reaction of spring barley varieties.   *Phytopath.* **34**:572–585.   1944.

———— and J. G. DICKSON.   Variation in hyphal tip cultures from conidia of *Helminthosporium gramineum*.   *Phytopath.* **24**:559–560.   1934.

———— and C. W. SCHALLER.   Response of spring barley varieties to floral loose smut inoculation.   *Phytopath.* **36**:534–548.   1946.

SHANDS, R. G.   Chevron, a barley variety resistant to stem rust and other diseases.   *Phytopath.* **39**:209–211.   1939.

————.   An apparent linkage of resistance to loose smut and stem rust in barley.   *Jour. Am. Soc. Agron.* **38**:690–692.   1946.

————. Inheritance of covered smut resistance in two barley crosses. *Agron. Jour.* **48**:81–86. 1956.

SIMMONDS, P. M. Detection of the loose smut fungi in embryos of barley and wheat. *Sci. Agr.* **26**:51–58. 1946.

SMITH, G. The haustoria of the Erysiphaceae. *Bot. Gaz.* **29**:153–184. 1900.

SMITH, L. Cytology and genetics of barley. *Bot. Rev.* **17**:1–51, 133–202, 285–355. 1951.

SMITH, N. J. G. Observations of the Helminthosporium diseases of cereals in Britain. *Ann. Appl. Biol.* **16**:236–260. 1929.

———— and J. M. RATTRAY. Net blotch, spot blotch and leaf-stripe diseases of barley in South Africa. *So. African Jour. Sci.* **27**:341–351. 1930.

SOROKIN, H. Ueber einege Krankheiten der Kulturpflanzen im Sudussurischchen Gebeit. *Trudy Obslich. Estest Imp. Kazan Univ.* **22**:1–32. 1890.

SPRAGUE, R. Septoria disease of Gramineae in Western United States. *Oregon State Col. Monogr.* 6. 1944.

————. Diseases of cereals and grasses in North America. The Ronald Press Company, pp. 249–251. New York. 1950.

STAKMAN, E. C. Spore germination of cereal smuts. *Minn. Agr. Exp. Sta. Tech. Bul.* 133. 1913.

STRAIB, W. Auftreten und Verbreitung biologischer Rassen des Gelbrostes (*Puccinia glumarum* (Schm.) Eriks. und Henn.) im Jahre 1934. *Arb. biol. Reichs Land-und Forstw.* **21**:455–466, 467–481. 1935. **22**:43–63. 1936.

SUNESON, C. A. Physiologic and genetic studies with the stripe disease of barley. *Hilgardia* **20**:29–36. 1950.

TAPKE, V. F. A study of the cause of variability in response of barley loose smut to control through seed treatment with surface disinfectants. *Jour. Agr. Research* **51**:491–508. 1935.

————. Pathogenic strains of *Ustilago nigra*. *Phytopath.* **26**:1033–1034. 1936.

————. Physiologic races of *Ustilago hordei*. *Jour. Agr. Research* **55**:683–692. 1937.

————. Studies on the natural inoculation of seed barley with covered smut (*Ustilago hordei*). *Jour. Agr. Research* **60**:787–810. 1940.

————. Occurrence, identification, and species validity of the barley loose smuts, *Ustilago nuda, U. nigra* and *U. medians*. *Phytopath.* **33**:194–209. 1943.

————. Physiologic races of *Ustilago nigra*. *Phytopath.* **33**:324–327. 1943.

————. New physiologic races of *Ustilago hordei*. *Phytopath.* **35**:970–976. 1945.

THREN, R. Über Zustandekommen und Erhaltung der Dikaryophase von *Ustilago nuda* (Jensen) Kellerm. et Sw. und *Ustilago tritici* (Persoon) Jensen. *Zeitschr. Bot.* **36**:449–498. 1941.

TIDD, J. S. Studies concerning the reaction of barley to two undescribed physiologic races of barley mildew, *E. graminis hordei* Marchal. *Phytopath.* **27**:51–68. 1937.

TINLINE, R. D. Studies on the perfect stage of *Helminthosporium sativum*. *Can. Jour. Bot.* **29**:467–478. 1951.

TISDALE, W. H., and V. F. TAPKE. Infection of barley by *Ustilago nuda* through seed inoculation. *Jour. Agr. Research* **29**:263–287. 1924.

TRANZSCHEL, V. A. Kulturversuche mit Uredineen in den Jahren 1911–1913. *Mycol. Centralbl.* **4**:70–71. 1914.

WALLIN, J. R. Parasitism of *Xanthomonas translucens* (J. J. and R.) Dowson on grasses and cereals. *Iowa State Col. Jour. Sci.* **20**:171–193. 1946.

WATERHOUSE, W. L.   Studies in the inheritance of leaf rust, *Puccinia anomala* Rostr., in crosses of barley.   *Jour. Proc. Roy. Soc. N.S. Wales* **61**:218–247.   1928.

WEAVER, J. C.   Climatic relations of American barley production.   *Geograph. Rev.* **33**:569–588.   1943.

WEHMEYER, L. E.   Studies in the Genus Pleospora.   I.   *Mycologia* **41**:565–593. 1949.

YARWOOD, C. E.   Effect of mildew and rust infection on dry weight and respiration of excised clover leaflets.   *Jour. Agr. Research* **49**:549–558.   1934.

———.   Tolerance of *Erysiphe polygoni* and certain other powdery mildews to low humidity.   *Phytopath.* **26**:845–859.   1936.

———.   Copper sulphate as an eradicant spray for powdery mildews.   *Phytopath.* **35**:895–909.   1945.

CHAPTER 4

CORN DISEASES

Corn, or maize, is represented by the single species, *Zea mays* L. All the cultivated varieties of corn are included in this species, and no wild or uncultivated forms are known. Botanical varieties or subspecies of *Zea mays* are pod corn, *Z. mays* var. *tunicata;* flour corn, *Z. mays* var. *amylacea;* flint corn, *Z. mays* var. *indurata;* popcorn, *Z. mays* var. *everta;* sweet corn, *Z. mays* var. *saccharata;* and dent corn, *Z. mays* var. *indentata.* Dent corn comprises the major acreage in the United States and flint corn in South America, Europe, and Asia. The United States grows approximately three-fourths of the world's supply of corn.

The basic chromosome number in corn is 10 pairs. The inheritance of some 350 genes is known, as reported by Emerson, Beadle, and Fraser (1935), Jenkins (1936), Mangelsdorf and Reeves (1939), and others. The extensive development of inbred lines of corn stimulated by the practical program of hybrid-corn production has resulted in the isolation and sta-bilization of many characters, including some conditioning disease resistance. Practical disease control for seedling blights, root and stalk rot diseases, ear rots, smut, and a few other diseases is being accomplished through resistance in the development of inbred lines. In some instances, diseases of minor importance have become epiphytotic in areas where susceptible hybrids have been distributed, as in the case of bacterial blight and *Helminthosporium* blight.

Corn is a warm-climate annual. However, it is grown under a wide range of environmental conditions because of its great adaptability. Changes in the length of growing period enable the crop to be grown economically from the frost-free southern tropics to an 80-day, frost-free period in the Northern United States and southern Canada. The use of hybrid corn is increasing the productivity for both grain and silage over this wide geographical range.

Corn is damaged by a relatively large number of diseases. The devel-opmental anatomy and physiology of the plant play an important role in many of these diseases. Estimated annual losses in the United States for the 10-year period 1930 to 1939 were 12.5 per cent of the annual grain crop, or over 250 million bushels (Plant Disease Survey). The losses in sweet corn, an important canning and garden crop in the United States, averaged 9.5 per cent, or over 38,000 tons of canning corn. These losses are being reduced appreciably in recent years with the development of better-adapted disease-resistant hybrids.

1. Nonparasitic Diseases. The nonparasitic diseases in corn are manifest by various combinations of symptoms. Usually stalk and leaf symptoms are closely associated. The symptoms under field conditions are frequently similar to those caused by pathogenic organisms. The group of maladies is divided into two general classes based upon cause: (1) temperature and moisture and (2) mineral deficiencies.

Low temperatures, especially during the period of seedling growth, produce chlorophyll disturbances and retarded growth. Many of the virescent types are expressed more commonly at low light intensity and at low temperatures (Demeric, 1924, Smith, 1935). Continued low temperatures result in browning of leaf tissue, depletion of endosperm reserves, and blighting of seedlings by soil-borne organisms. Warm weather usually results in the formation of chlorophyll in the younger leaves and recovery, although the plants frequently are retarded in growth. Low temperatures and frost before the mature corn is dried fully cause a bleached wrinkled pericarp and low or weakened germination (Kiesselbach and Ratcliff, 1918).

High temperatures and drought cause firing of upper leaves and tassels. The affected tissues are conspicuous as they bleach and dry out.

Corn is relatively sensitive to certain mineral deficiencies or balance of inorganic elements. Leaf spotting, blotching, or chlorosis is frequently accompanied by necrosis of stalk tissue below the apical growing point. In other instances, deposits of inorganic-organic complexes occur in the basal stalk tissues. External symptoms are usually reduced internodal elongation, yellow to brown leaf lesions, and barren stalks or nubbin ears. This condition occurs commonly on marsh or peat soils or on soils high in soluble aluminum and iron compounds. Low potash, magnesium, and other elements are associated with the condition. Abbott *et al.* (1913), Hartwell and Pember (1918), Hoffer and Carr (1923), Jones (1929), and Magistad (1925) have investigated this type of injury in corn.

2. Mosaics, Stripe, and Streak. Viruses transmitted mechanically or by aphids and leaf hoppers are incitants of these diseases on corn. Several of the viruses are also of economic importance on other crop plants, such as sugarcane, sorghum, wheat, and several grasses. Some of these viruses are transmitted to a number of crop plants in these groups, while others are restricted in range of suscepts. Damage from these diseases generally is limited to the warmer climates. These maladies are common on corn in the sugarcane-producing areas of the world. Others are restricted to specific geographic areas.

Corn mosaics incited by the celery and lily strains of the cucumber virus occur in some areas. Leaf flecking, mottling, striping, yellowing, and severe dwarfing of plants in early infections are the common symptoms. The corn aphid *Rhopalosiphum* (*Aphis*) *maidis* (Fitch), the cotton aphid *Aphis gossypii* (Glover), and probably others transmit the virus in nature (Wellman, 1934).

Sugarcane mosaics occur on corn in areas where the two crops are grown in association (Brandes, 1920, Brandes and Klaphaak, 1923). Symptoms are similar to the other mosaics on corn (see Chap. 10).

Corn mosaic or stripe, described by Stahl (1927) in Cuba as distinct from sugarcane mosaic and chlorotic streak, probably is similar to the mosaic on corn in Hawaii (Kunkel, 1922) and in the Philippines (Carter, 1941, Lawas and Fernandes, 1949). Chlorotic, elongate blotches and stripes, intracellular bodies, phloem necrosis, and poorly developed ears are the characteristic symptoms. The virus is transmitted by the corn lanternfly, *Peregrinus maidis* (Ashm.), which apparently is not a vector of the other mosaic viruses on corn. Symptoms of stripe mosaic frequently are similar to others on corn.

Leaf fleck, or mottle mosaic, occurs on corn and some grasses in local areas. Stoner (1952) reports this virus as specific and transmitted by three aphid species, *Rhopalosiphum* (*Aphis*) *maidis* (Fitch), *R. fitchii* (Sand.) [*R. prunifoliae* (Fitch)], and *Myzus persicae* (Sulz.).

A number of mosaic viruses are transmitted to corn. Seedlings of susceptible inbreds or single hybrids occasionally are used for identification of viruses. The sugarcane mosaic viruses and strains and barley yellow dwarf virus are examples. Tobacco ring spot virus, tobacco necrosis virus, some cucumber mosaic virus strains, wheat streak virus, and others are reported to incite symptoms on corn seedlings.

At least three virus diseases of the yellows type occur on corn. The vectors of this group of maladies are leaf hoppers. Generally the virus has an incubation period in the insect vector and frequently persists for some time in the insect. The host range of this group of viruses is more restricted than in the mosaics. Strains of the virus frequently are associated with specific host groups.

Corn streak apparently is restricted to Africa and India (McClean, 1947, Storey, 1925, 1929, 1951). Corn, sugarcane, and some grasses are hosts; strain A of the virus is severe on wheat and barley (Gorter, 1947) (see Chap. 11). Small chlorotic spots; narrow, interrupted, yellow to translucent streaks; and severe stunting of plants are the general symptoms. The virus strains are transmitted by at least three species of leaf hoppers, *Cicadulina mbila* (Naude), *C. zeae* (China), and *C. storeyi* (China). Storey and associates (McClean, 1947, Storey, 1925, 1928, 1933, 1934, 1937, Storey and McClean, 1930) have studied the insect-vector biological relationships and transmitted the virus mechanically into the insect. Resistant varieties, especially from Peru, are used in the control of the disease.

A *mottle virus* associated with streak and transmitted by the same leaf hoppers shows transitory symptoms on corn (Storey, 1937).

Corn stunt occurs in Central America, Mexico, and the Southern United States, and at least a similar disease is reported in Italy and adjacent areas (Kunkel, 1946, 1948, Niederhauser and Cervantes, 1950, Scossiroli, 1951). Corn and teosinte are the only hosts known. Yellowing followed by stunting and necroses of all plant parts are characteristic in severe early infection. Bud proliferation, excessive tillering, stunting, and poor development of tassel and ear are common symptoms. Chlorotic spotting, streaking, and banding of leaves, husks, and stalks, and undeveloped or poorly developed roots and ears occur in late infections. At least two leaf hoppers, *Dulbulus (Baldulus) maidis* (De L. & W.) and *D. elimatus* (Ball), are vectors. Maramorosch (1951) studied the virus in the insect vector and transmitted the virus mechanically into the insect. Varieties differ widely in susceptibility, and resistant lines of corn have been studied.

Corn wallaby-ear virus occurs in Australia (Schindler, 1942). Symptoms are dark-green, faintly mottled leaves rolling inward and narrow, elongate white galls on the underside of leaf veins. Early infections result in stunting and poor ear development. The leaf hopper *Cicadula bimaculata* (Evans) is a vector.

These diseases are controlled best by resistant varieties as in the sugarcane. However, relatively little breeding for resistance in corn is reported to date. Certain Peruvian flint varieties are resistant to the streak disease (see also Chap. 10).

3. Bacterial Wilt (Stewart's Disease), *Bacterium stewartii* E. F. Sm. The bacterial wilt of corn is a destructive disease of susceptible varieties of sweet, flint, and dent corn. Before the use of resistant varieties, large acreages of sweet corn in the United States were devastated by epiphytotics of this disease. The wilt earlier was largely responsible for the northern shift in the canning-corn acreage in this country. The use of the resistant single-cross hybrid Golden Cross Bantam and other resistant hybrids has reduced this hazard and made possible the economical production of the Bantam-type sweet corn in the corn-belt area (Smith, 1933).

The disease is common in the warmer areas. It occurs infrequently in the northern tier of states in the United States. It has been reported from most corn-producing areas (Elliott, 1941). The major damage is confined to the United States, where sweet corn is an important commercial crop.

SYMPTOMS AND EFFECTS. The disease is a typical bacterial vascular wilt. The bacteria develop in a gelatinous matrix inside the vascular bundles of the stalks, inflorescences, and leaves. The bacterial mass breaks out of the bundles in the later stages of disease development. In susceptible varieties of sweet corn, the disease is a typical wilt, especially in the early stages of plant growth. Plants not killed are stunted, form tassels and nubbin ears early, and show considerable leaf necrosis. Later leaf infections from insect injuries result in pale-green wilted streaks, which frequently extend into the stalk (Fig. 16). On dent corn the leaf lesions are the characteristic symptom. Abundant lesions result in the killing of the leaf blade. The yellow bacterial exudate is conspicuous in the vascular bundles of stalks and leaves. In dent corn, stalk infection and the formation of the exudate are not so extensive as in the more susceptible sweet corn varieties.

THE BACTERIUM

Bacterium stewartii E. F. Sm.
[*Phytomonas stewartii* (E. F. Sm.) Bergey *et al.*]
(*Pseudomonas stewartii* E. F. Sm.)
[*Aplanobacter stewartii* (E. F. Sm.) McCull.]

The nonmotile rods with rounded ends develop in a gelatinous matrix.

ETIOLOGY. The corn plants are infected either from the seed or through the feeding injuries of insects carrying the bacteria. Apparently, infection direct from the soil is not common. The corn flea beetle is the common insect carrier of the inoculum both in primary and secondary infections (Elliott, 1941, and Rand and Cash, 1924). Once inside the tissues, the bacteria develop in the conductive tissues, frequently filling the vessels with the bacterial mass. The bacteria spread through the vascular bundles to tassels and ears. Kernel infection occurs under favorable conditions. The bacteria embedded in the matrix survive the winter in diseased stalk tissues. This source of inoculum apparently is not important in initiating the infection the following year, as infested beetles carry the bacteria for long periods inside their bodies. From 10 to 20 per cent of the beetle populations carry the bacteria when they come out of hibernation in the spring. The infestation and infection increase simultaneously during the summer. Bacterial lesions, following beetle feeding on dent corn, develop abundantly after the plants are in tassel. This is the period when beetle infestation usually is highest.

Resistance to bacterial wilt is the best means of control. Usually the early-maturing varieties of corn are more susceptible. The majority of the resistant inbred lines and hybrids possess a type of resistance associated with height and lateness of the plants. A few highly resistant

F IG. 16. Bacterial wilt of Golden Bantam sweet corn (A) and bacterial lesions on a blade of dent corn (C). Bacterial exudate in bundles is shown in (B). (*Courtesy of Cereal Crops Section, Agricultural Research Service, U.S. Department of Agriculture*)

lines have a type of resistance not correlated with either height or lateness (Ivanoff and Riker, 1936). A few dent corn hybrids are very susceptible to bacterial wilt.

4. Cobb's Disease of Sugar Cane, *Xanthomonas vasculorum* (Cobb) Dows. [*Phytomonas vasculorum* (Cobb) Bergey *et al.*]. Ivanoff (1935) inoculated corn, sugarcane, and sorghum with *Xanthomonas vasculorum*. Symptoms and development of the disease were similar to wilt, although the organism was sufficiently different to retain the species. The discussion of this disease is given in Chap. 10.

5. Bacterial Blights. Several bacterial blights have been described on corn. Certain of these are more common on sorghums and related crops than on corn. Local water-soaked lesions occur on leaf sheath and culm tissues. The discussion of this group of diseases is included in Chap. 9.

6. Physoderma Brown Spot, *Physoderma maydis*, Miyabe. Corn and teosinte are the only known hosts of this parasite. The disease is of minor importance in the extreme southern portion of the corn belt of the United States and other warm humid regions of corn culture (Eddins, 1933, Tisdale, 1919, 1920). The lesions occur on leaf and stalk tissues. The oblong to round lesions are slightly water-soaked, light green, later turning reddish brown. The small spots coalesce to form brown blotches, especially at the base of the leaf blade and adjacent sheath tissue. The lesions also occur on the stalk beneath the leaf sheath. Later the epidermis and parenchyma collapse to form small pockets of dusty-brown sporangia in the leaf sheath and stalk tissues. The stalks frequently break at these infection centers.

THE FUNGUS

Physoderma maydis Miyabe
(*Cladochytrium maydis* Miyabe)
(*Physoderma zeae-maydis* Shaw)

The sporangia are smooth, brown, flattened on one side, and measure 18–24 by 20–30 microns. The lid of the sporangia opens on germination, liberating the uniciliate zoospores, 3 to 4 microns wide by 5 to 7 microns long. The fusion of gametes and the formation of a true sporangium are described by Sparrow (1934, 1947), and Sprague (1950).

ETIOLOGY. The fungus develops as an obligate parasite on corn, and the spores persist in the old stalks. The sporangia are distributed by wind and other agencies. Those falling within the leaf whorl and leaf axils germinate to form zoospores. The zoospores become attached to the young corn tissues, usually within the leaf whorl, and form an infection hypha that enters the mesophyll or parenchymatous cells. Larger vegetative cells develop from the fine mycelium, and sporangia are formed which ultimately fill the cell. Sporangia are discharged when the tissues disintegrate, or they are carried over unfavorable conditions in the corn tissues. Abundant moisture and high temperature are essential to the development of the disease (Tisdale, 1919, Voorhees, 1933).

Control is largely through sanitation and the use of resistant hybrids, according to Eddins and Voorhees (1935).

7. Pythium Root Rot, *Pythium arrhenomanes* Drechsl. and *P. graminicola* Subrm. Species of *Pythium* are parasitic on germinating seed and the rootlets of many of the grasses as well as other crops throughout the world. The behavior on this former group of suscepts is somewhat different from that on the dicotyledonous crop plants, mainly because of plant structure and development. Root systems are depleted from the seedling stage until maturity. Under favorable conditions, seed rotting and seedling blight of corn are more common than damping-off, the latter being more general in seedlings with small delicate hypocotyls. This group of fungi is associated with crop sequences and soils high in moisture.

The disease is common on corn, although not so destructive as it is on sugarcane, sorghums, and related plants. The root rot of corn is widely distributed, from the tropical to northern limits of corn culture and extending into the heavier soils of even the semihumid areas (Branstetter, 1927, and Rands and Dopp, 1934).

SYMPTOMS AND EFFECTS. The disease is primarily a seed rot and rootlet rot. Light-brown water-soaked lesions develop on the finer rootlets and at the root ruptures. In the earlier stages of germination, the rot is in close proximity to the seedling tissues; at later stages it may be some distance from the crown. The rot advances into the main roots and seedling or crown tissues in wet, cold soils. Lobulate sporangia occur in the cortex of the rotted tissues, usually near the surface. Later, oospores are formed throughout the invaded tissues. In wet, cold soils seed rot and seedling blight occur, as reported by Hoppe (1949), Johann *et al.* (1928), and others (Fig. 17). Plating on tissue media is useful in identification of the species involved (Johann, 1928).

THE FUNGUS

Pythium arrhenomanes Drechsl.
(*Pythium butleri* Subrm.)
[*Pythium aphanidermatum* (Eds.) Fitzp.]
(*Rheosporangium aphanidermatum* Eds.)

Drechsler (1936) retains the species *P. graminicola* Subrm. as morphologically distinct from *P. arrhenomanes*, although they are similar morphologically. Both attack corn, sugarcane, sorghum, etc.: the former is more common on the small grains.

The differentiating morphology of these two species is given by Drechsler (1936) and Middleton (1943). *Pythium arrhenomanes* forms numerous large lobulate sporangia in tissues or culture. Oogonia are formed in tissue with crook-necked antheridia and remote connections between antheridia and oogonial stalks. Numerous antheridia are attached to the oogonium. The antheridial walls do not persist and in the mature oospore are nearly indiscernible (Fig. 17). This is in contrast to the close mycelial connection between antheridium and oogonium, fewer antheridia attached to an oogonium, and the persistence of the antheridial membranous parts on the wall of the mature oospore in *P. graminicola*. Elliott (1943) described a

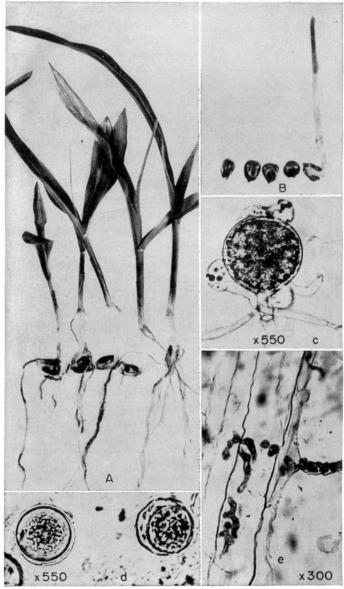

Fig. 17. (*A*) Pythium root rot of corn in a cold dry soil and (*B*) seed rot and seedling blight in a cold wet soil.   The lobulate sporangia in the roots (*e*), oogonium (*c*), and oospores (*d*) of *Pythium arrhenomanes* are shown, highly magnified, in the inserts. (*Courtesy of Helen Johann.*)

stalk rot of corn caused by the similar species, *P. aphanidermatum* (Eds.) Fitzp. (*P. butleri* Subrm.).   See Middleton (1943) and Sprague (1950) for other species on corn.

ETIOLOGY.   The mycelium develops in the soils in association with crop residue and invades the plant roots whenever environmental conditions are favorable (Carpenter, 1934).   The fungus develops in root tissues throughout the growing season.

CONTROL.   Crop sequence and soil preparation, including balanced fertility, are remedial measures.   The use of resistant varieties and hybrids is becoming an important control measure, as reported by Bowman *et al.* (1937), Johann *et al.* (1928), and Rands and Sherwood (1927). Early seedling and root infections in infested soils are prevented by seed treatment with Arasan and Orthocide or organic sulfur compounds.

8. Downy Mildews, *Sclerospora* spp.   Corn is attacked by several species of the genus *Sclerospora*.   The tropical downy mildews on corn are of major economic importance especially in the South Pacific islands, southern Asia, and South Africa.   The increasing prevalence of the grass downy mildew, *Sclerospora macrospora* Sacc. or *Sclerophthora macrospora* (Sacc.) Thir, Shaw, & Naras. (1953), on corn in southern Europe and in the United States (Ullstrup, 1952) causes some loss in corn in humid areas.   The age of the corn plant, the species of the pathogen, and environment all influence the symptom expression.   In the tropical downy mildews, white to yellow streaking of the leaves and development of the downy mildew on the streaks, followed by necrosis and browning of the streaks, are the characteristic leaf symptoms.   Dwarfing or reduced elongation of the upper internodes, suckering, retarded development of the plants, and poorly formed ears are typical where the infection occurs early, as reported by Butler (1913) and Weston (1920) for the disease incited by *S. indica* Butl. & Bisby, *S. maydis* (Rac.) Butl., *S. sacchari* Miyake, and *S. philippinensis* Weston.   Only the imperfect stage of the pathogens develops on the infected plants.   Symptoms on susceptible corn incited by the closely related species *S. graminicola* (Sacc.) Schroet. and *S. sorghi* (Kulk.) Weston & Uppal are similar to those described above for the tropical downy mildews.   In addition, oospores in the latter two species are formed sparsely in the lesion.   The two pathogens do not cause shredding of lesioned corn tissue as in millets and sorghum, respectively (Butler, 1907, Kulkarni, 1913, Melhus *et al.*, 1928, Patel, 1949, Safeeulla and Thirumalachar, 1955, Weston and Uppal, 1932).

Symptoms on corn incited by *S. macrospora* Sacc. are characterized by bud proliferation; axillary crown buds when infection occurs in the seedling stage and inflorescences when infection occurs after ear and tassel primordia are differentiated.   These modified leaf-like inflorescences are described as "crazy top" of corn.   Linear yellow to brown stripes occur in the narrow leaves and leaf-like structures. Infected tissues show no shredding in corn or other grasses. The sporangial stage occurs on the leaf lesions, but these nocturnally formed structures are evanescent and difficult to find. The large oospores are numerous or sparse, apparently dependent upon host-pathogen reaction (Cugini and Traverso, 1902, Ippolito and Traverso, 1903, Ullstrup, 1952).

THE FUNGI. Several species occur on corn in various parts of the world.

Sclerospora philippinensis Weston. Philippine species on corn

S. maydis (Rac.) Butl. Java species on corn

(*S. javanica* Palm)

S. indica Butl. & Bisby. Indian species on corn

S. spontanea Weston. Philippine on *Saccharum spontaneum* and corn

S. sacchari Miyake. South Pacific on sugarcane and corn

S. sorghi (Kulk.) Weston & Uppal. Asia, Africa on sorghum and corn

S. graminicola (Sacc.) Schroet. General on *Setaria* spp. and corn

S. macrospora Sacc. or *Sclerophthora macrospora* (Sacc.) Thir., Shaw, & Naras. General on Gramineae

According to Weston (1920, 1921), the first three species are characterized by the conidial stage only having been found. An oospore stage of a Sclerospora occurs on *Saccharum spontaneum* L., but it has not been connected definitely with *S. spontanea*. The morphology of the conidial stage of *S. philippinensis*, *S. maydis*, and *S. sacchari* is similar. Conidiophores are erect, varying slightly in length and shape for the different species, and all three have a conspicuous basal cell. They are dichotomously branched, with the conidia borne on rather long conoid-subulate curved sterigmata. Conidia are ellipsoid to ovoid, slightly rounded at the apex, and vary in size in the different species (see Table 5). The conidia germinate by forming a germ tube (Uppal and Weston, 1936). Leece (1941), Lyon (1915), and Miyake (1911) reported oospores of *S. sacchari* on sugarcane. According to Weston (1921), the conidia of *S. spontanea* are narrower and longer than in the other species. The species *S. sorghi*, occurring on sorghum and corn in southern Asia and South Africa, is intermediate in morphology between *S. graminicola*, described in Chap. 5, and the tropical group of species, as suggested by Weston and Uppal (1932). The conidiophores are erect, with a bulbous basal cell and branch in regular close succession. The sterigmata are tapering and long (16 microns). The conidia are broadly rotund, the ends are bluntly rounded, and they germinate by the formation of a germ tube. The oogonial stage is similar to *S. graminicola*. *S. macrospora* or *Sclerophthora macrospora* forms *Phytophthora*-like sporangia borne on hyphoid sporangiophores emerging through the stomata. Sporangia are large, lemon-shaped, apically poroid, and germinate in water by the division of cell contents into biflagilate zoospores. Oogonial stage is like that of other *Sclerospora* except that the oospore germination in place in the tissue is indirect by the formation of a sporangium similar in shape to the asexual sporangium (McDonough, 1946, Peglion, 1930, Peyronel, 1929, Tanaka, 1940, and Thirumalacher *et al.*, 1953).

ETIOLOGY. The group of species, occurring in tropical climates, persists in the mycelial and conidial stage. Primary and secondary infections from conidia occur when weather conditions are favorable. Epiphytotics are common on corn and less common on sugarcane. The sorghum downy mildew extends into the drier areas, as the oospores are long-lived and produce primary infections on the new crop. No very satisfactory control measures have been reported. Some differences in response of varieties are evident.

A summary of the *Sclerospora* species on the cereals and grasses is given in Table 6.

9. Kernel Mold, Scutellum Rot, and Seedling Injury, *Rhizopus, Aspergillus,* and *Penicillium* spp. This type of damage to corn is common in the more humid areas throughout the world. The disease is associated with the maturation and moisture loss in the field and with storage conditions after harvest. Losses from this type of damage affect yield somewhat, but more especially the quality of the grain. The organisms involved are generally secondary, as they are semiparasitic in nature. The primary cause is immature corn of high moisture content left in the field too long or stored with high moisture.

The symptoms of this type of damage are not easily differentiated from kernel damage by *Gibberella, Fusarium,* and *Diplodia.* The mold damage in the kernel is more superficial, with less rotting of the kernel and a characteristic moldy odor. Frequently, the two groups of diseases are combined. Damaged corn under the Federal grain standards includes both types. Koehler and Holbert (1930, 1938) and others have discussed the scutellum rot in which these organisms invade the scutellum and adjacent endosperm to damage the stored reserves and weaken the embryo.

Damage of this type in seed corn is reduced by early harvesting and artificial drying. Better drying and storage facilities for the commercial corn crop are necessary and economical in preventing this type of loss. Corn hybrids that mature and dry rapidly in the field are important in reducing the damage.

10. Gibberella Ear Rot, Kernel Rot, Stalk Rot, and Seedling Blight, *Gibberella* and *Fusarium* Spp. Two species of *Gibberella* are common on corn, and they produce different symptoms as well as differing in their distribution. *Gibberella zeae* (Schw.) Petch or *G. roseum f. cerealis* (Cke.) Snyder & Hansen causes a pink ear rot, stalk rot, and seedling blight on corn. This fungus also occurs commonly on the other cereal crops where it causes more damage perhaps than on corn. In the United States, *G. zeae* is distributed on corn and wheat, barley, and rye through the eastern and central sections of the corn belt, and it is less common in the southern and western corn-producing sections. *G. fujikuroi* (Saw.) Wr. produces a kernel rot and stalk rot of corn throughout the entire corn belt and southward into the subtemperate and tropical zones. The dis-

TABLE 6. THE FUNGI CAUSING DOWNY MILDEW ON THE GRAMINEAE*

	S. or Sclerophthora macrospora	Sclerospora spp.						
		S. graminicola	S. sorghi	S. noblei	S. spontanea	S. philippinensis	S. maydis	S. sacchari
Plants reported suscepts	Triticum spp. Hordeum spp. Zea mays Avena spp. Secalis sp. Oryza sativa Bromus commutatus Phalaris spp. Phragmites spp. Glyceria spp. Andropogon spp. Panicum sp. Lolium spp. Alopecurus spp.	Setaria viridis S. italica S. magna S. glauca S. verticillata Zea mays Euchlaena mexicana Sorghum vulgare Saccharum officinarum Pennisetum spp.	Sorghum vulgare S. arundinaceum Zea mays Euchlaena mexicana	Sorghum plumosum Andropogon australis	Saccharum spontaneum S. officinarum Zea mays Euchlaena luxurians Miscanthus japonicus Sorghum sp.	Zea mays Euchlaena luxurians Sorghum vulgare Saccharum spontaneum Miscanthus japonicus	Zea mays Teosinte × Corn Hybrids	Saccharum officinarum Zea mays Euchlaena spp.
Common suscepts in North America	Wheat, Bromus, Avena	Millets	Unknown	Unknown	Unknown	Unknown	Unknown	Unknown
Distribution	General	General	South Asia, Africa	Australia	Philippines	Philippines	Java	South Pacific
Conidiophores: length, microns	Hyphoid	268	180–300	300–450	350–550	150–400	150–300	190–280
Basal cell	Absent	Absent	Present	Present	Present	Present	Present	Present
Conidia, microns	Sporangia	14–23 × 11–17	15–29 × 15–27	21–37 × 13–29	39–45 × 15–17	17–57 × 11–27	28–45 × 16–22	25–41 × 15–23
Conidial germination	Zoospores	Zoospores	Germ tube	Germ tube	Germ tube	Germ tube	Germ tube	Germ tube
Oospores, microns	60–65	30–60	31–69	19–35	Connection not certain	Unknown	Unknown	Oogonia 11–23
Germination	Sporangium	Germ tube	Germ tube	Germ tube				

* In addition to the species listed in the table, the following species have been described: *Sclerospora miscanthi* Miya, on species of *Miscanthus* and *Saccharum* from Japan and the Philippines with oospores (47 to 49 microns) only described; *S. farlowii* Griff. on *Chloris elegans* from the Southwestern United States with oospores (28 to 45 microns) the only stage known; *S. magnusiana* Sorok. on *Equisetum* from Russia, *S. butleri* Weston on *Eragrostis aspera* from Nyasaland, *S. northi* Weston on *Erianthus maximum* from Fiji, and *S. oryzae* Brizi on rice from Formosa.

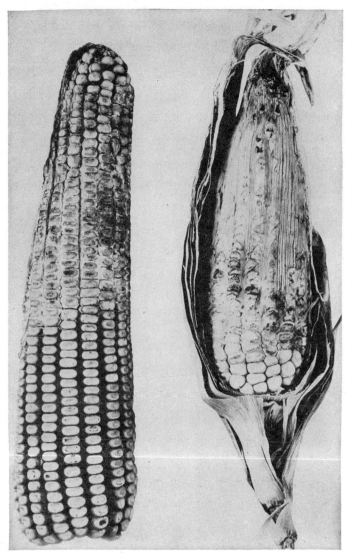

FIG. 18. *Gibberella* or pink ear rot of corn caused by *Gibberella zeae* or *Fusarium roseum*
f. cerealis var. Graminearum; left, tip portion of the ear rotted; right, rot involving
the entire ear.

tribution of the two species is similar in the other corn-producing areas
of the world, the former species being common in the humid temperate
zones, the latter extending over the humid and semihumid temperate
zones and into the subtemperate and tropical zones (Blattny, 1931,
Edwards, 1937, Maher, 1931, Mendiola, 1930, Nishikado, 1933, Stevens
and Wood, 1935, and others).

Fig. 19. *Fusarium* or pink kernel rot of corn caused by *Fusarium moniliforme*, the conidial stage of *Gibberella fujikuroi* or *G. moniliforme*.

SYMPTOMS AND EFFECTS.   The symptoms of the ear rots vary with the fungus and severity of attack.   The *Gibberella* ear rot caused by *G. zeae* is typically a pink or reddish rot progressing from the tip of the ear downward.   A small portion of the tip to all the ear is rotted, depending upon age of the plant when infection occurs and the environmental conditions (Fig. 18).   The *Fusarium* kernel rot caused by *G. fujikuroi* or *F. moniliforme* is typically a rot of individual or groups of kernels.   The symp-

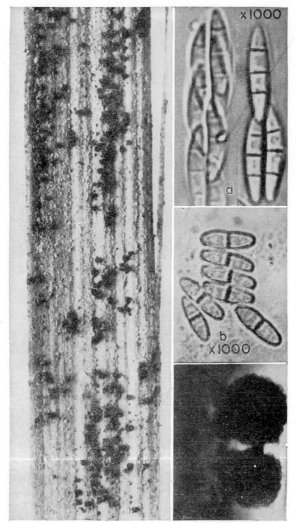

FIG. 20. Perithecia of *Gibberella zeae* or *G. roseum f. cerealis* on cornstalk tissue.   Highly magnified ascospores of *G. zeae* (*a*) and *G. fujikuroi* or *G. moniliforme* (*b*) are shown in the inserts.

toms produced by *G. fujikuroi* and *G. fujikuroi subglutinans* Ed. are similar.   The color of the rotted kernels is pink to reddish brown or gray, depending upon the general prevalence of mycelium of the fungus and weather conditions (Fig. 19).   In both types of rots, the symptoms are evident after the husks are removed.

The diseased kernels are not all distinguishable in the shelled grain. The badly rotted kernels are pink to reddish brown, with mycelium conspicuous on the surface. The less damaged kernels are evident by a lusterless pericarp surface and shrunken, sometimes brown, area over the embryo. Many kernels infected with these fungi show no external symptoms. Plating the kernels on agar or germinating the kernels in a moist atmosphere soon shows the pinkish-white mycelium on the kernel surface. Grain infected with *Gibberella zeae* causes emesis in pigs and humans.

The stalk rots, more commonly, are associated with the roots, crown, and lower nodes of the stalk. They cause reddish lesions, premature ripening, and stalk breaking. Local sheath and node lesions, pinkish in color, occur frequently with stalk rotting. *Gibberella zeae* frequently causes premature ripening of the stalks, but both fungi develop aggressively on the dead stalks late in the autumn or in the following spring. Bluish-black round perithecia develop abundantly on these stalks in the late autumn in mild climates or in the following spring in the main corn belt of the United States (Fig. 20). The perithecia of *G. fujikuroi* and the variety *G. fujikuroi subglutinans* Ed. are not so common in the United States as they are in more tropical areas (Edwards, 1935, 1936, Ullstrup, 1936).

The seedling blight of corn is characteristically a water-soaked rotting of the cortical tissues. The blight occurs before emergence or when the seedlings are in the first to third leaf stage, rarely later. Brown water-soaked lesions on the subcrown internode and roots are characteristic, as described by Dickson (1923), Pearson (1931), Voorhees (1933). Environmental conditions, especially low temperatures, influence the amount of blighting and symptoms, according to Dickson (1923) and Dufrenoy and Fremont (1931). Seedling blight is common with *Gibberella zeae*, whereas root lesions and slender weakened plants occur more frequently with *G. fujikuroi* infections.

THE FUNGI

1. *Gibberella zeae* (Schw.) Petch
   [*Gibberella saubinetti* (Mont.) Sacc.] Used until recently. Or
   Fusarium graminearum Schw. Conidial stage. Or
   G. roseum f. cerealis (Cke.) Snyder & Hansen
   F. roseum f. cerealis (Cke.) Snyder & Hansen var. Graminearum

See wheat, Chap. 11 for detailed synonymy and morphology.

Macroconidia are produced in sporodochia or pseudopionnotes, 3- to 5-septate, curved with gradual tapering toward tip. Microconidia and vegetative resting cells (chlamydospores) are not produced.

Perithecia are borne superficially, are globose, smooth, and blue black. Asci are oblong-clavate, contain eight spores, arranged obliquely in one row. Ascospores are 3-septate, slender, tapering uniformly to the ends, and slightly curved (Fig. 20).

2. *Gibberella fujikuroi* (Saw.) Wr.
   [*Gibberella moniliformis* (Sheld.) Wine.]
   Fusarium moniliforme Sheld.   Conidial stage

Macroconidia are produced sparingly, 3- to 5-septate, and curved toward the tip. Microconidia are borne in chains or false heads on branches of the hyphae, usually nonseptate except when germinating.   Perithecia are similar to former species.   Asci are less clavate, more oblong than in former species, contain eight spores, arranged in two irregular rows.   Ascospores are straight, tapering to tips, 1- to 3-septate, usually 1-septate.

3. *Gibberella fujikuroi* (Saw.) Wr. var. *subglutinans* Ed.
   Fusarium moniliforme Sheld. var. *subglutinans* Wr. & Reinking.
   Conidial stage

Macroconidia are similar to the species, but less curved toward the tip, usually 3-septate.   Microconidia are borne singly or in false heads, never in chains.   Asci are long, narrow, subclavate, contain usually eight spores, less commonly four or six arranged in one oblique row.   Ascospores are straight, rounded at tips, short and thick, and 1-septate (Fig. 20).

The Snyder and Hansen (1945) nomenclature for this species and variety.

Gibberella moniliforme (Sheld.) Snyder & Hansen
Fusarium moniliforme (Sheld.) Snyder & Hansen

ETIOLOGY.   The fungi develop on crop residue of the cereal crops that remain in and on the surface of the soil.   Spore development is confined largely to residue on the soil surface.   Mycelium, conidia, and ascospores are produced during the growing season.   Primary and secondary infections occur when environmental conditions are favorable.   Seed infection, even in artificially dried corn, is frequently high, especially with *Gibberella fujikuroi*.   Seedling blight occurs during germination and the early seedling stage when the soil is cold while the seed is germinating.   Root and stalk rots become evident soon after pollination and increase in severity as the plants mature.   Ear infections occur through and around the silks as these tissues decline in physiological activity after pollination.   Wind-borne spores apparently are the main source of inoculum for ear infection.

CONTROL.   Crop rotation and plowing under crop residue aid in reducing the volume of inoculum.   Resistant hybrids are important in reducing seedling blight and stalk rots.   Commercial hybrids also vary in susceptibility to ear rot, according to Hayes *et al.* (1933), Holbert *et al.* (1924, 1926, 1929), McIndoe (1931).

11. Diplodia Ear Rot, Stalk Rot, and Seedling Blight, *Diplodia zeae* (Schw.) Lév. and *D. macrospora* Earle and *Physalospora* spp.   The *Diplodia* rot is of major importance in economical corn production.   The disease occurs rather generally wherever the crop is grown intensively. In the United States, the disease is important in the corn belt and diminishes in amount and severity in the Western drier sections and in the Northern, cooler areas. The disease produced by the less common species *D. macrospora* occurs in the more humid, warmer climates. The distribution of the disease and damage from it vary considerably, depending upon climatic conditions.

SYMPTOMS AND EFFECT. Seedling blight is generally less prevalent than in the *Fusarium* diseases. The cortical lesions and blighted seedlings resulting from infected seed are characterized by a brown dry rot, especially below the soil surface.

The stalk rot and leaf-sheath lesions are generally not conspicuous until after pollination of the corn plant. Reddish-purple to dark-brown blotches occur on the leaf sheath and extend into the nodes and basal portion of the internodes. Mycelial development frequently is extensive between the leaf sheath and stalk; however, other organisms, including saprophytes, develop beneath the leaf sheath. Stalk rot is initiated from lesions within the leaf sheaths and from rotted roots. Stalk rot more frequently develops from the adventitious roots and crown upward into the stalk, causing premature ripening and chaffy or rotted ears. The brown discoloration of the internodes and nodes is concealed by the dead bleached leaf sheaths (Fig. 21). The rotted stalks break over as the plants mature.

The ear rot varies from an inconspicuous infection of the kernels to a complete rotting of the ear and husks. The age of the ear tissues when infection occurs and environmental conditions influence the ear rot symptoms. The ear rot usually progresses upward from the base of the ear. The white to grayish-brown mycelium occurs between the kernels on the rotted ears (Fig. 21). The dark-brown pycnidia occur late in the season in the invaded husks, the floral bracts, and the pericarp of rotted kernels. Pycnidial development on the rotted stalks occurs in late autumn, and they are abundant on the old stalk tissues the following spring and summer.

THE FUNGI

1. *Diplodia zeae* (Schw.) Lév.
   [*Diplodia maydis* (Berk.) Sacc.]
   (*Sphaeria striaeformis* var. 4 Schw.)
   (*Sphaeria zeae* Schw.)
   (*Sphaeria maydis* Berk.)
   [*Sphaeria Hendersonia zeae* (Schw.) Curr.]
   [*Hendersonia zeae* (Curr.) Hazsl.]
   (*Diplodia zeae* Lév.)
   [*Dothiora zeae* (Schw.) Benn.]
   [*Macrodiplodia zeae* (Schw.) P. & S.]
   [*Phaeostagonosporopsis zeae* (Schw.) Wor.]

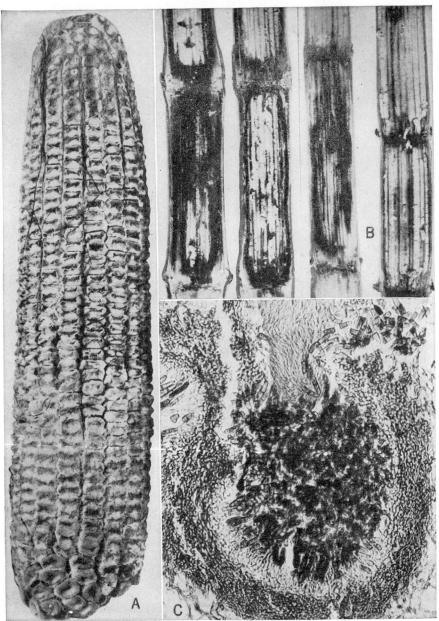

FIG. 21. Ear rot (*A*) and stalk rot (*B*) of corn caused by *Diplodia zeae*.   (*C*) Pycnidium and conidia highly magnified.

Shear and Stevens (1935) discuss the confused nomenclature for this fungus and suggest the continued use of *Diplodia zeae* on the basis of long usage, although *D. maydis* is the more correct binomial.

Globose flask-shaped or irregular pycnidia develop below the surface with a well-defined beak protruding through the epidermis (Fig. 21*C*). The conidiophores are simple, short, and pointed. The conidia are ovate, two-celled, straight to slightly curved with rounded to bluntly tapered ends, measure 25–30 by 6 microns, and are olivaceous in color. Slender, thread-like hyaline scolecospores are described by Butler (1913) and Johann (1939).

2. *Diplodia macrospora* Earle.
   [*Dothiora zeae* var. *macrospora* (Schw.) P. & S.]

The morphology of this species is similar to the former except that the mycelium is coarser and the conidia are about twice the length of *Diplodia zeae*, or 70–80 by 6–8 microns. Scolecospores are reported associated with this species.

ETIOLOGY. These fungi develop saprophytically on dead corn tissue on which inoculum is produced in abundance. The physiologically active corn tissues are infected more generally by contact infections from seed-borne mycelium, mycelium developing on pollen or organic materials collected within the leaf sheaths, and mycelium in crop residue in the soil. Mechanical injuries also offer avenues of entrance into susceptible tissues. As the corn plants approach maturity, the fungus established in the parenchymatous tissues is capable of advancing intercellularly through the stalk and into the ear, although most of the ear rot results from local infections. According to Clayton (1927), Durrell (1923), Holbert (1935), Koehler and Holbert (1930), McNew (1937), and Young (1926), the aggressive development of the fungus is associated with tissues approaching physiological maturity. Conidia and mycelia in the old stalk tissues are the important sources of inoculum. Infected seed is the principal cause of seedling blight.

The disease is reduced by the use of stalk rot–resistant hybrids, sanitation and rotation, and seed treatment (Hoppe and Holbert, 1936, Smith and Trost, 1934). The organic mercury dusts are very effective in controlling the seed infection (Hoppe, 1945, Raleigh, 1930), but they do not control the stalk and ear rots. An inhibitor or antibiotic substance produced by the fungus is suggested by Kent (1940).

Physalospora zeicola Ell. & Ev., *Diplodia frumenti* Ell. & Ev., and *P. zeae* Stout, *Macrophoma zeae* Tehon & Daniels occur on corn in the Southeastern and Central United States (Eddins and Voorhees, 1933, and Ullstrup, 1946). These fungi also occur on other crops and crop residues. Stalk lesions and a minor ear rot are produced by the former, and leaf lesions, ear rot, and stalk and tassel lesions on corn are caused by the latter fungus. Early symptoms on the stalks and ears are similar to the more widely distributed *Diplodia* rot of corn. The darker-colored mycelium, the slate-gray color of the rotted ears, and the frequent presence of sclerotia differentiate the *Physalospora* rot from the *Diplodia* rot. Sterile mycelium is more commonly plated from the tissues rotted by *P. zeae* than by *P. zeicola* and *Diplodia* spp., which usually produce pycnidia in culture.

3. *Physalospora zeicola* Ell. & Ev.
   Diplodia frumenti Ell. & Ev.

The black perithecia are gregarious, covered by the epidermis, with conical necks terminating in glossy black ostioles protruding through the cuticle. The asci are clavate-cylindrical, nearly sessile, measure 95–140 by 10–13 microns, are double-walled, and appear white in section. Ascospores, usually eight and arranged biseriately, are ellipsoid in shape, unicellular, hyaline, and measure 20–30 by 8–9 microns. Pycnidia are submerged, black, and contain dark-brown striate-septate spores. Pycnidia are produced in cultures on media.

4. *Physalospora zeae* Stout
   Macrophoma zeae Tehon & Daniels

The black perithecia with minute papillate ostioles through the cuticle are formed in the mesophyll of the leaves and on tassel necks and branches. The asci are long, clavate to cylindrical, stalked, double-walled, and measure 85–175 by 17–22 microns. Ascospores, usually eight and arranged subbiseriately, are hyaline to dilute amber, narrow ellipsoid, unicellular, and measure 19–25 by 6.5–8 (mean 26.8 by 10) microns (Ullstrup, 1946). Pycnidia of the imperfect stage occurring in association with perithecia (Ullstrup, 1946) are submerged, black, globose, protruding, usually with a short neck, and produce unicellular ellipsoid tapering conidia. Pycnidium-like structures that produce numerous nongerminating, hyaline, unicelled microconidia extruded in droplets in a mucus-like matrix are produced in nature and occasionally in culture.

12. Nigrospora Cob and Stalk Rot, *Nigrospora oryzae* (Berk. & Br.) Petch and *N. sphaerica* (Sacc.) Mason. Corn is the common suscept, although the above species and *Nigrospora sacchari* (Speg.) Mason occur on a group of monocotyledonous plants. The disease causes losses of corn in some years; however, it is of minor importance on corn and other cereals in North America, Europe, and Asia, as reported by Durrell (1925), Mason (1927), Petch (1924), Savulescu and Rayss (1930, 1931), and Standen (1945).

SYMPTOMS AND EFFECTS. The stalk and cob rots are not conspicuous until about harvest. The thin-walled cells of the stalk and cob of the ear are rotted away, leaving the vascular and sclerenchyma tissue. The kernels are chaffy, and the cob is shredded and easily broken (Fig. 22). The stalks break over at any point below the ear. The cob rot is the most important manifestation of the disease, as local losses frequently occur in the northern section of the corn belt.

FIG. 22. Cob rot of corn caused by *Nigrospora sphaerica*. The chaffy light-colored ear (*A*) is the characteristic appearance. (*B*) The rotted basal portion of the cob and dark masses of spores on the glumes. (*C*) Surface mycelium, less common.

THE FUNGI

Nigrospora oryzae (Berk. & Br.) Petch and *Nigrospora sphaerica*
   (Sacc.) Mason
(*Basisporium gallarum* Moll.)
(*Coniosporium gacevi* Bubak.)

The two species were differentiated tentatively on spore size: 13.5 to 14.9 microns for *Nigrospora oryzae* and 16.5 to 17.8 microns for *N. sphaerica* (Mason, 1927). Standen (1943) studied the variability of cultures from corn and other crops and demonstrated the difficulty of differentiating by spore size. If combined, the binomial *N. oryzae* should be used. The conidia are spherical, 10 to 20 microns in diameter, black, and borne on short lateral branches.

Durrell (1925), Savulescu and Rayss (1932), and Standen (1944, 1945) suggest that the disease develops under conditions of slow or checked development of the corn plant, such as light frosts and sudden and acute drought.

13. Helminthosporium Leaf Blight and Leaf Spots, *Helminthosporium turcicum* Pass., *Cochliobolus heterostrophus* Drechsl. or *H. maydis* Nishikado & Miyake, and *H. carbonum* Ullstrup. Three *Helminthosporium* diseases occur rather generally on corn. These are associated predominantly with the corn leaves, and they differ in geographic distribution, suscept range, and damage.

The northern leaf blight caused by *H. turcicum* is distributed widely over the world on corn, Sudan grass, Johnson grass, and other sorghums. Epidemics on corn occur more generally in the Eastern United States, and the fungus extends north and west on Sudan grass. The characteristic symptoms are large linear to irregular, somewhat elliptical lesions on the leaf blades and extending into the leaf sheaths. The lesions are first water-soaked, then light olivaceous to brown, and finally black to straw color as the tissues dry out (Drechsler, 1923, Koehler and Holbert, 1938, Ullstrup, 1943). Narrow bands of pigmentation occur along the margins of the lesions on Sudan grass (Fig. 23). Under favorable environment, the entire leaf blade is killed. Conidia are produced abundantly on the older portions of the lesions. Tassel infection on corn is less conspicuous, and ear infection is rare, although infection of the floral bracts in Sudan grass is more common.

The southern leaf spot caused by *Cochliobolus heterostrophus* or *Helminthosporium maydis* is distributed widely over the world on corn and teosinte in the warmer climatic zones. The disease is common in the Southern United States, and it occurs associated with the former disease in the southern edge of the corn belt. The lesions on the leaves are distinct from those caused by *H. turcicum*, as they are smaller, more definite, and different in pattern and color (Fig. 23). The numerous spots are elongated between the veins with limited and parallel margins, usually less than 0.5 mm. long, buff to reddish brown in color, with a zonated or target-like color pattern (Drechsler, 1925). Less conspicuous spots occur on the floral bracts of the tassels. Ear infection is not common.

The leaf spot caused by *Helminthosporium carbonum* occurs on corn in the United States. The leaf spots are similar to the southern leaf spot in race 1 of the parasite, and narrow irregular chocolate-brown spots are produced by race 2 (Fig. 23). Both races of the parasite produce a black moldy growth over the kernels of susceptible varieties and result in a charred appearance of the infected ear (Ullstrup, 1941, 1943, 1944).

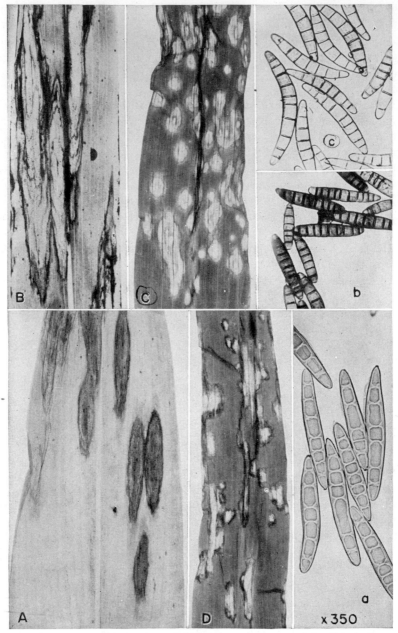

FIG. 23. Leaf blights of corn (A) and Sudan grass (B) caused by *Helminthosporium turcicum* and of corn caused by *H. carbonum* (C) and *H. maydis* or *Cochliobolus heterostropus* (D). Inserts, at same magnification, show the conidia of (a) *H. turcicum* (b) *H. carbonum,* and (c) *H. maydis.* (*Courtesy of A. J. Ullstrup.*)

The Fungi

1. *Helminthosporium turcicum* Pass.
(*Helminthosporium inconspicuum* Cke. & Ell.)

Conidial production is abundant on the leaf lesions. Conidiophores emerge two to six from the stomata, are olivaceous in color, 2- to 4-septate, and measure 7–9 by 150–250 microns. The cells of the conidiophore are longer than in most species. Conidia vary greatly in size and shape, ranging from 45–132 by 15–25 microns. The spores are straight or slightly curved, pale olivaceous, widest near the middle and tapering toward both ends, usually 3- to 8-septate with a protruding hilum (Fig. 23). Germination is characteristically by polar germ tubes. Mycelium and conidia develop on culture media. Specialized races occur on corn, some of which infect Sudan grass, and on Sudan grass; the latter do not generally infect corn. The fungus is highly variable in both cultural characters and pathogenicity (Roberts, 1952). The use of heavily diseased leaves, dried and pulverized as inoculum, overcomes some of the variability.

2. *Cochliobolus heterostrophus* Drechsl.
(*Ophiobolus heterostrophus* Drechsl.)
Helminthosporium maydis Nishik. & Miy.   Conidial stage

The conidiophores arise two to three from the stomata on the older portion of the spot, 120 to 170 microns long, olivaceous. Conidia are light olivaceous, 30 to 115 microns long by 10 to 17 microns wide, often curved, tapering toward the rounded ends (Fig. 23). They germinate by polar germ tubes. Perithecia are numerous in old tissue, ellipsoidal, about 0.4 mm. in diameter, black, with conidiophores and mycelium on the surface, ostiolate beak well defined. Asci are numerous, short stipitate, with a rounded end, are 160 to 180 microns long, and contain typically four spores. Ascospores are filamentous, arranged in parallel multiple coils, usually four coils. Drechsler (1925, 1934) gives the details of the morphology and the taxonomic significance of the new genus *Cochliobolus*.

3. *Helminthosporium carbonum* Ullstrup.

This species occurs on a few susceptible corn inbreds and varieties. Originally described by Ullstrup (1943, 1944) as *Helminthosporium maydis* Nishik. & Miy., it is relatively similar in conidial morphology to the above species. Conidia are straight to slightly curved, widest in the center tapering toward the rounded ends, measure 25–100 by 7–18 microns, are 2- to 12-septate, have an inconspicuous hilum, and germinate from polar cells (Fig. 23). *H. zeicola* Stout and *H. rostratum* Drechsl. are reported on corn. The former pathogen occurs in warm climates (Sprague, 1950).

ETIOLOGY. The leaf infections of all three species occur under favorable environmental conditions throughout the growing season. The source of primary inoculum is largely that produced on the crop residues. Secondary conidial infection is general after the leaf infections are established. Mitra (1923), Nishikado and Miyake (1926), and Nishikado (1927) discuss the general etiology and distribution .of the former two diseases. The southern leaf spot disease develops at somewhat higher temperatures than the leaf blight. Apparently only *Helminthosporium carbonum* is seed-borne on corn (Ullstrup, 1943, 1944). *H. turcicum* is seed-borne on Sudan grass (Chilton, 1940).

Control measures consist largely of sanitation, rotation, proper covering of crop residue, and use of resistant corn hybrids. The inbred lines NC 34 and Mo 21A each carries a dominant factor pair for resistance and combine well in producing resistant hybrids. Inbred R4 contains a recessive factor for susceptibility to *H. turcicum* (Jenkins and Robert, 1952). Other lines are resistant to some isolates of the pathogen.

Many of the commercial hybrids in use are relatively resistant or tolerant to all three diseases. Resistance to the destructive race 1 of *H. carbonum* is inherited as a monogenic recessive; therefore hybrids are resistant unless susceptible inbreds are combined (Roman and Ullstrup, 1951). Seed treatment of Sudan grass seed with mercury dusts reduces the infection.

14. Rhizoctonia Rot, *Rhizoctonia* spp. *Rhizoctonia*, especially in the warmer climates, causes a root, stalk, and ear rot of corn. Many of the species of *Rhizoctonia* are capable of attacking corn. The disease is of minor importance on this crop. Voorhees (1934) described an ear rot of corn in which the rotted ears are covered by a salmon pink and later a gray mycelium, and brown to black sclerotia are formed on the husks. He described the fungus on corn as a new species, *Rhizoctonia zeae* Voorhees.

15. Smut, *Ustilago maydis* (DC.) Cda. The common corn smut is one of the most widely distributed diseases on this crop. Teosinte is the only other plant infected. Corn smut has been distributed with the spread of the crop; yet the disease commonly is not carried over on the grain as in many other cereal smuts. Evidently sufficient inoculum is carried on the seed to introduce the parasite. Once the parasite is introduced and established, the extremely hardy spores, produced in countless numbers, perpetuate the fungus. The corn smut eradication program in Australia is still uncertain, as new infections occur even after 1 or 2 years of apparent absence of the disease. The use of the mechanical corn harvester increases materially the amount of inoculum carried on the grain.

The losses from corn smut average high each year. The losses are generally greater in the warmer and somewhat drier areas. According to Garber and Hoover (1928), Immer and Christensen (1928, 1931), Johnson and Christensen (1935), and Jorgensen (1929), the location of the galls on the plant and the time of development and size influence the effect on yield. Hurd-Karrer's (1926, 1927) reports on the effect of smut infection on sap concentration and sugar content partly explain the location effect of the gall on yield.

SYMPTOMS. The smut galls occur on any part of the corn plant where embryonic tissues are exposed. Gall formation is induced by the fungus. As shown by Knowles (1889), the mycelium developing between the cells of the embryonic thin-walled tissues induces hyperplasia and hypertrophy and the excessive development of the phloem elements of the bundles. The galls are common in axillary buds, individual flowers of ear and tassel, and leaves and are less common on stalk tissue unless mechanical injuries occur (Fig. 24). The white membrane of modified corn epidermal tissue ruptures during gall enlargement, releasing the black mass of spores. Galls on corn seedlings usually result in the dwarfing and blighting of the plants. The later gall development rarely kills the plant or plant parts.

THE FUNGUS

Ustilago maydis (DC.) Cda.
(*Uredo zeae-mays* DC.)
(*Uredo maydis* DC.)
(*Uredo zeae* Schw.)
[*Ustilago zeae* (Schw.) Ung.]
(*Ustilago maydis* Cda.)
[*Ustilago mays-zeae* (DC.) Magn.]

The spores are spherical to ellipsoidal, 8 to 11 microns in diameter, black, with heavy spine-like echinulations. The germination is typically by the formation of a basidium (promycelium) and sporidia. According to Chilton (1940), the type of germination and sporidial development is variable between strains of the fungus. Sporidia increase by budding. According to Bowman (1946), Christensen and Rodenhiser (1940), Sampson (1939), and Stakman (1940), fusions occur between compatible sporidia.

ETIOLOGY. Infection of the corn plant occurs during the period of vegetative development. Davis (1936) and Melhus and Davis (1931) describe a systemic infection of the plant occurring in the late seedling stage and sori development as the tissues differentiate throughout the season. Local infections of embryonic tissue occur at any time during the vegetative development of the corn plant. Injuries, such as hail injury and detasseling, frequently result in late localized infections. The source of inoculum is largely air-borne sporidia from chlamydospores on the soil and crop residue. The spores are very resistant to extremes of environment. Delayed germination is common in a fair proportion of the spore population; this greatly lengthens the period of abundant inoculum. Sporidial fusion and the development of the dicaryotic infection hyphae are characteristic, although Chilton (1940) and Christensen (1931) report exceptions. Gall formation and spore dissemination are continuous during the summer growing period. The use of the mechanical corn harvester increases the spread of spores. The smut galls are

FIG. 24. Corn smut galls on tassel and ear (*B*) and axillary buds (*A*). Infection before the tissues are differentiated fully results in a sorus involving the entire ear (*C*). Chlamydospores of *Ustilago maydis*, highly magnified, are shown in insert.

not deleterious to animals other than increasing the dust content of dry fodder.

CONTROL. In areas where corn is grown continuously or in short rotations on contiguous acreages, crop rotation and sanitation do not control the disease. This is especially true where the ripe corn is harvested mechanically and the stalks are left on the fields. Under such conditions, resistant hybrids offer the only satisfactory means of control. Cutting the green corn and ensiling it reduce the spore inoculum. Under such conditions with crop rotation, it is possible greatly to reduce the damage from corn smut. |Seed treatment with the mercury dusts is important where corn is introduced into smut-free areas.

Smut-resistant hybrids offer the best means of control. Hayes *et al.* (1924), Hoover (1932), Immer (1927), Jones (1918), and others summarize the problems and the reaction of hybrids and inbreds. Many of the more recently developed hybrids for the corn belt are moderately resistant to many of the variants of this parasite. Some few inbreds in both sweet and dent corn are resistant over a wide range of conditions. The variability of the parasite, as shown by Christensen and Stakman (1926) and Stakman *et al.* (1929, 1933, 1940), makes the control problem more difficult.

16. Head Smut, *Sphacelotheca reiliana* (Kuehn) Clint. The head smut occurs on corn and sorghums. In the Central United States, the disease is more common on the latter crop group. In the United States, the disease on corn is localized in limited sections in the Western Inter-mountain and Southwestern areas. It is the more prevalent smut on corn in southern Russia, India, and South Africa, according to Bressman (1933), McAlpine (1910), Pole-Evans (1911), and Potter (1914). Losses from this smut are high in areas favorable for its development.

SYMPTOMS. The large smut sori replace the tassel and ear in corn and the panicle in the sorghums (Fig. 25). The sorus consists of the conductive tissues of the suscept surrounded by the spore mass and the fragile exterior fungal membrane. Frequently, only part of the tassel or panicle is replaced by the sorus. In such cases the floral bracts frequently develop into leaf-like proliferations. The black dusty spore mass is conspicuous soon after the tassels or panicles emerge.

THE FUNGUS

Sphacelotheca reiliana (Kuehn) Clint.
[*Sorosporium reilianum* (Kuehn) McAlp.]
(*Ustilago reiliana* Kuehn)
(*Ustilago pulveracea* Cook)
[*Ustilago reilana f. zeae* (Kuehn) Pass.]
(*Cintractia sorghi* de Toni)

The sorus is composed of loosely united spores and the conductive tissues of the suscept at first enclosed by a fragile fungal membrane. The reddish-brown to black chlamydospores are finely echinulate, irregular to spherical, and 9 to 12 microns in diameter. The spores germinate by forming basidia and lateral sporidia. Germination is frequently irregular with branching promycelia.

ETIOLOGY. Infection occurs in the seedling or young plant, resulting in a systemic distribution of the mycelium in the apical primordial tissues. Sori develop in the floral structures. Potter (1914) demonstrated the prevalence of seedling rather than floral infection. Christensen (1926) and Reed *et al.* (1927) have demonstrated the importance of environmental conditions to the seedling infection. Zehner and Humphrey (1929) produced infection by introducing the spore suspension

FIG. 25. Head smut of sorghum and corn incited by *Sphacelotheca reiliana*.

through the leaf whorl into the growing point in seedlings. Spores are distributed widely during the latter part of the growing season. Soilborne inoculum is the important source of natural infection. Reed (1927) has shown that separate physiologic races occur on corn and the sorghums.

CONTROL. Sanitation and rotation are important in preventing seedling infection. In areas where the disease is distributed extensively, these measures are not sufficient to prevent its occurrence. Seed treatment with the mercury dusts helps prevent seedling infection. Differences in susceptibility in corn occurs in south Russia and in Kansas. Reed (1927) has shown that the sweet sorghum and sorgo varieties are relatively susceptible and that the feterita, milo, broom corn, kaffir, and kaoliang varieties are resistant.

17. Leaf Rust, *Puccinia sorghi* Schw. The rust of corn is of importance over much of the corn-producing area of the world. In the American corn belt many of the commercial hybrids of dent corn are moderately resistant in the mature plant stage, and therefore reduction in yields due to rust is not great. The long-established native corn varieties in Central and South America carry genes for resistance and again are not damaged severely by rust when in the heterozygous condition. Most of the sweet corn varieties and especially some of the recent single crosses are susceptible, and losses are heavy. Susceptible lines of dent corn soon are eliminated in the breeding programs by severe rust losses. Niederhauser and Barnes (1955) place rust and *Helminthosporium* blight as the major diseases of corn in tropical America. The round to elongated uredia occur on the leaves and less frequently on the husks and the floral bracts of the tassel. The accumulation of uredia results in chlorosis and death of the leaf followed by the production of open and covered telia. Corn and teosinte are the uredial and telial hosts. Several species of *Oxalis*, chiefly *O. corniculata* L., *O. europaea* Jord. (*O. cymosa* Small), and *O. stricta* L. are the aecial hosts of this and related rust pathogens. The aecial infection is limited in nature although reported occasionally in the United States, Mexico, Europe, and South Africa. Allen (1933, 1934), Arthur (1904), LeRoux (1954), Mains (1934), Pole-Evans (1923), Tranzschel (1907), Zogg (1949), and others have reported on the life cycle and heterothallism of this pathogen.

THE FUNGUS

Puccinia sorghi Schw.
(*Puccinia maydis* Bér.)
(*Puccinia zeae* Bér.)
(*Aecidium oxalidis* Thüm.)
(*Dicaeoma sorghi* Ktze.)

Pycnia are on upper surface of leaflets, young stems, and petioles, few to numerous; aecia develop on undersurface of leaflet and surrounding pycnia on stems of *Oxalis* spp. Aeciospores are globoid or ellipsoid, finely verrucose, and pale yellow. Uredia are round, oblong or elongate, pulverulent, frequently on both upper and lower leaf surfaces, yellowish brown to brown. Urediospores are round to ellipsoid, 23 to 32 microns, moderately echinulate, yellowish brown, pores approximately equatorial. Telia are scattered or in groups, oblong or irregular in mass, naked or covered, blackish or brown. Teliospores are oblong to ellipsoid, slightly constricted at septum, rounded, sometimes flattened, at apex, wall dark brown, pedicels long to frequently short, colorless. Mesospores present in abnormal telia.

Puccinia polysora Underw., described on *Erianthus divaricatus* (L.) Hitch. and *Tripsacum* spp. (Chap. 12), was reported on corn by Cummins

(1941) and more recently reported as the major rust on corn in tropical areas. This species on corn is included under *P. sorghi*. The covered telia, the irregular-shaped, thin-walled teliospores with shortened pedicels, and the range in morphology of the urediospores occur within the isolates and pure lines of *P. sorghi* on corn lines and hybrids. Uredial cultures of the *P. polysora* morphological type on corn from the tropical areas produce typical *P. sorghi* on many corn lines, and telial cultures from corn produce the aecial stage on *Oxalis* spp. Morphologically these two species and *P. purpurea* Cke. on *Sorghum* spp. are very similar. The aecial stages develop on the several *Oxalis* spp. These species in nature show specialization on host groups in the uredial stage. A few pure-line segregates of *P. purpurea* infect corn. This group of species probably comprises specialized varieties of *P. sorghi* rather than species.

A second species, *Angiospora zeae* Mains (1934, 1938), occurs sparingly on corn in the tropical sections of Central and South America and the Caribbean Islands. The uredia and telia occur in limited groups on corn leaves; urediospores are 16 to 30 microns and pale brown. The telia are covered, and teliospores are unicellular, catenulate in sessile chains of usually two spores. Germination of the teliospores is not described.

SPECIALIZATION AND RESISTANCE. Specialization in *P. sorghi* has been reported by Mains (1924) and Stakman *et al.* (1928). LeRoux *et al.* (1954), using eight inbred lines of corn originating in North and South America, differentiated 15 races, including Mains race 3 differentiated on Golden Glow 208R. Two sister lines, Cuzco 1 and 1b from Peru, are resistant to all isolates of the pathogen studied. The distribution of the major races of *P. sorghi* extends from south central Mexico to southern Canada, indicating wind distribution as in *P. graminis* (see wheat, Chap. 11).

Mains (1924, 1926, 1931) reported resistant varieties and the inheritance of resistance on a single factor pair basis. Rhoades (1935) and Rhoades and Rhoades (1939) reported the location of the gene pair conditioning resistance in G.G. 208R on the tenth or shortest chromosome. Resistance to the specific race groups in the inbred lines used in differentiating the specialized races of the pathogen is conditioned by single dominant factor pairs in each line. On the basis of hybrid reaction the resistance to rust appears to be due to dominant factors in 38 strains and to recessive factors in 10.[1]

REFERENCES

ABBOTT, J. B., *et al.* The reclamation of an unproductive soil of the Kankakee Marsh region. *Ind. Agr. Exp. Sta. Bul.* 170. 1913.

[1] Unpublished data from A. L. Hooker, Wisconsin, and G. F. Sprague, Iowa Agricultural Experiment Stations and Section Cereal Crops, Agricultural Research Administration, U.S. Department of Agriculture.

ALLEN, R. F.   The spermatia of corn rust, *Puccinia sorghi*.   *Phytopath.* **23** :923–925. 1933.

———.   A cytological study of heterothallism in *Puccinia sorghi*.   *Jour. Agr. Research* **49** :1047–1068.   1934.

ARTHUR, J. C.   The aecidium of maize rust.   *Bot. Gaz.* **38** :64–67.   1904.

BOWMAN, D. H., *et al.*   Inheritance of resistance to Pythium root rot in sorghum. *Jour. Agr. Research* **55** :105–115.   1937.

———.   Sporidial fusion in *Ustilago maydis*.   *Jour. Agr. Research* **72** :233–243. 1946.

BRANDES, E. W.   Mosaic disease of corn.   *Jour. Agr. Research* **19** :517–521.   1920.

———.   Mechanics of inoculation with sugar cane mosaic by insect vectors.   *Jour. Agr. Research* **23** :279–283.   1923.

——— and P. J. KLAPHAAK.   Cultivated and wild hosts of sugar cane or grass mosaic.   *Jour. Agr. Research* **24** :247–262.   1923.

BRANSTETTER, B. B.   Corn root rot studies.   *Mo. Agr. Exp. Sta. Res. Bul.* 113. 1927.

BUTLER, E. J.   Some diseases of cereals caused by *Sclerospora graminicola*.   *Mem. Dept. Agr. India Bot. Ser.* **2** :1–24.   1907.

———.   The downy mildew of maize [*Sclerospora maydis* (Racc.) Butl.]   *Dept. Agr. India Bot. Ser. Mem.* **5** :275–280.   1913.

——— and A. H. KHAN.   Some new sugar cane diseases.   *Dept. Agr. India Bot. Ser. Mem.* **6** :191–203.   1913.

CARPENTER, C. W.   Predisposing factors in Pythium root rot.   *Hawaiian Planters' Record* **38** :279–338.   1934.

CARTER, W.   *Peregrinus maidis* Ashm. and the transmission of corn mosaic.   *Ann. Ent. Soc. Am.* **34** :551–556.   1941.

CHILTON, ST. J. P.   Delayed reduction of the diploid nucleus in promycelia of *Ustilago zeae*.   *Phytopath.* **30** :622–623.   1940.

———.   The occurrence of *Helminthosporium turcicum* in the seed and glumes of Sudan grass.   *Phytopath.* **30** :533–536.   1940.

CHRISTENSEN, J. J.   The relation of soil temperature and soil moisture to the development of head smut of sorghum.   *Phytopath.* **16** :353–357.   1926.

———.   Studies on the genetics of *Ustilago zeae*.   *Phytopath. Zeitschr.* **4** :129–188. 1931.

——— and H. A. RODENHISER.   Physiologic specialization and genetics of the smut fungi.   *Bot. Rev.* **6** :389–425.   1940.

——— and E. C. STAKMAN.   Physiologic specialization and mutation in *Ustilago zeae*.   *Phytopath.* **16** :979–999.   1926.

CLAYTON, E. E.   Diplodia ear rot disease of corn.   *Jour. Agr. Research* **34** :357–371. 1927.

COOK, M. T.   Phloem necrosis in the stripe disease of corn.   *Jour. Agr. Porto Rico* **20** :684–688, 699–700.   1936.

CUGINI, G., and G. B. TRAVERSO.   La *Sclerospora macrospora* Sacc. parassita della *Zea Mays* L.   *Staz. Sper. Agr. Ital.* **35** :46–49.   1902.

CUMMINS, G. B.   Heterothallism in corn rust and effect of filtering the pycnial exudate.   *Phytopath.* **21** :751–753.   1931.

———.   Identity and distribution of three rusts of corn.   *Phytopath.* **31** :856–857. 1941.

DAVIS, G. N.   Some of the factors influencing the infection and pathogenicity of *Ustilago zeae* (Beckm.) Unger on *Zea mays* L.   *Iowa Agr. Exp. Sta. Res. Bul.* 199.   1936.

DEMERIC, M.   Genetic relations of five factor pairs for virescent seedlings in maize. *N.Y.* (*Cornell*) *Agr. Exp. Sta. Mem.* 84.   1924.

DICKSON, J. G.   The influence of soil temperature and moisture on the development of seedling blight of wheat and corn caused by *Gibberella saubinetii*. *Jour. Agr. Research* **23**:837–870. 1923.

DRECHSLER, C.   Some graminicolous species of Helminthosporium. *Jour. Agr. Research* **24**:641–740. 1923.

———.   Leafspot of maize caused by *Ophiobolus heterostrophus*, n. sp., the ascigerous stage of a Helminthosporium exhibiting bipolar germination. *Jour. Agr. Research* **31**:701–726. 1925.

———.   Phytopathological and taxonomic aspects of Ophiobolus, Pyrenophora, Helminthosporium, and a new genus, Cochliobolus. *Phytopath.* **24**:953–983. 1934.

———.   *Pythium graminicolum* and *P. arrhenomanes*. *Phytopath.* **26**:676–684. 1936.

DUFRENOY, J., and T. FREMONT.   Influence de la température sur les réactions du maïs à l'infection fusarienne. *Phytopath. Zeitschr.* **4**:37–41. 1931.

DURRELL, L. W.   Dry rot of corn. *Iowa Agr. Exp. Sta. Res. Bul.* 77. 1923.

———.   Basisporium dry rot of corn. *Iowa Agr. Exp. Sta. Res. Bul.* 84. 1925.

EDDINS, A. H.   Infection of corn plants by *Physoderma zeae-maydis* Shaw. *Jour. Agr. Research* **46**:241–253. 1933.

——— and R. K. VOORHEES.   *Physalospora zeicola* on corn and its taxonomic and host relationship. *Phytopath.* **23**:63–72. 1933.

——— and ———.   Breeding for resistance to brown spot of corn. *Fla. Agr. Exp. Sta. Rept.* (1934) **48**:73. 1935.

EDWARDS, E. T.   Studies on *Gibberella fujikuroi* var. *subglutinans*, the hitherto undescribed ascigerous stage of *Fusarium moniliforme* var. *subglutinans*, and on its pathogenicity on maize in New South Wales. *Dept. Agr. N.S. Wales Sci. Bul.* 49. 1935.

———.   Root and basal stock rot of maize. *Agr. Gaz. N.S. Wales* **47**:259–261. 1936.

ELLIOTT, C.   Bacterial wilt of corn. *U.S. Dept. Agr. Farmers' Bul.* 1878. 1941.

———.   A Pythium stalk rot of corn. *Jour. Agr. Research* **66**:21–39. 1943.

EMERSON, R. A., *et al.*   A summary of linkage studies in maize. *N.Y. (Cornell) Agr. Exp. Sta. Mem.* 180. 1935.

GARBER, R. J., and M. M. HOOVER.   The relation of smut infection to yield in maize. *Jour. Am. Soc. Agron.* **20**:735–746. 1928.

GORTER, G. J. M. A.   Wheat stunt—a new cereal disease. *Fmg. So. Africa* **22**:29–32. 1947.

HARTWELL, B. L., and T. R. PEMBER.   Aluminum as a factor influencing the effect of acid soils on different crops. *Jour. Am. Soc. Agron.* **10**:45–47. 1918.

HAYES, H. K., *et al.*   Reactions of selfed lines of maize to *Ustilago zeae*. *Phytopath.* **14**:268–280. 1924.

———.   Reaction of maize seedlings to *Gibberella saubinetii*. *Phytopath.* **23**:905–911. 1933.

HOFFER, G. N., and R. H. CARR.   Accumulation of aluminum and iron compounds in corn plants and its probable relation to root rots. *Jour. Agr. Research* **23**:801–823. 1923.

HOLBERT, J. R., *et al.*   The corn root, stalk, and ear rot diseases and their control through seed selection and breeding. *Ill. Agr. Exp. Sta. Bul.* 255. 1924.

——— *et al.*   Some factors affecting infection with and the spread of *Diplodia zeae* in the host tissue. *Phytopath.* **25**:1113–1114. 1935.

——— and J. G. DICKSON.   Corn varieties resistant to rot disease. *U.S. Dept. Agr. Yearbook*, pp. 254–259. 1926.

——— and ———. The development of disease resistant strains of corn. *Int. Cong. Plant Sci. Proc.* **1**:155–160. 1929.

HOOVER, M. M. Inheritance studies of the reaction of selfed lines of maize to smut (*Ustilago zeae*). *W. Va. Agr. Exp. Sta. Bul.* 253. 1932.

HOPPE, P. E. Differences in Pythium injury to corn seedlings at high and low soil temperatures. *Phytopath.* **39**:77–84. 1949.

——— and J. R. HOLBERT. Methods used in the determination of relative amounts of ear rot in dent corn. *Jour. Am. Soc. Agron.* **28**:810–819. 1936.

HURD-KARRER, A. M. Effect of smut on sap concentration in infected corn stalks. *Am. Jour. Bot.* **13**:286–290. 1926.

——— and H. HASSELBRING. Effect of smut (*Ustilago zeae*) on the sugar content of cornstalks. *Jour. Agr. Research* **34**:191–195. 1927.

IMMER, F. R., and J. J. CHRISTENSEN. Determination of losses due to smut infections in selfed lines of corn. *Phytopath.* **18**:599–602. 1928.

——— and ———. Further studies on reaction of corn to smut and effect of smut on yield. *Phytopath.* **21**:661–674. 1931.

IPPOLITO, G., and C. B. TRAVERSO. La *Sclerospora macrospora* Sacc. parassita delle inflorescenze virescenti di *Zea mays* L. *Staz. Sper. Agr. Ital.* **36**:975–995. 1903.

IVANOFF, S. S. Studies on the host range of *Phytomonas stewartii* and *P. vascularum*. *Phytopath.* **25**:992–1002. 1935.

——— and A. J. RIKER. Resistance to bacterial wilt of inbred strains and crosses of sweet corn. *Jour. Agr. Research* **43**:927–954. 1936.

——— and ———. Genetic types of resistance to bacterial wilt of corn. *Phytopath.* **26**:95–96. 1936.

JENKINS, M. T. Corn Improvement. *U.S. Dept. Agr. Yearbook*, pp. 455–522. 1936.

——— and A. L. ROBERT. Inheritance of resistance to the leaf blight of corn caused by *Helminthosporium turcicum*. *Agron. Jour.* **44**:136–140. 1952.

JOHANN, H., *et al.* A Pythium seedling blight and root rot of corn. *Jour. Agr. Research* **37**:443–464. 1928.

———. Scolecospores in *Diplodia zeae*. *Phytopath.* **29**:67–71. 1939.

JOHNSON, I. J., and J. J. CHRISTENSEN. Relation between number, size, and location of smut infections to reduction in yield of corn. *Phytopath.* **25**:223–233. 1935.

JONES, D. F. Segregation of susceptibility to parasitism in maize. *Am. Jour. Bot.* **5**:295–300. 1918.

JONES, J. P. Deficiency of magnesium the cause of chlorosis in corn. *Jour. Agr. Research* **39**:873–892. 1929.

JORGENSEN, L. R. Effect of smut infection on the yield of selfed lines and F₁ crosses in maize. *Jour. Am. Soc. Agron.* **21**:1109–1112. 1929.

KENT, G. C. An inhibitor produced by *Diplodia zeae* (Schw.) Lév. *Iowa Agr. Exp. Sta. Res. Bul.* 274. 1940.

KIESSELBACH, T. A., and J. A. RATCLIFF. Freezing injury of corn. *Nebr. Agr. Exp. Sta. Bul.* 163. 1918.

KNOWLES, E. L. A study of the abnormal structures induced by *Ustilago zeae-maydis*. *Jour. Mycology* **5**:14–18. 1889.

KOEHLER, B., and J. R. HOLBERT. Corn diseases in Illinois, their extent, nature and control. *Ill. Agr. Exp. Sta. Bul.* 354. 1930. *Cir.* 484. 1938.

KULKARNI, G. S. Observations on the downy mildew *Sclerospora graminicola* (Sacc.) Schroet. of bajri and jowar. *Mem. Dept. Agr. Bot. Ser.* **5**:268–273. 1913.

KUNKEL, L. O. A possible causative agent for the mosaic disease of corn. *Bul. Exp. Sta. Hawaiian Sugar Planters' Assn.* **3**:44–58. 1921.

————. Insect transmission of yellow stripe disease. *Hawaiian Planters' Record* **26**:58–64. 1922.

————. Leafhopper transmission of corn stunt. *Proc. Nat. Acad. Sci.* **32**:246–247. 1946.

————. Studies on a new corn virus disease. *Arch. Ges. Virusforsch.* **4**:24–46. 1948.

LAWAS, O. M., and W. L. FERNANDEZ. A study of the transmission of the corn mosaic and some of the physical properties of its virus. *Philipp. Agr.* **32**:231–238. 1949.

LEECE, C. W. Downy-mildew disease of sugar cane and other grasses. *Queensland Bur. Sugar Exp. Stas. Tech. Com.* **5**:111–135. 1941.

LE ROUX, P. M. Investigations on spore germination and host-pathogen reactions in corn rust incited by *Puccinia sorghi*. Thesis. University of Wisconsin, pp. 1–161. 1954.

————, J. C. DICKSON, A. L. HOOKER, and G. F. SPRAGUE. A genetic basis for rust reaction on corn. *Phytopath.* **44**:496. 1954.

LYON, H. L. The Australian leaf stripe disease of sugar cane. *Hawaiian Planters' Record* **4**:257–265. 1915.

MAGISTAD, O. C. Aluminum content of the soil solution and its relation to soil reaction and plant growth. *Soil Sci.* **20**:181–225. 1925.

MAINS, E. B. Studies in rust resistance. *Jour. Heredity* **17**:313–325. 1926.

————. Inheritance of resistance to rust, *Puccinia sorghi* in maize. *Jour. Agr. Research* **43**:419–430. 1931.

————. *Angiospora*, a new genus of rusts on grasses. *Mycologia* **26** :122–132. 1934.

————. Host specialization of *Puccinia sorghi*. *Phytopath.* **24**:405–411. 1934.

————. Two unusual rusts of grasses. *Mycologia* **30**:42–45. 1938.

MANGELSDORF, P. C., and R. G. REEVES. The origin of Indian corn and its relatives. *Texas Agr. Exp. Sta. Bul.* 574. 1939.

MARAMOROSCH, K. Mechanical transmission of corn stunt virus to an insect vector. *Phytopath.* **41**:833–839. 1951.

MASON, W. E. On species of the genus Nigrospora Zimm. recorded on Monocotyledons. *Trans. Brit. Myc. Soc.* **12**:152–165. 1927.

McALPINE, D. The smuts of Australia. Department of Agriculture, Victoria. 1910.

McCLEAN, A. P. D. Some forms of streak virus occurring in maize, sugarcane and wild grasses. *Dept. Agr. So. Africa Sci. Bul.* 265. 1947.

McDONOUGH, E. S. A cytological study of the development of the oospore of *Sclerospora macrospora* Sacc. *Trans. Wis. Acad. Sci., Arts, Letters* **38**:211–218. 1946.

McINDOE, K. G. The inheritance of the reaction of maize to *Gibberella saubinetti*. *Phytopath.* **21**:615–639. 1931.

McNEW, G. L. Crown infection of corn by *Diplodia zeae*. *Iowa Agr. Exp. Sta. Res. Bul.* 216. 1937.

MELHUS, I. E., and G. N. DAVIS. Nodal infection with the corn smut organism. *Phytopath.* **21**:129. 1931.

————, F. H. VAN HALTER, and D. E. BLISS. A study of *Sclerospora graminicola* (Sacc.) Schroet. on *Setaria viridis* (L.) Beauv. and *Zeae mays* L. *Iowa Agr. Exp. Sta. Res. Bul.* 111. 1928.

MENDIOLA, V. B. The Fusarium disease of corn. *Philipp. Agr.* **19**:79–106. 1930.

MIDDLETON, J. T. The taxonomy, host range and geographic distribution of the genus *Pythium*. *Mem. Torrey Bot. Club* **20**:1–171. 1943.

MITRA, M. Helminthosporium species on cereals and sugar cane in India. I. *Dept. Agr. India Bot. Ser. Mem.* **11**:219–242. 1923.

MIYAKE, T. On a fungus disease of sugar cane caused by a new parasitic fungus, *Sclerospora sacchari* T. Miy. *Rept. Sugar Exp. Sta. Govt. Formosa Div. Path. Bul.* 1. 1911. (In Japanese.)

NIEDERHAUSER, J. S., and D. BARNES. Use of pesticides on basic food crops in the tropics. *Adv. in Chemistry Ser.* **13**:3–8. 1955.

NISHIKADO, Y. Temperature relations to the growth of graminicolous species of Helminthosporium. I. *Ber. Ōhara. Inst. Landw. Forsch. Kuraschiki* **3**:349–377. 1927.

—— and C. MIYAKE. Studies on two Helminthosporium diseases of maize caused by *H. turcicum* Pass. and *Ophiobolus heterostrophus* Drech. (*H. maydis* N. and M.). *Ber. Ōhara Inst. Landw. Forsch. Kuraschiki* **3**:221–266. 1926.

PATEL, M. K. Production of oospores by *Sclerospora sorghi* on maize. *Indian Phytopath.* **2**:52–57. 1949.

PEARSON, N. L. Parasitism of *Gibberella saubinetti* on corn seedlings. *Jour. Agr. Research* **43**:569–596. 1931.

PEGLION, V. La formazione dei conidi e la germinazione della oospore della *Sclerospora macrospora* Sacc. *Boll. R. Staz. Patol. Veg. Roma* (*N.S.*) **10**:153–164. 1930.

PETCH, T. *Monotospora oryzae* B. et Br. *Jour. Indian Bot. Soc.* **4**:21–22. 1924.

PEYRONEL, B. Gli zoosporangi nella *Sclerospora macrospora*. *Boll. R. Staz. Patol. Veg. Roma* (*N.S.*) **9**:353–357. 1929.

POLE-EVANS, I. B. Maize smut or "Brand" *Sorosporium reilianum* (Kühn) McAlp. *Agr. Jour. Union So. Africa* **1**:697. 1911.

POLE-EVANS, MARY. Rusts in South Africa. II. A sketch of the life-cycle of the rust on mealie and Oxalis. *Union So. Africa Div. Bot. Sci. Bul.* 2, pp. 1–8. 1923.

POTTER, A. A. Head smut on sorghum and maize. *Jour. Agr. Research* **2**:339–372. 1914.

RALEIGH, W. P. Infection studies on *Diplodia zeae* (Schw.) Lév. and control of seedling blights of corn. *Iowa Agr. Exp. Sta. Res. Bul.* 124. 1930.

RAND, F. V., and L. C. CASH. Further evidence of insect dissemination of bacterial wilt of corn. *Science* **59**:67–69. 1924.

RANDS, R. D., and E. DOPP. Variability in *Pythium arrhenomanes* in relation to root rot of sugar cane and corn. *Jour. Agr. Research* **49**:189–221. 1934.

—— and S. F. Sherwood. Yield tests of disease resistant sugar cane in Louisiana. *U.S. Dept. Agr. Cir.* 418. 1927.

REED, G. M., *et al.* Experimental studies on head smut of corn and sorghum. *Bul. Torrey Bot. Club* **54**:295–310. 1927.

RHOADES, M. M., and V. H. RHOADES. Genetic studies with factors in the tenth chromosome in maize. *Genetics* **24**:302–314. 1939.

RHOADES, V. H. The location of a gene for disease resistance in maize. *Nat. Acad. Sci. Proc.* **21**:243–246. 1935.

ROBERT, ALICE L. Cultural and pathogenic variability in single-conidial and hyphal-tip isolates of *Helminthosporium turcicum* Pass. *U.S. Dept. Agr. Tech. Bul.* 1058, pp. 1–18. 1952.

ROMAN, H., and A. J. ULLSTRUP. The use of A-B translocations to locate genes in maize. *Agron. Jour.* **43**:450–454. 1951.

SAFEEULLA, K. M., and M. J. THIRUMALACHAR. Resistance to infection by *Sclerospora sorghi* of sorghum and maize varieties in Mysore, India. *Phytopath.* **45**:128–131. 1955.

SAMPSON, K.   Life cycles of smut fungi.   *Brit. Myc. Soc. Trans.* **23**:1–23.   1939.

ȷAVULESCU, T., and T. RAYSS.   Une nouvelle maladie du maïs en Roumaine provoquée par *Nigrospora oryzae* (B. et Br.) Pet.   *Arch. Roumaines path. exp. Microbiol.* **3**:41–53.   1930.

—— and ——.   Contribution à la connaissance de la biologie de *Nigrospora oryzae* (B. et Br.) Pet. parasite du maïs.   *Trav. cryptogam. Louis Mangin Mus. Nat. Hist. Paris,* pp. 233–240.   1931.

—— and ——.   Influence des conditions extérieures sur le développement de *Nigrospora oryzae* (B. et Br.) Pet. parasite de maïs en Roumaine.   *Acad. Sci. Paris compt. rend.* **194**:1262–1265.   1932.

SCHINDLER, A. J.   Insect transmission of wallaby ear disease of maize.   *Jour. Aust. Inst. Agr. Sci.* **8**:35–37.   1942.

SCOSSIROLI, R.   Per la conoscenza del manismo del mais.   *Ann. Sper. Agr.* (N.S.) **5**:157–177.   1951.

SHEAR, C. L., and N. E. STEVENS.   *Sphaeria zeae* (*Diplodia zeae*) and confused species.   *Mycologia* **27**:467–477.   1935.

SMITH, G. M.   Golden cross bantam sweet corn.   *U.S. Dept. Agr. Cir.* 268.   1933.

—— and J. F. TROST.   Diplodia ear rot in inbred and hybrid strains of sweet corn.   *Phytopath.* **24**:151–157.   1934.

SMITH, O. F.   The influence of low temperature on seedling development in two inbred lines of corn.   *Jour. Am. Soc. Agron.* **27**:467–479.   1935.

SPARROW, F. K.   The occurrence of true sporangia in the Physoderma disease of corn.   *Science* (N.S.) **79**:563–564.   1934.

——.   Observations on Chytridiaceous parasites of phanerogams.   II.   A preliminary study of the occurrence of ephemeral sporangia in the Physoderma disease of maize.   *Am. Jour. Bot.* **34**:94–97.   1947.

SPRAGUE, R.   Diseases of cereals and grasses in North America.   The Ronald Press Company, pp. 11–13.   New York.   1950.

STAHL, C. F.   Corn stripe disease in Cuba not identical with sugar cane mosaic.   *Trop. Res. Found. Bul.* 7.   1927.

STAKMAN, E. C., *et al.*   Physiologic specialization in *Puccinia sorghi*.   *Phytopath.* **18**:345–354.   1928.

——.   Mutation and hybridization in *Ustilago zeae*.   *Minn. Agr. Exp. Sta. Tech. Bul.* 65.   1929.

——.   The constancy of cultural characters and pathogenicity in variant lines of *Ustilago zeae*.   *Bul. Torrey Bot. Club* **60**:565–572.   1933.

——.   The genetics of pathogenic organisms.   *Am. Assn. Adv. Sci. Pub.* 12.   1940.

STEVENS, N. E., and J. I. WOOD.   Losses from ear rots in the United States.   *Phytopath.* **25**:281–283.   1935.

STONER, W. N.   Leaf fleck, an aphid borne persistent virus disease of maize.   *Phytopath.* **43**:683–689.   1953.

STOREY, H. H.   Streak disease of sugar-cane.   *Union S. Africa Dept. Agr. Sci. Bul.* 39, pp. 1–30.   1925.

——.   The transmission of streak disease of maize by the leafhopper *Balclutha mbila* Naude.   *Ann. Appl. Biol.* **12**:422–439.   1925.

——.   Transmission studies of maize streak disease.   *Ann. Appl. Biol.* **15**:1–25.   1928.

——.   A mosaic virus of grasses not virulent to sugar-cane.   *Ann. Appl. Biol.* **16**:525–532.   1929.

——.   Investigations of the mechanism of the transmission of plant viruses by insect vectors.   I.   *Proc. Roy. Soc. London* B **113**:463–485.   1933.

————. Studies on the mechanism of the transmission of plant viruses by insects. *Arch. exper. Zellforsch.* **15**:457–458. 1934.

————. A new virus of maize transmitted by *Cicadulina* spp. *Ann. Appl. Biol.* **24**:87–94. 1937.

————. Basic research in agriculture: a brief history of research at Amani 1928–1947. *E. Africa Agr. For. Res. Org. Kenya.* February, 1951.

———— and A. P. D. McCLEAN. The transmission of streak disease between maize, sugar cane and wild grasses. *Ann. Appl. Biol.* **17**:691–719. 1930.

TANAKA, I. *Phytophthora macrospora* (Sacc.) S. Ito et I. Tanaka on wheat plant. *Ann. Phytopath. Soc. Japan* **10**:127–138. 1940.

THIRUMALACHAR, M. J., C. G. SHAW, and M. J. NARASIMHAN. The sporangial phase of the downy mildew on *Eleusine coracana* with a discussion of the identity of *Sclerospora macrospora* Sacc. *Bul. Torrey Bot. Club* **80**:299–307. 1953.

TISDALE, W. H. Physoderma disease of corn. *Jour. Agr. Research* **16**:137–154. 1919.

————. The brown spot of corn with suggestions for its control. *U.S. Dept. Agr. Farmers' Bul.* 1124. 1920.

TRANZSCHEL, W. Beitrage zur Biologie der Uredineen. II. *Trudy. Bot. Mus. Imp. Akad. Nank. St. Petersbourg* **3**:37–55. 1907.

ULLSTRUP, A. J. The occurrence of *Gibberella fujikuroi* var. *subglutinans* in the United States. *Phytopath.* **26**:685–693. 1936.

————. Inheritance of susceptibility to infection by *Helminthosporium maydis* race 1 in maize. *Jour. Agr. Research* **63**:331–334. 1941.

————. Two physiologic races of *Helminthosporium maydis* in the corn belt. *Phytopath.* **31**:508–523. 1941.

————. Diseases of dent corn in the United States. *U.S. Dept. Agr. Cir.* 674. 1943.

————. Further studies on a species of Helminthosporium parasitizing corn. *Phytopath.* **34**:214–222. 1944.

————. An undescribed ear rot of corn caused by *Physalospora zeae*. *Phytopath.* **36**:201–212. 1946.

————. Observations on crazy top of corn. *Phytopath.* **42**:675–680. 1952.

VALLEAU, W. D. Seed transmission of Helminthosporium of corn. *Phytopath.* **25**:1109–1112. 1925.

VOORHEES, R. S. Effect of certain environmental factors on the germination of the sporangia of *Physoderma zeae-maydis*. *Jour. Agr. Research* **47**:609–615. 1933.

————. *Gibberella moniliformis* on corn. *Phytopath.* **23**:368–378. 1933.

————. Histological studies of the seedling disease of corn caused by *Gibberella moniliformis*. *Jour. Agr. Research* **49**:1009–1051. 1934.

————. Sclerotial rot of corn caused by *Rhizoctonia zeae* n. sp. *Phytopath.* **24**:1290–1303. 1934.

WALTER, J. M. The mode of entrance of *Ustilago zeae* into corn. *Phytopath.* **24**:1012–1020. 1934.

WELLENSIEK, S. J. The nature of resistance in *Zea mays* L. to *Puccinia sorghi* Schw. *Phytopath.* **17**:815–825. 1927.

WELLMAN, F. L. Infection of *Zea mays* and various other Gramineae by the celery virus in Florida. *Phytopath.* **24**:1035–1037. 1934.

WESTON, W. H. Philippine downy mildew of maize. *Jour. Agr. Research* **19**:97–112. 1920.

————. Another conidial Sclerospora of Philippine maize. *Jour. Agr. Research* **20**:669–684. 1921.

———— and B. N. UPPAL. The basis of *Sclerospora sorghi* as a species. *Phytopath.* **22**:573–586. 1932.

YOUNG, P. A. Penetration phenomena and facultative parasitism in Alternaria, Diplodia, and other fungi. *Bot. Gaz.* **81**:258–279. 1926.

ZEHNER, M. G., and H. B. HUMPHREY. Smuts and rusts produced in cereals by hypodermic injection of inoculum. *Jour. Agr. Research* **38**:623–627. 1929.

ZOGG, H. Untersuchungen über die Epidemiologie des Maisrostes *Puccinia sorghi* Schw. *Phytopath. Zeitschr.* **15**:143–190. 1949.

CHAPTER 5

MILLET DISEASES

The millets are not grown extensively in the United States. Two species, *Setaria italica* (L.) Beauv. [*Chaetochloa italica* (L.) Scrib.] and *Panicum miliaceum* L., comprise the major acreage. Several additional species are grown less commonly in the United States, but extensively elsewhere: *Echinochloa crusgalli* var. *frumentacea* (Rozb.) Wright, *E. colonum* (L.) Link, *Pennisetum glaucum* (L.) R. Br., and *Eleusine coracana* (L.) Gaertn. The cultivated millets are important food crops in parts of southern Europe, Asia, and Africa.

In both *Setaria* and *Panicum*, polyploid series exist. The basic chromosome number in both genera is apparently nine pairs. The cultivated *Setaria* millet is in the nine-chromosome pair, and the *Panicum* millet is in the 18-pair group, based on the reports of Arenkova (1940), Avdulov (1931), Kishimoto (1938), and Rau (1929).

The cultivated millets are summer annuals adapted to warm climates. Most of the species are drought-resistant. Downy mildew and smuts are the more important diseases causing losses in this crop.

1. Bacterial Blights. Two bacterial blights are reported on the millets. *Bacterial spot* caused by *Pseudomonas alboprecipitans* Rosen [*Phytomonas alboprecipitans* (Rosen) Bergey *et al.*] occurs on the *Setaria* millet and some wild species. Small grayish-green spots with brown pigmentation are the common symptoms, as described by Rosen (1924).

A *bacterial stripe* of the *Panicum* millet caused by *Xanthomonas panici* (Elliott) Savul. [*Phytomonas panici* (Elliott) Bergey *et al.*] was reported by Elliott (1923). The brown stripes on the leaves are water-soaked and later show scales of exudate on the surface of the lesions.

2. Pythium Root Rot, *Pythium* spp. Rootlet rot and blighting of seedlings occurs under certain soil conditions. The graminicolous species of *Pythium* parasitize these crop plants.

3. Downy Mildew, *Sclerospora graminicola* (Sacc.) Schroet. The cultivated millets as well as many wild millet-like grasses are damaged by the disease, as described by Melhus, Van Haltern, and Bliss (1928). The disease is world-wide in its distribution and causes heavy losses, notably in Asiatic countries where the millets constitute an important cereal crop.   The disease is generally distributed in the United States on the weed grass, *Setaria viridis* (L.) Beauv.

SYMPTOMS AND EFFECTS.   The symptoms are very characteristic on these millets.   The plants are dwarfed chiefly through reduced internodal elongation of the culms.   Excessive tillering from the crown and development of branches from the axillary buds along the culm are generally characteristic.   The development of leaf-like malformations of the floral bracts and failure of kernel development are further common symptoms. The downy mass of conidiophores and conidia are usually common on suscepts grown in a humid climate and are less common under dry conditions.   Oospores develop abundantly in the mesophyll and parenchyma.   Leaf necrosis and browning are followed by splitting and shredding of the invaded tissues, especially as the plants approach maturity.   Axillary bud development and shredding of the leaves are less pronounced in the *Pennisetum* spp.   The excessive proliferation of buds and inflorescences during vegetative development combined with little or no kernel development causes a serious reduction in yield where infection is high (Fig. 26).

THE FUNGUS

Sclerospora graminicola (Sacc.) Schroet.
(*Protomyces graminicola* Sacc.)

The conidiophores emerge singly or in groups through the stomatal openings.   They are short (average 268 microns), thickened, without a basal cell, and produce numerous short branches near the apex.   The conidia are borne apically on short (8 microns) sterigmata.   The conidia are elliptical, slightly pointed with a conspicuous dehiscent papilla, smooth-walled, hyaline, and range from 14 to 23 microns long by 11 to 17 microns wide.   Occasional large conidia occur.   The conidia germinate to form three or more kidney-shaped two-ciliate zoospores.   Oospores are produced in large number in the cells of the mesophyll and parenchyma.   The oospores are irregular to round (30 to 60 microns in diameter), thick-walled, have a smooth outer wall, and are reddish brown in color.   They germinate by the formation of a germ tube, as described by McDonough (1937).

ETIOLOGY.   The oospores remain viable in the soil and crop refuse for long periods.   The oospores frequently are carried with the seed.   Only a small percentage of the spores germinate at one time; therefore the oosporic inoculum is present over long periods.   Infection of young plants occurs from the spores in the soil.   Wind-borne spores also may serve as inoculum when they come in contact with embryonic tissues. Conidia cause secondary spread when the plant tissues are still susceptible and weather favorable.   The mature plant tissues are relatively resistant.

CONTROL.   Control of the disease is difficult in areas where these crops are grown continuously over large areas, because of the general soil infestation.   In the United States where millets are not grown exten-

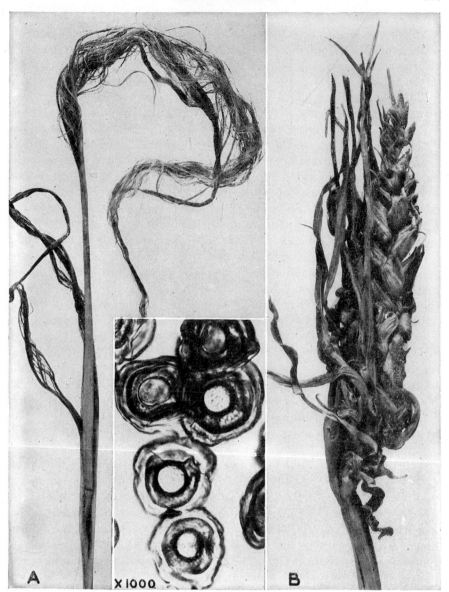

Fig. 26. Two of the more common downy mildews of the Gramineae. (A) *Sclerospora graminicola* incites disease in the millets and related grasses. Oospores, highly magnified, are shown in insert. Plant tissues shred, releasing the oospores. (B) *Sclerospora macrospora* incites disease in wheat, other cereals, and grasses. Plant tissues proliferate into leaf-like structures, and oospores remain in the diseased tissues.

sively, the control is easier unless the crop is sown in areas where the wild *Setaria viridis* is infected. Seed treatment with formaldehyde, sulfuric acid, and organic mercury compounds is reported as the best treatment. Tasugi and Akaishi (1933, 1935) report resistant varieties of the millets.

The downy mildew incited by *Sclerospora macrospora* Sacc. or *Scleroph-thora macrospora* (Sacc.) Thir., Shaw, & Naras. occurs on *Eleusine coracana* in India (Thirumalachar *et al.*, 1953). The proliferated tissues do not shred and release the spores as in the former species (Fig. 26).

4. Helminthosporium Leaf Spots. Several species have been reported on the millets and closely related species. Drechsler (1923) described *Helminthosporium monoceras* Drechsl. on *Echinochloa crusgalli* (L.) Beauv. in the United States. Sprague (1950) lists *H. giganteum* Heald & Wolf, *H. halodes* Drechsl., *H. nodulosum* Berk. & Curtis, *H. sacchari* (Breda de Haan) Butl., *H. sativum* P. K. B., *H. setariae* Saw. on the millet-like grasses and millets in North America. Nishikado (1929) reported *H. panici-miliacei* Nishikado and *H. yamadai* Nishikado on the *Panicum* millet and Ito (1930) *H. setariae* Saw. with the perithecial stage, *Cochlio-bolus setariae* (Ito & Kuribay.) Drechsl. on *Setaria* millets in Japan. Mitra and Mehta (1934) reported *H. nodulosum* (Berk. & Curt.) Sacc. and *H. leucostylum* Drechsl. on millet-like grasses in India.

5. Long Smuts, *Tolyposporium penicillariae* Bref. This smut is not reported in the United States but occurs in Asia and Africa, according to Britton-Jones (1922) and Butler (1918). The pear-shaped brown to black sori protrude from the floral bracts. The membrane ruptures exposing the greenish-brown spores. *T. bullatum* (Schröt.) Schröt. occurs on the millets also.

6. Kernel Smuts, *Ustilago crameri* Koern. This smut is common on the cultivated *Setaria* millets in Asia and Africa and in the United States on *Setaria italica* (L.) Beauv. and *S. viridis* (L.) Beauv. It is similar in appearance to the kernel smut caused by *Ustilago neglecta* Niessl. on *S. lutescens* (Weigel) Hubb. [*S. glauca* (L.) Beauv.] and *S. viridis* common weed grasses (Fig. 27). The disease is severe in some of the Asiatic areas, especially where the spores persist in the soil. The ovaries are replaced by the spore mass. The sori are enclosed in the floral bracts and are similar in shape to an enlarged kernel. The brittle floral bracts break, releasing the loose spore mass as the plants reach maturity, or they persist, enclosing the spores, and appear in the threshed grain.

THE FUNGI

1. *Ustilago crameri* Koern.

The sori in the spikelet destroy the kernel and basal portion of the spikelet. Spores are subglobose to ellipsoid and irregular, smooth, yellowish brown to olive green, 8 to

FIG. 27. (A) Head smut of millet caused by *Sphacelotheca destruens;* (B) kernel smut of *Setaria* caused by *Ustilago neglecta.*

11 microns wide and up to 16 microns long. The spores germinate to form a basidium with lateral branching or sporidia.

2. *Ustilago neglecta* Niessl.

Forms sori in the spikelets that usually rupture to release the purplish-brown spore mass. Spores are reddish brown to dark olive brown, irregularly globose to elongate, 7 to 11 microns wide and 9 to 12 microns long, walls echinulate. Germination as in

U. crameri. This species occurs commonly on *Setaria lutescens,* less commonly on *S. viridis* and *Bouteloua* spp. (Fischer, 1950).

Control of the smuts of millet is difficult, especially in dry areas where the spores remain in the soil for long periods. Seed treatment with volatile mercury compounds controls the smuts when all smut balls are removed first and reduces somewhat the infection from spores in the soil. Crop rotation is advisable. Resistant varieties are limited, but not fully explored.

7. Head Smuts, *Sphacelotheca destruens* (Schlect.) Stevenson & A. G. Johnson. The head smut of the *Panicum* millets is widespread on this crop. Similar-appearing smuts occur on several of the wild species. The smut sori are first evident as the panicles emerge. The entire inflorescence is modified into a sorus enclosed by a grayish-white false membrane. The membrane ruptures as the plants mature, exposing the dark-brown spore mass and the vascular tissues of the smutted panicle (Fig. 27).

THE FUNGI

1. *Sphacelotheca destruens* (Schlect.) Stevenson & A. G. Johnson

The sori are formed in the inflorescence, completely destroying all but the vascular elements. The sorus is covered by a grayish-white false membrane of fungus origin. Hyaline angular sterile cells of false membrane adhere in masses to the surface of the sorus. Spores are reddish brown, spherical to subspherical, 7 to 10 microns in diameter, and mostly smooth. The chlamydospores germinate to form basidia and sporidia (Fischer, 1953).

2. *Ustilago crusgalli* Tr. & Earle

Occurs on *Echinochloa crusgalli* (L.) Beauv., including the cultivated varieties. The sori are formed in a series of gall-like swellings on nodes and inflorescence and remain covered by the peridium of host tissue. Spores are globose to irregularly elongate, 9 to 13 microns, yellow to olivaceous, bluntly and sparsely echinulate.

The etiology and control are similar to the kernel smut.

REFERENCES

ARENKORA, D. N. Polyploid races in millet (*Panicum miliaceum* L.). *Acad. Sci. U.S.S.R.* **29**:332–335. 1940. (In Russian with English summary.)

AVDULOV, N. P. Karyo-systematische Untersuchung der Familie Gramineen. *Bul. Appl. Bot. Gen. and Plant Breed.* Suppl. 43. 1931. (In Russian with German summary.)

BRITTON-JONES, H. R. The smuts of millets. *Egypt. Min. Agr. Tech. Sci. Service Bul.* 18. 1922.

BUTLER, E. J. Fungi and disease in plants. Thacker, Spink and Co. Calcutta and Simla. 1918.

DRECHSLER, C. Some graminicolous species of Helminthosporium. *Jour. Agr. Research* **24**:641–739. 1923.

ELLIOTT, C.   A bacterial stripe disease of proso millet.   *Jour. Agr. Research* **26**:151–159.   1923.

FISCHER, G. F.   Manual of the North American smut fungi.   The Ronald Press Company, p. 343.   New York.   1953.

ITO, S.   On some new ascigerous stages of the species Helminthosporium parasitic on cereals.   *Imp. Acad. Tokyo Proc.* **6**:352–355.   1930.

KISHIMOTO, E.   Chromosomenzahlen in den Gattungen Panicum und Setaria.   I.   *Cytologia* **9**:23–27.   1938.

McDONOUGH, E. S.   The nuclear history of *Sclerospora graminicola*.   *Mycologia* **29**:151–173.   1937.

MELHUS, I. E., F. VAN HALTERN, and D. E. BLISS.   A study of *Sclerospora graminicola* (Sacc.) Schroet. on *Setaria viridis* (L.) Beauv. and *Zeae mays* L.   *Iowa Agr. Exp. Sta. Res. Bul.* 111.   1925.

MITRA, M., and P. R. MEHTA.   Diseases of *Eleusine coracana* G. and E. and *E. aegyptiaca* D. caused by species of Helminthosporium.   *Indian Jour. Agr. Sci.* **4**:943–975.   1934.

NISHIKADO, Y.   Studies on the Helminthosporium diseases of Gramineae in Japan.   *Ber. Ōhara Inst. Landw. Forsch.* **4**:111–126.   1929.

PORTER, R. H., *et al.*   Seed disinfectants for the control of kernel smut of foxtail millet.   *Phytopath.* **18**:911–919.   1929.

RAU, N. S.   On the chromosome numbers of some cultivated plants of South India.   *Jour. Indian Bot. Soc.* **8**:126–128.   1929.

ROSEN, H. R.   A bacterial disease of foxtail (*Chaetochloa lutescens*).   *Ark. Agr. Exp. Sta. Bul.* 193.   1924.

SPRAGUE, R.   Diseases of cereals and grasses in North America.   The Ronald Press Company.   New York.   1950.

TASUGI, H., and Y. AKAISHI.   Studies of the downy mildew (*Sclerospora graminicola setarae-italicae*) on Italian millet in Manchuria.   *So. Manch. Ry. Co. Agr. Exp. Sta. Res. Bul.* 11.   1933.   *Bul.* 15.   1935.

THIRUMALACHAR, M. J., C. G. SHAW, and M. J. NARASIMHAN.   The sporangial phase of the downy mildew on *Eleusine coracana* with a discussion of the identity of *Sclerospora macrospora* Sacc.   *Bul. Torrey Bot. Club* **80**:299–307.   1953.

TU, C., and H. W. LI.   Breeding millet resistant to smut in North China.   *Phytopath.* **25**:648–649.   1935.

VASEY, H. E.   Millet smuts and their control.   *Colo. Agr. Exp. Sta. Bul.* 242.   1918.

YU, T. F., *et al.*   Seed treatment experiments for controlling kernel smut of millet.   *Nanking Col. Agr. Forest. Bul.*   (N.S.) 14.   1934.

CHAPTER 6

OAT DISEASES

The cultivated oats are derived mainly from two wild species. The common oat, *Avena sativa* L., apparently is the principal commercial species originating from the wild oat, *A. fatua* L.; and *A. byzantina* Koch. originates from the wild red oat, *A. sterilis* L. The varieties of *A. sativa*, the common oat, grown in the cooler climates comprise the largest acreage devoted to this crop. The red oat varieties are grown chiefly in the warmer climates. The basic chromosome number in *Avena* is seven pairs. The wild and cultivated species mentioned above and *A. nuda* L. and *A. orientalis* Schreb. all have 21 pairs of chromosomes. *A. barbata* Brot. and *A. abyssinica* Hochst have 14 pairs, and *A. brevis* Roth, *A. nudibrevis* Vav., and *A. strigosa* Schreb. 7 pairs. Species in all three groups are used in disease studies and in the production of hybrids. Kihara and Nishigama (1932), Malzew (1930), Nishiyama (1951, 1953), Stanton (1936), and others discuss the genetics and cytology of this genus.

The more recent oat hybrids, combining disease resistance and high yielding capacity, represent a significant advance in disease control. Murphy (1942) and Stanton (1936) list the source of these selections and their reaction to the smuts and rusts. This parental material represents an available collection of superior germ plasm for rapid advance in oat breeding.

The cultivated oats are grouped into two classes—winter and spring—based on character of growth. Varieties of the common oat and hybrids including it and closely related species comprise the main group of spring types. Varieties of the red oat or combinations with it comprise the more important winter types grown. In oats, like barley, the winter types are distributed in the milder climates and the spring varieties occupy the larger acreage in the temperate zones. Fall-sown spring oats are grown on limited acreages in the extreme Southern United States.

Oats occupy an extensive acreage, and they are grown chiefly for feed grain. Oats rank third in acreage devoted to the cereal grains in the United States. Limited acreages are produced for pasture and hay. The oat plant is predominantly a low-temperature crop, especially during the seedling and early vegetative period. The red oat type will develop at somewhat higher temperature than the common oat, particularly in the later vegetative stages.

Oat diseases cause large losses in the United States. These losses are being reduced rapidly by the use of disease-resistant varieties. There is still, however, a large acreage of older varieties susceptible to one or more of the major oat diseases. Estimated average annual losses in the United States for the 10-year period 1930 to 1939 amounted to 10 per cent of the crop, or over 103 million bushels annually (Plant Disease Survey). Losses from the oat smuts and crown and stem rusts have been appreciably lower, since the general use of the newer disease-resistant varieties.

1. Nonparasitic. Two nonparasitic maladies are common on oats in limited areas. The blasting of spikelets of the panicle is conspicuous and causes a reduction in yield. The gray speck or dry leaf spot disease is important on oats and other cereals and grasses.

Blast. Blast of oats is caused by a number of environmental conditions. Species and varieties vary greatly in their tendency to produce white empty spikelets, especially near the base of the panicle. The blasting apparently is associated with a disturbed plant metabolism either when the panicle tissues are differentiating or near the period of pollination. Derick and Forsyth (1935), Elliott (1925), and Huskins (1931) discuss the influence of moisture, unbalanced fertility, and other factors upon the development of blast. These factors influence blasting when they occur at the tillering stage or later near the period of flower pollination. Derick and Hamilton (1939) show the influence of blast on yield. There is an indication that the conditions associated with blasting of the spikelets tend to reduce the potential yield of the remaining spikelets. The better-adapted varieties are relatively free from blast.

Gray Speck, or Dry Leaf Spot. The gray speck is not common in the major oat-growing sections of North America, although it occurs in alkaline organic soils (Hageman *et al.*, 1942, MacLachlan, 1941, 1943, and Sherman and Harmer, 1941). The malady is severe in parts of Europe and Australia, as reported by Davies and Jones (1931), Lundegårdh (1931, 1932), Rademacker (1935), Samuel and Piper (1928), and Stiles (1948). The disease occurs with less severity on wheat, barley, and some grasses.

Symptoms and Effects. Light-green to gray irregular to oblong flecks occur on the leaves, especially the leaf blades. The areas enlarge, dry out, and change to a buff or light-brown color (Fig. 28). The size and extent of the spots are modified by variety, severity of the manganese deficiency, and soil moisture. The plants are reduced in height; the leaf blades are narrow and more erect; and the plants are chlorotic as well as showing leaf spots in severe manifestations of the disease. Yields are reduced greatly when plant development is retarded. The dry leaf

Fig. 28. Gray speck and dry leaf spot caused by a deficiency of manganese and other minor elements. (A) Gray speck and dry spot on oat leaves and (B) a similar disturbance on winter wheat.

spot stage symptoms are easily differentiated from the heritable blotch described by Ferdinandsen and Winge (1929).

The disease is associated with certain soil types. Alkaline soils, low in soluble manganese and frequently high in organic matter, are conducive to leaf spot development. Unbalanced soil nutrients and their direct or indirect effect upon soluble or available manganese appear to be among the basic contributing factors. In organic soils, certain bacteria are associated with the conversion of soluble manganese compounds to insoluble oxides (MacLachlan, 1941, 1943). According to Lundegårdh (1931) all factors that cause a decrease in manganese absorption by the plant influence the development of the disease. Indirectly then, the addition of lime, humus or colloids, nitrates, or alkaline phosphates increases the development of gray speck. In addition to the manganese deficiency, a disturbed ion balance in the plant, especially extremely low or high potassium-calcium ratios, increases the manifestation of gray speck. There is apparently no distinct relation between the absolute manganese content of the leaves and the appearance of the necrosis. However, the direct cause of the disease is a deficiency of manganese. Fertilization with ammonium salts facilitates the solution of the manganese in the soil. Blatty (1932), Gerretsen (1937), Hiltner (1924), Samuel and Piper (1928, 1929), and others report control by the use of manganese sulfate, although in certain soils the results are temporary as the added manganese is changed quickly to insoluble oxides. MacLachlan (1941, 1943) obtained quicker and more permanent results by the use of 1 per cent manganese sulfate spray with bentonite and soap. Leach et al. (1954), MacLachlan (1941, 1943), Rademacker (1935), and Sherman and Harmer (1941) report differences in the tolerance of various crop plants.

Deficiencies in boron, copper, zinc, molybdenum, and other elements frequently occur in soils low in available manganese. This is associated with soils high in calcium and low in magnesium or with soils in which these elements do not occur in the substrates from which the soils are formed. Apparently, under conditions of high calcium ratio, the application of small quantities of soluble magnesium salts to the soils aids in correcting the deficiency in minor elements. Plant symptoms usually are chlorosis and necrosis of leaf blade, reduced plant growth, and low yields. Oats and legumes are affected more than many other crops (Kline, 1955, Quinlan-Watson, 1953).

2. Mosaics, Viruses Soil-borne and Insect-transmitted. Several mosaic diseases occur on oats throughout the world.

Soil-borne mosaics are present chiefly in the winter oat areas. Two types are described, leaf mottling and chlorosis and eyespot or ring spot (McKinney, 1946). These occur in limited areas and, once established, persist for some time.   The use of rotation and resistant varieties represents the best means of control.

Yellow dwarf virus is distributed widely in both spring and winter oat areas.   Damage is severe in susceptible varieties, especially in seasons of high aphid infestations.   The disease frequently is called red leaf. The symptoms vary somewhat with time of infection and oat variety. Red or yellow blotches, extending into linear lesions and red or yellow discoloration of the foliage, are the common symptoms.   Early infections cause plant dwarfing as in barley.   Aphids are the vectors, and the virus is not soil-borne nor transmitted mechanically (Oswald and Houston, 1953).

Blue dwarf occurs on oats independently or in association with yellow dwarf.   The plants appear blue-green in color, the leaf blades are shortened and thickened, and the plants are dwarfed.   The degree of dwarfing depends upon the stage of plant development when infection occurs. Apparently this is the same or a similar virus to that inciting yellow dwarf.   See Virus Diseases of barley, Chap. 3, and of wheat, Chap. 11.

The *brome mosaic* and *wheat streak mosaic* are transmitted to oats and probably occur in nature.   The symptoms incited by these viruses differ in minor details from those described.

Resistant varieties appear to offer the best means of control of this group of virus diseases.   The use of systemic insecticides early in the season indicates another means of control through the elimination of the vectors.

3. Halo Blight, *Pseudomonas coronafaciens* (Elliott) Stevens.   The cultivated oats and several grasses (*Agropyron, Avena, Bromus* spp.) are susceptible in varying degrees.   The disease is common in North America, and it is reported in most areas of the world.   The severe attack early in the season apparently does not cause an appreciable reduction in yield of grain.   Severe infection just previous to emergence of the panicles reduces yield.

DESCRIPTION.   The lesions are more common on the leaf blades, but they occur on the leaf sheaths and floral bracts in severe late infections. The initial lesion is an oval to oblong water-soaked small spot, changing gradually from green to buff or light brown.   The initial infection is associated with the stomata or more frequently aphid or other insect punctures.   The tissues surrounding the small spot gradually lose the green color and become slightly water-soaked and light yellow in color. The light-yellow zone forms a halo area around the restricted brown lesion.   As the number of infections increases, the lesions coalesce, forming an irregular halo area (Fig. 29).   The tissues dry out and fade to light-brown and straw-colored mottling.   No exudate is present on the lesion.   The spots on the floral bracts are less conspicuous, owing chiefly to the reduction in size of the halo area.

The bacterial colony is restricted to the stomatal cavity or mechanical puncture and between the adjacent mesophyll cells.   Spread between

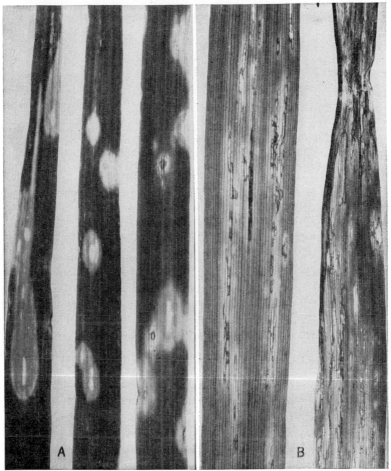

Fig. 29. Halo blight (A) and stripe blight (B) of oats caused by *Pseudomonas coronafaciens* and *P. striafaciens*, respectively, showing the progressive stages in lesion development.

the cells in the tissues is limited.   The cells of the adjacent tissue change in composition and function.   Chlorophyll regeneration is stopped; the permeability of the membranes is changed; and the intercellular spaces become water-soaked.   The bacterial colony apparently modifies the physiology of the adjacent cells without rapid necrosis of the tissue and in this manner establishes a nutritive balance with the suscept tissues without rapid advance between the cells. The pathological histology in this type of infection is in marked contrast to that of the bacteria in the gelatinous matrix advancing between the cells of the suscept, as in the bacterial blight of barley caused by *Xanthomonas translucens* and the stripe blight of oats caused by *Pseudomonas striafaciens* (Elliott) Starr & Burk.

THE BACTERIUM

Pseudomonas coronafaciens (Elliott) Stevens
[*Phytomonas coronafaciens* (Elliott) Bergey *et al.*]
(*Bacterium coronafaciens* Elliott)
(*Pseudomonas avenae* Manns)

The motile rods with rounded ends and one or more polar flagella develop without a gelatinous matrix. The colony is white on nutrient media. See Elliott (1920, 1951) for the detailed description of the organism. Tessi (1953) showed that *P. coronafaciens* and *P. striafaciens* are similar morphologically, culturally, and serologically. *P. coronafaciens* in media containing glucose produces a toxin causing the halo on a large number of grasses. On media without glucose, the halo is incited only on the same host group as in *P. striafaciens*.

ETIOLOGY. The bacteria enter the tissues through natural openings or injuries. Insects are important factors in the distribution and infection. Abundant moisture is necessary for the rapid development and spread of the disease. Johnson (1937) discusses the importance of high moisture content of the tissues (water-soaking) and the development of the disease. Secondary infections occur when conditions are favorable. Spikelet infection frequently occurs as the panicle is emerging from the leaf whorl. The organism is capable of existing considerable periods in crop residue. Infections of the hull and pericarp carry the bacteria over on the seed.

CONTROL. Seed treatment, sanitation, and rotation reduce the general abundance of the disease. The disease occurs rather generally in the humid areas despite the practice of these control measures. Apparently, insects play an important role in the general spread and establishment of the disease. Varieties show differences in susceptibility. Victoria × Richland selections are intermediate to susceptible in reaction, and Bond hybrid selections, especially D69 × Bond and Fulghum, are resistant (Tessi, 1949, 1952).

4. Bacterial Stripe Blight, *Pseudomonas striafaciens* (Elliott) Starr & Burk. The stripe blight occurs sparingly on oats in various sections of North America. The lesions first appear as sunken water-soaked minute spots that coalesce to form long water-soaked stripes (Fig. 29). Bacterial exudate is apparent on the surface of the lesion. The disease is similar in appearance, and the etiology is essentially the same as the bacterial blight of barley.

THE BACTERIUM

Pseudomonas striafaciens (Elliott) Starr & Burk.
[*Phytomonas striafaciens* (Elliott) Bergey *et al.*]
(*Bacterium striafaciens* Elliott)

The small rods with rounded ends and polar flagella are smaller than *Pseudomonas coronafaciens*. The colony is white on media. See Elliott (1927, 1951) for the detailed description.

5. Downy Mildew, *Sclerospora macrospora* Sacc. or *Sclerophthora macrospora* (Sacc.) Thir., Shaw & Naras. The downy mildew occurs on oats in scattered locations in Europe, Australia, and the United States. See Chap. 11, Downy Mildew on wheat.

6. Powdery Mildew, *Erysiphe graminis avenae* E. Marchal. The powdery mildew is not common on most of the cultivated varieties of oats. Reed (1920) demonstrated that many of the older varieties of oats are susceptible to certain specialized races of the fungus. This variety of the fungus is specialized on species of *Avena* and *Arrhenatherum*. See Powdery Mildew of barley (Chap. 3) for the detailed discussion of the disease.

7. Fusarium Blight, *Gibberella* and *Fusarium* Spp. The seedling blight occurs on all varieties of oats, especially in the Northern United States, Canada, and northern Europe. According to Greany *et al.* (1938) and Simmonds (1928), foot rot is also common. Losses from the kernel blight are limited, largely because of the open panicle and closed, generally pendent, spikelets and flowers. Occasionally, individual spikelets or kernels are infected. The blighted kernels are straw-colored and generally have a pinkish cast from the mycelial and conidial masses. See Fusarium Blight of barley (Chap. 3) and wheat (Chap. 11) for details of the disease.

8. Helminthosporium Leaf Blotch, *Pyrenophora avenae* Ito & Kuribay. Conidial Stage *Helminthosporium avenae* Eidam. The disease is distributed generally on the cultivated oats, although generally it is of minor importance, as reported by Dennis (1935), Drechsler (1923), O'Brien and Dennis (1933), Ravn (1901), and Turner and Milliard (1931).

DESCRIPTION. The blotches are oblong to linear with an irregular margin. They are light reddish brown, frequently with a sunken center, and conidia are fairly abundant on the older portion of the lesion (Fig. 30). The blotches are generally on the leaf blade. The infected leaf blades turn yellow and dry out as the infection advances.

THE FUNGUS

Pyrenophora avenae Ito & Kuribay.
Helminthosporium avenae Eidam. Conidial stage
(*Helminthosporium teres forma avenae-sativae* Bri. & Cav.)
(*Helminthosporium avenae-sativae* Bri. & Cav.)

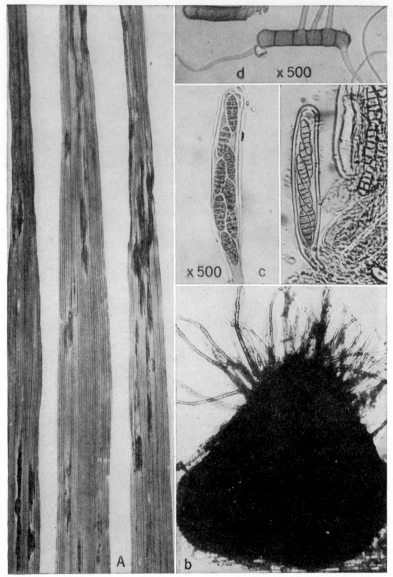

FIG. 30. Helminthosporium leaf spot of oats caused by *Helminthosporium avenae* or *Pyrenophora avenae* and the perithecium, ascospores, and conidia of the fungus. (*A*) Typical reddish-brown leaf lesions, (*b*) perithecium, (*c*) asci and ascospores, and (*d*) germinating conidium.

The morphology of the conidial stage is similar to *P. teres* (Drechsler, 1923, Sprague, 1950). According to Dennis (1935), Dickson (1946), and Ito and Kuribayashi (1931), the perithecial stage is not extensively distributed. The perithecia are partly submerged, irregular in shape, and less than 0.5 mm. in diameter. Setae and conidiophores are common on the surface. Asci, when fully developed, are clavate to cylindrical, slightly curved, rounded at the apex, characteristically eight-spored, many with two to four spores. Many asci are without organized spores. Ascospores are light brown, 5-septate, constricted at the septa with the center two or three cells divided longitudinally in the mature spores (Fig. 30).

ETIOLOGY. The primary infection occurs on the coleoptile or seedling leaves from seed-borne inoculum or mycelium, conidia, or ascospores from crop refuse. Secondary infections occur throughout the growing season. Ascosporic inoculum is a possible source of secondary infection as well as conidia from lesions on the leaves and crop refuse. Seed infection is common where the blotch develops abundantly, according to O'Brien and Dennis (1933) and Ravn (1900).

9. Helminthosporium Blight, *Helminthosporium victoriae* Meehan & Murphy. The disease in severe form occurs only on oat varieties with the Victoria parentage. Victoria from South America, resistant to leaf or crown rust and smut, was crossed with Richland. Leaf rust, stem rust, and smut-resistant selections were increased in 1933 (Stanton *et al.*, 1934). Selections were used extensively in both the spring and winter oat breeding, and some 30 varieties were distributed, placing this group as the dominant varieties by 1946. Soon after extensive distribution, the blight appeared in 1944 in epiphytotic form in the United States, and soon after in other countries where seed was introduced. The symptoms and the pathogen were different than the long-known leaf blotch disease (Meehan and Murphy, 1946, Murphy and Meehan, 1946).

The disease is characteristically a seedling blight and culm necrosis of varieties from Victoria. Reddening and necrosis of the seedling leaves without invasion of the pathogen into the leaf tissue are the early symptoms. Root and crown rot progress during the growing season. During the period the kernels are filling to ripening, culm invasion and necrosis develop rapidly. The nodes and basal portion of the lower internodes darken, and the culms break in the necrotic internodal section. Some leaf spotting occurs at this late stage of plant development. Later dark-olive-colored conidia are produced abundantly on the necrotic areas of the culm.

THE FUNGUS

Helminthosporium victoriae Meehan & Murphy

Conidiophores are dark olivaceous, slightly curved, rounded at the base. Conidia are widest near the center, tapering to a rounded apex 70 by 15 microns, 4- to 11-septate, average 8 septa. Germination is generally bipolar. The fungus produces a dark-gray mycelium in culture similar to *H. sativum*, and the conidia of the two are somewhat similar (Luttrell, 1955, Meehan, 1951).   As *H. sativum* incites a root and crown rot on oats under environmental conditions unfavorable for the host, conidia of the two species frequently are confused.

ETIOLOGY.   The fungus is seed-borne and persists in crop residue. Once established the disease appears each season on the susceptible varieties.   Seed treatment with the volatile organic mercury compounds and crop rotation control the disease on the susceptible varieties except under conditions where extensive acreages of susceptible varieties are grown in the area.

Apparently most other oat varieties are resistant.   One of three factor pairs conditioning resistance to leaf rust in Victoria is the same or closely associated with a simple dominant factor pair conditioning susceptibility to the blight.   However, the factor pairs for leaf rust resistance and many of the desirable agronomic characters of Victoria have been obtained free from blight susceptibility in several recent hybrids (Welsh *et al.*, 1954).

10. Scolecotrichum Leaf Blotch, *Scolecotrichum graminis* var. *avenae* Eriks.   This leaf blotch is common throughout the world on a large group of the grasses and less general on oats.   The disease is distributed widely on the grasses.   The oblong to linear reddish-brown to brownish-purple blotches with regular margins develop on the leaves.   The necrotic area is dry and sunken with conspicuous rows of the tufts of conidiophores that emerge through the stomata.   Conidial production on most of the oat varieties is sparse.   For the full discussion of the disease refer to Chap. 12.

11. Anthracnose, *Colletotrichum graminicolum* (Ces.) Wils.   The disease generally is not so common on oats as on the other cereals.   As reported by Sanford (1935), the disease develops on the root, crown, and basal culm tissues.   In severe attacks the plants are killed prematurely.   The disease is associated with dry soils low in fertility.   See Chap. 8 for the detailed discussion.

12. Septoria Leaf Blotch, *Septoria avenae* Frank.   The disease occurs on oats and a number of grasses.   The leaf blotch phase is distributed widely in both the winter- and spring-oat areas.   The blotches on the leaves are mottled light and dark brown, restricted irregular areas on the blade or spreading over the entire blade and sheath depending upon the susceptibility of the variety.   The lesions in the leaf sheath extend into the culm resulting in necrosis and blackening, frequently followed by lodging.   Small irregular brown blotches occur on the floral bracts and frequently result in brown discoloration of the enclosed kernel (Fig. 31). Pycnidial development on the extensive-type lesion generally is sparse and late in development.   The disease causes severe damage in areas where susceptible varieties are grown.

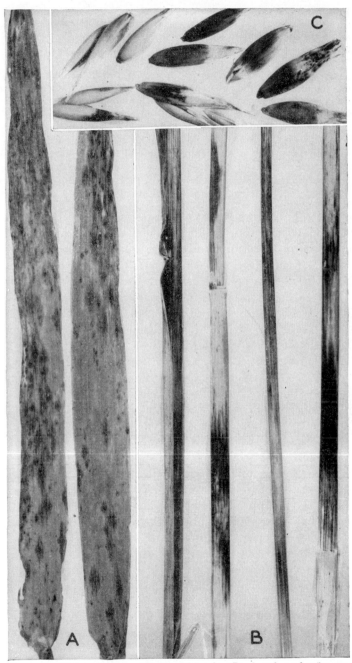

Fig. 31. Septoria blight of oats. (A) Leaf blotch, (B) sheath and culm necrosis and blackening, and (C) kernel necrosis and blackening.

The Fungus

Leptosphaeria avenaria G. F. Weber
Septoria avenae Frank.   Conidial stage

Pycnidia are scattered, subepidermal, globose, average 120 microns, smooth-walled, and distinct ostiole.   Spores are rod-shaped, usually straight, 3-septate, hyaline and guttulate, and 25–45 by 3–4 microns.   Perithecia are subepidermal on dead leaves and culms, globose to subglobose, walls smooth, black, ostiole does not protrude.   Asci are narrowly clavate with rounded apex, hyaline, thin-walled, and 30–100 by 10–18 microns.   Ascospores are fusoid, straight or slightly curved rounded ends, 3-septate, constricted at the septa, light yellow to olivaceous, and 23–48 by 4–6 microns (Sprague, 1950, Weber, 1922).   Races of this species on many of the grasses and the variety *Septoria avenae* Frank var. *triticea* T. Johnson, with a perithecial stage similar to *L. avenaria* and common on several grasses, do not infect oats.[1]

ETIOLOGY.   The fungus persists from one season to the next as mycelium and pycnidia on crop residue.   On susceptible varieties it is seed-borne.   Ascosporic inoculum probably is not important in the etiology of the disease.   Leaf infections occur especially during cool, wet weather. The culm rot is associated with the spread of the lesions into the leaf sheath.   The pathogen produces a toxic compound that kills tissue in advance of extensive mycelial invasion and pycnidial development. Some of the varieties, especially with Bond parentage developed for crown rust resistance, are susceptible.

13. Black Loose Smut, *Ustilago avenae* (Pers.) Rostr.   Two species of *Ustilago* cause black loose smuts on the cereals and grasses.   The black loose smut on oats and some grasses is caused by *Ustilago avenae* (Pers.) Rostr.   The morphology and life cycle of this fungus are similar to *U. nigra* Tapke (*U. medians* Biedenkopf) on barley.   A physiologic race of *U. avenae* formerly classified as *U. perennans* Rostr. occurs on certain grasses.   Fischer (1953) combines these species under *U. avenae*.

See Chap. 3, Black Semiloose Smut of barley, for detailed discussion.

The black loose smut is world-wide in its distribution on the cultivated oats and wild *Avena* spp. and related grasses.   This smut is less prevalent perhaps than the covered smut in the major oat-producing areas of the United States.   Separate estimates of losses from the two oat smuts are difficult to make because of variations in symptoms of the two smuts on different oat varieties (Fig. 32).   The combined losses, however, are high but are being reduced rapidly in the past few years by the use of resistant varieties.

DESCRIPTION AND EFFECT.   The individual flowers of the oat panicle are replaced, in large part, by the spore mass.   The smut sori vary from the loose powdery black spore mass replacing the floral structures, except the rachilla, to a semiloose spore mass enclosed within the lemma and

---

[1] Personal communication Drs. Dorothy Shaw and T. Johnson, Winnipeg, Canada.

Fig. 32. Oat panicles and spikelets infected with the black loose smut caused by *Ustilago avenae A* and *a* and covered smut caused by *U. kolleri* or *U. hordei* (*B* and *b*), showing the characteristic differences in symptoms and spores. The symptoms produced by the two species vary with oat varieties. Chlamydospores of *U. avenae* right of (*a*) and of *U. kolleri* right of (*b*).

palea. Oat variety and physiologic race of the parasite largely determine the difference in symptoms. The sori at first are covered by delicate gray membranes as the panicles emerge, but these membranes soon rupture, releasing the black spore masses. The smutted panicles are the first conspicuous evidence of the disease as they appear simultaneously with the emergence of the healthy inflorescences.

THE FUNGUS

Ustilago avenae (Pers.) Rostr.
(*Uredo segetum* subsp. *avenae* Pers.)
(*Uredo carbo* var. *avenae* DC.)
(*Ustilago segetum* var. *avenae* Jens.)
(*Ustilago avenae* Jens.)
[*Ustilago avenae* (Pers.) Jens.]

The fungus is distinguished by the finely echinulate chlamydospores that germinate to form a promycelium (basidium) and sporidia. The echinulations vary from distinct to very fine thickenings on the exospore wall. See Chap. 3, Black Semiloose Smut of barley, for the morphology.

ETIOLOGY. The cycle of development is similar in both of the black loose smut fungi. Seedling infection from seed-borne chlamydospores occurs during the early stages of germination and seedling development. Systemic invasion of the growing point of the seedling results, and spores are formed in the individual flowers. Environmental factors play an important role in smut infection and development, as reported by Bartholomew and Jones (1923), Johnson (1927), Jones (1923), Reed and Faris (1924), and Reed (1938). The spores are distributed from anthesis through harvesting of the crop. The spores within the floral bracts are the best situated for seedling infection and the most difficult to reach by seed treatment. The spores germinating beneath the hulls frequently are in close contact with the young coleoptile of the seedling that the smut fungus penetrates most readily before the coleoptile is 1 cm long. The chlamydospores located between the hulls and the pericarp are protected from the direct action of fungicides in solution as trapped air prevents complete wetting of the inner surfaces.

CONTROL. Seed treatment, especially with volatile fungicidal compounds, controls the loose and covered smuts of oats. Use of smut-resistant oat varieties is the most effective and economical method of control. Seed treatment is advised by many pathologists and plant breeders even though smut-resistant varieties are used. This advice is based upon two main premises: (1) seed treatment to prevent the possible establishment and accumulation of races of the smut fungi capable of attacking the resistant variety, and (2) seed treatment to protect the seedlings from soil-borne and other seed-borne organisms parasitic on oat seedlings. Seed-treatment experiments with smut-resistant varieties in the North Central humid area indicate increased stand in many instances, but relatively few show significant increases in yield at the standard rate of seeding (2 bushels per acre). Economizing on seed by reducing the rate of seeding, combined with seed treatment, is probably desirable in many areas.

TABLE 7. THE REACTION OF 10 DIFFERENTIAL VARIETIES OF OATS TO 15 PHYSIOLOGIC RACES OF *Ustilago avenae*

Physiologic race number	Smut reaction on differential varieties									
	Anthony (C.I. 2143)	Black Diamond (C.I. 1878)	Victory (C.I. 560)	Gothland (C.I. 1898)	Monarch (C.I. 1876)	Fulghum (C.I. 708)	Black Mesdag (C.I. 1877)	Camas (C.I. 2965)	Nicol (C.I. 2925)	Lelina (C.I. 3404)
A										
1	S	S	S	R	R	R	R	R	R	R
2	S	S	S	R	R	R	R	S	R	R
3	S	S	S	R–	S	R	R	R	R	R
4	S	S	S	R	S	R	S	R	R–	R
5	S	S	S	S	R	R	R	R	R	R
6	S	S	S	S	S	R	R	R	R	R
7	S	S	S	S	S	R	R	S	R–	R
8	S	S	R	R	R	R	R	R	R	R
9	S	S	R	R–	R	S	R	R	R	R
10	S	R	S	S	R	R	R	R	R	R
11	S	R	S	S	R	R	R	R	S	R
12	R	S	R	R–	R	S	R	R	R	R
13	R–	R	R	R	S	R	R	R	R	R
14	S	S	S	R	R	S	R	R	R	S
15	R	S	R	S	S	S	R	R	R	S

R = resistant; mean, 0 to 5 per cent. R– = resistant to susceptible; mean, 5 to 10 per cent. S = susceptible; 10 per cent or above.

SPECIALIZATION AND RESISTANCE. Physiologic specialization has been investigated extensively since Reed (1924) first described the two physiologic races for each species. Reed (1940) differentiated 29 races of *Ustilago avenae;* of these, 2 races (10 and 11) infected a few species and varieties, 6 races attack Fulghum (C.I. 3211), 6 races are unable to attack Fulghum but infect Gothland (C.I. 1898), and 15 races are unable to infect either Fulghum or Gothland. Roemer *et al.* (1937) and Vaughan (1938) reported a race capable of attacking Black Mesdag (C.I. 1877), and Sampson (1929, 1938) one infecting *Avena brevis.* Reed and Stanton (1942) reported a subrace 30A infecting Victoria (C.I. 4201) and Lee × Victoria selections. Races 30 and 31 infect Victoria but not the Victoria

× Richland varieties; however, Hansing *et al.* (1946) reported a race infecting this latter group of varieties. Holton and Rodenhiser (1946) have reviewed the literature and reported on extensive experiments with the specialized races collected in the United States and Canada. Using 10 differential varieties of oats, they report 15 races of *U. avenae* and 7 races of *U. kolleri* or *U. hordei*. These include many of the races differentiated by others, although the equivalents are indicated in only a few. Table 7 is condensed from the data presented by Holton and Rodenhiser (1946).

The practical application of information on specialization is in relation to determining smut resistance. Holton and Rodenhiser (1946) list the following oat varieties and hybrid selections, which were smut-free in field tests at Pullman, Wash., with the 22 individual races of *Ustilago avenae* and *U. kolleri* or *U. hordei*.

Variety or selection with pedigree	C.I. number
Benton (D69 × Bond)	3910
Boone (Victoria × Richland)	3305
Clinton (D69 × Bond)	3971
Huron (Markton × Victory)	3756
Marion (Markton × Rainbow)	3247
Markton	2053
Marvic (Markton × Victory)	2597
Neosho (Fulghum × Markton × Victoria-Richland)	4141
Rangler (Nortex × Victoria)	3733
Bond × Anthony Sel	4004
D69 × Bond Sel	3662
D69 × Bond Sel	3663
D69 × Bond Sel	3841
D69 × Bond Sel	3846
D69 × Bond Sel	4285
D69 × Bond Sel	4272
Fulghum-Markton × Victoria-Richland Sel	4001
Markton × Rainbow Sel	3350
Red Rustproof × (Victoria-Richland) Sel	3720
Richland × Fulghum Sel	3966
Victoria-Richland × Markton-Rainbow Sel	3609
Victoria-Richland × Morota-Bond Sel	4301

They report many other of the newer varieties and selections as resistant or with only small percentages of smut produced by a few physiologic races. Certain varieties, however, were highly susceptible to specific races.

Resistance to the two oat smuts is conditioned by several factors and modifiers or inhibitors acting independently or in combinations. The oat varieties and physiologic races of the parasites determine the factors functioning in the expression of resistance. Single factor pairs, single

factors and modifiers, and multiple factors are reported by Hayes *et al.*
(1939), Reed *et al.* (1934, 1935, 1937, 1938, 1941, 1942), Stanton *et al.*
(1934, 1943), Torrie (1939), and many others. The nature of smut
resistance and invasion of the resistant plants are discussed by Western
(1936, 1937) and Zade and Arland (1933). The genetics of the oat smut
fungi and inheritance of characters in crosses are summarized by Christensen and Rodenhiser (1940), Holton (1931, 1932, 1936), and Sampson
(1939).

14. Covered Smut, *Ustilago kolleri* Wille or *Ustilago hordei* (Pers.)
Lagerh. The covered smuts of oats and barley are similar in symptoms,
morphology of the fungi, and etiology. The persistence of the membrane
enclosing the sorus varies with oat variety. Hybridization between *U.
avenae* and *U. kolleri* results in normal segregation for both fungus characters and disease symptoms, according to Holton (1931, 1932, 1936).

The covered smut is world-wide in its distribution on cultivated oats
and wild species of *Avena*. Losses in general probably are greater than
those caused by the black loose smut, owing to the greater prevalence of
the covered smut. Varieties resistant to the covered and black loose
smuts are effective in reducing losses from these diseases in the major
oat-producing regions of the world.

DESCRIPTION AND EFFECT. The smut sori replacing the kernels are
enclosed in a fairly permanent membrane composed of pericarp and
floral bracts. The smutted panicles are the first conspicuous evidence of
the disease. In varieties in which the floral bracts are not modified
greatly by the smut, the smut sori are not conspicuous until the crop is
mature. The bleached lusterless lemma and palea then appear gray
because of the spore mass within. The smut sori are broken during
ripening of the grain and threshing, releasing the spores over the surface
of the healthy kernels. In many varieties of oats, the sori develop in the
lemma and palea as well as in the kernel, and in such varieties the smut
is more apparent (Fig. 32).

THE FUNGUS

Ustilago kolleri Wille or
Ustilago hordei (Pers.) Lagerh.
(*Ustilago avenae* var. *levis* Kell. & Swing.)
[*Ustilago levis* (Kell. & Swing.) Magn.]

The fungus is distinguished by the small smooth chlamydospores which germinate
to form a promycelium (basidium) and sporidia. See Chap. 3, Covered Smut of
barley, for morphology and detailed discussion.

ETIOLOGY. Seed-borne chlamydospores and seedling infection occur
as in the black loose smut.

SPECIALIZATION AND RESISTANCE.   Reed (1940) differentiated 14 physiologic races of *Ustilago kolleri* on 10 species and varieties of *Avena*. Holton and Rodenhiser (1946) differentiate 7 races, using the same 10 differential varieties employed for determining the races of *U. avenae*, as condensed in Table 8.

TABLE 8. THE REACTION OF 10 DIFFERENTIAL VARIETIES OF OATS TO 7 PHYSIOLOGIC RACES OF *Ustilago kolleri* OR *U. hordei*

Physiologic race numbers	Smut reaction on differential varieties									
	Anthony	Black Diamond	Victory	Gothland	Monarch	Fulghum	Black Mesdag	Camas	Nicol	Lelina
K										
1	S	S	S	R	R	R	R	R	R	R
2	S	S	S	R	S	R	R	R	R	R
3	S	S	S	S	R–	R	R	R	R	R
4	S	S	R	R	S	S	S	R	R	R
5	S	S	S	R	S	R	S	R	R	R
6	S–	S	R	R	R	R	R	R	R	R
7	S	S	S	S	R	R	R	R	R	S

R = resistant; mean, 0 to 5 per cent.
R– = resistant to susceptible; mean, 5 to 10 per cent.
S = susceptible; 10 per cent or above.

The list of smut-free varieties and selections given under the black loose smut (page 138) indicates the wide range of selections resistant to both species.   In addition, some of the newer oat varieties are reported by Holton and Rodenhiser (1946) as highly resistant to black loose smut and free from covered smut, notably Marida (C.I. 2571) (Markton × Idamine), Sac (C.I. 3907) (D69 × Bond), Bannock × Victoria-Richland Sel. (C.I. 4181), and D69 × Bond Sel. (C.I. 4532), whereas a number of the newer varieties showed a low percentage of smut produced by race 7 of *Ustilago kolleri* or *U. hordei*.

15. Stem Rust, *Puccinia graminis avenae* Eriks. & Henn.   Stem rust occurs on some varieties of all *Avena* spp. and on many related grasses. The disease is world-wide in its distribution on oats, and it reduces the forage value and yield of grain.   Stem rust–resistant varieties are reducing the importance of this disease on oats.   According to Stakman and Loegering (1944), physiologic races 8 and 10 of the fungus are a potential danger to the many rust-resistant varieties originating from the Victoria × Richland cross.   However, the control of stem rust through the use of resistant varieties in oats as in wheat is a continuous battle against a biologically dynamic pathogen (Koo *et al.*, 1955, Stakman and Loegering, 1944, Welsh and Johnson, 1951, 1954).

The red rust stage of the disease is conspicuous on the leaves and culms. In the northern spring-oat area, the disease occurs relatively late on most oat varieties. The black rust stage is apparent late in the season on the leaf sheaths and culms. See Chap. 11, Stem Rust of wheat, for the detailed discussion of the disease.

SPECIALIZATION AND RESISTANCE. Specialization occurs within the physiologic variety *Puccinia graminis avenae*. Levine and Smith (1937), Newton *et al.* (1940), and Stakman *et al.* (1935) differentiated 12 physiologic races on 3 groups of oat varieties. Fischer and Clausen (1944) reported race 14 obtained from *Poa ampla* Merr. Koo *et al.* (1955), Welsh *et al.* (1953), and Welsh and Johnson (1951, 1954) regrouped the races and 7A on the basis of the several factor pairs conditioning resistance to stem rust. Table 9 on race reactions is compiled from the investigations reported.

Four dominant factor pairs conditioning stem rust resistance in oats have been studied in some detail. Smith (1934) was unable to combine the White Tartar (White Russian) and the Richland factor pairs, and he concluded that the two genes appeared to be allelic. Later investigations seemed to confirm this conclusion (Cochran *et al.*, 1945, Litzenberger 1949, and others). The Canadian investigators using the Hajira and Hajira-Joanette type of stem rust resistance found two and three factor pairs in crosses involving Hajira. The two factors from Hajira were associated closely and conditioned resistance to all races except 7A. These two factors and the Richland factor conditioned resistance to all known races (Welsh and Johnson, 1954). They suggested the possibility of the one factor pair from Hajira selection conditioning resistance to all races except 7A. Their postulated C factor possibly is similar to the "Canadian" factor studied at Minnesota (Kehr *et al.*, 1950). The inheritance in the cross [Landhafer × (Mindo × Hajira-Joanette)] × Andrew in combination with the White Tartar and the Richland factor pairs and Gopher susceptible to the races studied suggest that these two genes, White Tartar and Richland, are combined and linked in the coupling phase in this multiple cross (Koo *et al.*, 1955). This combination of what appears to be the White Tartar and Richland factors and the two closely associated factor pairs from Hajira selections, inherited independently of these former two genes, provides a relatively high degree of protection of oats from stem rust over a wide range of environmental conditions. The expression of resistance when conditioned by some of the factor pairs, notably those from Hajira-Joanette, is influenced by temperature and light (Gordon, 1933, Koo *et al.*, 1955). Konzak (1954) obtained the Richland or similar factor pair from Mohawk (C.I. 4237) by exposure to thermal neutrons.

TABLE 9. REACTION IN THE SEEDLING STATE OF 7 GROUPS OF OAT VARIETIES TO THE UREDIAL STAGE OF 13 RACES OF *Puccinia graminis avenae*

Source of resistance by groups and varieties	C.I. no.	Seedling reaction to races of *P. graminis avenae*												
		1	2	5	3	7	12	7A	8	9	10	11	4	6	13
Group 1:															
White Tartar (White Russian)*	551														
	1614	R	R	R	S	S	S	S	R	R	R	R	S	S	S
Green Russian	1978														
Marion	3247														
Rainbow	2345														
Group 2:															
Richland	787	R	R	R	R	R	R	R	S	MS	S	S	S	S	S
Iogold	2329														
Tama	3502														
Vicland	3611														
Group 3:															
Hajira, original	1001	R	R	R	R	R	R	R	Sg	Sg	Sg	Sg	Sg	Sg	Sg
Group 4:															
Jostrain	2660	R	S	MR	R	S	MR	S	S	S	MS	R	R	S	MR
Roxton	4134	R	S	S	S	S	S	S	S	R	R	R	S	S	S
Sevnothree	3251	R	S	S	R	S	S	S	S	S	S	S	R	S	S
Group 5:															
Hajira Sel. (C.A.N. 810)															
Canuck	4024	R	R	R	R	R	R	S	R	R	R	R	R	R	R
Rodney	6661														
Minland	6765														
Group 6:															
Hajira Sel. + Richland															
Garry	4801	R	R	R	R	R	R	R	R	R	R	R	R	R	R
Ransom	5927														
Group 7:															
Combined resistance, groups 1, 2, etc. [Landhafer × (Mindo × Hajira-Joanette)]															
× Andrew	....	R	R	R	R	R	R	R	R	R	R	R	R	R	R

R = resistant (0; −1), MR = moderately resistant (1–2), MS = moderately susceptible (2–3), S = susceptible (4), and Sg = segregating.
* Two C.I. numbers of this variety with same reaction.

16. Crown Rust, *Puccinia coronata* (Pers.) Cda. The crown rust or leaf rust occurs on most species of *Avena* and many related grasses. The aecial stage occurs more commonly on *Rhamnus cathartica* L., *R. dahurica* Pallas, and *R. lanceolata* Pursh., according to Dietz (1926), Melhus *et al.* (1922), and Tranzschel (1934). The distribution of the rust is worldwide in temperate humid and semihumid areas. The disease causes heavy losses that are probably larger than those from stem rust in most oat-growing areas.

DESCRIPTION AND EFFECT. All stages of this heteroecious long-cycle rust are common symptoms in the temperate zones. The aecial infections on the *Rhamnus* spp. are common and conspicuous, the orange-yellow elevated lesions occurring on leaves, young stems, and fruits. The uredia develop on the leaves and floral structures of oats and grasses. The uredia are round to oblong but soon spread and coalesce to form irregular orange-yellow patterns. The epidermis of the suscept is not turned back as in the stem rust uredia. The telia frequently form a dark border around the uredia and develop independently, especially on the leaf sheath in linear dark-brown spots covered by the epidermis of the suscept (Fig. 33). Oats heavily infected with crown rust lodge badly and ripen prematurely.

THE FUNGUS

Puccinia coronata (Pers.) Cda. or
Puccinia coronata var. *avenae* (Cda.) Fraser & Led.
(*Puccinia coronifera* Eriks.)
(*Puccinia lolii* Niel.)
(*Puccinia coronifera Kleb.*)

The aecial stage consists of the raised pycnia (spermagonia) with exudate usually on the upper surface of the leaf and the aecia with large peridia on the lower surface (Fig. 34). The aeciospores are subglobose, finely verrucose, and light orange-yellow. The urediospores are globose or ovate, markedly echinulate, have three to four germ pores located irregularly, and are orange-yellow in color. The telia are covered by the epidermis except when forming directly in the uredia. The teliospores are constricted slightly at the septum, the apex is thickened with several blunt processes forming a crown-like apex, the pedicels are short and thickened, the color is dark brown (Fig. 34).

McGinnis (1954) reported three chromosome pairs in *P. coronata* based on the study of germinating teliospores.

ETIOLOGY. In the major spring oat areas, the aecial host plays an important role in the cycle of development of the fungus. Heavy aecial infection occurs in the early spring usually when the spring-sown oats are in the late seedling or early tillering stage of development. Aeciospore spread to the oats and grasses occurs early. The development of the uredial stage from primary infections from these wind-borne spores is scattered considerable distances from the *Rhamnus*. In agricultural areas where the *Rhamnus* is used as a hedge plant, as occurred some years ago in Rock County, Wisconsin, the early spread of crown rust to the oat plants is general. The uredial stage develops rapidly, and secondary infections from urediospores occur whenever weather is favorable. Telia develop later as the oats and grasses mature. The teliospores overwinter

FIG. 33. Crown rust reaction of oat varieties under field conditions. (A) Susceptible reaction, numerous uredia and telia, (B) necrosis and minute uredia characteristic of Victoria and Santa Fe types of resistance, and (C) absence of necrosis and uredia typical of Bond.

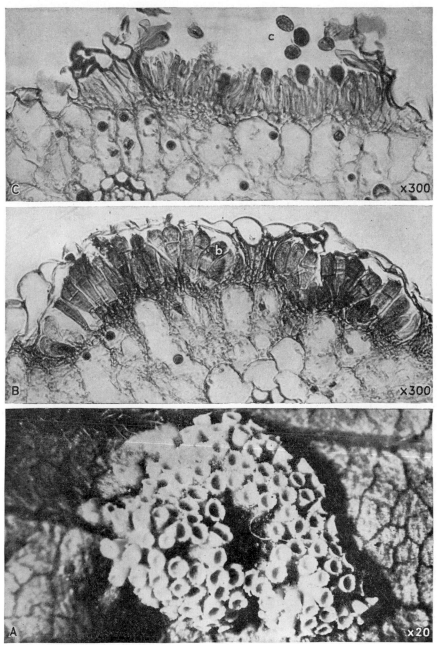

Fig. 34. The spore forms of *Puccinia coronata*. The aecium on *Rhamnus cathartica* showing the peridia of the aecial cups (*A*). The urediospores (*c*) and the teliospores (*b*) are produced in the uredium (*C*) and telium (*B*) shown in transections of the oat leaf.

on the old leaves and germinate to form the sporidia that infect the *Rhamnus* spp. the following spring.   Allen (1932) and Buller (1941) studied the cytology and sexual fusion of the aecial stage.   Murphy (1935) investigated physiologic races from *Rhamnus* spp. inoculations.   In the southern oat areas, the uredial stage develops continuously on the oats and grasses. Northern spread of the urediospores occurs from these southern sections. The spring infection from this source of inoculum is usually much later in the season than where the aecial infection occurs.   Extensive investigations on the overwintering of the uredial stage in the northern spring oat area of North America and Europe indicate the rarity of its occurrence.

CONTROL.   Eradication of the *Rhamnus* is practiced in some sections. Some states combine *Rhamnus* and *Berberis* eradication.   *Rhamnus* hedges around fields are generally removed by farmers when they realize the connection between the *Rhamnus* and the crown rust in their oat fields.   Rust-resistant varieties are rapidly replacing the older susceptible varieties.

SPECIALIZATION AND RESISTANCE.   Specialized races of *Puccinia coronata avenae* are numerous.   The earlier identification of races of this pathogen employed several sets of differential oat varieties in various parts of the world (Murphy *et al.*, 1942, Straib, 1937, Vallega, 1942). The confusion in race designation and in their association with resistance of the various oat varieties increased with the addition of new factors for leaf rust resistance.   Recently, races were regrouped on the reaction of 10 species and varieties containing single genes or gene groups conditioning resistance in the seedling stage (Simons and Murphy, 1955, Welsh *et al.*, 1953, 1954).   Former race numbers were replaced by new designations above 200, and the history of race designation in the pathogen is reviewed by Simons and Murphy (1955).   Many of the factor pairs conditioning resistance to crown rust have not been studied in detail for all of the races of the pathogen.   Therefore the present new classification of races 201 to 275 is subject to further change.   The 10 varieties used as differentials represent some of the major factor pairs used in breeding for crown rust resistance.

Resistance to crown rust from the races studied is conditioned by single dominant factor pairs, but some resistant varieties contain two or more factors.   The Victoria type of crown rust resistance is expressed in necrosis and no sporulation of the pathogen.   The Victoria reaction is not expressed fully in the presence of other factors giving a zero-type reaction at least for races studied.   Finkner (1954) and Litzenberger (1949) report a single factor pair conditioning resistance intermediate on the scale of dominance or epistasis to other genes for resistance to races 1 and 57 (old race numbers) and linked with susceptibility to *Helminthosporium* blight.   Poehlman and Kingsolver (1950) and Welsh *et al.*

(1954) report Victoria has more than one major factor for resistance to crown rust. Welsh *et al.* (1954) suggest that a dominant factor pair conditions resistance to races 4, 5, 34A, and 57, and the same factor or one linked with this factor conditions susceptibility to *Helminthosporium* blight. The resistance to race 45 and races 1, 2, 3, 6, 24, 34, and 38 associated with 45 is dominant, but resistance is governed by the factor pairs inherited independently of *Helminthosporium* blight reaction. Different varieties were used with Victoria by the several investigators, and possibly different races were used based on the new race classification (Simons, 1955). Bond resistance is conditioned by two complementary factor pairs (Cochran *et al.*, 1945, Hays, 1941, Ko *et al.*, 1946, and Torrie, 1939). Landhafer resistance involves one factor pair with less dominance expressed than in some others (Finkner, 1954, Kehr and Hayes, 1950, Litzenberger, 1949). Santa Fe varies in its genotype for crown rust resistance and perhaps contains one or both of the factor pairs allelic to the Ukraine-linked duplicate factors (Finkner, 1954). Trispernia varies in resistant reaction to race 57; parental plants with highest-type resistance indicated more factor pairs were involved than in plants with lower-type resistance. Trispernia may have any combination of these factor pairs. Finkner (1954) suggests a genetic pattern for the inheritance of resistance to crown rust and the relationship of the factors involved. Ascencao (P.I. 186603) moderately resistant to race 263 (new race number) of South America and recently reported in North America contains a factor pair for resistance to this race that is not allelic to the factors carried by Santa Fe and Landhafer (Simons, 1954). Brown and Shands (1954) discuss the resistance of *A. strigosa* and its combination with varieties in the higher chromosome groups.

REFERENCES

ALLEN, R. F. A cytological study of heterothallism in *Puccinia coronata*. *Jour. Agr. Research* **45**:513–541. 1932.

BARTHOLOMEW, L. K., and E. S. JONES. Relation of certain soil factors to the infection of oats by loose smut. *Jour. Agr. Research* **24**:569–575. 1923.

BROWN, C. M., and H. L. SHANDS. Behavior of the interspecific hybrid and amphiploid of *Avena abyssinica* × *A. strigosa*. *Agron. Jour.* **46**:557–559. 1954.

BULLER, A. H. R. The flexuous hyphae of *Puccinia graminis* and other Uredinales. *Phytopath.* **31**:4. 1941.

CHRISTENSEN, J. J., and H. A. RODENHISER. Physiologic specialization and genetics of the smut fungi. *Bot. Rev.* **6**:389–425. 1940.

COCHRAN, G. W., C. O. JOHNSTON, E. G. HEYNE, and E. D. HANSING. Inheritance of reaction to smut, stem rust and crown rust in four oat crosses. *Jour. Agr. Research* **70**:43–61. 1945.

DAVIES, D. W., and E. T. JONES. Grey speck disease of oats. *Welsh Jour. Agr.* **7**:349–358. 1931.

DENNIS, R. W. G. Notes on the occurrence of *Pyrenophora avenae* Ito in Scotland. *Trans. Brit. Myc. Soc.* **19**:288–290. 1935.

DERICK, R. A., and J. L. FORSYTH. A study of the cause of blast in oats. *Sci. Agr.* **15**:814–824. 1935.

———— and D. G. HAMILTON. Further studies on oat blast. *Sci. Agr.* **20**:157–165. 1939.

DIETZ, S. M. The alternate hosts of crown rust, *Puccinia coronata* Corda. *Jour. Agr. Research* **33**:953–970. 1926.

DRECHSLER, C. Some graminicolous species of Helminthosporium. I. *Jour. Agr. Research* **24**:641–739. 1923.

ELLIOTT, C. Halo-blight of oats. *Jour. Agr. Research* **19**:139–172. 1920.

————. Oat blast. *Phytopath.* **15**:564–567. 1925.

————. Bacterial stripe blight of oats. *Jour. Agr. Research* **35**:811–824. 1927.

————. Manual of bacterial plant pathogens, 2d ed. Chronica Botanica Company, Waltham, Mass. 1951.

FERDINANDSEN, C., and O. WINGE. A heritable blotch leaf in oats. *Hereditas* **13**: 164–176. 1929–1930.

FINKNER, V. C. Genetic factors governing resistance and susceptibility of oats to *Puccinia coronata* Corda var. *avenae*, F. and L., race 57. *Iowa Agr. Exp. Sta. Res. Bul.* 411, pp. 1041–1063. 1954.

FISCHER, G. W. Manual of the North American smut fungi. The Ronald Press Company. New York. 1953.

———— and C. E. CLAASSEN. Studies of stem rust (*Puccinia graminis*) from *Poa ampla*, *Avena fatua*, and *Agropyron spicatum* in the Pullman, Washington region. *Phytopath.* **34**:301–314. 1944.

GERRETSEN, F. C. Manganese deficiency of oats and its relation to soil bacteria. *Ann. Bot.* (N.S.) **1**:207–330. 1937.

GORDON, W. L. A study of the relation of environment to the development of the uredinial and telial stages of the physiologic forms of *Puccinia graminis avenae* Eriks. and Henn. *Sci. Agr.* **14**:184–237. 1933.

GREANEY, F. G., *et al.* Varietal resistance of wheat and oats to root rot caused by *Fusarium culmorum* and *Helminthosporium sativum*. *Sci. Agr.* **18**:500–523. 1938.

HAGEMAN, R. H., *et al.* The production of grey speck of oats in purified sand cultures. *Jour. Am. Soc. Agron.* **34**:731–735. 1942.

HANSING, E. D., *et al.* A new race of *Ustilago avenae*. *Phytopath.* **36**:400. 1946.

HAYES, H. K., *et al.* Studies of inheritance in crosses between Bond, *Avena byzantina*, and varieties of *A. sativa*. *Minn. Agr. Exp. Sta. Tech. Bul.* 137. 1939.

HILTNER, E. Die Dörrfleckenkrankheit des Hafers und ihre Heilung durch Mangan. *Landw. Jahrb.* **60**:689–769. 1924.

HOLTON, C. S. Hybridization and segregation in the oat smuts. *Phytopath.* **21**:835–842. 1931.

————. Studies in the genetics and cytology of *Ustilago avenae* and *U. levis*. *Minn. Agr. Exp. Sta. Tech. Bul.* 87. 1932.

————. Inheritance of chlamydospore characters in oat smut fungi. *Jour. Agr. Research* **52**:535–540. 1936.

————. Origin and production of morphologic and pathogenic strains of the oat smut fungi by mutation and hybridization. *Jour. Agr. Research* **52**:311–317. 1936.

———— and H. A. RODENHISER. Physiologic specialization in the oat smut fungi with relation to breeding oats for smut resistance. *U.S. Dept. Agr. Tech. Bul.* (Manuscript.)

HUSKINS, C. L. Blindness or blast of oats. *Sci. Agr.* **12**:191–199. 1931.

ITO, S., and K. KURIBAYASHI. The ascigerous forms of some graminicolous species of Helminthosporium in Japan. *Jour. Fac. Agr. Hokkaido Imp. Univ. Sapporo* **29**:85–125. 1931. (In English.)

JOHNSON, J. Relation of water-soaked tissues to infection by *Bacterium angulatum* and *Bact. tabacum* and other organisms. *Jour. Agr. Research* **55**:599–618. 1937.

JOHNSTON, C. O. Effects of soil moisture and temperature and of dehulling on the infection of oats by loose and covered smuts. *Phytopath.* **17**:31–36. 1927.

JONES, E. S. Influence of temperature, moisture and oxygen on spore germination of *Ustilago avenae*. *Jour. Agr. Research* **24**:577–591. 1923.

KEHR, W. R., H. K. HAYES, M. B. MOORE, and E. C. STAKMAN. The present status of breeding rust resistant oats at the Minnesota Station. *Agron. Jour.* **42**:356–359. 1950.

KIHARA, H., and I. NISHIYAMA. The genetics and cytology of certain cereals. III. Different compatibility in reciprocal crosses of Avena with special reference to tetraploid hybrids between hexaploid and diploid species. *Japan. Jour. Bot.* **6**:245–305. 1932.

KLINE, C. H. Molybdenum and lime in the treatment of acid soils. *Jour. Soil and Water Conserv.* **10**:63–69. 1955.

KO, S. Y., J. H. TORRIE, and J. G. DICKSON. Inheritance of reaction to crown rust and stem rust and other characters in crosses between Bond, *Avena byzantina*, and varieties of *A. sativa*. *Phytopath.* **36**:226–235. 1946.

KONZAK, C. F. Stem rust resistance in oats induced by nuclear radiation. *Agron. Jour.* **46**:538–540. 1954.

KOO, F. K. S., M. B. MOORE, W. M. MYERS, and B. J. ROBERTS. Inheritance of seedling reaction to races 7 and 8 of *Puccinia graminis avenae* Eriks. and Henn. at high temperatures in three oat crosses. *Agron. Jour.* **47**:122–124. 1955.

LEVINE, M. N., and D. C. SMITH. Comparative reaction of oat varieties in the seedling and maturing stages to physiologic races of *Puccinia graminis avenae* and the distribution of these races in the United States. *Jour. Agr. Research* **55**:713–729. 1937.

LITZENBERGER, S. C. Inheritance of resistance to specific races of crown and stem rust, to Helminthosporium blight, and to certain agronomic characters of oats. *Iowa Agr. Exp. Sta. Res. Bul.* 370, pp. 451–496. 1949.

LUNDEGÅRDH, H. Studier over Stråsädens näringsupptagande samt dettas betydelse för tillväxten och för uppkomsten ave icke-parasitara sjukdomar. *Centralanst Försoksväs. Jordbruk. Avdel. Lantbruksbot. Medd.* 403. 1931.

―――. Om grafläcksjuka och liknande bristsjukdomar hos Kulturväxter. *Landt. männer Tidskr. Landt. Nr.* **15**:775–777. 1932.

LUTTRELL, E. S. A taxonomic revision of *Helminthosporium sativum* and related species. *Am. Jour. Bot.* **42**:57–68. 1955.

MACLACHLAN, J. D. Manganese deficiency in soils and crops. I. Control in oats by spraying: studies of the role of soil micro-organisms. *Sci. Agr.* **22**:201–207. 1941.

―――. Manganese deficiency in soils and crops. II. The use of various compounds to control manganese deficiency in oats. *Sci. Agr.* **24**:86–94. 1943.

MALZEW (MALTZEV), A. I. Wild and cultivated oats. Section Euavena Griseb. *Bul. Appl. Bot., Gen., Plant Breeding (Leningrad). Suppl.* 38. 1930. (In Russian with English summary.)

McGINNIS, R. C. Cytological studies of chromosomes of rust fungi. II. The mitotic chromosomes of *Puccinia coronata*. *Can. Jour. Bot.* **32**:213–214. 1954.

McKINNEY, H. H. Mosaics of winter oats induced by soil-borne viruses. *Phytopath.* **36**:359–369. 1946.

MEEHAN, F. L.   *Helminthosporium victoriae* and other graminicolous species.   *Iowa State Col. Jour. Sci.* **25**:292–294.   1951.

—— and H. C. MURPHY.   A new Helminthosporium blight of oats.   *Science* (N.S.) **104**:413–414.   1946.

MELHUS, I. E., *et al.*   Alternate hosts and biological specialization of crown rust in America.   *Iowa Agr. Exp. Sta. Res. Bul.* 72.   1922.

MURPHY, H. C.   Physiologic specialization in *Puccinia coronata avenae*.   *U.S. Dept. Agr. Tech. Bul.* 433.   1935.

—— *et al.*   Breeding for disease resistance in oats.   *Jour. Am. Soc. Agron.* **34**:72–89.   1942.

—— and F. L. MEEHAN.   Reaction of oat varieties to a new species of Helminthosporium.   *Phytopath.* **36**:407.   1946.

NEWTON, M., *et al.*   Seedling reactions of wheat varieties to stem rust and leaf rust and of oat varieties to stem rust and crown rust.   *Can. Jour. Research* **C 18**:489–506.   1940.

NISHIYAMA, I.   A genetic analysis of the grain character of oats by means of the monosomic inheritance.   *Res. Inst. Food Sci. Kyoto Univ. Bul.* 4.   1951.

——.   Cytogenetic studies in Avena.   V. Genetic studies of steriloids found in the progeny of a triploid Avena hybrid.   *Mem. Res. Inst. Food Sci. Kyoto Univ.* 5.   1953.

O'BRIEN, D. B., and R. W. G. DENNIS.   Helminthosporium disease of oats.   *Scottish Agr. Col. Res. Bul.* 3.   1933.

OSWALD, J. W., and B. R. HOUSTON.   The yellow-dwarf virus disease of cereal crops.   *Phytopath.* **43**:128–136.   1953.

POEHLMAN, J. M., and C. H. KINGSOLVER.   Disease reaction and agronomic qualities of oat selections from a Columbia × Victoria-Richland cross.   *Agron. Jour.* **42**:498–502.   1950.

QUINLAN-WATSON, F.   The effect of zinc deficiency on the aldehase activity in the leaves of oats and clover.   *Biochem. Jour.* **53**:457–460.   1953.

RADEMACHER, B.   Genetisch bedingte Unterschiede in der Neigung zu physiologischen Slorungen beim Haffer.   *Zeitschr. Zücht.* **A 20**:210–250.   1935.

RAVN, F. K.   Nogle Helminthosporium-arter ogde af dem fremkalte sygdomme hos byg og havre.   *Bot. Tidsskr.* **23**:101–322.   1900.

——.   Ueber einige Helminthosporium-arten und die von denselben hervorgerufenen Krankheiten bei Gerste und Hafer.   *Zeitschr. Pflanzenkr.* **11**:1–26. 1901.

REED, G. M.   Varietal resistance and susceptibility of oats to powdery mildew.   *Mo. Agr. Exp. Sta. Res. Bul.* 37.   1920.

——.   Inheritance of resistance to loose and covered smut in hybrids of Black Mesdag with Hulless, Silvermine, and Early Champion oats.   *Am. Jour. Bot.* **21**:278–291.   1934.

——.   Inheritance of resistance to loose smut in hybrids of Fulghum and Black Mesdag oats.   *Bul. Torrey Bot. Club* **62**:177–186.   1935.

——.   Physiologic specialization of the parasitic fungi.   *Bot. Rev.* **1**:119–137. 1935.

——.   Influence of the growth of the host on oat smut development.   *Proc. Am. Philos. Soc.* **79**:303–326.   1938.

——.   Physiologic races of oat smuts.   *Am. Jour. Bot.* **27**:135–143.   1940.

——.   Inheritance of smut resistance in some oat hybrids.   *Am. Jour. Bot.* **28**:451–457.   1941.

——.   Inheritance of smut resistance in hybrids of Navarro oats.   *Am. Jour. Bot.* **29**:308–314.   1942.

———— and J. A. FARIS. Influence of environmental factors on the infection of sorghums and oats by smuts. II. *Am. Jour. Bot.* **11**:579–599. 1924.

———— and T. R. STANTON. Inheritance of resistance to loose and covered smuts in oat hybrids. *Jour. Am. Soc. Agron.* **29**:997–1006. 1937.

———— and ————. Inheritance of resistance to loose and covered smuts in Markton oat hybrids. *Jour. Agr. Research* **56**:159–176. 1938.

———— and ————. Susceptibility of Lee × Victoria oat selections to loose smut. *Phytopath.* **32**:100–102. 1942.

ROEMER, T., *et al.* Die Züchtung resistenter Rassen der Kultur pflanzen. *Kühn-Archiv.* **45**:1–427. 1937.

SAMPSON, K. The biology of oat smuts. III. *Ann. Appl. Biol.* **20**:258–271. 1933.

————. Life cycles of smut fungi. *British Myc. Soc. Trans.* **23**:1–23. 1939.

SAMUEL, G., and C. S. PIPER. Grey speck (manganese deficiency) disease of oats. *Jour. Dept. Agr. So. Aust.* **31**:696–705. 1928.

———— and ————. Manganese as an essential element in plant growth. *Ann. Appl. Biol.* **16**:494–524. 1929.

SANFORD, G. B. *Colletotrichum graminicolum* (Ces.) Wils. as a parasite of the stem and root tissues of *Avena sativa.* *Sci. Agr.* **15**:370–376. 1935.

SHERMAN, G. D., and P. M. HARMER. Manganese deficiency of oats on alkaline organic soils. *Jour. Am. Soc. Agron.* **33**:1080–1092. 1941.

SIMMONDS, P. M. Seedling blight and foot-rots of oats caused by *Fusarium culmorum* (W.G.Sm.) Sacc. *Can. Dept. Agr. Bul.* (N.S.) 105. 1928.

SIMONS, M. D. Inheritance of resistance of the oat selection P. I. 186603 to specific races of crown rust. *Am. Soc. Agron. Abstr.*, p. 74. 1954.

————. The relationship of temperature and stage of growth to the crown rust reaction of certain varieties of oats. *Phytopath.* **44**:221–223. 1954.

———— and H. C. MURPHY. A comparison of certain combinations of oat varieties as crown rust differentials. *U.S. Dept. Agr. Tech. Bull.* 1112, p. 22. 1955.

SMITH, D. C. Correlated inheritance in oats of reaction to diseases and other characteristics. *Minn. Agr. Exp. Sta. Tech. Bul.* 102. 1934.

SPRAGUE, R. A physiologic form of *Septoria tritici* on oats. *Phytopath.* **24**:133–143. 1934.

————. Diseases of cereals and grasses in North America. The Ronald Press Company. New York. 1950.

STAKMAN, E. C., *et al.* Die Bestimmung physiologischer Rassen Pflanzenpathogenerpilze. *Nova Acta Leopoldina* **3**:281–336. 1935.

———— and W. Q. LOEGERING. The potential importance of race 8 of *Puccinia graminis avenae* in the United States. *Phytopath.* **34**:421–425. 1944.

STANTON, T. R. Superior germ plasm in oats. *U.S. Dept. Agr. Yearbook*, pp. 347–414. 1936.

———— *et al.* Field studies on resistance of hybrid selections of oats to covered and loose smuts of oats. *U.S. Dept. Agr. Tech. Bul.* 422. 1934.

———— and F. A. COFFMAN. Grow disease resistant oats. *U.S. Dept. Agr. Farmers' Bul.* 1941. 1943.

———— and ————. Disease resistant and hardy oats for the South. *U.S. Dept. Agr. Farmers' Bul.* 1947. 1943.

STILES, W. Trace elements in plants and animals. The Macmillan Company. New York. 1948.

STRAIB, W. Die Bestimmung der physiologischen Rassen von *Puccinia coronata* Cda. auf Hafer in Deutschland. *Arb. biol. Reich Land.-Forstw. Berlin-Dahlam* **22**:121–157. 1937.

TESSI, J. L. Presencia de "*Pseudomonas coronafaciens*" en la Argentina y reacción de algunas variedades de Avena frente a este parásito. *Rev. Invest. Agric. Buenos Aires.* **3**:319–334. 1949.

———. Razas de distinto poder patogeno dentro de *"Pseudomonas striafaciens"* en Argentina y reacción de variedades de Avena a su ataque. *Rev. Invest. Agric.* **6**:235–246. 1952.

———. Estudio comparativo de dos bacterios patogenos en Avena y determinacion de una toxina que origina sus differencias. *Rev. Invest. Agric.* **7**:131–145. 1953.

TORRIE, J. H. Correlated inheritance in oats of reaction to smuts, crown rust, stem rust, and other characters. *Jour. Agr. Research* **59**:783–804. 1939.

TRANZSCHEL, V. Alternate hosts of the cereal rusts and their distribution in U.S.S.R. *Bul. Plant Prot. Leningrad.* **5**:4–40. 1934. (In Russian.)

TURNER, D. M., and W. A. MILLARD. Leaf spots of oats, *Helminthosporium avenae* (Bri. and Cav.) Eidam. *Ann. Appl. Biol.* **18**:535–558. 1931.

VALLEGA, J. Especializacion fisiologica de *Puccinia coronata avenae* en Argentina. *Anales Inst. Fitot. Santa Catalina* **2**:53–84. 1942.

VAUGHAN, E. K. A race of *Ustilago avenae* capable of infecting Black Mesdag oats. *Phytopath.* **28**:660–661. 1938.

WEBER, G. F. Speckled blotch of oats. *Phytopath.* **12**:449–470. 1922.

WELSH, J. N. The synthetic production of oat varieties resistant to race 6 and certain other physiologic races of oat stem rust. *Can. Jour. Research* **C15**:58–69. 1937.

———, R. B. CARSON, W. J. CHEREWICK, W. A. F. HAGBORG, B. PETURSON, and H. A. H. WALLACE. Oat varieties—past and present. *Can. Dept. Agr. Pub.* 891, pp. 1–51. 1953.

——— and T. JOHNSON. The source of resistance and the inheritance of reaction to 12 physiologic races of stem rust, *Puccinia graminis avenae* (Erikss. and Henn.). *Can. Jour. Bot.* **29**:189–205. 1951.

——— and ———. Inheritance of reaction to race 7A and other races of oat stem rust, *Puccinia graminis avenae*. *Can. Jour. Bot.* **32**:347–357. 1954.

———, B. PETURSON, and J. E. MACHACEK. Associated inheritance of reaction to races of crown rust, *Puccinia coronata avenae* Erikss., and to victoria blight, *Helminthosporium victoriae* M. and M., in oats. *Can. Jour. Bot.* **32**:55–68. 1954.

WESTERN, J. H. The biology of oat smuts. IV. The invasion of some susceptible and resistant oat varieties, including Markton, by selected biologic species of smut *Ustilago avenae* (Pers). Jens. and *U. kolleri* Wil. *Ann. Appl. Biol.* **23**:245–263. 1936.

———. Sexual fusion of *Ustilago avenae* under natural conditions. *Phytopath.* **27**:547–553. 1937.

ZADE, A., and A. ARLAND. The relation of host and pathogene in *Ustilago avenae*. *Bul. Torrey Bot. Club.* **60**:77–78. 1933.

CHAPTER 7

RICE DISEASES

The cultivated rice varieties belong to the one species *Oryza sativa* L. The rice plant is a warm-climate annual with both spring and intermediate to winter types of growth response or, more correctly expressed in the case of rice, long- and short-day photoperiodic responses are represented. The crop is grown without irrigation (upland) and with irrigation (lowland). The major commercial crop is grown under irrigation or where the plants are grown in water the greater part of the growing season. The basic chromosome number in the cultivated rice varieties and most wild varieties of this species is 12 pairs. Multiples of this basic number occur in the wild *Oryza* spp. According to Jones (1936), haploid, triploid, and tetraploid rice plants occur.

The rice plant is adapted to relatively warm climates. Jones (1936) gives a mean temperature of 70°F. or above during the entire growing season as essential for successful rice culture. The physiological anatomy of the plant is similar to the other Gramineae. The high silicon content of the cell membranes especially of leaf and floral bracts is notable for rice and closely related species. Attempts have been made to correlate silicon content with disease resistance in some instances. The anatomy of the leaf is different than many of the grasses, as discussed by Tullis (1935). Long-, medium-, and short-grain types are represented in the commercial varieties.

Diseases are important in the economical production of rice. Seedling blight and culm rot maladies are of major importance. Estimated average annual losses from rice diseases in the United States are not tabulated. The total reduction in yield and quality is probably less than in the other cereal crops. Losses in the more intensive rice areas of the Pacific are frequently higher than those in the United States. Padwick (1950) summarizes the literature on rice diseases and estimates losses in various rice-growing countries.

1. Nonparasitic. *Straighthead* of rice is prevalent in certain sections of the United States, Mexico, and other areas. The disease apparently occurs on new rice land or on soils where large amounts of organic material are plowed under. Tisdale and Jenkins (1921) reported it as one of the

153

most destructive diseases of irrigated rice in the southern United States. According to Tullis (1940), more recently the occurrence of the trouble is limited and practical control is practiced. The leaves are dark green and stiff. Frequently one or both glumes are absent, and the grain does not fill. Large roots, without branches and root hairs, develop rather than the fine well-branched fibrous root system. The plants remain green with erect panicles as the crop matures. The disease is prevented under most conditions by draining off the irrigation water after 6 weeks and allowing the soil to dry thoroughly, after which the water is applied and left on for the remainder of the season. Proper crop rotation and avoidance of excessive amounts of raw organic material turned into the soil constitute the best means of preventing soil conditions conducive to straighthead.

Deficiency diseases of rice are similar to those occurring on other Gramineae. Symptoms vary with the mineral elements deficient, but the minor elements manganese, boron, copper, etc., when nonavailable, result in white leaf tips, gray spotting of the leaves, and incomplete panicle emergence as in oats and other cereals (Martin, 1939, Martin and Allstat, 1940, Tullis and Cralley, 1940). The balance between calcium and magnesium as affecting the availability of the nutrient elements as well as controlling magnesium toxicity is important (see oats, Chap. 6). The symptoms of this mineral-deficiency complex in rice earlier were confused with the somewhat similar symptoms of the seed-borne nematode, *Aphelenchoides oryzae* Yokoo., developing in the apical culm tissues of rice in Japan and in the United States. This nematode is different from the culm gall nematode *Ditylenchus angustus* (But.) Filipjev of Bengal.

2. Dwarf, or Stunt, *Insect-transmitted Virus.* The stunt disease of rice occurs only in the Japanese Islands and eastern Asia in so far as present distribution is recorded. The disease is important historically both from the early records of damage and the first recorded insect transmission of a virus. As summarized by Fukushi (1934, 1940) and Katsura (1936), the insect-transmission studies on dwarf started by Takami in 1883 and reported from 1895 to 1908 antedated the report of Ball in 1905 on the insect transmission of the virus causing curly top of beets.

DESCRIPTION AND EFFECT. The dwarf disease of rice is a streak and rosette combination of symptoms. The plants are dwarfed, tillering excessively, leaves and stunted panicles are erect and late, and leaves are streaked. The leaves frequently show white specks along the veins. These develop into narrow interrupted to continuous light-yellow streaks on a dark-green leaf background. Intracellular bodies are present in the cells of the chlorotic tissues and in the cells of crown and roots. According to Fukushi (1934, 1940), *Panicum miliaceum* L., *Echinochloa crusgalli* var. *edulis* Hon., *Alopecurus fulvus* L., and *Poa pratensis* L. are subject to attack but not extensively diseased. Rye, wheat, and oats are slightly susceptible.

The major insect vector is the rice leaf hopper, *Nephotettix apicalis* var. *cincticeps* (Uhl.). *Deltocephalus dorsalis* (Motsch.) also transmits the virus. Fukushi (1933, 1935, 1937, 1939) has studied the insect and virus relationship in great detail. Control is accomplished in part by destruction of the leaf hopper. The insets overwinter on *Astragalus sinicus* L., which is common as a winter crop and weed in the rice areas.

Kuribayashi (1931) described a stripe virus on rice that is characterized by yellowish stripes on the leaf and some stunting of the plant. The virus is transmitted by another leaf hopper, *Delphacodes striatellus* (Fall.).

3. Bacterial Diseases

Bacterial Blight, Xanthomonas oryzae (Uyeda & Ishiyama) Dowson (*Pseudomonas oryzae* Uyeda & Ishiyama). The bacterial blight of rice occurs in the Philippines, Japan, Mexico, and probably other areas. Linear stripes develop between the veins of the leaf blade and sheath. The stripes are first water-soaked, later light brown or bleached, with exudate on the surface (Elliott, 1951, Ishiyama, 1928, Reinking, 1918).

Kernel Rot, Xanthomonas itoana (Toch.) Dowson (*Pseudomonas itoana* Toch.). The disease is reported in Japan as a black rot of rice kernels. The necrotic dark areas extend into the endosperm and therefore reduce the quality of the milled grain (Tochinai, 1932).

4. Fusarium and Gibberella Blight, or "Bakanae Disease." *Fusarium moniliforme* (Sheld.) Snyder & Hansen, *Gibberella moniliforme* (Sheld.) Snyder & Hansen, or *G. fujikuroi* (Saw.) Wr. and *Fusarium roseum f. cerealis* (Cke.) Snyder & Hansen, *Gibberella roseum f. cerealis* (Cke.) Snyder & Hansen or *G. zeae* (Schw.) Petch. Rice in the more humid Asiatic regions and in the humid areas of the Western Hemisphere is damaged by this group of fungi. Grain spoilage, especially reduced germination, seedling blight, and kernel blight are reported by Seto (1935) including a good summary of the literature. The disease on rice is similar in general symptoms and etiology to the disease on the other cereal crops. Certain cultures, notably of *Gibberella fujikuroi*, stimulate internodal and root elongation. Extracts of these fungus cultures produce the same effect not only on rice but also on many other species of the Gramineae and Leguminosae. The Japanese investigators have studied this phase of the problem extensively.

Tullis (1936, 1940) reported the isolation of *Fusarium* from pink and yellow discolored rice kernels and from those with a chalky endosperm from the humid rice-producing areas in the United States. *Fusarium* species ranked second in frequency of occurrence in the discolored grain. Germination and vigor of seedling development is reduced in the discolored kernels, according to Tisdale (1922).

Although the *Fusarium* blight on corn, wheat, and barley has been studied extensively in the United States, the disease on rice has received little attention. Probably the pathogens occur commonly on this crop in the humid rice sections of the United States as is the case in Asiatic rice areas.

5. False Smut, *Ustilagoinoidea virens* (Cke.) Tak. The sclerotia develop and replace the flowers similar to ergot. Usually only a few flowers are infected in a panicle, and the sclerotia are large enough to be removed in cleaning the threshed rice. According to Butler (1918) and Fulton (1908), the disease is widely distributed but of little economic importance.

THE FUNGUS

Ustilaginoidea virens (Cke.) Tak.
[*Ustilaginoidea oryzae* (Pat.) Bref.]

According to Brefeld (1895), this parasite as well as *Ustilaginoidea setariae* Bref. on *Setaria* are related closely to *Claviceps*. Gäumann and Dodge (1928) state that the genus, based on *U. setariae*, differs from *Claviceps* chiefly in the imperfect forms. The conidia are borne over the surface of the pseudomorphs replacing the ovary. The conidia are spherical, echinulate, and 4 to 6 microns in diameter. They germinate to form secondary spores. The sclerotia, protruding from the floral bracts, are spherical, 5 to 8 min. in diameter, and olive green in color because of the layer of conidia. Germination of the sclerotia of *U. virens* is undescribed. Yoshino and Yamamoto (1952) produced the diseased kernels by placing the spores in the leaf whorl at the booting stage.

The etiology is indefinite. The scattered sclerotia in individual flowers indicate a cycle similar to that of *Claviceps purpurea*.

6. Culm Rot, *Leptosphaeria salvinii* Catt. The culm rot is an important disease in most rice-producing sections of the world. The disease causes lodging and lightweight grain. The lesions, on the leaf sheath near the water line, appear first, a month to 6 weeks before the plants head. The black discolored areas spread around the culm and inward into the culm tissue. Numerous black sclerotia develop inside the leaf sheath and later in the culm. The fungus mycelium forms dark threads on the epidermis of the culm. The lesion extends in dark-brown streaks above the older black necrotic area. As the panicles fill, the stalks break over in the necrotic area, and early infection results in poorly filled grain. Butler (1913) reported excessive late tillers as a symptom. Tullis (1933) indicated that excessive tillering is not associated with the disease in the United States. The characteristic symptoms are shown in Fig. 35.

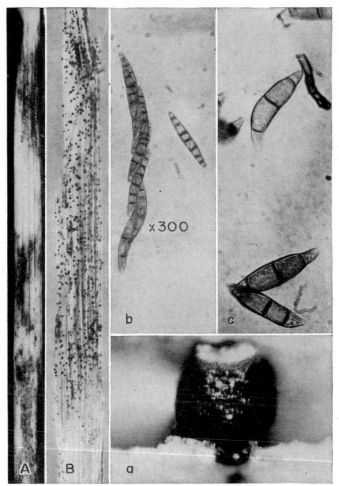

Fig. 35. Culm rot of rice caused by *Leptosphaeria salvinii* showing the lesion on the culm (*A*) and the sclerotia on the old straw (*B*). The perithecium (*a*), the ascus and eight ascospores (*b*), and conidia (*c*) are shown in the inserts. (*Courtesy of E. C. Tullis.*)

THE FUNGUS

Leptosphaeria salvinii Catt.
Sclerotium oryzae Catt. Sclerotial stage
Helminthosporium sigmoideum Cav. Conidial stage
Helminthosporium sigmoideum var. *irregulare* Cralley & Tullis
(*Helminthosporium sigmoideum* var. *microsphaeroides* Nakata)

The mycelium is white inside the tissues and olivaceous on the surface of the tissues with numerous olivaceous irregular appresoria. The spherical sclerotia are black, nearly smooth, and mostly 230 to 270 microns in diameter. Conidiophores are dark, septate, simple or sparsely branched. The conidia are borne singly on a sharp-pointed sterigmata. Conidia are fusiform, slightly curved, typically 3-septate. The two intercalary cells are dark brown; apical cells are lime green. Germination of the apical cells is characteristic. Perithecia are dark, globose, embedded in the outer tissues of the sheath, with a short beak flush with the epidermis of the sheath, and average 381 microns in diameter. Asci are narrowly clavate, short-stalked, with delicate wall. Ascospores are arranged biseriately, fusiform, 3-septate with the center septum constricted when mature, brown with two end cells lighter brown, mostly 44 to 48 microns long by 8 microns wide (Fig. 35).

ETIOLOGY. The sclerotial stage occurs commonly on rice and certain wild grasses. This stage has been described from most rice-growing sections. The conidial stage is usually associated with the sclerotial stage, but it is known to occur in fewer areas. The ascigerous stage is less common and is reported from fewer locations than the sclerotial stage. The fungus persists in the sclerotial stage and develops mycelium and conidia from these structures. Primary infections are commonly from the sclerotia.

CONTROL. According to Tullis (1940) the most satisfactory control is to drain the water from infected fields before the infections reach the rice culm. Only enough water is added, after draining to maturity, to keep the soil saturated. This control measure results in a slight reduction in yield of healthy fields, but it prevents lodging in heavily infected fields. No highly resistant commercial varieties are available. Cralley (1936) and Tullis and Cralley (1933) noted the difference in susceptibility of varieties and that the Japanese short-grained varieties are more resistant than the long-grained types. According to Cralley (1939), a complete fertilizer containing some excess of potassium increased yields without increasing stem rot. According to Cralley and Tullis (1935), a similar culm rot is caused by *Helminthosporium sigmoideum* var. *irregulare* Cralley & Tullis. The disease is not so severe as in the former. The sclerotia are numerous and irregular in shape and size and smaller than those of *H. sigmoideum*. The conidia are similar in both, except that those of the variety *H. sigmoideum irregulare* commonly germinate before maturity. No perfect stage of this latter fungus is described.

7. **Black Sheath Rot,** *Ophiobolus oryzinus* Sacc. Tullis (1933) has reviewed the literature on the several species of *Ophiobolus* reported on rice. Certain of these more common on wheat and other cereals are discussed in Chap. 11. Miyake (1910) described *Ophiobolus oryzae* Miyake, which apparently is different morphologically from *O. oryzinus* and *Cochliobolus* (*O.*) *miyabeanus* (Ito & Kuribay.) Dickson. The latter was described by Ito and Kuribayashi (1927) as the ascigerous stage of *Helminthosporium oryzae* Breda de Haan.

The black sheath rot is of minor importance as the culm tissues are rarely invaded. The disease appears as a black rotted lesion on the sheath tissues near the water line. The fungus perithecia are formed on rice straw and stubble and on the sheaths of *Typha latifolia* L., the common cattail.

8. Helminthosporium Blight, *Cochliobolus* (*Ophiobolus*) *miyabeanus* (Ito & Kuribay.) Dickson, *Helminthosporium oryzae* Breda de Haan. The disease is apparently world-wide in distribution on rice. Damage is frequently severe. Losses in stand due to seedling blight, in yield due to leaf and culm infection, and in quality and yield by kernel infection are frequently high.

DESCRIPTION AND EFFECT. The seedling blight is similar in appearance to that caused by *Helminthosporium sativum* on wheat and barley. The brown cortical lesions appear on the coleoptile, subcrown internode, and seminal roots. Seedling blight occurs before or after emergence. Brown leaf spots and leaf-sheath lesions develop on the less severely infected seedlings. Circular to elongate brown leaf spots are first small without marked water soaking and later spread with reddish-brown margins and gray centers (Fig. 36). On severely infected plants the leaves dry out before the plants are mature. Conidiophores and conidia develop on the lesions. Lesions on the culm, below the panicle, or at the base of the panicle are common when weather conditions are favorable. The brown necrotic areas result in shriveled kernels and broken panicles. Blighted kernels result from early flower infections. Small brown lesions on the floral bracts and pericarp cause discoloration of the milled rice.

THE FUNGUS

Cochliobolus (*Ophiobolus*) *miyabeanus* (Ito & Kuribay.) Dickson
Helminthosporium oryzae Breda de Haan. Conidial stage
(*Helminthosporium macrocarpum* Thuem)
(*Helminthosporium oryzae* Miy. & Hori)

Ito and Kuribayashi (1927) described the ascigerous stage produced in culture from single conidia and named it *Ophiobolus miyabeanus* Ito & Kuribay. They state (1931) that perithecia were secured only in culture and have not been collected in the field. Tullis has collected the perfect stage under field conditions in the United States. The author found the perfect stage in old straw and stubble in southern Mexico in 1953.

The mycelium forms grayish-brown to dark-brown mycelial mats in and on the plant parts and in culture. The conidiophores form in mats or singly and are light brown to olivaceous and vary greatly in both width and length. The conidia are brown, slightly curved, tapering toward the round apex and toward the base, and vary greatly in shape and septation. The peripheral wall is moderately thin, and the hilum is inconspicuous. Spores germinate characteristically from the two apical cells. Perithecia in culture are globose, pseudoparenchymatous, black, and form an ostiolar beak. Asci are cylindrical to long, fusiform, slightly curved, and contain mostly four to six ascospores. Ascospores are filiform and form in a close helix in the ascus (Fig. 37). *Helminthosporium sativum* Pamm., King, & Bakke also produces a seedling blight on rice.

FIG. 36. Leaf and kernel lesions and seedling blight of rice caused by *Cochliobolus miyabeanus* or *Helminthosporium oryzae*.

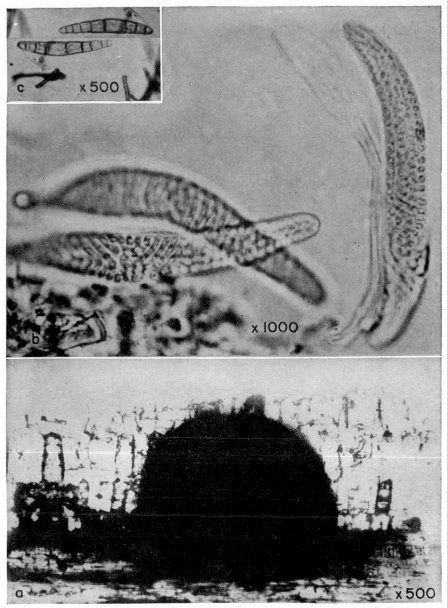

FIG. 37. Perithecium (a), asci and coiled ascospores (b), and conidia (c) of *Cochliobolus miyabeanus, Helminthosporium oryzae,* highly magnified.

ETIOLOGY.   The conidia and mycelium are common on crop residue. The fungus persists over unfavorable periods in both the conidial and mycelial phases.   Primary infection is from one or the other of these sources.   Ocfemia (1924), Nishikado (1923), and Wei and Lin (1936) have studied the influence of temperature and moisture on seedling blight and conidial morphology.   Seedling blight is more prevalent in the cooler soils.   Lesions on the aerial parts develop rapidly at the higher temperatures and moistures.   Ocfemia (1924) and Nishikado (1927) have reported on the regional variation in the fungus.   Tochinai and Sakamoto (1937) found specialization in pathogenicity and differences in physiology and morphology.

CONTROL.   Control by sanitation and crop rotation is unsatisfactory, as the fungus develops abundantly on crop residue.   Seed treatment with mercury or copper compounds will help reduce seedling blight, but it is not effective alone in the control of the disease.   Resistant varieties offer the best means of control.   Wei and Lin (1936) reported that all Chinese rice varieties tested were infected, although there was a difference in relative susceptibility.   Several Japanese varieties are moderately resistant, according to Adair (1941) who has used the resistant variety, Mubo Aikoku, and others in breeding for resistance.   He reported that resistance is governed by at least two factor pairs and that suscept reaction is similar in the seedling and mature plant.   Tullis (1935) studied the leaf invasion by *Helminthosporium oryzae* in moderately resistant and susceptible strains of rice.   The physiological response of the cells, the relation of large intercellular spaces in the mesophyll, and the sclerenchyma in the bundle sheath determine the type of fungus infection and spread in the leaf tissue.   Leaf structure plays an important role in preventing excessive leaf necrosis and large lesions.   Martin and Alstatt (1940) found both *H. oryzae* and the small-spored (*Helminthosporium*) *Curvularia lunata* (Wakk.) Boed. associated with the black kernels.

9. Blast, *Piricularia oryzae* Cav.   The disease is serious in most of the humid rice-producing areas of the world.   Rice grown under irrigation in the areas of low relative humidity, as in California, is not damaged by blast.   Abe (1933) and Hemmi and Imura (1939) demonstrated that conidia are not produced below 88 per cent relative humidity.

DESCRIPTION AND EFFECT.   The disease occurs on the leaves, culms, branches of the panicle, and the floral structures.   Rotten neck, *i.e.*, lesions on the neck of the culm and on the panicle branches near the base of the panicle, is the most conspicuous symptom.   Rotten neck also causes the greatest damage, as the lesions prevent kernel filling.   The dark-brown lesions on the culms and branches of the panicles are indicative of severe necrosis of the tissues.   The panicle or panicle branches break over at the lesioned area.   Brown lesions at the crown and nodes also occur. The leaf spots on young leaves are linear, and on older leaves they are small and circular. The leaf spots caused by this fungus are somewhat similar in appearance to *Helminthosporium* leaf spots. The difference in conidiophores and conidia differentiate the two diseases. Leaf lesions occur on the seedlings and as the plants approach maturity. Severe infections result in the leaves and sheaths drying out and browning. The coalescing of lesions at the base of the leaf blade is a characteristic symptom of blast. Small circular brown lesions occur on the kernels.

THE FUNGUS

Piricularia oryzae Cav.
[*Piricularia grisea* (Cke.) Sacc.]

The conidiophores are simple, rarely branched, grayish, septate, and slender. Conidia are borne terminally. They are ovate, 2-septate when mature, the apex is pointed to blunt depending upon race, according to Tochinai and Shimamura (1932).

ETIOLOGY. The fungus on rice and the one on numerous grasses are similar morphologically. Sprague (1950) suggests *P. grisea* for the pathogen on the grasses, especially in the temperate zones, and retains *P. oryzae* for the fungus on tropical grasses, mostly rice. Apparently races of the pathogen on grasses in the tropics do not infect rice. The conidia are wind-borne and frequently are distributed widely by this means as in the rusts. The distribution of races of the pathogen on rice in North America suggests the importance of wind distribution. The mycelium and conidia persist on crop residue, and conidial infections occur when environmental conditions are favorable for conidial production, distribution, and infection. Terui (1940) described the formation of conidia on the plant surface and in internal cavities of the leaves and culms. The pathogen is seed-borne. See Padwick (1950) for a summary of the literature. Abe (1933), Henry and Anderson (1948), and Hemmi and Imura (1939) studied the relation of environment to the development of conidia. The physiological condition of the rice plant is associated with development of the disease (Abe, 1936, Hemmi, 1933, Sakamoto, 1940).

CONTROL. Injury from early leaf infections and the production of conidial inoculum are reduced by flooding as soon as leaf spots appear (Tullis, 1940). Hemmi (1933) and others report that maintaining the irrigation water on the crop until near maturity helps reduce damage. Proper balance of fertilizers, especially avoiding excess nitrate, helps prevent damage from lodging and rotten neck during kernel filling and maturation. Seed treatment and the use of disease-free seed are advisable, especially in preventing the introduction and spread of new races of the pathogen.

FIG. 38. Leaf spot of rice caused by *Cercospora oryzae*, showing resistant reactions and susceptible reaction (*right*).

SPECIALIZATION AND RESISTANCE. Specialization of the fungus is summarized by Anderson *et al.* (1947), Aoki (1935), Inoue (1939), and Padwick (1950). Races of the pathogen in Florida attack varieties that are resistant in the South Central United States and Mexico. Similar differences in varietal reaction to the disease are reported by Padwick (1950).

10. Cercospora Spot, *Cercospora oryzae* Miyake. The disease is distributed widely with the rice crop, according to Tullis (1937). The narrow linear spots are reddish brown to dark reddish brown, depending upon the rice variety.   On susceptible varieties, the lesion fades to a lighter brown along the margins, and it is light gray brown in the older center.   On the more resistant varieties, the lesions are smaller and uniform in color.   The lesions are generally more abundant on the leaves, although spots on the sheath, culm, and floral bracts are present in heavy infections.   Spots on the floral bracts spread laterally to form oblong lesions (Fig. 38).

THE FUNGUS

Cercospora oryzae Miyake

The brown conidiophores emerge from the stomata singly or two to three and range from 88–140 microns long by 4–5 microns thick.   The conidia are cylindrical to tapering toward the apex and vary in length and septation; septation is 3 to 10, length 20–60 microns by 5 microns wide, as reported in the original description by Miyake (1910).

The pathogen persists on crop residue and the weed, red rice.   The fungus invades the floral bracts and therefore is seed-borne.   Although seedling infection is not reported, this group of pathogens frequently incites leaf lesions during the seedling stage of plant development.

Specialization of the fungus and screening out new races on resistant rice selections represent major problems in control.   Ryker and Coward (1948) summarize the reaction of the 10 physiological races based on rice selections and red rice.   Resistant varieties represent the chief means of control (Adair, 1941, Jodon *et al.*, 1944, Jodon and Chilton, 1946, Ryker and Jodon, 1940, Tullis, 1937).   Resistance is conditioned by several single dominant factor pairs with at least one factor dominant for susceptibility in Supreme Blue Rose.   Some linkage of factors is indicated.

11. Sheath and Culm Blight, *Rhizoctonia* spp.   The Rhizoctonia sheath blight occurs on many crop plants in addition to rice.   The symptoms vary somewhat on the cereals and grasses.   On rice, the lesions are large, irregular, elliptical with reddish-brown margin, straw color, and light-ochre-yellow or greenish-yellow center (Fig. 39).   The lesions occur more commonly just below the ligule.   The lesions on the culms are smaller than those on the sheath tissues.   Ryker and Gooch (1938) reported that culms are not infected.   White or light-tan mycelial strands develop under moist conditions.   Salmon or tan irregular-shaped sclerotia frequently develop on the surface mycelium and within the leaf sheath under and adjacent to the lesion.   Seedlings and mature plants are blighted under favorable conditions for disease development.

THE FUNGI

1. *Rhizoctonia oryzae* Ryker & Gooch

Main mycelial strands are 6 to 10 microns in width, branching at an acute angle, with a slight constriction at the point of branching and a septation a short distance from the point of constriction. Later, short-celled much-branched hypae emerge at right angles from the main branches; certain of these form thickened short hyphae. These anastomose, forming masses of sclerotial cells of various shape and size. The

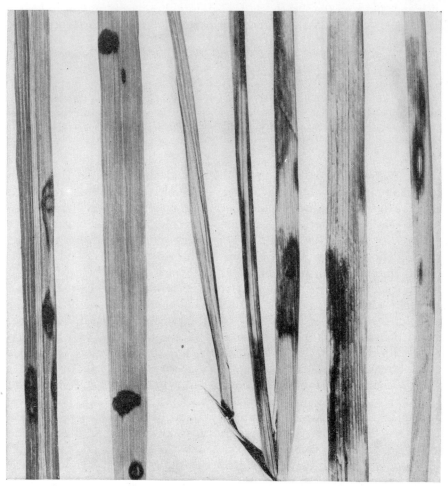

FIG. 39. Leaf and sheath spots on rice caused by *Rhizoctonia oryzae*.

sclerotia are salmon-colored. Sclerotial masses are not formed on rice plants. No basidial stage is found.

Ryker and Gooch (1938) have reviewed the literature and discussed the other species found on rice. *Rhizoctonia oryzae* is not pathogenic on the wide range of plants as occurs in *R. solani* Kuehn.

2. *Pellicularia filamentosa* (Pat.) Rogers or *Corticium solani* (Prill. &
   Del.) Bourd. & Galz.
   (*Corticium vagum* Berk. & Curt.)
   (*Corticium vagum* var. *solani* Burt.)
   (*Hypochnus filamentosa* Pat.)
   (*Hypochnus solani* Prill. & Del.)
   (*Corticium solani* Prill. & Del.)
   (*Corticium botryosum* Bresa.)
   (*Rhizoctonia solani* Kuehn)

This species occurs on rice and many other crop plants. The fungus strains on rice
in Asia and the Pacific islands apparently produce larger sclerotia than is given in the
species description, as discussed by Matsumoto (1934), Palo (1926), Park and Bertus
(1934), Ryker and Gooch (1938), Wei (1934), and others.

The mycelium and branching are similar to the former species. Brown to black
sclerotial structures form from intertwined anastomosed thickened cells rather than
the separately branched filaments as in the above species. The sclerotia are tan,
brown, or black and irregular in shape and size. The hymenophore consists of a dark
loose weft of hyphae, turning grayish white when sporulating. The outer branches
function as basidia and form two to six apical sterigmata bearing ovate to ovate-
oblong sporidia. The sporidia germinate to form the *Rhizoctonia* mycelium. Rogers
(1943) transferred the parasitic forms of *Hypochnus filamentosa* Pat., H. *solani* Prill. &
Del., *Corticium vogum* Bert. & Curt., and *C. solani* Prill. & Del. to the genus *Pellicularia*.

Matsumoto (1934) compared *Hypochnus sasakii* Shirai with *Corticium*
vagum or probably *C. solani*. The hyphal cells in the former are smaller
in diameter, mostly 6 to 8 microns, than in the latter, mostly 8 to 12
microns. The sclerotial cells near the periphery of the sclerotium show
a corresponding difference in size. Hyphal fusions do not occur between
the two fungi. The basidial stages of the two are similar. Growth of the
mycelium on different media and temperatures differs in the two. He
concludes, "It would be better to name our fungus *Corticium sasakii*
(Shirai) T. Matsumoto instead of *Hypochnus sasakii* until further altera-
tion is needed."

Voorhees (1934) reported a stalk rot of corn caused by *Rhizoctonia*
zeae Voorh. Ryker and Gooch (1938) reported this fungus on rice in
Louisiana.

ETIOLOGY. The mycelium and sclerotia, where they are formed, persist on straw and stubble of rice and grasses. Infection occurs on the rice
seedlings and plants when conditions are favorable. Thick stands of
plants are conducive to infection. Proper balance of fertility is desirable
to prevent excessive vegetative growth. The disease is of minor importance in the United States.

12. Kernel Smut, *Tilletia horrida* Tak. [*Neovossia horrida* (Tak.)
Pad. & Kahn.]. The kernel smut of rice is distributed widely. The
disease ordinarily causes little damage, as few kernels in a panicle are

smutted and the smutty grain can be removed in part by cleaning equipment, for when completely smutted they are lighter in weight than the
healthy rice kernels.   Anderson (1899) reported as high as 25 per cent
smutted grains in rice fields in a limited area in South Carolina.   Samples

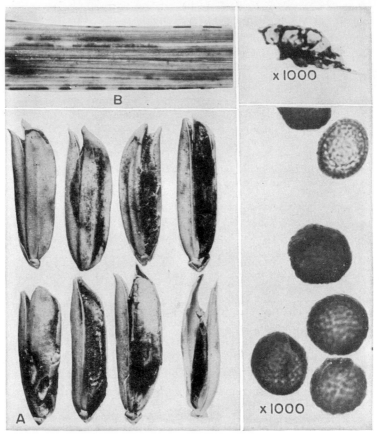

FIG. 40. Kernel smut of rice (A), caused by *Tilletia horrida*, and leaf smut (B), caused
by *Entyloma dactylidis* or *E. oryzae*, with chlamydospores of the two fungi at same
magnification.

of threshed rice showed as high as 3.7 per cent smutty grains.   These
apparently were unusual cases.

The smutted kernels are not conspicuous in the field or in the threshed
grain unless the smutty grains are broken.   The hulls of the smutted
kernels appear dull gray.   The smut sorus is enclosed almost completely
by the lemma and palea (Fig. 40).

THE FUNGUS

Tilletia horrida Tak.
[*Neovossia horrida* (Tak.) Padw. & Kahn]

Anderson (1899) incorrectly reported this fungus as *Tilletia corona* Scribn. (*Neovossia corona* Scribn.) which occurs on several species of grasses. Fischer (1953) places this fungus in *Tilletia* on the basis that the conspicuous sheath and short apiculus of the spore are not the investing sheath and long pedicel-like appendage characteristic of the genus *Neovossia*.

The sori replace the ovaries and are concealed by the floral bracts. Sterile cells are globose to angular, hyaline to dull yellow, thin-walled, smooth. Spores are globose to subglobose, yellowish brown to olive brown, deeply reticulate with areolae varying in size, and 21 to 30 microns in diameter. The spores germinate by the formation of basidia and apical whorls of sporidia. Padwick and Kahn (1944) and others describe the sporidia as not fusing in place.

The etiology is different from that of the *Tilletia* spp. on the cereals and grasses. Takahashi (1896) and Teng (1931) described the germination after soaking in water. Tullis (1940) reported that the embryo is not invaded and the seed will germinate even when the entire endosperm is replaced by the spore mass. Floral infection by sporidia apparently accounts for the few scattered infections in the rice panicles. Varieties appear to differ in susceptibility and in the number of smutted kernels in a panicle.

13. Leaf Smut, *Entyloma dactylidis* (Pass.) Cif. or *Entyloma oryzae* Syd. The disease occurs in most of the rice-producing areas of the world (Butler, 1913, Tullis, 1934). A similar leaf smut occurs sparingly on *Zizania aquatica* L. and on several other genera of the Gramineae. Specimens of "straight head" of rice collected by Tisdale in Arkansas in 1918 show abundant sori of this fungus on the leaves, indicating its presence in the United States for a long period. The disease is of minor importance.

The small black spots on the leaves are angular to linear with the epidermis persisting over the sorus (Fig. 40). Symptoms are similar on the other grasses.

THE FUNGUS

Entyloma dactylidis (Pass.) Cif.
Entyloma oryzae Syd.
Entyloma lineatum (Cke.) Davis

Fischer (1953) places the morphologically similar leaf smut pathogens including *E. oryzae* under the earlier-described *E. dactylidis*. Tentatively *E. lineatum* on *Zizania* is retained as a separate species on the basis of the more linear sori and smaller spores.

Spores are subglobose, angular, and adhere in irregular groups, light olivaceous to dark smoky brown, smooth wall without a sheath, 7–11 by 8–17 microns. The spores germinate, usually in place, to form a promycelium with apical sporidia.

REFERENCES

ABE, T. On the relationship of atmospheric humidity to the infection of the rice plant by *Piricularia oryzae* B. et C. *Forsch. Geb. Pflanzenkrankh.* (*Kyoto*) **2**:98–124. 1933.

———. Comparison of pathogenicity in different culture strains of *Piricularia oryzae* and varietal susceptibility of the rice plant to the blast disease. *Ann. Phytopath. Soc. Japan* **6**:15–26. 1936.

ADAIR, R. C. Inheritance in rice of reaction to *Helminthosporium oryzae* and *Cercospora oryzae*. *U.S. Dept. Agr. Tech. Bul.* 772. 1941.

ANDERSON, A. L., B. W. HENRY, and E. C. TULLIS. Factors affecting infectivity, spread and persistence of *Piricularia oryzae* Cav. *Phytopath.* **37**:94–110. 1947.

AOKI, Y. On the physiologic specialization in the rice blast fungus, *Piricularia oryzae* Br. et Cav. *Ann. Phytopath. Soc. Japan* **5**:107–120. 1935.

BUTLER, E. J. Diseases of rice. *Agr. Research Inst. Pusa Bul.* 34. 1913.

———. Fungi and disease in plants. Thacker, Spink and Co. Calcutta and Simla. 1918.

CRALLEY, E. M. Resistance of rice varieties to stem rot. *Ark. Agr. Exp. Sta. Bul.* 329. 1936.

———. Effect of fertilizers on stem rot of rice. *Ark. Agr. Exp. Sta. Bul.* 383. 1939.

——— and E. C. TULLIS. A comparison of *Leptosphaeria salvinii* and *Helminthosporium sigmoideum irregulare*. *Jour. Agr. Research* **51**:341–348. 1935.

FISCHER, G. W. Manual of the North American smut fungi. The Ronald Press Company, pp. 1–343. New York. 1953.

FUKUSHI, T. Transmission of the virus through the eggs of an insect vector. *Proc. Imp. Acad. Japan* **9**:457–460. 1933.

———. Studies on the dwarf disease of rice plant. *Jour. Fac. Agr. Hokkaido Imp. Univ.* **37**:41–164. 1934.

———. Multiplication of virus in its insect vector. *Proc. Imp. Acad. Japan* **11**:301–303. 1935.

———. An insect vector of the dwarf disease of rice plant. *Proc. Imp. Acad. Japan* **13**:328–331. 1937.

———. Retention of virus by its insect vectors through several generations. *Proc. Imp. Acad. Japan* **15**:142–145. 1939.

———. Further studies on the dwarf disease of rice plant. *Jour. Fac. Agr. Hokkaido Imp. Univ.* **55**:83–154. 1940.

FULTON, H. R. Diseases affecting rice. *La. Agr. Exp. Sta. Bul.* 105. 1908.

GÄUMANN, E. A., and C. W. DODGE. Comparative morphology of fungi. McGraw-Hill Book Company, Inc. New York. 1928.

HEMMI, T. Experimental studies on the relation of environmental factors to the occurrence and severity of blast disease in rice plants. *Phytopath. Zeitschr.* **6**:305–324. 1933.

——— and J. IMURA. On the relation of air-humidity to conidial formation in the rice blast fungus, *Piricularia oryzae*, and the characteristics in the germination of conidia produced by the strains showing different pathogenicity. *Ann. Phytopath. Soc. Japan* **9**:147–156. 1939. (Japanese-English summary.)

HENRY, B. W., and A. L. ANDERSON. Sporulation by *Piricularia oryzae*. *Phytopath.* **38**:265–278. 1948.

INOUE, Y.   Comparison of the cellulose-decomposition by cultural strains of the rice blast fungus, *Piricularia oryzae* Br. et Cav.   *Ann. Phytopath. Soc. Japan* **9**:33–40.   1939.   (Japanese-English summary.)

ISHIYAMA, S.   Bacterial leaf blight of the rice plant.   *Proc. 3d Pan-Pacific Sci. Congr. Tokyo* (1926), p. 2112.   1928.

ITO, S., and K. KURIBAYASHI.   Production of the ascigerous stage in culture of *Helminthosporium oryzae.   Ann. Phytopath. Soc. Japan* **2**:1–8.   1927.

—— and ——.   The ascigerous forms of some graminicolous species of Helminthosporium in Japan.   *Jour. Fac. Agr. Hokkaido Imp. Univ. Sapporo* **29**:85–125.   1931.

JODON, N. E., and S. J. P. CHILTON.   Some characters inherited independently of reaction of physiologic races of *Cercospora oryzae* in rice.   *Jour. Am. Soc. Agron.* **38**:864–872.   1946.

——, T. C. RYKER, and S. J. P. CHILTON.   Inheritance of reaction to physiologic races of *Cercospora oryzae* in rice.   *Jour. Am. Soc. Agron.* **36**:497–507.   1944.

JONES, J. W.   Improvement in rice.   *U.S. Dept. Agr. Yearbook*, pp. 415–454.   1936.

KATSURA, S.   The stunt disease of Japanese rice, the first plant virosis shown to be transmitted by an insect vector.   *Phytopath.* **26**:887–895.   1936.

KURABAYASHI, K.   Studies on the stripe of rice plant.   *Nagano Agr. Exp. Sta. Bul.* 2.   1931.   *Jour. Plant Prot. Japan* **18**:565–571, 636–640.   1931.   (Japanese-English summary.)

MARTIN, A. L., and G. E. ALSTATT.   Black kernel and white tip of rice.   *Texas Agr. Exp. Sta. Bul.* 584.   1940.

MATSUMOTO, T.   Some remarks on the taxonomy of the fungus *Hypochnus sasakii* Shirai.   *Sapporo Nat. Hist. Soc.* **13**:115–120.   1934.

MIYAKE, I.   Studien über die Pilze des Reispflanze in Japan.   *Jour. Col. Agr. Imp. Univ. Tokyo* **2**:237–276.   1910.   (Good list of fungi.)

NAKATOMI, S.   On the variability and inheritance of the resistance of the rice plants against the rice blast disease.   *Japanese Jour. Gen.* **4**:31–38.   1927.

NISHIKADO, Y.   Effect of temperature on the growth of *Helminthosporium oryzae* Breda de Haan.   *Ann. Phytopath. Soc. Japan* **1**:20–30.   1923.   (In Japanese.)

——.   Comparative studies on Helminthosporium diseases of rice in the Pacific regions.   *Ber. Ōhara Inst. Landwirts. Forsch.* **3**:425–440.   1927.

OCFEMIA, G. O.   The Helminthosporium disease of rice occurring in the southern United States and in the Philippines.   *Am. Jour. Bot.* **11**:385–408.   1924.

——.   The relation of soil temperature to germination of certain Philippine upland and lowland varieties of rice and infection by the Helminthosporium disease.   *Am. Jour. Bot.* **11**:437–460.   1924.

PADWICK, G. W.   Manual of rice diseases.   Commonwealth Mycological Institute, pp. 1–198.   Kew, Surrey.   1950.

—— and AZMATULLAH KAHN.   Notes on Indian fungi.   *Gr. Brit. Imp. Mycol. Inst. Mycol. Pap.* 10.   1944.

PALO, M. A.   Rhizoctonia disease of rice.   *I. Philippine Agr.* **15**:361–375.   1926.

PARK, M., and L. S. BERTUS.   Sclerotial diseases of rice in Ceylon.   *Ceylon Jour. Sci. Botany* **A12**:1–36.   1934.

RAMIAH, K., and K. RAMASWAMI.   Breeding for resistance to *Piricularia oryzae* in rice (*O. sativa*).   *Proc. Indian Acad. Sci.* **3**:450–458.   1936.

REINKING, O. A.   Philippine economic-plant diseases.   *Philipp. Jour. Sci.* **A13**:165–274.   1918.

ROGERS, D. P.   The genus Pellicularia (Thelephoraceae).   *Farlowia* **1**:95–118.   1943.

RYKER, T. C., and L. E. COWART.   Development of Cercospora-resistant strains of rice.   *Phytopath.* **38**:23.   1948.

———— and F. S. Gooch. Rhizoctonia sheath spot of rice. *Phytopath.* **28**:233–246. 1938.

———— and N. E. Jodon. Inheritance of resistance to *Cercospora oryzae* in rice. *Phytopath.* **30**:1041–1047. 1940.

Sakamoto, M. On the facilitated infection of the rice blast fungus, *Piricularia oryzae* Cav. due to the wind. *Ann. Phytopath. Soc. Japan* **10**:119–126. 1940. (Japanese-English summary.)

Seto, F. Beitrage zur Kenntnis der "Bakane" Krankheit der Reispflanze. *Col. Agr. Kyoto Imp. Univ. Mem.* 36. 1935.

Sprague, R. Diseases of cereals and grasses in North America. The Ronald Press Company, pp. 1–538. New York. 1950.

Suzuki, H. Studies on the relations between the anatomical characters of the rice plant and its susceptibility to blast disease. *Jour. Col. Agr. Tokyo* **14**:181–264. 1937.

Takahashi, Y. On *Ustilago virens* Cooke and a new species of *Tilletia* parasitic on rice plant. *Bot. Mag. Tokyo* **10**:16–20. 1896.

Teng, S. C. Observations on the germination of the chlamydospores of *Tilletia horrida* Tak. *Sci. Soc. China Biol. Lab. Cont. Bot.* **6**:111–114. 1931.

Terui, M. Internal formation of conidia of the rice blast fungus. *Ann. Phytopath. Soc. Japan* **10**:265–267. 1940. (Japanese-English summary.)

Tisdale, W. H. Seedling blight and stack-burn of rice and the hot water seed treatment. *U.S. Dept. Agr. Bul.* 1116. 1922.

———— and J. M. Jenkins. Straighthead of rice and its control. *U.S. Dept. Agr. Farmers' Bul.* 1212. 1921.

Tochinai, Y. The black rot of rice grains caused by *Pseudomonas itoana* n. sp. *Ann. Phytopath. Soc. Japan* **2**:453–457. 1932.

———— and M. Sakamoto. Studies on the physiologic specialization in *Ophiobolus miyabeanus* Ito et Kurib. *Jour. Fac. Agr. Hokkaido Imp. Univ.* **51**:1–96. 1937.

———— and M. Shimamura. Studies on the physiologic specialization in *Piricularia oryzae* Br. et Cav. *Ann. Phytopath. Soc. Japan* **2**:414–441. 1932.

Tullis, E. C. *Leptosphaeria salvinii*, the ascigerous stage of *Helminthosporium sigmoideum* and *Sclerotium oryzae*. *Jour. Agr. Research* **47**:675–687. 1933.

————. Leaf smut of rice in the United States. *Phytopath.* **24**:1386. 1934.

————. Histological studies of rice leaves infected with *Helminthosporium oryzae*. *Jour. Agr. Research* **50**:81–90. 1935.

————. Fungi isolated from discolored rice kernels. *U.S. Dept. Agr. Tech. Bul.* 540. 1936.

————. *Cercospora oryzae* on rice in the United States. *Phytopath.* **27**:1005–1008. 1937.

————. Diseases of rice. *U.S. Dept. Agr. Farmers' Bul.* 1854. 1940.

———— and E. M. Cralley. Laboratory and field studies on the development and control of stem rot of rice. *Ark. Agr. Exp. Sta. Bul.* 295. 1933.

Voorhees, R. K. Sclerotial rot of corn caused by *Rhizoctonia zeae* n. sp. *Phytopath.* **24**:1290–1303. 1934.

Wei, C. T. Rhizoctonia sheath blight of rice. *Col. Agr. For. Uni. Nanking Bul.* 15. 1934.

———— and C. K. Lin. Studies on Helminthosporiose of rice. *Col. Agr. For. Univ. Nanking Bul.* 44. 1936.

Yoshino, M., and T. Yamamoto. Pathogenicity of the chlamydospores of the rice false smut. *Agr. and Hort. Tokyo,* **27**:291–292. 1952.

CHAPTER 8

RYE DISEASES

Cultivated rye, *Secale cereale* L., is chiefly a winter cereal grown extensively in Europe and Asia and on the lighter soils of North and South America. Spring varieties of rye occur; however, their use is limited as the extreme winter hardiness of rye permits the growing of the higher-yielding winter varieties under most conditions. Cultivated rye has the basic chromosome number of seven pairs.

Rye is cross-pollinated and, therefore, heterozygous and relatively variable in the expression of characters. Inbreeding is possible, although progress in stabilizing a large number of characters is slow owing to general self-sterility. Self-fertile lines are obtainable, and considerable progress is possible in securing stable disease-resistant lines. Such lines are not in general use commercially in the production of hybrid rye seed, although their use is possible.

Rye diseases, in general, are of less importance economically than in the other cereal crops. The average annual loss from 1930 to 1939 in the United States is reported at 1.8 per cent of the crop, or 497,000 bushels (Plant Disease Survey).

1. Mosaic, Virus. The mosaic on rye is found sparingly in the United States and in Asia. The virus is perhaps one of several that occurs on wheat (see wheat, Mosaic, Chap. 11).

2. Bacterial Blight, *Xanthomonas translucens f.sp. secalis* (Reddy, Godkin & A. G. Johnson) Hagb. The disease occurs in North America and Australia, according to Noble (1935) and Reddy, and Johnson (1924). The lesions on rye are less stripe-like than on barley, and they generally develop an irregular margin (see Bacterial Blight of barley, Chap. 3).

3. Powdery Mildew, *Erysiphe graminis secalis* E. Marchal. Light infections of powdery mildew are common on rye although of no economic importance. Specialization of the physiologic variety on rye is not pronounced, owing largely to the heterozygous condition of the plant. According to Germar (1934) and Mains (1926), mildew-resistant rye selections are not difficult to secure.

The symptoms and etiology are similar to those on barley (Chap. 3).

173

4. Fusarium Blight or Scab, *Gibberella* and *Fusarium* spp.   The disease is common on rye in the humid and semihumid rye areas.   The losses are frequently high in both reduced yields and in the damaged quality. The "drunken bread" malady described as formerly common in the humid regions of Europe and Asia is attributed largely to *Gibberella-* and *Fusarium*-infected grain.   The appearance of the disease on rye is similar to that on wheat (Chap. 11).

5. Snow Mold, Foot Rot, and Head Blight, *Fusarium nivale* (Fr.) Ces., *Calonectria graminicola* (Berk. & Br.) Wr. or *F. nivale f. graminicola* (Berk. & Brme.) Snyder & Hansen.   The disease is common on rye, wheat, and many grasses.   Its occurrence is associated with heavy snow covering in many areas where the disease is severe, as in the northern sections of North America and Europe and Asia.   The symptoms of the disease are similar to those on wheat (Chap. 11).

6. Ergot, *Claviceps purpurea* (Fr.) Tul.   Ergot is common on species of many genera of the Gramineae.   The disease is world-wide in the temperate, humid, and semihumid areas.   Ergot is usually more severe on rye, although certain varieties of barley and wheat are infected heavily.   *Bromus, Agropyron, Poa* spp., and other cultivated grasses are damaged, especially where grown for seed production.

Grain marketed through the Federal grading system of the United States is designated "ergoty" when it contains more than 0.3 per cent of ergot sclerotia by weight.   In the past such grain was discounted heavily. The perfection of the gravity-type separator and other grain-cleaning equipment greatly facilitates the removal of the sclerotia to within the tolerance of the Federal grades and pure-food laws.   Many other cereal-producing countries apply a similar low tolerance for grains used for feed or food purposes.

The low tolerances of ergot sclerotia are necessary because these fungus bodies contain compounds harmful to the circulatory system of animals.   The sclerotia contain varying amounts of compounds known as ergosterol, ergotoxin, ergotamin, ergostetrine, ergoclarin, etc.   Some of these when properly purified are valuable medicinals.   According to Barger (1931), Dudley and Moir (1935), Krebs (1936), Kussner (1934), Thompson (1935), Trabucchi (1934), and others, specific compounds or groups of these compounds cause constriction of the capillary blood vessels in specific tissues or generally.   Large dosages or continuous smaller amounts of ergot result in constriction of the capillaries of the placental tissue and abortion.   The lactation of animals is reduced or prevented by continuous small amounts of ergot.   The effect is cumulative and frequently causes reduced circulation and breakdown of tissues especially in the extremities, such as fingers, toes, hoofs, and ears. Extract of ergot is used extensively in obstetrical medicine.   Sound sclerotia of suitable composition demand high prices, especially during periods when the regular supply from the Mediterranean area is cut off from commerce.   Many rye growers and processors receive high prices for ergot sclerotia during these periods of shortage.   Environmental conditions, especially excess moisture, influence the infection of the sclerotia with *Fusarium* spp. and the composition of the sclerotia; therefore all sclerotia are not suitable for this use.

DESCRIPTION AND EFFECT.   The ergot infection is conspicuous from blossoming of the cereals and grasses to maturity of the crop.   Infection is evident first in the conidial "honeydew" stage when the conidia of the fungus are produced on a folded stromatic surface and released in a sugary exudate.   The exudate and conidia accumulate in droplets or adhere to the surface of the floral structures, depending upon the concentration of the exudate which is influenced by moisture.   Insects feed upon the exudate and are conspicuous around infected spikes.   The sclerotial stage soon follows the conidial; however, under some conditions only the conidial stage develops (see grass diseases, Chap. 12).   The blue-black sclerotial body replaces the kernel in the infected flowers. As the sclerotia develop, the floral bracts spread apart, and the bodies when fully developed are evident as they usually protrude beyond the floral bracts.   Few to many sclerotia occur on a spike or panicle (Fig. 41). These sclerotia are conspicuous in the threshed grain or seed, and they can be detected by specific color tests in milled-grain products.

THE FUNGUS

Claviceps purpurea (Fr.) Tul.

Synonyms arranged under the three stages in the development of the fungus.

Sclerotial Stage	Conidial Stage
(*Clavaria solida* Munch.)	(*Spermoedia clavis* Fries)
(*Clavaria clavus* Schr.)	(*Sphacelia segetum* Lév.)
(*Clavaria secalina* Paul.)	(*Fusarium heterosporum* Nees)
[*Sclerotium clavus* (Tode) DC.]	(*Oidium abortifaciens* Berk. & Bro.)

Ascigerous Stage

(*Sphaeropus fungorum* Paul.)
(*Sphaeria purpurea* Fries)
(*Kentrosporium purpurea* Wallr.)
(*Cordyceps purpurea* Fries)
(*Cordiliceps purpurea* Tul.)

The morphology of the fungus comprises a series of developmental stages involving specific types of mycelium and spores.   A mass of tightly compacted large hyphae form an absorbing structure in the vascular bundle of the rachilla in the base of the

flower.   The hyphal cells above this foot-like structure are shorter and remain in an
active state of division, or they constitute an embryonic or generative hyphal mass
situated in the position of the young ovary and are frequently associated with the

FIG. 41. Rye spikes showing the conidial (*A*) and sclerotial (*B*) stages of ergot caused
by *Claviceps purpurea*.

residual tissues of the ovary wall of the grass flower.   The latter structure is replaced
completely or in part by this compact mass of actively dividing hyphae.   The conid-
ial-bearing mycelial stroma is differentiated first on the top of the generative hyphae.
It is composed of a thin layer of compacted hyphae producing a stromatal surface of

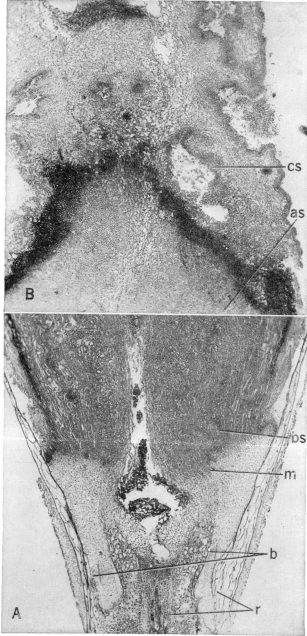

FIG. 42. Median longisection of rye flower, showing young sclerotium (*A*) and conidial stroma (*B*) of *Claviceps purpurea*. (*r*) Rachilla and flower tissue, (*b*) attaching and absorbing mycelium, (*m*) region of mycelial differentiation, (*bs*) basal and (*as*) apical sections of sclerotium, and (*cs*) conidial stroma. × 100.

palisade-like single-celled conidiophores on the undulating and folding outer surface of the sphacelial stroma. Small spheroid hyaline conidia are produced successively from the conidiophores, and they are held in a sugary exudate secreted by the stroma. The generative mass of hyphae soon differentiate sclerotial hyphae below the conidial stroma, and they in turn supply the contact with the conidial stroma during the remaining period of activity of the conidial stage. The sclerotium continues to elongate from the base by the differentiation of sclerotial cells from the generative mass of hyphae. As the sclerotium elongates, the hyphae included differentiate into the harder outer surface layer, the fertile hyphal mass, and the central larger storage cells. At the maturity of the suscept cereal or grass, the fungus development consists of the foot-like attachment in the rachilla, the sclerotium similar in shape to the caryopsis of the suscept it replaces and enclosed by the floral bracts, and the dry conidial stroma still attached to the apex of the sclerotium (Fig. 42). The size of the sclerotia vary depending upon the development of the suscept and the number of infections in each spike.

The fertile hyphae within the sclerotium develop into stromata and stipes (Fig. 43). Sclerotial germination occurs only after a suitable conditioning period, such as overwintering in the soil. The stipes are cylindrical, the length and thickness varying depending upon the depth of the sclerotium below the soil surface. The stromatal head (sphaeridium) is spherical, pale-fawn-colored, and the upper surface is covered with minute elevations, the projecting ostioles of the submerged peripheral perithecia. Perithecia are flask-shaped, sunken below the surface with the pronounced ostiole protruding. The asci are long, somewhat curved, and surrounded by numerous paraphyses similar in form to the asci. Each ascus contains a bundle of eight slender filiform hyaline ascospores. The ascospores are ejected from the apex of the ascus through the ostiole of the perithecium.

ETIOLOGY. Infection occurs through the young flowers, from windborne ascospores first and later from conidia. Penetration of the young ovary tissues occurs, and an absorbing organ of mycelium forms in the conductive tissue at the base of the ovary or in the rachilla. All or portions of the ovary are destroyed. The conidial and sclerotial mycelial masses develop within or above the remaining ovary tissues. The specialized fungus structures are attached to the flower by the absorbing mycelium until the sclerotium is mature. The sclerotia fall to the ground or are harvested with the grain. The sclerotia overwintering in the soil form perithecial stomatal heads that emerge from the soil by elongation of the stipe (Fig. 43). The stromata emerge from the soil shortly before blossoming of the grain or grass plants. Ascospores are produced in large numbers and are extruded from the asci at the period the suscepts are blossoming. The primary infection occurs from the wind-borne ascospores. In the spring of 1944, 10 to 15 germinating sclerotia per square foot were found under the plants in a nursery of Bromus inermis L. at Madison, Wis. Secondary infections from meteoric water or insect-borne conidia are extensive over a period of 1 to 2 weeks, depending upon environmental conditions.

Lewis (1945) discussed methods of preserving conidia in sugar solutions and using them in large-scale inoculations in the commercial production

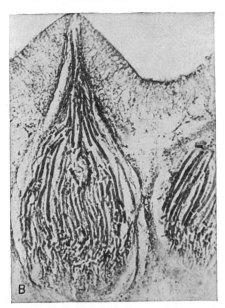

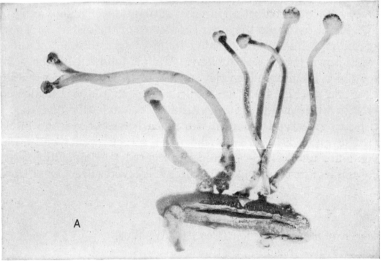

FIG. 43. Sclerotium and stromatal heads containing the perithecia of *Claviceps purpurea* (*A*) and section through perithecium (*B*).

of ergot. Atanasoff (1920) and McFarland (1921) have summarized the literature and data on infection.

CONTROL. Crop rotations and removal or cutting of weed grasses prevent the formation and spread of sclerotia and reduce the amount of primary inoculum. Heavy infection of spring barley in the North Central United States is usually associated with weed grasses in or adjacent to the field. According to McFarland (1921), the sclerotia formed on *Agropyron*, *Arrhenatherum*, *Bromus*, *Glyceria*, *Phleum*, and *Poa* spp. infect the cereal crops. The strain on *Lolium* spp. tested did not infect the cereals (see Chap. 12). Modern seed-cleaning equipment removes the sclerotia from the seed grain. Sclerotia do not remain viable when stored over 1 year.

Some varieties of the cereals and grasses are less susceptible to ergot than others. Investigations on ergot resistance, however, indicate to date no resistance in selections in rye or in most of the cultivated grasses (see Chap. 12).

7. **Anthracnose,** *Colletotrichum graminicolum* (Ces.) G. W. Wils. The disease is distributed widely on the cereals and grasses. In North America according to Sanford (1935), Selby and Manns (1909), and Wilson (1914), the disease is important especially in the soft red winter wheat area and in the North Central rye area. Barley, oats, and many grasses are damaged in this same general area. The disease is associated with conditions of low or unbalanced soil fertility, open coarse soils, and continuous grass-cereal culture.

DESCRIPTION AND EFFECT. The disease is apparent toward maturity of the crop. General reduction in vigor of plant development and premature ripening or dying are the gross symptoms of the malady. The presence of the mycelium and bleaching followed by browning of the culm bases and crown tissues appear first. This is followed by the development of the black acervuli on the surface of the lower leaf sheaths and culms especially (Fig. 44). Acervuli develop on the leaf blades of the dead plants when moisture is plentiful. Pronounced round to oblong lesions bearing acervuli occur on the green leaves of rye and of Sudan grass. The grain is shriveled when the attack occurs early. Spike infection and superficial grain infection occurs less commonly. Seedling and crown infection occur where the disease is severe (Bruehl and Dickson, 1950).

THE FUNGUS

Colletotrichum graminicolum (Ces.) G. W. Wils.
(*Colletotrichum cereale* Manns)

The acervuli are superficial, circular to oval, with dark mycelium forming the basal stroma. The black to dark-brown setae form through or surrounding the conidial-forming stroma of the mycelium. Setae are septate and tapering at the apex. Conidia are spindle-shaped, slightly curved, hyaline, and one-celled.

ETIOLOGY. The fungus is saprophytic on crop residues. Primary infection is commonly from the mycelium and conidia on the crop residue. Infected seed is a possible source of seedling root and crown infection.

x500

Fig. 44. Numerous acervuli, bearing black setae and hyaline conidia of *Colletotricum graminicolum*, formed on the basal leaf sheath and culm tissues of rye. Acervulus from Sudan grass, highly magnified, is shown in the insert.

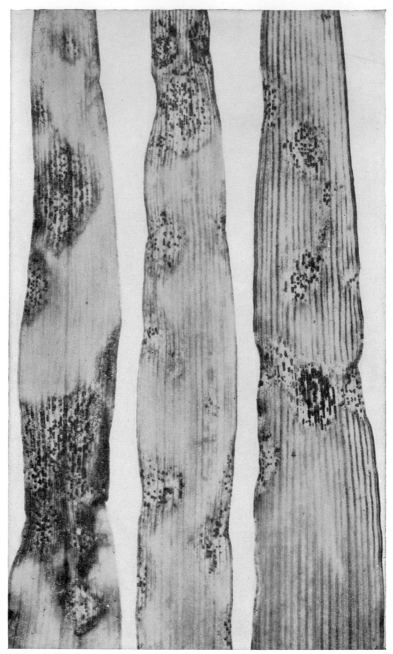

Fig. 45. Leaf blotch of rye caused by *Septoria secalis*, showing the abundant dark-brown pycnidia on the blotches.

Secondary spread late in the growing season is general from conidia as well as mycelium on the crop residue.

CONTROL. The disease is reduced by the rotation of crops and improved soil fertility. The use of legumes and other dicotyledonous crops in the rotation, plowing under cover crops, and proper balance of phosphate and potash greatly reduce infection and damage.

8. Rhynchosporium Leaf Scald, *Rhynchosporium secalis* (Oud.) J. J. Davis. The disease is distributed widely on rye, but it causes less damage on this crop than on barley and smooth brome grass. See Chap. 3 for a detailed discussion of the disease.

9. Septoria Leaf Blotch, *Septoria secalis* Prill. & Del. The *Septoria* leaf blotch occurs generally in the rye-producing areas. The lesions are characteristic of this group of fungi on the cereals and grasses. The light-brown irregular blotches are differentiated readily by the development of the pycnidia so typical of this genus (Fig. 45).

THE FUNGUS

Septoria secalis Prill & Del.

The pycnidia are subepidermal, black, globose, and smooth. The hyaline conidia are slender, cylindrical, slightly curved, rounded at the ends, and 3-septate. They average 2.7 microns wide by 35 microns long. See Chap. 11 for the full discussion of the disease.

10. Stalk Smut, *Urocystis occulta* (Wallr.) Rab. This fungus apparently is restricted to rye, and it is widely distributed throughout the world with the crop. The general distribution of this smut is in striking contrast to the rather limited distribution of the flag smut of wheat. The symptoms of the disease on rye and wheat are different, as indicated by the common names. In rye, the sori form in the parenchymatous tissue between the bundles of the culm tissue and less commonly in the leaf tissue. The sori form in the flowers and usually prevent the spike from emerging. Sori occur in the mesophyll of the leaf sheaths and less commonly in the leaf blade. The sori are covered by the epidermis in the early stages of development. The epidermis ruptures and the tissues shred, releasing the dark-brown spore mass (Fig. 46).

THE FUNGUS

Urocystis occulta (Wallr.) Rab.
(*Erysibe occulta* Wallr.)

The spore balls are oblong to spherical, 16 to 32 microns in diameter, and generally incompletely covered by the sterile cells. The sterile cells are light yellow to tan with irregular margins. The spores are reddish brown, oblong to subspherical, with flattened sides, and occur 1 to 2, rarely 3 to 4, in a smut ball. The spores germinate in place to form a promycelium (basidium) with usually four sporidia borne in an apical whorl and promycelium with irregular branching. Stakman *et al.* (1934) studied the

cycle of development of the fungus. Fischer (1953) retains this species and combines
Urocystis tritici Koern. and *U. agropyri* (Preuss) Schroet. on the basis of similar
morphology.

ETIOLOGY. Seedling infection occurs commonly from seed-borne spores. The fungus develops with the culm primordia of the seedlings, and sori are produced in the parenchymatous and mesophyll tissues. In the Northern United States and Canada, soil infestation is uncommon. The source of inoculum is largely seed-borne spores. The spores remain viable in dry soil to infect the fall-sown rye, as reported by Tisdale and Tapke (1927). Seed treatment controls this smut in most areas. According to Ling and Moore (1937) and Stakman and Levine (1916), temperature and moisture influence smut development and control of the disease. See Chap. 11 for a detailed discussion.

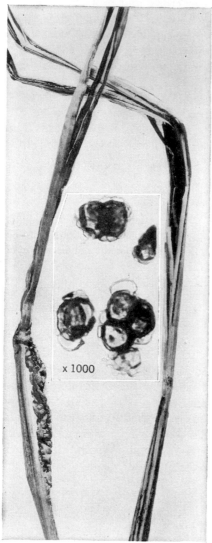

11. Head and Kernel Smuts, *Ustilago* and *Tilletia* spp. Other smuts are reported occasionally on rye. Several of the smuts have been designated by specific binomials, probably without adequate study of the disease. According to Humphrey and Tapke (1925), *Ustilago nuda* (Jens.) Rostr. or *U. tritici* (Pers.) Rostr. is capable of infecting rye under favorable conditions, and it has been reported on rye several times. *U. vavilovii* Jacz. is reported on rye in Asia (Gunther, 1941, pp. 87–89). Bunt caused by the species of *Tilletia* common on wheat is found occasionally on rye, and certain races of these parasites are capable of infecting rye, as reported by Gaines and Stevenson (1923), Lobik (1930), Nieves (1935), and others.

FIG. 46. Stalk smut of rye caused by *Urocystis occulta* and chlamydospores of the fungus, highly magnified.

12. Stem Rust, *Puccinia graminis secalis* Eriks. Stem rust on rye is distributed more generally, and the physiologic races of this variety of the parasite occur more commonly on the weed grasses than those on wheat or oats, as discussed by Stakman *et al.* (1934). Stem rust-resistant varieties and self-pollinated lines of rye occur, as reported by Cotter and Levine (1932), Mains (1926), Stakman *et al.* (1934, 1935), and Waterhouse (1936). Over 14 physiologic races occur on rye in the United States, according to Stakman *et al.* (1935); however, the heterozygous condition of the cross-pollinated rye makes determination of races unreliable. The detailed discussion of the disease is given in Chap. 11.

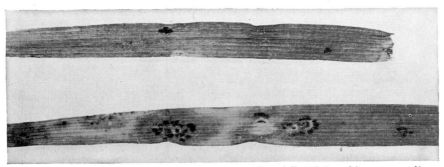

Fig. 47. Leaf blade of rye showing the new uredia of *Puccinia rubigo-vera secalis* or *P. recondita* formed early in the spring from mycelium overwintering in the tissues adjacent to the center uredium, which was formed the previous autumn. Contrasted with spring infections, upper leaf.

13. Stripe Rust, *Puccinia glumarum* (Schm.) Eriks. & Henn. The stripe rust is less common on rye in both the Pacific and Intermountain areas of North America, Europe, and Asia (see Chap. 11).

14. Leaf Rust, *Puccinia rubigo-vera secalis* (Eriks. & Henn.) Carleton or *P. recondita* Rob. The leaf rust of rye is widely distributed on rye and many wild species of the genus *Secale.* The aecial stage of the fungus occurs on species of *Anchusa*, although the natural aecial infections are rare in the United States, according to Arthur (1934), Mains (1933), and Mains and Jackson (1924).

The leaf rust causes a reduction in tillering and lower yields when heavy infection occurs early. Losses occur more commonly in the southern range of rye culture where the fungus overwinters in greater abundance.

DESCRIPTION. The round to ovate orange-brown uredia develop on the leaves and leaf sheaths. The telia are covered by the epidermis. The uredial development in the spring furnishes an excellent demonstration of overwintering of the uredial mycelium in the leaf tissue and the development of new uredia early in the spring (Fig. 47).

THE FUNGUS

Puccinia rubigo-vera secalis (Eriks. & Henn.) Carleton or *P. recondita*
Rob. ex Desm.
[*Puccinia rubigo-vera* (DC.) Wint.]
(*Puccinia dispersa secalis* Eriks. & Henn.)
(*Puccinia secalina* Grove)
(*Puccinia dispersa* Eriks. & Henn.)

Cummins and Caldwell (1956) review the synonymy and show that *P. recondita*
Rob. ex Desm. becomes the oldest valid binomial applicable to the leaf rust fungi
of the "*rubigo-vera* complex."

The morphology of the fungus and etiology are similar to the leaf rust of wheat
(Chap. 11). Specialization and resistance are reported by Gassner and Kirchhoff
(1934), Mains (1926), and Vavilov (1919).

REFERENCES

ARTHUR, J. C. Manual of the rusts in the United States and Canada. Purdue
Research Foundation, Lafayette, Ind. 1934.

ATANASOFF, D. Ergot of grains and grasses. *U.S. Dept. Agr. Div. Cereal Crops and
Diseases Stenciled Pub.* 1920. (Good literature list.)

BARGER, G. Ergot and ergotism. A monograph based on the Dohme lectures, Johns
Hopkins University. Gurney and Jackson. London. 1931.

BRUEHL, G. W., and J. G. DICKSON. Anthracnose of cereals and grasses. *U.S.
Dept. Agr. Tech. Bul.* 1005. 1950.

COTTER, R. U., and M. N. LEVINE. Physiologic specialization in *Puccinia graminis
secalis*. *Jour. Agr. Research* **45**:298–315. 1932.

CUMMINS, G. B., and R. M. CALDWELL. The validity of binomials in the leaf rust
fungus complex of cereals and grasses. *Phytopath.* **46**:81–82. 1956.

DUDLEY, H. W., and C. MOIR. The substance responsible for the traditional clinical
effect of ergot. *Brit. Med. Jour.* **3871**:520–523. 1935.

FISCHER, G. W. Manual of the North American smut fungi. The Ronald Press
Company. New York. 1953.

GAINES, E. F., and F. J. STEVENSON. Occurrence of bunt in rye. *Phytopath.* **13**:210–
215. 1923.

GASSNER, G., and H. KIRCHHOFF. Einige Versuche zum Nachweis biologischer
Rassen innerhalb des Roggenbraunrostes, *Puccinia dispersa* Eriks. & Henn.
Phytopath. Zeitschr. **7**:479–486. 1934.

GERMAR, B. Über einige Wirkungen der Kieselsäure in Getreidepflazen insbesondere
auf deren Resistenz gegenüber Mehltau. *Zeitschr. Pflanzenern. Dungung Bodenk.*
A35:102–115. 1934.

GUTNER, L. S. The smut fungi of the U.S.S.R. State Printing Office. Moscow-
Leningrad. 1941.

HUMPHREY, H. B., and V. F. TAPKE. The loose smut of rye *Ustilago tritici*. *Phytopath.* **15**:598–606. 1925.

KREBS, J. Untersuchungen über den Pilz des Mutterkorns, *Claviceps purpurea* Tul.
Ber. schweiz. bot. Gesel. **47**:71–165. 1936.

KUSSNER, W. Ergoclavin ein neues specifiches Alkoloid des Mutterkorns. *Arch.
Pharm. Ber. deutsch. pharm. Gesel.* **44**:503–504. 1934.

LEWIS, R. W. The field inoculation of rye with *Claviceps purpurea*. *Phytopath.*
35:353–360. 1945.

LING, L., and M. B. MOORE. Influence of soil temperature and soil moisture on infection of stem smut of rye. *Phytopath.* **27** :633–636. 1937.

LOBIK, V. I. On the occurrence of bunt [*Tilletia foetens* (Berk. and Cort.) Trel.] on rye (*Secale cereale*). *Bul. No. Caucasian Plant Prot. Sta. Rostoff on Don* **7–6** :165–166. 1930. (In Russian.)

MAINS, E. B. Rye resistant to leaf rust, stem rust, and powdery mildew. *Jour. Agr. Research* **32** :201–221. 1926.

———. Studies on rust resistance. *Jour. Heredity* **17** :313–325. 1926.

———. Host specialization in the leaf rust of grasses, *Puccinia rubigo-vera. Mich. Acad. Sci., Arts, Letters* **17** :289–394. 1933.

——— and H. S. JACKSON. Aecial stages of the leaf rust of rye, *Puccinia dispersa* Eriks. and barley, *P. anomala* Rostr., in the United States. *Jour. Agr. Research* **28** :1119–1126. 1924.

MCFARLAND, F. T. Infection experiments with Claviceps. *Phytopath.* **11** :41. 1921. Thesis. University Wisconsin. Madison, Wis.

NIEVES, R. Infection experimental del centeno de Pelkus (*Secale cereale* var. *vulgare*) por las caries del trigo: *Tilletia tritici* y *Tilletia levis. Phytopath.* **25** :503–515. 1935.

NOBLE, R. J. Notes on plant diseases recorded in New South Wales for the year ending 30th June 1935. *Int. Bul. Plant Prot.* **9** :270–273. 1935.

REDDY, C. S., and A. G. JOHNSON. Bacterial blight of rye. *Jour. Agr. Research* **28** :1039–1040. 1924.

SANFORD, G. B. *Colletotrichum graminicolum* (Ces.) Wils. as a parasite of the stem and root tissues of *Avena sativa. Sci. Agr.* **15** :370–376. 1935.

SELBY, A. D., and T. F. MANNS. Studies in diseases of cereals and grasses. *Ohio Agr. Exp. Sta. Bul.* 203. 1909.

STAKMAN, E. C., and M. N. LEVINE. Rye smut. *Minn. Agr. Exp. Sta. Bul.* 160. 1916.

——— et al. The cytology of *Urocystis occulta. Phytopath.* **24** :874–889. 1934.

———. Relation of barberry to the origin and persistence of physiologic forms of *Puccinia graminis. Jour. Agr. Research* **48** :953–969. 1934.

———. Die Bestimmung physiologischer Rassen pflanzenpathogener Pilze. *Nova Acta Leopoldina* **3** :281–336. 1935.

THOMPSON, M. A. The active constituents of ergot: a pharmacological and chemical study. *Jour. Ann. Pharm. Assn.* **24** :24–38, 185–196. 1935.

TRABUCCHI, E. Ricerche sui principii attivi *Segale cornuta.* I. *Boll. Soc. Ital. Biol. Sper.* **9** :501–507. 1934.

VAVILOV, N. I. Immunity of plants to infectious diseases. *Bul. Petrovsk. Agr. Acad. Moscow.* 1918.

WATERHOUSE, W. L. Some observations on cereal rust problems in Australia. *Proc. Linn. Soc. N.S.W.* **41** :5–38. 1936.

WILSON, G. W. The identity of the anthracnose of grasses in the United States. *Phytopath.* **4** :106–112. 1914.

CHAPTER 9

SORGHUMS, SUDAN GRASS, AND
JOHNSON GRASS DISEASES

The cultivated sorghums are largely annuals of the species *Sorghum vulgare* Pers. Martin (1936) included the perennial Johnson grass, *S. halepense* (L.) Pers., in the sorghum group. The following key is from Martin's grouping (1936, p. 528), illustrating the relationship of the different sorghum types.

1. Annual sorghums, *Sorghum vulgare*
   A. Sorgo (sweet or saccharine sorghum)
   B. Grain sorghum
      1. Milo
      2. Kaffir
      3. Feterita
      4. Durra
      5. Misc. (hegari, darso, shallu, kaoliang, etc.)
   C. Broom corn
   D. Grass sorghum; Sudan grass, *S. vulgare* var. *sudanense* (Piper) Hitchc.
2. Perennial sorghum or Johnson grass, *S. halepense* (L.) Pers.

The crop is cultivated for grain, forage, and juice in the drier climates and on limited acreage in the humid sections of the United States. Like corn, the sorghum is sufficiently diverse so that varieties are grown in all locations in the United States.

The basic chromosome number of the genus is 5 chromosome pairs. The annual sorghums included in the above key all have 10 pairs; the perennial Johnson grass has 20 pairs of chromosomes. Certain of the sorghum varieties have been crossed with sugarcane hybrids. So far as known, such hybrids are pollen-sterile and only a few fertile embryos develop.

The physiological anatomy of the sorghums is similar to corn. The seedling development and susceptibility to disease is very similar in the two. The development of the inflorescence, however, is different, as the sorghums develop panicles or heads of perfect flowers. The caryopsis of the sorghums threshes free from the floral bracts in most commercial varieties. The development of prussic acid, especially in the leaves, and the formation of hydrocyanic acid under some conditions are characteristic of the group.

Diseases cause relatively large losses in this crop. Seedling blights and root rots are important in reducing stands and plant vigor. The sorghum smuts are severe where control methods are neglected.

1. Nonparasitic Leaf Spotting and Weak Neck. *Leaf spotting* and other types of pigmentation of the leaves and floral bracts as well as chlorophyll deficiencies occur in this group of plants. Purple, brown, and yellow spots and blotches are characteristic of the sorghums and Sudan grass. The color of the spots is determined in part by the genetic composition of the variety or selection. The nonparasitic spots are differentiated from the parasitic maladies as follows: (1) general absence of water-soaked areas in any portion of the pigmented spot, (2) uniform pigmentation over the spot and rather regular margin, (3) absence of necrosis, especially in the early stages of spot development, and (4) the absence of fungi and bacteria associated with the spots. These pigmented areas on mature leaves, however, frequently show necrosis followed by secondary organisms invading the tissues. Environmental conditions influence the expression of the spotting. Pigmentation of this type is associated especially with progenies from hybrids or inbreds. Freed's sorgo is one of the very few sorghums that has never been observed to develop leaf spots. The various types of chlorophyll deficiencies—albino, virescent, and other chlorophyll-deficient manifestations—are associated frequently with material in the segregating progenies growing under environmental conditions favorable for expression of the defects. Chlorosis due to soil conditions, temperatures, etc., is present in some areas and varieties (Quinby and Martin, 1954).

Weak neck is common in the major grain sorghum areas of the United States where the malady causes reduction in yield and plumpness of grain. The damage is pronounced in relation to combine harvesting, as the grain must stand in the field until completely dry. The culm breaks below the head, resulting in poorly developed heads with lightweight seed. Apparently the tissues of the peduncle and rachis do not develop sufficient thick-walled tissue, and these tissues ripen too early to support the developing head. The early maturation of these tissues is accompanied by physiological decline of the tissues; the upper culm tissues dry out and bleach to straw color; a soft spongy condition of the upper culm develops; and frequently the decline is accompanied by water soaking and the accumulation of a sticky exudate during wet weather. Secondary organisms such as *Fusarium, Helminthosporium,* and *Alternaria* spp. later are associated with the dead tissues (Swanson, 1938, and Leukel *et al.,* 1943). The dwarf varieties developed for combining are derived largely from the milo types. These and the feterita types are early-maturing and develop a dry stem and a relatively thin rind, conditions conducive to weak neck. The development of dwarf varieties with culms more like the sorgo and kaffir types in which the culms remain green longer apparently is the best means of control (Hansing *et al.*, 1950).

2. Mosaics, Infectious Viruses. Certain of the grass virus maladies affect the sorghums. These are discussed in detail in Chap. 10.

3. Bacterial Blights. The bacterial stripe and streak diseases are common on the sorghums. Symptoms are more pronounced and defoliation is more severe than with the similar diseases on sugarcane. Burrill (1887, 1889) in Illinois was probably the first to describe the bacterial blights of sorghums in his early studies of bacterial diseases of plants. He described an organism that he considered to be the cause of the complex and named it *Bacillus sorghi* Burr. Kellerman (1888) and Kellerman and Swingle (1889) conducted a parallel study of this disease complex on the sorghums of Kansas. The red lesions on the sorghums were described and attributed to several distinct organisms by other investigators during this same period, as reviewed by Elliott and Smith (1929). Smith and his associates (1905, 1911) redescribed the symptoms of "Burrill's bacterial disease of broom corn" but attributed it to another bacterium, nonsporiferous, white on culture media, and with one to three polar flagella. They named the organism *Bacterium andropogoni* Smith & Hedges. Elliott (1929, 1930) continued the study of these bacterial diseases and differentiated the bacterial stripe and streak diseases of the sorghums.

Bacterial Stripe, Pseudomonas andropogoni (E. F. Sm.) Stapp [*Phytomonas andropogoni* (E. F. Sm.) Bergey *et al.*]. The disease occurs on many sorghums, including Sudan grass and Johnson grass, and it is distributed chiefly in the grain sorghum areas of the United States and similar areas in other countries. Inoculations on corn and sugarcane result in lesions on some varieties. The disease is widely distributed and under favorable conditions causes leaf killing.

The lesions are linear stripes with the color continuous throughout the lesion. The stripes are narrow, water-soaked, and bounded by the veins when young. Later the stripes fuse and cover a large part of the leaf surface and extend into the leaf sheath. Stalks and floral structures show similar but more restricted lesions. The color ranges from dark purplish red to brown, depending upon the sorghum variety and genetic factor pair conditioning pigment. Abundant exudate, pigmented similar to the color of the stripe, forms in droplets and scales over the lesions (Fig. 48).

Bacterial Streak, Xanthomonas holcicola (Elliott) Starr & Burk. [*Phytomonas holcicola* (Ell.) Bergey *et al.*]. The disease is widely distributed in the United States and abroad on the sorghums, including Sudan grass and Johnson grass. Apparently this disease occurs more commonly on the sorghums, especially in the cooler climates, than the stripe. Streak causes considerable defoliation in the grain and grass sorghums.

The young lesions are narrow water-soaked streaks with red to brown margins and irregular blotches of color interrupting the continuity of the streak.   Irregular oval blotches with tan centers and red margins, inter-

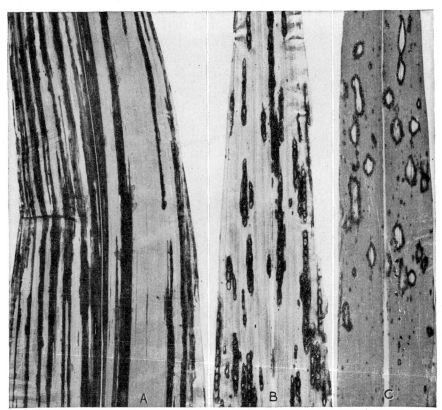

Fig. 48. Bacterial blights of Sorghum and Sudan grass.   (A) Bacterial stripe, (B) bacterial streak, and (C) bacterial spot.

mingled with the streaks, develop as the disease advances and lesions coalesce (Fig. 48).   Exudate is abundant as yellow to cream-colored droplets or scales.

Holcus Spot, Pseudomonas syringae v. Hall [*Phytomonas syringae* (v. Hall) Bergey *et al.*], [*P. holci* (Kend.) Bergey *et al.*].   Kendrick (1926) described this leaf spot of corn and sorghum.   The holcus spot is characterized by tan, red-bordered, round to elliptical lesions on the leaves. The spots coalesce to form irregular blotches, but do not elongate to form streaks or stripes.   Exudate is not present on the lesion.

The various bacteria associated with the sorghum blights are similar in gross morphology (rods with polar flagella) but vary in physiology. All are seed-borne to some extent and are carried over on crop residue. Seed treatment, crop rotation, and the use of resistant varieties are the principal means of control. The prevalence of these diseases and the degree of control are dependent on seasonal conditions.

4. Seed Rot, Seedling Blight, and Root Rot. *Pythium* spp. and Other Fungi. The sorghums, like corn, are warm-weather plants. Seed rotting and seedling blight in cold soils reduce stand (Leukel and Martin, 1943). *Pythium debaryanum* Hesse, *P. graminicola* Subrm., and other species are associated with the seed rotting. *Penicillium oxalicum* Currie & Thom also causes seedling blight. *Pythium arrhenomanes* Drechsl. incites seedling blight and rootlet rotting. This pathogen earlier was associated with milo disease. Seed treatment with protective fungicides such as Arasan and Orthocide increases stands in cold soils. Later planting of sorghums, after the soil has become warm, greatly reduces the damage from this disease. See corn, Chap. 4.

5. Downy Mildews, *Sclerospora sorghi* (Kulk.) Weston & Uppal and Other Species. The sorghums are not damaged extensively by the downy mildews, although several species of *Sclerospora* occur on them in Asia and Africa (see Table 6, page 86, The Fungi Causing Downy Mildew on the Gramineae). *Sclerospora sorghi* is limited apparently to southern Asia, chiefly India, where it is destructive on this crop and occurs on a few varieties of corn. The symptoms on sorghum are similar to those of the oriental downy mildews on corn and sugarcane, according to Butler (1917), Melchers (1931), Uppal and Desai (1932), and Weston and Uppal (1932).

THE FUNGUS

Sclerospora sorghi (Kulk.) Weston & Uppal
(*Sclerospora graminicola* var. *andropogonis-sorghi* Kulk.)

Weston and Uppal (1932) described the conidial and oogonial stages of the fungus and compared them with *S. graminicola* as the basis for the new species. The conidia of *S. sorghi* lack an apical dehiscence papilla and germinate by the formation of a germ tube. The conidiophore has a definite basal cell, an extensive branching, and consequent arrangement of the conidia in a hemispherical plane on the long sterigmata of the conidiophores. The oospores are similar in morphology to *S. graminicola* and germinate by the formation of a germ tube.

Conidial production is nocturnal in nature; however, Safeeulla and Thirumalachar (1955) demonstrated that low temperature and saturated atmosphere are the critical factors rather than periodic response. Infection occurs from both conidia and oospores. Some sorghum varieties are resistant, and only a few varieties of corn are susceptible.

6. Gibberella Seedling Blight and Stalk Rot, *Gibberella zeae* (Schw.) Petch and *G. fujikuroi* (Saw.) Wr. or *Fusarium roseum* f. *cerealis* (Cke.) Snyder & Hansen and *Fusarium moniliforme* (Sheld.) Snyder & Hansen. Seedling blight in cooler soils is frequently an important factor in reducing stands. Seed treatment with the mercury dusts increases stands, according to Leukel (1943) and Leukel and Martin (1943). The symptoms are similar to those on corn (Chap. 4).

7. Helminthosporium Leaf Blight, *Helminthosporium turcicum* Pass. The corn leaf blight occurs sparingly on the sorghums and very extensively on Sudan and Johnson grasses. Considerable defoliation results on these two grass sorghums (Fig. 49). The lesions are similar to those on corn except for the development of more pigmentation around the margin of the lesions in the sorghum group. According to Lefebvre and Sherwin (1945) and Mitra (1923) as well as inoculation experiments by Allison, specialization occurs within the species. Certain of the physiologic races of the parasite on corn infect the sorghum group. The sorghum races of the fungus apparently do not infect corn naturally. The extensive severe development of the disease on Sudan grass in the Northern United States and the relatively sparse occurrence of the leaf blight on corn are explained in part on the basis of specialization of the parasite.

The morphology of the fungus and etiology and control of the disease are discussed in Chap. 4. According to Chilton (1940), the fungus is seed-borne as well as carried over on crop residue.

8. Anthracnose, *Colletotrichum* spp. The anthracnose is distributed widely on the sorghums, especially on Sudan and Johnson grasses. The disease is of little economic importance on the sweet and grain sorghums, but causes damage on Sudan grass (Allison and Chamberlain, 1946) and broom corn. *Colletotrichum graminicolum* (Ces.) G. W. Wils. causes root and crown lesions and small ovate to irregular zonate spots on the leaves of Sudan grass. The central area of the leaf spot is tan; the border is red to brown. Acervuli develop on the older portion of the spot (Fig. 49). Root and crown rot occurs on Sudan grass and broom corn, causing lodging and reduced yields. The interior of the base of the stalk is rotted, and numerous acervuli develop on the surface of the rotted portion. *C. graminicolum* isolates from *Sorghum* spp. develop at higher temperatures than isolates from cool-temperature cereals and grasses. Seedling blight develops at temperatures between 16 and 32°C. (Bruehl and Dickson, 1950). *C. falcatum* Went or *Physalospora tucumanensis* Speg. occurs on *Sorghum* spp., especially in the warmer climates. *C. lineola* Corda is listed also as the species on the sorghums, although Wilson (1914) includes this under *C. graminicolum*.

The disease is abundant from mid-summer to maturity of the crop.

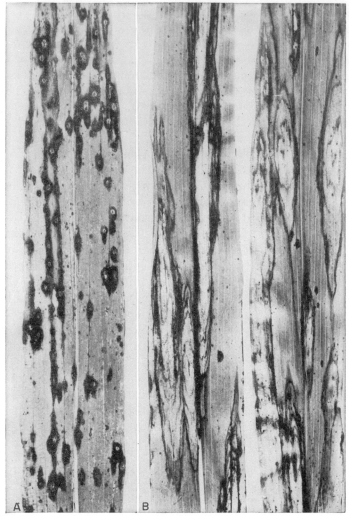

Fig. 49. Leaf spots of Sudan grass produced by *Colletotrichum graminicolum* (*A*) compared with the leaf lesions caused by *Helminthosporium turcicum* (*B*). These are two serious diseases of this crop.

Resistant inbreds and hybrids are the best means of control on Sudan grass. The disease is discussed in detail in Chaps. 8 and 10.

9. Milo Root Rot, *Periconia circinata* (Mangin) Sacc. The disease occurs principally on the milo-type grain sorghums across the Southern United States and adjacent Mexico. Milo and its derivatives developed for combine harvesting were damaged severely, especially where grown continuously on the same fields. Resistant milo plants were observed occasionally in the badly diseased fields and experimental plots. Selections from such plants led to the development of resistant varieties and information on the inheritance of resistance to the disease long before the cause was known (Bowman *et al.*, 1937, Heyne *et al.*, 1944, Kendrick and Briggs, 1939). The reaction of the resistant milos was an important factor in the determination of the pathogen involved.

FIG. 50. The milo root rot disease incited by *Periconia circinata*. The disease is controlled by using resistant varieties; resistant, left, susceptible, center, and F_1, right, growing in infested soil. (*Courtesy of L. F. Melchers.*)

Earlier reports attributed the disease to *Pythium arrhenomanes*, ubiquitously associated with the diseased plants (Elliott *et al.*, 1937). However, in the investigations with this pathogen the resistant milos were diseased. Finally, by the process of elimination by testing the reaction of resistant milos to the fungi isolated from diseased plants, the pathogen *Periconia circinata* was discovered (Leukel 1948) (Fig. 50).

THE FUNGUS

Periconia circinata (Mangin) Sacc.
(*Aspergillus circinatus* L. Mangin)

Conidiophores are filiform erect, fuliginous, lobed at the apex. Conidia borne in short chains in dense black heads, young conidia smooth, mature spiny, spherical, black, and 15 to 27 microns. Thick-walled, mycelial cells remain viable in soil and root residues.

The pathogen is associated with the rotted, blackened roots of many grasses. A similar species incites a leaf spot on rubber plants. The fungus invades the roots and crown tissues, later causing wilting and firing of the leaves and gradual death of the susceptible plants (Fig. 50).

Apparently a toxic product is produced by the pathogen, causing slow death of the plant tissues similar to the *Helminthosporium* blight of oats. The *Pythium* rootlet rot frequently is an associated factor in the severe damage.

Crop rotation and resistant varieties are the best means of control. Most of the sorghums are resistant (Wagner, 1942). Reaction to the disease is determined by a single factor pair with susceptibility to the disease partly dominant (Bowman *et al.*, 1937, Heyne *et al.*, 1944, Melchers and Lowe, 1943, Quinby, 1954).

10. Gloeocercospora Leaf Spot, *Gloeocercospora sorghi* Bain & Edg. Bain and Edgerton (1943) described a zonate leaf spot on sorghums, corn, and sugarcane. The fungus also occurs on the bent grasses, *Agrostis* spp. (Sprague, 1950). The fungus was placed in a new genus in the Tuberculareaceae, as the conidia are borne in a sporodochium-like structure and in a slimy matrix on short conidiophores. The spots appear first as small red to brown water-soaked lesions. As the spots enlarge they become dark-colored and elongate and form large zonate, semicircular or irregular lesions frequently extending across the leaf blade. Both leaves and floral bracts are infected. The fungus is seed-borne (Bain, 1950).

The Fungus

Gloeocercospora sorghi Bain & Edg.

Sporodochia bearing hyaline, septate, short, single or branched conidiophores form on the lesion. Conidia are hyaline, elongate to filiform, variable in length, the longer ones tapering at the apex, and measure 20–195 by 1.4–3.2 microns. The conidia are borne in a slimy matrix and are salmon pink in mass. Sclerotia are 0.1 to 0.2 mm. in diameter, lenticular to spherical, black, and borne inside the necrotic tissue.

11. Sooty Stripe, *Ramulispora sorghi* (Ell. & Ev.) Olive & Lefebvre. A fungus, somewhat similar in conidial morphology to the above, produces sooty stripe on the sorghums in the Southern United States and Asia. The lesions are elongate elliptic, regular in outline, first gray to tan in the center with a red margin, later sooty-colored as the black loosely attached sclerotia develop on the lesions.

Ramulispora sorghi (Ell. & Ev.) Olive & Lefebvre in the Tuberculareaceae, with synonyms (*Septorella sorghi* Ell. & Ev.), (*Ramulispora andropogonis* Miura), [*Titaeospora andropogonis* (Miura) Tai], apparently is the accepted binomial for this parasite.

The sclerotia are amphigenous, scattered, superficial, subglobose, coarsely tuberculate, and black. Sporodochia are amphigenous, erumpent from subepidermal stromata. Fasiculate conidiophores form from the sporodochia and the sclerotia. Conidia are filiform, one- to three-branched, hyaline, curved, tapering toward the apex, 3- to 8-septate, and measure 38–86 by 1.9–3 microns (Olive *et al.*, 1946).

Cercospora sorghi Ell. & Ev. occurs on the sorghums and corn in the Southern United States, according to Seymour (1929) and Stevenson (1926).

12. Charcoal Rot, *Macrophomina phaseoli* (Maubl.) Ashby or *Botryodiplodia phaseoli* (Maubl.) Thir. The root, crown, and stem rot is distributed widely especially in warm, dry climates on many crop plants. Sorghum, corn, millets, cotton, jute, hemp, soybean, castor bean are some of the field crops damaged by this pathogen under low moisture and high temperature. The symptoms are similar on all crops. The spongy rot and eventual breakdown of parenchymatous tissues leaving vascular bundles, fibers, and other mechanical tissue intact in roots, crown, and basal stem tissues followed by lodging are characteristic symptoms. Numerous black sclerotia and frequently subcarbonaceous pycnidia give the charred appearance to the diseased plant parts. The diseased plants frequently occur in local areas in the field. The pathogen on the various hosts has been described under various binomials.

THE FUNGUS

Macrophomina phaseoli (Maubl.) Ashby

Macrophomina phaseolina (Tassi) G. Goid. According to Sprague (1950)

Botryodiplodia phaseoli (Maubl.) Thir. According to Thirumalachar (1953)

Dothiorella phaseoli (Maubl.) Petrak & Sydow.

Sclerotium bataticola Taubenhaus

(*Macrophoma phaseolina* Tassi)

The fungus, as the binominals indicate, has several spore forms. Sclerotia are generally small to variable in size, black, numerous in tissues. *Rhizoctonia bataticola* (Taub.) Butler is considered the same as *Sclerotium bataticola* by some mycologists (Sprague, 1950, Thirumalachar, 1953). Pycnidia are subcarbonaceous, mostly 100 to 200 microns with small ostioles. Conidia are ovate-elongate, hyaline nonseptate or ovate-elongate, reddish brown, two-celled depending on temperature and age (Thirumalachar, 1953).

Apparently the pathogenicity and morphological development of this fungus are influenced greatly by temperature, moisture, and association with other microflora, as summarized by Norton (1953) and Sprague (1950). The pathogen persists in crop residues in the soil and attacks subterranean plant tissues when the crop is growing under conditions causing physiological stress, especially temperatures of 30°C. and above and low soil moisture. Apparently the fungus is held in balance in the soil by other microflora until high soil temperature and low moisture inhibit their development.

Crop rotation, balanced fertility, and proper management to ensure adequate soil moisture and avoid high soil temperatures are important control measures (Livingston, 1945, Norton, 1953, Wadsworth and Sieglinger, 1950). All varieties are relatively susceptible under conditions favorable for the development of the disease. Hoffmaster and Tullis[1] rank the sorghums in the following order: milos, darso, Sudan grass, feterita, Legari, kaffir, and sweet sorghums.

Tullis[2] described a stalk rot of sorghum in Texas with symptoms similar to charcoal rot which was caused by *Spicaria elegans* (Cda.) Harz. and *Fusarium moniliforme*.

13. Ascochyta spot, *Ascochyta sorghina* Sacc. The leaf and sheath spot occurs in many of the warmer areas of the world. The spots coalesce to form reddish-purple to brown blotches on leaf blades and sheaths. The large pycnidia protrude through the stomata to give the surface a characteristic roughened texture (Leukel *et al.*, 1951, Sprague, 1950). Yields are reduced when foliage damage is severe. The pycnidia are globose, depressed, papillate. Conidia are oblong-ellipsoid, 1-septate, hyaline, 20 by 8 microns. The several species listed on corn, *Ascochyta zeae* Stout, *A. zeicola* Ell. & Ev., *A. zeina* Sacc., *A. maydis* Stout, and *A. sorghina*, are similar morphologically.

14. Loose Kernel Smut, *Sphacelotheca cruenta* (Kuehn) Potter. The loose kernel smut of the sorghums is not distributed so widely in the United States as the covered kernel smut. The disease is common in Asia and Africa, according to Butler (1918). Apparently the disease is not common on Sudan grass. A similar loose kernel smut occurs on Johnson grass in South Central United States, Mexico, South America, and Africa. Both grain and fodder yields are reduced by the smut.

SYMPTOMS AND EFFECTS. The smut is apparent by the early appearance of the smutted heads and dwarfing of the plants in most varieties. Generally all the flowers in a head are smutted. The floral bracts tend to elongate and proliferate. Frequently the lemma and palea as well as the ovary contain smut sori. The membrane ruptures early, releasing the powdery black spore mass. The elongate central columella of the sorus persists after the spores are discharged. The symptoms range from a loose to covered-type kernel smut, depending upon the sorghum variety (Fig. 51).

THE FUNGUS

Sphacelotheca cruenta (Kuehn) Potter
(*Ustilago cruenta* Kuehn)
(*Sphacelotheca holci* Jacks.)

The sori are formed in the ovaries and floral bracts. The chlamydospores are enclosed in a fungal membrane (peridium) composed of loosely joined rounded gray

---

[1] *U.S. Dept. Agr. Plant Dis. Reptr.* **28**:1175–1184. 1944.
[2] *Ibid.*, p. 1100.

FIG. 51. Sorghum spikelets, magnified, showing the loose (A) and covered (B) kernel smuts of sorghums caused by *Sphacelotheca cruenta* and *S. sorghi*, respectively. Chlamydospores, highly magnified, are shown in the inserts.

cells about twice the diameter of the spores.   In many sorghum varieties the membrane is fragile and ruptures early.   The chlamydospores are formed in elongated irregular clumps, not spore balls, that separate as they mature.   The spores are round to elliptical, light brown, with indistinct pits to reticulations on the surface, and 5 to 10 microns in diameter.   They germinate to form characteristically a four-celled basidium (promycelium) with laterally borne sporidia.   Secondary sporidia are produced on media.   See Fischer (1953) for synonymy.

ETIOLOGY.   Infection occurs in the early stages of seedling development.   The fungus becomes established in the primordial tissues of the developing shoot, and spores are produced in the kernels and adjacent floral tissues.   This systemic type of infection is characteristic of many of the smut fungi.   As shown by Melchers and Hansing (1943), Reed (1923), and others, the development of *Sphacelotheca cruenta* in the sorghum plant causes dwarfing in most varieties and the early development of the head.   Smut infection occurs over the range of environmental conditions favorable for development of sorghum.   The chlamydospores are carried over on the seed and in the soil in the drier climates.   Infection of crown buds occurs in perennial hosts (Leukel and Martin, 1950).

Seed treatment and resistant varieties constitute the best means of control.   Leukel (1943) reported the mercury dusts and copper carbonate as suitable for the control of both loose and covered kernel smuts.   Stands were improved also in *Pythium*- and *Fusarium*-infested soils.   Melchers (1933) listed the milos, feteritas, hegari, and Dwarf Shantung kaoliang as resistant to moderately resistant to the two physiologic races of *Sphacelotheca cruenta*.   Race 1, originally from India, attacks kaffir × feterita and Pierce kaferita, while race 2 infection on these varieties is very light.   Race 2, found in Kansas, infects Red Amber × feterita and White Yolo, which are resistant to race 1.   Rodenhiser (1934) reported on crosses between races of *S. sorghi* and interspecific crosses between *S. sorghi* and race 1 of *S. cruenta*.   His results suggest (1) the origin of races by hybridization and (2) on the basis of sterility in the first generation interspecific hybrids, the two species are distinct rather than representing the extremes in a series of variants, as suggested by Tisdale *et al.* (1927).   According to Melchers (1940), Spur feterita (C.I. 623) is the most resistant to the known races of both loose and covered kernel smut parasites.

The loose kernel smut occurring on Johnson grass in the South Central United States is similar to the loose smut of sorghum.   Rodenhiser (1937) listed it as a third race of *Sphacelotheca cruenta*.   It infects some sorghums of feterita origin; the kafirs apparently are resistant.   Johnston *et al.* (1938) and Leukel and Martin (1950) suggest the binomial *S. holci*; Fischer (1953) includes it under *S. cruenta*.   Spur feterita (C.I. 623) is resistant to the smuts tested, and it is used extensively in breeding for smut-resistant varieties.

15. Covered Kernel Smut, *Sphacelotheca sorghi* (Lk.) Clint. The covered kernel smut is one of the most common diseases of sorghums in the United States and frequently reduces yield of grain. The disease is widely distributed in other countries on the sorghums and Sudan grass, as reviewed by Reed (1923) and Reed and Melchers (1925).

SYMPTOMS AND EFFECTS. Typically a kernel smut, the ovaries of the flowers are converted into smut balls. Sori rarely develop in the floral bracts. The outer membrane (peridium) of the sorus is tough and usually persists as the grain ripens. It varies in color with the different physiologic races of the fungus and with the sorghum varieties. The smut balls are elongated beyond the floral bracts and resemble somewhat the shape of the kernel (Fig. 51). The smutted plants are not reduced greatly in height and vegetative development as in the case of the loose smut, as shown by Melchers and Hansing (1943).

THE FUNGUS

Sphacelotheca sorghi (Lk.) Clint.
(*Sorosporium sorghi* Lk.)
[*Tilletia sorghi-vulgaris* Tul.)
(*Ustilago sorghi* (Lk.) Pass.]

The sori are formed chiefly in the ovaries. The chlamydospores are enclosed in a persistent fungal membrane that ruptures at the apex, exposing the dark spores and columella. The membrane is composed of round to elongate cells about the same diameter as the spores. Chlamydospores are globose to angular, olivaceous, brown, apparently smooth, but actually finely punctate to minutely echinulate, and 5 to 8 microns in diameter. They germinate to form characteristically a four-celled promycelium bearing lateral sporidia, although great variation occurs.

The etiology and control of the disease are essentially similar to the loose kernel smut. Standard and new chemical-dust seed treatments are described by Hansing and Melchers (1944). Environmental conditions influence infection, and some varieties, although infected, do not develop sori in the primary heads, according to Melchers (1933) and Melchers and Hansing (1938). Resistant varieties occur in most of the sorghum groups, as reported by Mehta *et al.* (1953), Melchers *et al.* (1932), Reed and Melchers (1925, 1940), and Tisdale *et al.* (1927), and others. Spur feterita (C.I. 623) is resistant to all known races, and dwarf yellow milo (K.B. 2515) and Red Amber × feterita (K.B. 33308) are resistant to one or more of the physiological races of *Sphacelotheca sorghi*. Specialization and variation in the fungus are reported by the above authors, as well as Ficke and Johnston (1930), Melchers *et al.* (1932), and Tyler (1938). Five physiologic races of the parasite are differentiated, as shown in Table 10.

16. Head Smut, *Sphacelotheca reiliana* (Kuehn) Clint. The disease occurs occasionally primarily on sorghums in the Central United States where it is of minor importance. In the Western United States and southeastern Europe and Asia, this smut is more generally distributed on corn. The smut is not evident until the panicle emerges. All or part of the floral structures are replaced by the smut sorus. The fragile membrane ruptures early, releasing the dark-brown dusty spore mass and exposing the vascular tissues of the inflorescence (Fig. 25). According to Reed (1923) the feterita, milo, kaffir, kaoliang, and broom corn groups of sorghums are resistant. Physiologic races are distinct on sorghum and corn. See Chap. 4 for the detailed discussion of the disease.

TABLE 10

Physio-logic race	Reaction on sorghum varieties					
	Dwarf yellow milo (C.I. 332)	White Yolo (K.B. 2525)	Pierce kaferita (K.B. 2547)	Feterita × kaffir (F.C.I. 8917)	Feterita (S.P.I. 51989)	Kaffir × feterita (H.C. 2423)
1	R	IR	R	R	R	R
2	S	S	R	R	R	R
3	R	R	S	IR	IR	S
4	R	S	R	R	R	R
5	R	R	R	R	R	S

R—resistant; I—intermediate; S—susceptible.

17. Long Smut, *Tolyposporium ehrenbergii* (Kuehn) Pat. This smut is not reported in the United States and according to Kamat (1933), Kulkarni (1918), and McAlpine (1910) causes little damage to this crop in other countries. The sori develop in the ovaries, frequently only a few in each panicle. The sori are long, cylindrical, slightly curved, and rupture at the apex to release the brownish-green spore balls. The spores remain united, more or less permanently in large groups. The exposed surfaces of the spores in the spore ball are covered by flattened echinulations. The spores germinate in place by the formation of an elongated promycelium, frequently branching. Numerous sporidia are produced singly or in chains.

Little experimental data are available on the etiology and control of the disease. Spores introduced into the leaf whorl of plants after the flower primordia are developed incited infection of individual kernels (Ramakrishnan and Reddy, 1949). Continuous culture in the same areas tends to increase the amount of infection.

18. Rust, *Puccinia purpurea* Cke. This rust occurs in limited amounts on all the sorghums, including many of the grass species of this genus. The disease is more common in the Southern United States, and it is distributed generally with the crop.   Butler (1918) reported the rust as one of the most common rusts of the cultivated crops of India.   The uredia develop on both surfaces of the leaves.   Pigmentation appears around the uredium.   Telia develop in and adjacent to the uredia as in the corn rust.   The aecial hosts are the same group of *Oxalis* spp. listed for *P. sorghi*.   See corn diseases, Chap. 4.   Resistant varieties occur in most of the sorghum groups.   The milos generally are resistant; a dwarf selection from Leoti-Atlas from Hays, Kans., is highly resistant (Johnston and Mains, 1931, Quinby and Martin, 1954).

REFERENCES

BAIN, D. C.   Fungi recovered from seed of *Sorghum vulgare* Pers.   *Phytopath.* **40**:521–522.   1950.

———— and C. W. EDGERTON.   The zonate leaf spot, a new disease of sorghum.   *Phytopath.* **33**:220–226.   1943.

BOWMAN, D. H., *et al.*   Inheritance of resistance to Pythium root rot in sorghum.   *Jour. Agr. Research* **55**:105–115.   1937.

BRUEHL, G. W., and J. G. DICKSON.   Anthracnose of cereals and grasses.   *U.S. Dept. Agr. Tech. Bul.* 1005.   1950.

BURRILL, T. J.   A disease of broom corn and sorghum.   *Soc. Prom. Agr. Sci. Proc.* **8**:30–36.   1887.

————.   Status of the sorghum blight.   *Jour. Mycol.* **5**:199.   1889.

BUTLER, E. J.   Fungi and disease in plants.   Thacker, Spink and Co.   Calcutta and Simla.   1918.

CHILTON, S. J. P.   The occurrence of *Helminthosporium turcicum* in the seed and glumes of Sudan grass.   *Phytopath.* **30**:533–536.   1940.

ELLIOTT, C.   Bacterial streak disease of sorghums.   *Jour. Agr. Research* **40**:963–976.   1930.

———— *et al.*   Pythium root rot of milo.   *Jour. Agr. Research* **54**:797–834.   1937.

———— and E. F. SMITH.   A bacterial stripe disease of sorghum.   *Jour. Agr. Research* **38**:1–22.   1929.

FICKE, C. H., and C. O. JOHNSTON.   Cultural characteristics of physiologic forms of *Sphacelotheca sorghi*.   *Phytopath.* **20**:241–249.   1930.

FISCHER, G. W.   Manual of the North American smut fungi.   The Ronald Press Company.   New York.   1953.

HANSING, E. D., and L. E. MELCHERS.   Standard and new fungicides for the control of covered kernel smut of sorghum and their effect on stand.   *Phytopath.* **34**:1034–1036.   1944.

————, L. E. MELCHERS, and J. C. BATES.   Weak neck of sorghum.   *Agron. Jour.* **42**:437–441.   1950.

HEYNE, E. G., *et al.*   Reaction of Fl sorghum plants to milo disease in the greenhouse and field.   *Jour. Am. Soc. Agron.* **36**:628–630.   1944.

JOHNSTON, C. O., *et al.*   Observations on the loose kernel smut of Johnson grass.   *Phytopath.* **28**:151–152.   1938.

———— and E. B. MAINS.   Relative susceptibility of varieties of sorghum to rust, *Puccinia purpurea*.   *Phytopath.* **21**:525–543.   1931.

KAMAT, M. N.   Observations on *Tolyposporium filiferum* cause of long smut of sorghum.   *Phytopath.* **23**:985–992.   1933.

KELLERMAN, W. A.   Preliminary report of sorghum blight.   *Kans. Agr. Exp. Sta. Bul.* 5, 1888.

—— and W. T. SWINGLE. Sorghum blight. *Kans. Agr. Exp. Sta. Rept.* (1888), pp. 281–302. 1889.

KENDRICK, J. B. Holcus bacterial spot of *Zea mays* and *Holcus* species. *Iowa Agr. Exp. Sta. Res. Bul.* 100. 1926.

KULKARNI, G. S. Smuts of jowar (sorghum) in the Bombay Presidency. *Agr. Res. Inst. Pusa Bul.* 78. 1918.

LEFEBVRE, C. L., and H. S. SHERWIN. Races of *Helminthosporium turcicum*. *Phytopath.* (Abstract) **35**:487. 1945.

LEUKEL, R. W. Chemical seed treatments for the control of certain diseases of sorghum. *U.S. Dept. Agr. Tech. Bul.* 849. 1943.

——. *Periconia circinata* and its relation to milo disease. *Jour. Agr. Res.* **77**:201–222. 1948.

—— et al. Weak neck in sorghum. *Jour. Am. Soc. Agron.* **35**:163–165. 1943.

—— and J. H. MARTIN. Seed rot and seedling blight of sorghum. *U.S. Dept. Agr. Tech. Bul.* 839. 1943.

—— and ——. Loose kernel smut of Johnson grass. *Phytopath.* **40**:1061–1070. 1950.

LIVINGSTON, J. E. Charcoal rot of corn and sorghum. *Nebr. Agr. Exp. Sta. Res. Bul.* 136. 1945.

MARTIN, J. H. Sorghum improvement. *U.S. Dept. Agr. Yearbook*, pp. 523–560. 1936.

McALPINE, D. The smuts of Australia. Government Printer. Melbourne, Australia. 1910.

MEHTA, P. R., B. SINGH, S. C. MATHUR, and S. B. SINGH. Varietal reaction of jowar to grain smut. *Nature* **172**:591–592. 1953.

MELCHERS, L. E. Downy mildew of sorghum and maize in Egypt. *Phytopath.* **21**:239–240. 1931.

——. Belated development of kernel smut (*Sphacelotheca sorghi*) in apparently healthy sorghum plants. *Jour. Agr. Research* **47**:343–350. 1933.

——. Physiological specialization of *Sphacelotheca cruenta* (Kühn) Potter. *Jour. Agr. Research* **47**:339–350. 1933.

——. The reaction of a group of sorghums to the covered and loose kernels smuts. *Am. Jour. Bot.* **27**:789–791. 1940.

——. On the cause of the milo disease. *Phytopath.* **32**:640–641. 1942.

—— et al. A study of the physiologic forms of kernel smut (*Sphacelotheca sorghi*) of sorghum. *Jour. Agr. Research* **44**:1–11. 1932.

—— and E. D. HANSING. The influence of environmental conditions at planting time on sorghum kernel smut infection. *Am. Jour. Bot.* **25**:17–28. 1938.

—— and A. E. LOWE. The development of sorghums resistant to milo disease. *Kans. Agr. Exp. Sta. Tech. Bul.* 55. 1943.

MITRA, M. *Helminthosporium* spp. on cereals and sugar cane in India. I. Diseases of *Zea mays* and *Sorghum vulgare* caused by species of *Helminthosporium*. *Mem. Dept. Agr. India Bot. Ser.* **11**:219–242. 1943.

NORTON, D. C. Linear growth of *Sclerotium bataticola* through soil. *Phytopath.* **43**:633–636. 1953.

OLIVE, L. S., et al. The fungus that causes sooty stripe of *Sorghum* spp. *Phytopath.* **36**:190–200. 1946.

QUINBY, J. R., and J. H. MARTIN. Sorghum improvement. *Adv. Agron.* **6**:305–359. 1954.

RAMAKRISHNAN, T. S., and G. S. REDDY. Artificial infection of sorghum with long smut. *Curr. Sci.* **18**:418. 1949.

REED, G. M. Varietal resistance and susceptibility of sorghum to *Sphacelotheca sorghi* (Link) Clinton and *Sphacelotheca cruenta* (Kühn) Potter. *Mycologia* **15**: 132–143. 1923.

———— *et al.* Experimental studies on head smut of corn and sorghum. *Bul. Torrey Bot. Club* **54**:295–310. 1927.

———— and L. E. MELCHERS. Sorghum smuts and varietal resistance in sorghums. *U.S. Dept. Agr. Bul.* 1284. 1925.

RODENHISER, H. A. Studies on the possible origin of physiologic forms of *Sphacelotheca sorghi* and *S. cruenta*. *Jour. Agr. Research* **49**:1069–1086. 1934.

————. Echinulation of chlamydospores and the pathogenicity of a previously undescribed physiologic race of *Sphacelotheca cruenta*. *Phytopath.* **27**:643–645. 1937.

SAFEEULLA, K. M., and M. J. THIRUMALACHAR. Resistance to infection by *Sclerospora sorghi* of sorghum and maize varieties in Mysore, India. *Phytopath.* **45**:128–131. 1955.

SEYMOUR, A. B. Host index of the fungi of North America. Harvard University Press. Cambridge, Mass. 1929.

SMITH, E. F. Bacteria in relation to plant diseases, *Carnegie Inst. Wash., D.C., Pub.* 27, vols. 1 and 2. 1905–1911.

———— and F. HEDGES. Burrill's bacterial disease of broom corn. *Science* (N.S.) **21**:502–503. 1905.

SPRAGUE, R. Diseases of cereals and grasses in North America. The Ronald Press Company. New York. 1950.

STEVENSON, J. A. Foreign plant diseases. *U.S. Dept. Agr. Off. Sec.* 1926.

SWANSON, A. F. "Weak neck" in sorghum. *Jour. Am. Soc. Agron.* **30**:720–724. 1938.

THIRUMALACHAR, M. J. Pycnidial state of charcoal rot inciting fungus with discussion of its nomenclature. *Phytopath.* **43**:608–610. 1953.

TISDALE, W. H., *et al.* Strains of kernel smuts of sorghum, *Sphacelotheca sorghi* and *S. cruenta*. *Jour. Agr. Research* **34**:825–838. 1927.

TYLER, L. J. Variation in *Sphacelotheca sorghi* (Link) Clinton. *Minn. Agr. Exp. Sta. Tech. Bul.* 133. 1938.

UPPAL, B. N., and M. K. DESAI. Two new hosts of the downy mildew of sorghum in Bombay. *Phytopath.* **22**:587–594. 1932.

WADSWORTH, D. F. and J. B. SIEGLINGER. Charcoal rot of sorghum. *Okla. Agr. Exp. Sta. Bul.* 355. 1950.

WAGNER, F. A. Reaction of sorghums to the root, crown, and shoot rot of milo. *Jour. Am. Soc. Agron.* **28**:643–654. 1936.

WESTON, W. H., and B. N. UPPAL. The basis for *Sclerospora sorghi* as a new species. *Phytopath.* **22**:573–586. 1932.

WILSON, G. W. The identity of the anthracnose of grasses in the United States. *Phytopath.* **4**:106–112. 1914.

CHAPTER 10

SUGARCANE DISEASES

The sugarcane varieties in cultivation are largely a group of inter-specific hybrids propagated vegetatively by stem cuttings. According to Brandes and Sartoris (1936), disease epidemics have played an important role in the development and propagation of the interspecific hybrids. Nobilizing or resorting to natural crossing between the cultivated varieties and primitive hardy disease-resistant forms was started during the sereh epidemic. However, hybrids superior to the former commercial varieties were not produced until much later. Hybridization really became important in sugarcane improvement after J. Jeswiet in Java produced the selection P.O.J. 2878. The sugarcane mosaic epidemics and better parental material and techniques stimulated further more systematic crossing and testing of larger populations of seedlings. Economic pressures and the better sugar yields of the latter hybrids also stimulated the breeding improvement. Table 11 gives the principal species used in crossing, their chromosome number, and disease reaction.

TABLE 11

Saccharum spp.	Number chromo-some pairs	Adapta-bility	Reactions to diseases			
			Mosaic	Sereh	Downy mildew	Smut
S. officinarum L...	40	Tropics	Susc.	Susc.	Susc.	Mod. susc.
S. sinense Rovb...	58–60	Wide	Some susc.	Immune	.....	Susc.
S. barberi Jesw....	42–46	Temp. and subtropics	Susc. tolerant	Immune	.....	Mod. susc.
S. spontaneum L..	56	Wide	Immune*	Immune	Susc.	Mod. susc.
S. robustum.......	42	Wide	Susc.			

* *S. spontaneum* var. Koelawa A. is susceptible to mosaic.

Disease-resistant hybrids are used in the control of most commercially important diseases. The propagation by clones, chiefly culm bud cuttings, facilitates the use of good-quality disease-resistant selections. Quarantines are effective in preventing the further distribution of diseases carried in the cuttings.

206

The history of sugarcane diseases and their relation to the breeding of better varieties is summarized by Edgerton (1955), Martin (1951), and Matsumoto (1950).

1. Virus Diseases. The virus diseases of sugarcane, as in rice and tobacco, were among the first recognized as incited by an infectious principle. The sugarcane virus disease complex was one of the early examples of disease resistance as a means of control. The "yellow stripe" and "sereh" reported in Java in 1890 by Musschenbrock and the series of reports summarized by Wakker and Went (1898) give the description of the disease and its alleviation by the propagation of resistant clones. The presence of the disease in Egypt in 1909 and in Hawaii in 1910 on canes imported from Java represents one of the early records of virus spread by means of clones. The repetition of the Java mosaic epidemic and the control of the disease and improvement of the crop by selecting resistant, high yielding varieties have occurred in every sugarcane-producing area of the world. The use of plant quarantine to prevent further dissemination of the disease was employed soon after its introduction into Hawaii in 1910.

Sugarcane Mosaic, Virus Transmitted by Aphids and Mechanically. Sugarcane, corn, sorghum, certain millets, and a number of tropical grasses are susceptible in varying degrees. The disease is world-wide in distribution and causes large losses in susceptible varieties (Brandes, 1919, Brandes and Klaphaak, 1923, Edgerton, 1950, Kunkel, 1929, Martin, 1951). The mosaic complex is manifest by light-green or yellow streaking and irregular mottled appearance of the leaves. The green or mild and the yellow or severe mosaics were differentiated in Louisiana and shown to be incited by strains of the virus (Forbes *et al.*, 1937, Tims *et al.*, 1935). Cross protection, *i.e.*, plants carrying one virus strain cannot be infected with the other, was demonstrated. The symptoms of the two in young plants are similar, but the stripes in the older leaves are predominantly light green in one and yellow in the other (Fig. 52). Yellowing and browning are later symptoms on susceptible varieties. Some varieties are dwarfed severely; others show only leaf symptoms, and these frequently are masked later. Certain resistant varieties carry the virus, but symptoms are inconspicuous. Culm cankers and internal necrosis occur in some varieties. The symptoms of the mosaic vary with host and age of plants, strain of the virus, and the environment (Kunkel, 1924, Martyn, 1946, Summers *et al.*, 1948).

Mechanical transmission of the virus on sugarcane and corn occurs readily (Matz, 1933, Sein, 1930, Wilbrink, 1929). The principal insect vector of the virus is *Rhopalosiphum* (*Aphis*) *maidis* (Fitch). Two other aphids, *Carolinaia cyperi* (Ainslie) and *Aphis* (*Carolinaia*) *setariae* (Thos.), are reported as vectors (Ingram and Summers, 1936, 1938, Tate

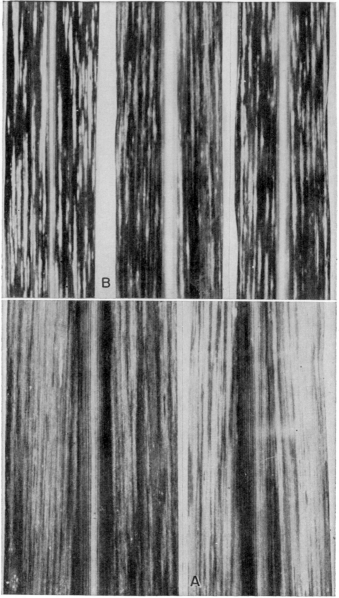

FIG. 52. The green (*A*) and yellow (*B*) mosaics of sugarcane caused by viruses transmitted by aphids and mechanically.   (*Courtesy of C. W. Edgerton.*)

and Vandenberg, 1939). Some physical properties of the virus are listed by Adsuar (1950). Strains of the virus are differentiated by symptoms on different hosts and host range and cross protection (Adsuar, 1954, Liu, 1950, Summers *et al.*, 1948). Costa and Penteado (1951) suggest the use of corn lines to determine virus content of strains. The virus persists in susceptible perennial hosts, sugarcane and related grasses. Aphids frequently increase on susceptible annual hosts, corn and sorghum, to spread the virus to sugarcane. Well-balanced fertility, resistant varieties, and elimination of susceptible hosts are the best control measures (Martyn, 1946). While the continued use of resistant sugarcane lines is dependent upon strains of the virus present, mosaic-resistant selections are the best means of control. Some resistant lines used in mosaic control are P.O.J. 2878, M28, M317, M336, M344, and the Kavangire variety.[1]

Streak Disease of Sugarcane and Corn, Virus Transmitted by Leaf Hoppers. The streak disease in Africa and Asia and possibly Brazil causes damage on susceptible sugarcane and corn varieties. The white leaf stripes are distributed uniformly over the leaves and vary in intensity, size, and length with the strain of the virus and host. The virus of sugarcane streak causes mild transitory symptoms on corn, and the virus from corn incites mild streak on sugarcane (Storey 1925, 1926, 1930, 1933).

The virus is transmitted by *Cicadulina mbila* (Nande), *C. zeae* (China), and *C. storeyi* (China) (Storey 1925, 1926). See corn diseases, Chap. 4. Resistant sugarcane varieties are reported by Storey and McClean (1930).

Chlorotic Streak, Virus Transmitted by Leaf Hopper. Chlorotic streak was first differentiated in Java and Hawaii in 1928 and 1929 and has since been reported as world-wide in distribution on sugarcane and several grasses (Abbott and Ingram, 1942, Abbott and Sass, 1945). The symptoms are distinct and differ from mosaic and streak. The chlorotic streaks are wide, frequently 1 to 3 cm., and continuous. The margins of the white streaks are irregular. Cellular modifications and necrosis of mesophyl occur (Fig. 53). Intracellular bodies are evident in the stalk tissues. The plants frequently are dwarfed when infection occurs early and growing conditions are unfavorable. The vector is *Draeculacephala portola* (Ball), widely distributed on the crop. Many of the better varieties are relatively susceptible. P.O.J. 2725, Co 281, Co 290, Co 421, and Kavangire are resistant in Puerto Rico.[2]

Roguing out diseased plants, balanced fertility with high nitrogen,

[1] Communication from Dr. J. A. B. Nolla, Diseases of Sugarcane in Puerto Rico, 1955.
[2] *Ibid.*

and hot-water treatment (30 minutes at 52 to 53°C.) of seed clones are major control methods (Abbott, 1947, Edgerton *et al.*, 1942).

Fiji Disease, Virus Transmitted by Leaf Hoppers. The disease is restricted to the South Pacific sugarcane areas where it is of major

FIG. 53. Chlorotic streak of sugarcane caused by a virus transmitted by leaf hoppers. (*Courtesy of C. W. Edgerton.*)

importance. The symptoms are short, stiff, erect, dark-green leaves with narrow, linear, smooth, white to brown gall-like raised lesions over the veins on the undersurface of the leaf blade. Internodal elongation of the canes is checked and the canes crack. Lower culm buds develop, forming shortened, distorted clumps. Root development from the lower nodes is characteristic on the shortened culms (Martin, 1947). Intracellular bodies are present in the cells of mesophyl and parenchyma (Kunkel, 1924). The vectors reported are *Perkinsiella vastatrix* (Bred.) and *P. saccharicida* (Kirk); probably other leaf hoppers are capable of transmitting the virus (Mungomery and Bell, 1933, Ocfemia *et al.*, 1933, 1934). Removing diseased plants, restricting ratooning, and use of virus-free cuttings with good management are suggested control methods. Some varieties are moderately resistant, but immune selections are not reported (Martin, 1947).

Other Virus Diseases. Ring mosaic in Java and the United States and probably Puerto Rico are listed (Martin, 1951).

Sereh, Cause Unknown. The sereh complex of Java includes the symptoms of many of the virus diseases discussed and, in addition, the red gum and staining of the vascular tissues and the development of adventitious roots from aerial culm nodes. The axillary buds of infected canes fail to develop after topping, and plants are generally unhealthy. The presence of bacteria in the tissues of diseased plants adds further complexity to the etiology. Undoubtedly in the early reports on the virus diseases of Java the sereh included mosaic (Lyon, 1921, 1923). The disease was controlled in Java for many years by growing seed canes at high altitudes. Resistant varieties and cuttings from healthy canes or treating cuttings 30 minutes in water at 52°C. are suggested as control measures (Wilbrink, 1923).

Ratoon Stunting, Virus Transmitted Mechanically and Carried in Cuttings. The disease was reported first in Australia (Bell, 1932). More recently it has been found in most sugarcane areas of the world.[1]

The symptoms are reduced growth of cuttings and ratoons, yellowing of the foliage, and discoloration and necrosis in the nodal tissues, especially near the culm apex. Plants are stunted severely or show only retarded growth. Large differences in yield between virus-free and virus-infected clones are the best means of establishing the presence of the virus (Hughes and Steindel, 1955). The virus is transmitted by cutting knives and by sap injection into healthy canes. No insect vector is known. Once established in the plant the virus persists indefinitely. Hot-water or hot-air treatment of cuttings appears to be the best means of control. Thirty minutes at 52°C. is approximately the time required to eliminate the virus. Destroying infected plants and disinfecting cutting knives are important in control. Differences in varietal reaction to the virus are evident, but resistant clones are not reported.

2. Gumming Disease, or Cobb's Bacterial Wilt, *Xanthomonas vasculorum* (Cobb) Dows. The disease causes serious loss in susceptible

[1] Communication from C. G. Hughes, Bureau of Sugar Experiment Stations, Queensland, Australia, 1955.

varieties in the Southern Pacific area, in the West Indies, in South America, and in other countries.   Apparently, the further general spread of the disease and introduction into the United States are preventable by quarantines and the use of resistant varieties.

In the early stages of development, the disease is primarily a leaf disease.   In this respect it is similar to the bacterial wilt of corn.   The pale-green to yellow stripes flecked with reddish dots, regular in outline when young and becoming diffused in outline as they become older, form along the margin and apex of the leaf blade.   The longitudinal streaks enlarge and turn red to brown as the leaves mature and tissue necrosis advances.   In the older canes, showing advanced stages of the disease, the inner leaves of the apical whorl develop linear stripes, while the older lower leaves show the red blotches and brown streaks.   As the disease advances in susceptible varieties, the organism in the vascular tissues of the leaves moves down into the stalk tissues.   The advanced leaf symptoms are associated with dwarfing of the plants and necrotic pockets in the stalk tissues.   The presence of the honey-yellow bacterial exudate in the conductive tissues of the stalk and veins constitutes the most characteristic symptom.   Hughes (1939), Matz (1922), North (1935), and Smith (1914) described the symptoms, the organism, and the etiology in detail.

THE BACTERIUM

Xanthomonas vasculorum (Cobb) Dows.
[*Phytomonas vasculorum* (Cobb) Bergey *et al.*]
(*Bacillus vasculorum* Cobb)
(*Bacillus sacchari* Speg.)
[*Bacterium vasculorum* (Cobb) Mig.]
[*Bacterium vasculorum* (Cobb) E. F. Sm.]
[*Pseudomonas vasculorum* (Cobb) E. F. Sm.]

ETIOLOGY.   The disease is confined largely to sugarcane in nature, although Orian (1939) reported it spreading to corn and a species of palm in Mauritius.   Inoculations into sorghums and corn produce symptoms similar to bacterial wilt of corn.   Infected cuttings spread the disease both locally and long distances.   In wet weather the exudate forms a slime on the surface of the infected plants.   The bacteria are spread by splashing rain and flies, according to reports by Leach (1940) and North (1935).   Some other insects, as in the case of Stewart's wilt of corn, may be associated with the spread of the disease.   Resistant varieties, elimination, and quarantine are the best means of control.

3. Leaf Scald, *Bacterium albilineans* Ashby.   The disease is distributed throughout the Southern Pacific sugarcane area, Hawaii, and Brazil. The scald so far is confined to these areas, and it is not so widespread in distribution as the similar gumming disease. Neither disease is reported in the continental United States. The two vascular bacterial diseases are about equally important in causing losses.

The leaf scald symptoms differentiate this disease from the gumming disease. The narrow definitely margined leaf streaks are creamy or grayish white in color. The streaks usually extend the full length of the leaf blade and the leaf sheath and later broaden from the tip downward, followed by withering of the leaf blade in the same direction (Arruda and Amaral, 1945). The streaks are associated with the infected vascular bundles. The older lesions turn brown or redden and finally wither and dry out. The severe infection results in rather sudden wilting and drying out of individual culms or the entire plant. In both cases, weakened shoots develop from the basal axillary buds. These show the characteristic white leaf streak symptoms. The vascular bundles of the stalk show red staining, especially in the nodes. The absence of the yellow exudate in scald also differentiates this disease from the gumming type. Ashby (1929), Bell (1929), and North (1926) discussed the disease and compared the symptoms of the two somewhat similar diseases.

THE BACTERIUM

Bacterium albilineans Ashby
[*Phytomonas albilineans* (Ashby) Bergey *et al.*]

Motile with a single polar flagellum, the rods are more slender than *Xanthomonas vasculorum*, and growth on most media is slow.

The distribution and development of the disease in fields vary from scattered plants with mild symptoms to rapid wilting of most plants in large areas. New infections appear suddenly in fields some distance from diseased areas. This behavior suggests the importance of environmental conditions in the development of scald and the possibility of insects playing an important role in both dissemination and introduction of the parasite into the cane tissues. The disease is controlled by the use of resistant varieties.

4. Bacterial Blights of Sugarcane. Two bacterial blights occur on sugarcane that are similar in symptoms to the two more common bacterial blights on sorghum and Sudan grass. The two diseases are widely distributed.

Red Stripe and Top Rot, Xanthomonas rubrilineans (Lee *et al.*) Starr & Burk. (*Phytomonas rubrilineans* Lee *et al.*). The disease is common through the sugarcane areas including the Southern United States, although it is of minor importance according to Christopher and Edgerton (1930). Cottrell-Dormer (1932) demonstrated that sorghums, corn, Sudan grass, and Johnson grass are infected.

The leaf stripes are at first water-soaked narrow lesions. They elongate rapidly and turn red to maroon in color. Usually the stripe is bordered by a water-soaked or yellow zone. The stripes are continuous with a uniformly necrotic and colored area. Later the stripes coalesce, forming alternate red and chlorotic stripes. Exudate is formed on the necrotic areas. Red staining occurs in the vascular bundles, progressing from the apex downward. Usually only portions of the culm interior are discolored by the bacteria. Later the apical culm tissues are rotted and the upper leaves are killed. Cottrell-Dormer (1932) and Wood (1927) associated the top rot and leaf stripe as the same disease complex. In the various studies in the Pacific area and in the United States, the disease is attributed to *Xanthomonas rubrilineans* (1924, 1925).

Streak or Mottled Stripe, Xanthomonas rubrisubalbicans (Christopher & Edg.) Salvu. (*Phytomonas rubrisubalbicans* Christopher & Edg.). The symptoms of the disease on sugarcane are similar to the streak disease of sorghum. Christopher and Edgerton (1930) compared the disease with the streak on Johnson grass. They produced the disease by inoculations on Johnson grass and sorghum but not on corn. Cottrell-Dormer (1932) reported the same disease in Queensland. The disease is of minor importance.

The leaf stripes are linear with less regular margins and centers. The color is predominantly red, although frequently white areas or white margins occur. Where the streaks coalesce, mottled red and white bands are formed across the leaf blade. Top rot is not associated with the disease. Moderately resistant varieties are reported by Christopher and Edgerton (1930).

5. Pythium Root Rot, *Pythium arrhenomanes* Drechsl., *P. graminicola* Subrm., and Others. The root rot complex is common on sugarcane in practically all areas. The practice of ratooning or growing several crops of cane on the same root system tends to increase the damage in comparison with an annual crop like corn. Soil type, drainage, fertility, and root damage from nematodes and insects influence the incidence and damage from this disease complex. Carpenter (1934) summarized the literature and experimental data on these predisposing factors. The rootlets and roots show a water-soaked brown to gray rot resulting in a depleted root system.

THE FUNGI. Essentially the same species of *Pythium* causing the root rots of corn, sorghums, and other grasses are associated with the sugarcane root rots. Edgerton *et al.* (1929) reported upon numerous isolations in Louisiana; *Pythium, Rhizoctonia,* and *Marasmius* spp. were the principal pathogenic fungi. Rands and Dopp (1934) reported on the variability of *Pythium arrhenomanes* and included nine of the species reported by Sideris (1931) in this species. The morphology of the two common graminicolous species of *Pythium* is given in Chap. 4.

Resistant varieties, soil drainage, and balanced fertility are essential in reducing losses from root rot.

6. Phytophthora rot, *Phytophthora erythroseptica* Pethyb. and Other Species. The soft rot of sugarcane cuttings in cold wet soils reduces stands. The disease is reported in Louisiana in the heavy soils subject to flooding (Steib and Chilton, 1950).

7. Downy Mildew, *Sclerospora sacchari* Miy. and Other Species. The disease on sugarcane caused by *Sclerospora sacchari* is the more common and destructive species on this crop in most of the Pacific area and in India. *S. macrospora*, common on many of the Gramineae, was reported on sugarcane in Peru in 1950 and in the United States in 1952. This is the common pathogen on sugarcane in Australia. Downy mildew is restricted to a few countries in the southwestern Pacific area. Other species of *Sclerospora* occur to a limited extent on this crop (see Table 6 on the Downy Mildew on the Gramineae, Chap. 4). According to Leece (1941), Lyon (1915), and Miyake (1911), conidia and oospores of *S. sacchari* occur on sugarcane. Symptoms on sugarcane are similar to those on corn.

The etiology and control of the disease on sugarcane, a perennial, vary somewhat from that on corn. Infection occurs on young tissues, especially those associated with the buds. Bud infection is probably more general in sugarcane than in corn. Disease-free cuttings and the use of resistant varieties constitute the best means of control (Hughes, 1950). Quarantine apparently has prevented the entry of the disease into the Caribbean area and the Southern United States.

8. Pokkah-bong, *Gibberella fujikuroi* (Saw.) Wr. and var. *subglutinans* Edwards, *Fusarium moniliforme* Sheldon and var. *subglutinans* Wr. & Reinking, or *G. moniliforme* (Sheld.) Snyder & Hansen, and Other Organisms. The disease is common on sugarcane and is widely distributed. Similar symptoms appear occasionally on corn. Damage is not severe except in very susceptible varieties of sugarcane.

The first symptoms of the disease are the light-colored twisted leaves as they come out of the terminal leaf whorl. Chlorotic areas especially on the base of the leaf and leaf sheath persist on plants with mild symptoms. Few to several leaves are affected. The chlorotic areas frequently are malformed, either narrow and twisted or stiff. In later symptoms on susceptible varieties, the top of the stalk is rotted and the growing point is killed. Pink necrotic areas are found in the leaves and culm.

Apparently *Gibberella fujikuroi* (Saw.) Wr. and the variety *subglutinans* Edwards are associated with the disease, as reported by Priode (1929) and others. Fawcett (1922) associated *Erwinia flavida* (G. Fawc.) Magrou (*Bacillus flavidus* G. Fawc.) with the disease (see Edgerton, 1955). The cause of the disease is still indefinite. Resistant varieties apparently are reducing the severity of the disease.

9. Red Rot, *Physalospora tucumanensis* Speg. or *Colletotrichum falcatum* Went. The red rot is prevalent in the major sugarcane areas. The disease apparently is more severe, especially in reducing stands, in the cooler climates. Perithecia were collected on the five species of *Saccharum* and the grass *Leptochloa filiformis* (Lam.) Beauv., common in some sugarcane areas, as reported by Carvajal and Edgerton (1944). It is possible to inoculate the sorghums including Sudan grass and Johnson grass with *Physalospora tucumanensis*, but the *Colletotrichum* found naturally on the sorghums is different, according to Edgerton (1911) and recent reports to the author. Abbott (1938) considered the same species on both groups of plants, based on morphology of the cultures isolated. *C. lineola* Cda., now considered a synonym of *C. graminicolum* (Ces.) G. W. Wils., is common on the sorghums, especially Sudan grass in the temperate zones (Edgerton, 1955).

SYMPTOMS AND EFFECTS. The disease is apparent as a red rot on the leaves, stalks, and stubble of the sugarcane (Fig. 54). The conspicuous red linear lesions are common on the midrib, especially during the latter part of the growing season. The stalk rot is less conspicuous from the exterior. In the split stalk, the longitudinal reddening of the internodal tissues interrupted by occasional white areas extending across the stalk tissue is the typical symptom. Stalk rotting is common in the cane after cutting, according to Edgerton and Carvajal (1944). Rotting of the young shoots on cuttings and reduction in stand are common during periods of cool weather (Steib and Chilton, 1951). Edgerton *et al.* (1937) reported reduction in stands and yields ranging from negligible in resistant to severe in susceptible varieties. Sugar yields are lowered by the stalk rot because of the inversion of sucrose.

THE FUNGUS

Physalospora tucumanensis Speg.
Colletotrichum falcatum Went.   Conidial stage.

Mycelium in culture is white to dark gray with a cottony texture in the light type and dark gray with compact texture in the dark type. Thick-walled vegetative resting cells (chlamydospores) are more abundant in the latter type. The conidia are borne in stromata or singly in culture. On the midrib lesions, they are formed in poorly defined acervuli intermingled with dark setae. Conidia are one-celled, mostly falcate, 17 to 33 microns long (16–48 by 4–8 microns, range reported by Abbott 1938), and hyaline to pinkish in mass. Perithecia are submerged, scattered, and irregular in shape, 100–260 microns wide by 85–250 high, with a small portion of the ostiole protruding. Asci are clavate, thickened at the apex. Ascospores are irregularly biseriate, single-celled, straight or fusoid, elliptical to ovate at maturity, and usually measure 18–22 by 7–8 microns. Paraphyses are abundant, septate, delicate, and extend to the ostiole.

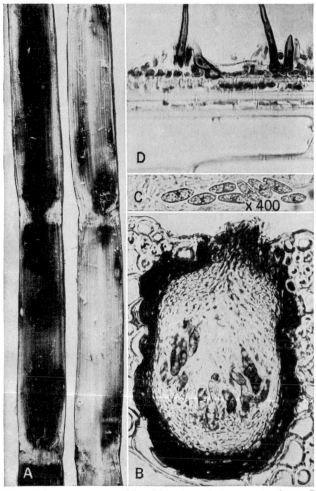

FIG. 54. Red rot of sugarcane caused by *Physalospora tucumanensis*, or *Colletotrichum falcatum* showing the stalk rot (*A*). The perithecium (*B*), ascospores (*C*), and acervulus (*D*) of the fungus are shown, highly magnified. (*Courtesy of C. W. Edgerton.*)

The present evidence indicates two species common on the Gramineae, *Physalospora tucumanensis* or *Colletotrichum falcatum* on the sugarcane and *C. graminicolum* on the small grains, Sudan grass, and many other grasses (see Anthracnose on rye, Chap. 8).

CONTROL. Resistant varieties reduce the loss from red rot. Soil preparation and fertilization assist in producing a strong young plant growth, especially during the cooler season. Edgerton and associates (1942) described a hot-water treatment useful in controlling the infection in cuttings, but apparently predisposing to infection and rotting when seed canes were inoculated after treatment.

   10. Helminthosporium Eye Spot and Brown Stripe, *Helminthosporium sacchari* (Breda de Haan) Butl. and *Cochliobolus stenospilus* (Carpenter) Matsu. & Yam., *H. stenospilum* Drechsl. The eye spot is widely distributed on sugarcane, lemon grass, *Cymbopogon citratus* (DC.) Stapf., and Napier grass, *Pennisetum purpureum* Schum. This is probably the common leaf spot of the group described on sugarcane. The brown stripe caused by *Helminthosporium stenospilum* Drechsl. is apparently less widely distributed and according to McRae (1933) is closely allied with the eye spot in both symptoms and morphology of the fungus, although the symptoms and the reporting of the ascigerous stage of the latter fungus by Carpenter differentiate the two fungi. Priode (1931) described a target blotch on sugarcane in Cuba differing somewhat in symptoms from the other leaf blights. Faris (1928), Lee and associates (1926), McRae (1933), Mitra (1930, 1931), Parris (1942, 1950), and others have reported on the symptomology of the leaf spots and the morphology of the fungi (Edgerton, 1955).

   The initial spots of eye spot are yellow to brown. As the lesions spread, they become oblong to linear, reddish brown to red, and show gray centers when the fungus is sporulating. Streaks or runners extend from the spots toward the leaf tip. The lesions spread rapidly under favorable conditions, causing death of the leaf blade and sheath. Stem cankers occur in Napier grass. The brown stripe apparently differs in less zonation and more restricted widening of the lesions and the absence of streaks or runners.

   THE FUNGI

   Helminthosporium sacchari (Breda de Haan) Butl.
   (*Cercospora sacchari* Breda de Haan)
   (*Helminthosporium ocellum* (Faris)

   The conidia are gray to brown, tapering toward the ends, straight or slightly curved, vary considerably in size, and germinate from the end cells. Specialization occurs within varieties and between species of the suscepts.

   Cochliobolus stenospilus (Carpenter) Matsu. & Yam.
   Helminthosporium stenospilum Drechsl. Conidial stage

   The fungus differs from *Helminthosporium sacchari* somewhat in conidial shape and is apparently restricted to sugarcane. The ascigerous stage with the coiled ascospores is similar to the other species of this genus (Paris, 1950).

Resistant varieties or hybrid selections offer the best means of control of this disease complex.

11. Brown Spot, *Cercospora longipes* Butl. The disease is world-wide in distribution and reduces yield and sugar content when severe. The leaf spots are first narrow and oval, later enlarging to elongate-oval brown rings with straw-colored centers. On susceptible varieties the leaf blade is killed by the numerous spots and appears grayish brown. Leaf stripes and sheath spots are described as symptoms for other species.

THE FUNGI

Cercospora longipes Butl.
Cercospora vaginae Krüger
Cercospora kopkei Krüger
Cercospora imperatae Sydow
Cercospora taiwanensis Mat. & Yam. or *Leptosphaeria taiwanensis* Yen. & Chu.

Several species differing chiefly in length and color of conidia and symptoms are described on sugarcane (Martin, 1951). Resistant varieties are reported in most areas where the disease is severe. In Puerto Rico P.O.J. 2725, Kavangire, M28, M275, M317, P.O.J. 36, M336, M338, Co 421 and S.C. 12/4 are resistant based on inoculations with the isolates of the pathogen occurring there (Nolla, 1955, see footnote, page 209).

12. Leaf Scorch, *Stagonospora sacchari* Lo & Ling. The disease is severe, especially on many of the improved varieties developed in the sugarcane breeding in various countries. It is restricted to Formosa (Taiwan) at present and represents a potential threat to sugarcane production in other areas. *Miscanthus* spp. are a common grass host of the pathogen. The fungus was collected on sugarcane in Taiwan in 1909, but was not considered of economic importance until the recent introduction and use of improved sugarcane varieties (Lo, Chu, and Chin, 1953, Lo and Dickson, 1956).

The lesions are narrow light-brown stripes that spread rapidly both in length and width to kill the entire leaf blade. Pycnidia develop in the mesophyll first on the initial lesions and later scattered over the necrotic tissue.

THE FUNGUS

Stagonospora sacchari Lo & Ling.

Pycnidia are globose to flattened, dark golden brown, ostiolate, 76–201 by 72–193 microns. Conidia are fusiform, 3-septate with restrictions at the septa, 24–58 by 8–13 microns.

Some varieties are moderately resistant as expressed by restricted narrow stripes and retarded necrosis.

13. Smut, *Ustilago scitaminea* Syd. The sugarcane smut is present in the Pacific area, Africa, Asia, and South America. The smutted plants are stunted and develop a whip-like smut sorus from the apex of the cane. The elongate sorus consisting of tissues of the apical bud or the inflorescence and the black spore mass is the typical symptom. The slender canes of the *Saccharum sinense* or slender Asiatic type and *S. spontaneum* are the most susceptible. The disease causes severe damage to sugarcane in the Tucuman district of Argentina, Uruguay, India, and South Africa (Chona and Gattani, 1950, Edgerton, 1955, Hirschhorn, 1949, Subramanian and Lakshmipati, 1954).

THE FUNGUS

Ustilago scitaminea Syd.
(*Ustilago sacchari* Rab.)

The morphology of the fungus and the description of similar species on *Erianthus* sp. are given by Hirschhorn (1941). Chlamydospores are olivaceous to brown, globose or irregular, 4 to 9 microns in diameter, smooth or with fine papilla on the surface. Many hyaline or brown single or joined sterile cells are present with the spores.

The pathogen infects terminal buds of the young shoots, develops systemically in the culm tissues, and forms spores in the inflorescences. The mycelium persists in the dormant bud tissue, and the fungus is spread with cuttings from infected plants. Eradication of infected plants, prevention of spore distribution, and resistant varieties offer the best means of control. Many of the improved varieties, especially from the United States, are susceptible (Mattos, 1949).

Sphacelotheca cruenta (Kuehn) Potter, the loose kernel smut of sorghum, is reported on sugarcane in India (Mundkur and Thirumalachar, 1952).

14. Rust, *Puccinia kuehnii* (Krueger) Butl. The sugarcane rust is distributed widely on sugarcane and related grasses. The disease is severe only on susceptible varieties such as Co 475 (Butler, 1918, Chona and Munjal, 1950, Martin, 1951, Patel *et al.*, 1950).

THE FUNGUS

Puccinia kuehnii (Krueger) Butler
(*Uromyces kuehnii* Krueger)

Frequently only uredia occur on sugarcane. Uredia are slightly elongate, orange brown, open frequently on both surfaces of the leaf blade. Paraphyses are present in the margin of uredia. Urediospores are ovate, orange, variable in size 48 by 27 microns, wall with short spines. Teliospores form in uredia or separate sori; they are oblong to pyriform, flattened or rounded at apex, smooth, and variable in size and formation of the septum. Pedicels are short on most hosts. Aecial host is unknown.

Races occur on the several hosts.   The species probably belongs to the group associated with corn, sorghum, etc., although the author has been unable to secure teliospore germination to date.   Patel *et al.* suggests *P. sacchari* as a binomial for the pathogen on sugarcane on the basis of very slight differences in morphology.

REFERENCES

ABBOTT, E. V.   Red rot of sugarcane.   *U.S. Dept. Agr. Tech. Bul.* 641.   1938.

———.   Influence of certain environmental conditions on chlorotic streak of sugarcane.   *Phytopath.* **37**:162–173.   1947.

——— and J. W. INGRAM.   Transmission of chlorotic streak of sugar cane by the leaf hopper *Draeculacephala portola*.   *Phytopath.* **32**:99–100.   1942.

——— and J. E. SASS.   Pathological histology of sugarcane affected with chlorotic streak.   *Jour. Agr. Research* **70**:201–207.   1945.

ADSUAR, J.   On the physical properties of sugar-cane mosaic virus.   *Phytopath.* **40**:214–216.   1950.

———.   Further studies on the mosaic of sugarcane variety Mayaguez 336.   *Jour. Agr. Univ. Puerto Rico* **38**:16–21.   1954.

ARRUDA, S. C., and J. F. Do AMARAL.   Leaf scald of sugar cane in Brazil.   *Phytopath.* **35**:135–137.   1945.

ASHBY, S. F.   Gumming disease of sugar cane.   *Trinidad Trop. Agr.* **6**:135–138.   1929.

BELL, A. F.   A key to the field identification of sugar cane diseases.   *Queensland Bur. Sugar Exp. Sta. Div. Path. Bul.* 2.   1929.

———.   Dwarf disease of sugar cane.   *Queensland Bur. Sugar Exp. Sta. Div. Path. Bul.* 3.   1932.

BRANDES, E. W.   The mosaic disease of sugar cane and other grasses.   *U.S. Dept. Agr. Bul.* 829.   1919.

——— and P. J. KLAPHAAK.   Cultivated and wild hosts of sugarcane or grass mosaic.   *Jour. Agr. Research* **24**:247–262.   1923.

——— and G. B. SARTORIS.   Sugarcane: Its origin and improvement.   *U.S. Dept. Agr. Yearbook*, pp. 561–623.   1936.

BUTLER, E. J.   Fungi and disease in plants.   Thacker, Spink and Co.   Calcutta and Simla.   1918.

CARPENTER, C. W.   Predisposing factors in Pythium root rot.   *Hawaiian Planters' Record* **38**:279–338.   1934.

CARVAJAL, F., and C. W. EDGERTON.   The perfect stage of *Colletotrichum falcatum*.   *Phytopath.* **34**:206–213.   1944.

CHONA, B. L., and M. L. GATTANI.   Kans grass (*Saccharum spontaneum*) a collateral host of sugarcane smut in India.   *Indian Jour. Agr. Sci.* **20**:359–362.   1950.

——— and R. L. MUNJAL.   *Puccinia kuehnii* (Krueg.) Butler on sugarcane in India.   *Curr. Sci.* **19**:151–152.   1950.

CHRISTOPHER, W. N., and C. W. EDGERTON.   Bacterial stripe diseases of sugarcane in Louisiana.   *Jour. Agr. Research* **41**:259–267.   1930.

COSTA, A. S., and M. P. PENTEADO.   Corn seedlings as test plants for the sugar-cane mosaic virus.   *Pytopath.* **41**:758–763.   1951.

COTTRELL-DORMER, W.   Red-stripe disease of sugar cane in Queensland.   *Queensland Bur. Sugar Exp. Sta. Div. Path. Bul.* 3.   1932.

EDGERTON, C. W.   The red rot of sugar cane: a report of progress.   *La. Agr. Exp. Sta. Bul.* 133.   1911.

———. Forty-two years of sugarcane disease research in Louisiana Agricultural Experiment Station. *La. Agr. Exp. Sta. Bul.* 448. 1950.

———. Sugarcane and its diseases. Louisiana Univ. Press. Baton Rouge, La. 1955.

——— *et al.* Relation of species of Pythium to the root-rot disease of sugar cane. *Phytopath.* **19**:549–564. 1929.

———. Investigations on sugar cane diseases in Louisiana in 1936–1937. *La. Agr. Exp. Sta. Bul.* 288. 1937.

——— *et al.* The hot water treatment of sugar cane. *La. Agr. Exp. Sta. Bul.* 336. 1942.

——— and F. CARVAJAL. Host-parasite relations in red rot of sugar cane. *Phytopath.* **34**:827–837. 1944.

FARIS, J. A. Three Helminthosporium diseases of sugar cane. *Phytopath.* **18**:753–774. 1928.

FAWCETT, G. L. Enfermedades de la caña de azúcar en Tucuman. I. "El polville" o podredumbre del brote terminal. *Rev. Indust. y Agr. de Tucuman* **13**:5–15. 1922.

FORBES, I. L., P. J. MILLS, and C. W. EDGERTON. Investigations on sugar cane diseases in Louisiana in 1936–1937. III. Immunity studies with sugar cane mosaic. *La. Agr. Exp. Sta. Bul.* 288. 1937.

HIRSCHHORN, E. Una especie de "Ustilago" nueva para la Argentina *Ustilago scitaminea. Revista Argentina de Agronomia* **8**:326–330. 1941.

———. Un nuevo método de infección artificial con el carbón de la caña de azúcar. *Rev. Invest. Agr. Buenos Aires* **3**:335–344. 1949.

HUGHES, C. G. Alternate hosts of *B. vasculorum,* the causal agent of gumming disease of sugar cane. *Queensland Bur. Sugar Exp. Sta. Tech. Com.* 3. 1939.

———. Downy mildew disease in North Queensland. *Proc. Conf. Queensland Soc. Sugarcane Tech.*, pp. 237–242. 1950.

——— and D. R. L. STEINDL. Ratoon stunting disease of sugar cane. *Queensland Bur. Sugar Exp. Sta. Tech. Com.* 2. 1955.

INGRAM, J. W., and E. M. SUMMERS. Transmission of sugarcane mosaic by the rusty plum aphid, *Hysteroneura setariae. Jour. Agr. Research* **52**:879–887. 1936.

——— and ———. Transmission of sugarcane mosaic by the green bug (*Toxoptera graminum* Rond.). *Jour. Agr. Research* **56**:537–540. 1938.

KUNKEL, L. O. Histological and cytological studies on the fiji disease of sugar cane. *Bul. Exp. Sta. Hawaiian Sugar Planters' Assn. Bot. Ser.* **3**:99–107. 1924.

———. Studies on the mosaic of sugarcane. *Bul. Exp. Sta. Hawaiian Sugar Planters' Assn. Bot. Ser.* **3**:115–167. 1924.

LEACH, J. G. Insect transmission of plant diseases. McGraw-Hill Book Company, Inc. New York. 1940.

LEE, H. A., *et al.* Red-stripe disease studies. *Hawaiian Sugar Planters' Assn. Exp. Sta. Bul.* 99. 1925.

——— *et al.* The history and distribution of eye spot. *Hawaiian Planters' Record* **30**:466–492. 1926. (A series of seven papers on the disease.)

——— and W. C. JENNINGS. Bacterial red stripe disease of tip canes. *Hawaiian Sugar Planters' Assn. Exp. Sta. Cir.* 42. 1924.

LEECE, C. W. Downy-mildew disease of sugar cane and other grasses. *Queensland Bur. Sugar Exp. Sta. Tech. Com.* **5**:111–135. 1941.

LIU, S-P. Studies on the sugarcane mosaic virus in Taiwan. I. Strains of the virus. *Rept. Taiwan Sugar Exp. Sta.* **5**:72–98. 1950.

LO, T. C., H. T. CHU, and R. J. CHIU. A comparative study of the fungi of Stagonospora causing leaf-scorch disease on sugarcane and *Miscanthus* spp. *Rept. Taiwan Sugar Exp. Sta.* **10**:105–112. 1953.

———— and J. G. DICKSON. Sporulation and variation in *Stagonospora sacchari*, the incitant of scorch disease of sugarcane. *Phytopath.* manuscript. 1956.

LYON, H. L. The Australian leaf stripe disease of sugar cane. *Hawaiian Planters' Record* **4**:257–265. 1915.

————. Three major cane diseases: Mosaic, sereh, and fiji disease. *Bul. Exp. Sta. Hawaiian Sugar Planters' Assn. Bot. Ser.* **3**:1–43. 1921.

————. A simple cure for sereh. *Hawaiian Planters' Record* **27**:352–356. 1923.

MARTIN, J. P. Sugar cane diseases in Hawaii. Hawaiian Sugar Planters' Association. Honolulu, T.H. 1938.

————. Fiji disease of sugarcane. *Hawaiian Planters' Record* **51**:103–118, 119–136. 1947.

————. Sugarcane diseases and their world distribution. *Proc. Cong. Int. Soc. Sugar Cane Tech.* (1950), 435–452. 1951.

MARTYN, E. B. Sugarcane mosaic in Jamaica. *Trop. Agr. Trinidad* **23**:123–129. 1946.

MATSUMOTO, T. An annotated list of sugarcane disease in Formosa and remarks on the taxonomy of the causal organisms. II. *Mem. Fac. Agr. Taiwan* **2**:1–50. 1950.

MATZ, J. Gumming disease of sugar cane. *Jour. Dept. Agr. Porto Rico* **6**:5–21. 1922.

————. Artificial transmission of sugar cane mosaic. *Jour. Agr. Research* **46**:821–839. 1933.

McMARTIN, A. Sugarcane mosaic disease and maize growing. *So. African Sugar Jour.* **31**:35, 81–82. 1947.

McRAE, W. Report of the imperial mycologist. *Sci. Rept. Imp. Inst. Agr. Res. Pusa* (1931–1932), pp. 129–130. 1933.

MITRA, M. A comparative study of species and strains of Helminthosporium on certain Indian cultivated crops. *Trans. British Myc. Soc.* **15**:254–293. 1930.

————. Saltation in the genus Helminthosporium. *Trans. British Myc. Soc.* **16**:115–127. 1931.

MIYAKE, T. On a fungus disease of sugar cane caused by new parasitic fungus, *Sclerospora sacchari* T. Miy. *Rept. Sugar Exp. Sta. Govt. Formosa Div. Path. Bul.* 1. 1911. (In Japanese.)

MUNDKUR, B. B., and M. J. THIRUMALACHAR. Ustilaginales of India. Commonwealth Mycological Institute. Key, Surrey. 1952.

MUNGOMERY, R. W., and A. F. BELL. Fiji disease of sugar cane and its transmission. *Queensland Bur. Sugar. Exp. Sta. Div. Path. Bul.* 4. 1933.

NORTH, D. S. Leaf-scald, a bacterial disease of sugar cane. *Colonial Sugar Refining Co. Sydney Agr. Tech. Rept.* 8. 1926.

————. The gumming disease of sugar cane. *Colonial Sugar Refining Co. Sydney Agr. Tech. Rept.* 10. 1935.

OCFEMIA, G. O. An insect vector of the Fiji disease of sugar cane. *Am. Jour. Bot.* **21**:113–120. 1934.

———— *et al.* Distribution of mosaic and Fiji diseases in sugar cane stalks. *Philippine Agr.* **21**:385–407. 1933.

ORIAN, G. Natural hosts of *Bacterium vascularium* (Cobb) E. F. Smith in Mauritius. *Proc. 6th Cong. Int. Soc. Sugar Cane Tech.*, pp. 437–447. 1939.

————. Artificial hosts of the sugar cane leaf scald organism. *Rev. de l'Ile Maurice* **21**:285–304. 1942.

PARRIS, G. K. Eye-spot of Napier grass in Hawaii caused by *Helminthosporium sacchari*. *Phytopath.* **32**:46–63. 1942.

———. The Helminthosporia that attack sugarcane. *Phytopath.* **40**:90–103. 1950.

PATEL, M. K., M. N. KAMAT, and Y. A. PADHYE. A new record of Puccinia on sugarcane in Bombay. *Curr. Sci.* **19**:121–122. 1950.

PRIODE, C. N. Pokkah-bong and twisted top disease of sugar cane in Cuba. *Phytopath.* **19**:343–366. 1929.

———. Target blotch of sugar cane. *Phytopath.* **21**:41–58. 1931.

RANDS, R. D., and E. DOPP. Variability in *Pythium arrhenomanes* in relation to root rot of sugar cane and corn. *Jour. Agr. Research* **49**:189–221. 1934.

SEIN, F., JR. A new mechanical method for artificially transmitting sugar cane mosaic. *Jour. Dept. Agr. Porto Rico* **14**:49–68. 1930.

SIDERIS, C. P. Taxonomic studies in the family Pythiaceae. I. Nematosporangium. *Mycologia* **23**:252–295. 1931.

SMITH, E. F. Bacteria in relation to plant disease. *Carnegie Inst. Wash., D.C., Pub.* **3**:1–71. 1914.

STAHL, C. F. Corn stripe disease in Cuba not identical with sugar cane mosaic. *Trop. Plant Research Found. Bul.* 7. 1927.

STEIB, R. J., and S. J. P. CHILTON. Phytophthora rot of sugarcane seed pieces in Louisiana. *Sugar Bul.* **29**:77–78. 1950.

—— and ———. Infection of sugarcane stalks by the red-rot fungus, *Physalospora tucumanensis* Speg. *Phytopath.* **41**:522–528. 1951.

STOREY, H. H. Streak disease of sugar cane. *Union So. Africa Dept. Agr. Sci. Bul.* 39. 1925.

———. The transmission of streak disease of maize by the leaf hopper, *Balclutha mbila* Naude. *Ann. Appl. Biol.* **12**:422–439. 1925.

———. Interspecific cross-transmission of plant virus diseases. *So. African Jour. Sci.* **23**:305–306. 1926.

———. Investigations of the mechanism of the transmission of plant viruses by insect vectors. I. *Proc. Roy. Soc. London.* B **113**:463–485. 1933.

SUBRAMANIAN, T. V., and V. L. RAO. Infection and development of *Ustilago scitaminea* Syd. in sugarcane. *Proc. Conf. Sugarcane Res. Wkrs. India* II-3, pp. 55–63. 1954.

SUMMERS, E. M., E. W. BRANDES, and R. D. RANDS. Mosaic of sugarcane in the United States, with special reference to strains of the virus. *U.S. Dept. Agr. Tech. Bul.* 955, pp. 1–124. 1948.

TATE, H. D., and S. R. VANDENBERG. Transmission of sugarcane mosaic by aphids. *Jour. Agr. Research* **59**:73–79. 1939.

TIMS, E. C., P. J. MILLS, and C. W. EDGERTON. Studies on sugarcane mosaic in Louisiana. *La. Agr. Exp. Sta. Bul.* 263. 1935.

WAKKER, I. H., and F. A. F. C. WENT. De Ziekten van het Suikerriet op Java. I. Ziekten die niet door dieren veroorzakt worden. E. J. Brill. Leiden. 1898.

WILBRINK, G. Warmwaterbehandeling van Stekken als Geneesmiddel Tegen de Serehziekte van het Suikerriet. *Arch. Suikerind. Nederlandsche-Indië Meded. het Proefsta. Java-Suikerind.* **31**:1–15. 1923.

———. Mechanical transmission of sugar cane mosaic. *Proc. 3d Cong. Int. Soc. Sugar Cane Tech.*, pp. 155–161. 1929.

WILSON, G. W. The identity of the anthracnose of grasses in the United States. *Phytopath.* **4**:106–112. 1914.

WOOD, E. J. F. Top rot in sugar cane. *Queensland Agr. Jour.* **28**:208–211. 1927.

CHAPTER 11

WHEAT DISEASES

The cultivated wheats of the world comprise several species. However, only three species are of major importance commercially, *i.e.*, common wheat, *Triticum aestivum* L. or *T. vulgare* Vill., and club wheat, *T. compactum* Host., both hexaploids, and the durum wheat, *T. durum* Desf., a tetraploid. Varieties of the common wheat represent the major acreages in most countries. In the genus *Triticum*, as in the other small grains, the basic chromosome number is seven pairs. The genus represents a polyploid series with the important cultivated species in the 14 and 21 chromosome pair groups. The following table includes the *Triticum* spp. divided into the three chromosome groups.

Diploid Series, 7 Chromosome Pairs	Tetraploid Series, 14 Chromosome Pairs	Hexaploid Series, 21 Chromosome Pairs
T. aegilopoides Bal.	*T. dicoccoides* Korn.	*T. spelta* L.
T. monococcum L.	*T. dicoccum* Schubl.	*T. vulgare* Vill. or
T. thaoudar Reu.	*T. durum* Desf.	*T. aestivum* L.
	T. turgidum L.	*T. compactum* Host.
	T. persicum Vav.	*T. sphaerococcum* Perc.
	T. orientale Perc.	*T. macha* Decap.
	T. pyramidale Perc.	*T. vavilovi* Jackub.
	T. timopheevi Zhuk.	

The species highly resistant to the major wheat diseases are found in the diploid and tetraploid series. Vavilov (1914) attempted to show that the *Triticum monococcum* wheats were the most resistant to the rusts and powdery mildew, the *T. durum* groups somewhat less resistant, and the *T. vulgare* group most susceptible. The emmer group (*T. dicoccum*) is resistant to many races of the fungi causing leaf rust, stem rust, powdery mildew, and bunt. *T. timopheevi*, an emmer-like wheat, is resistant to races of these same fungi and apparently possesses a genom not present in other tetraploid wheats or as yet detected in the hexaploid species. Interspecific crosses are made to transfer these resistance factors to the *T. vulgare* wheats. The earlier literature was summarized by Clark (1936). Ausemus (1943), Hayes *et al.* (1955), McFadden (1949), Sears and McFadden (1946), and Roemer *et al.* (1938) have discussed the resistance of wheat to the various important parasites.

During the early years, breeding for resistance to some of the major diseases was difficult. The common wheats were not resistant to a sufficient number of physiological races, and only resistant durum and emmer parental material was available. The hybrids between these latter resistant wheats and the susceptible common wheats resulted in high sterility and linkage of undesirable characters. McFadden (1930), in the development of Hope and H-44, accomplished the transfer of many of the factors for disease resistance into a hard red spring wheat from the cross Yaroslav emmer (*T. dicoccum*) × Marquis (*T. vulgare*). Hope and H-44 are used widely in producing disease-resistant wheats. The cross *T. timopheevi* × *T. vulgare* is difficult to use, owing to high sterility and failure in transferring the disease-resistance factors to the common wheats. The direct cross has been made by several breeders, but elimination of undesirable characters and the maintenance of all the factors for disease resistance have not been accomplished to date (Shands, 1941). Kostoff (1936, 1937) reported the production of an amphidiploid (*T. timococcum*) with the major disease-resistance factors of *T. timopheevi* and *T. monococcum* and with 21 chromosome pairs. The cytology and genetics of interspecific hybrids in the genus *Triticum* and intergeneric hybrids with *Aegilops, Haynaldia, Secale*, and *Agropyron* spp. have been studied extensively (Aase, 1930, Kihara, 1937, Percival, 1921, Sando, 1935, Thompson, 1934, and Watkins, 1930).

The wheats are grouped into three classes based on the character of growth: winter, intermediate, and spring. The winter and spring varieties comprise the major economic groups (Bayles and Clark, 1954, Clark and Bayles, 1935). The winter wheats, while more hardy than the other cereals excepting rye, are limited in their tolerance of winter conditions. The winter wheats are grown in North America south of approximately 45° latitude. Spring wheat culture starts about 42° latitude and extends northward to the northern cultivated areas of Canada. The intermediate types occur in localized areas in the Southern United States and Mexico. World distribution is on a somewhat similar basis, although more intermediate types are grown in the wheat areas of the Southern Hemisphere. The wheats of the world are grouped commercially on the basis of their general usage.

The wheats are recognized in commerce under five general classes. The official wheat standards established in the United States and operating in the world commerce under the "North American contract" are recognized as general wheat classes throughout most of the world. The five commercial classes in the United States are as follows: (1) hard red spring wheat grown in the North Central states and Canada; (2) durum wheat grown in limited areas in the drier sections of the same area as hard red spring wheat; (3) hard red winter wheat grown principally in the

South Central states; (4) soft red winter wheats grown in the more humid Eastern states; and (5) white wheats, chiefly intermediate types, grown in the Western states. The distribution of these general commercial classes throughout the world falls into similar ecological areas, but it is modified considerably by commercial demands, as, for example, the extensive durum wheat area in the Mediterranean region, the white wheats in Chile and Peru, etc.

Wheat is grown under a wide range of environmental conditions, principally for grain. The crop extends into the drier agricultural areas of the world as well as the humid sections (Carlton, 1920, Klages, 1942). Winter hardiness, drought resistance, and quality in addition to disease resistance are important in the breeding of wheats.

Diseases cause large losses in yield and quality. In the United States and Canada, reductions in yield and quality from stem rust, especially in 1904 and 1916, and again in 1935 and 1937, brought a demand for resistant varieties. The loss in North Dakota alone in 1935 was estimated at 100 million dollars. The aggregate loss in the United States and Canada caused by these stem rust epiphytotics was in the millions of dollars. Estimated average annual losses in the United States due to all wheat diseases for the 10-year period, 1930 to 1939, amounted to 10.5 per cent of the crop, or over 80 million bushels annually (Plant Disease Survey). Disease-resistant varieties are reducing these losses, stabilizing annual production, and materially reducing the cost of production.

1. Frost Injury and Winter Killing. Frost injury occurs in both winter and spring wheat. The damage usually is localized, although occasionally seasonal conditions develop that cause damage over extensive areas. The damage occurs more frequently in the northern sections of the wheat belts or at high altitudes, and it is usually during the spring growing period and again before the wheat is mature. Leaf, crown, and young crown roots are injured from spring frosts. The entire leaf blade, or bands of tissue are frosted, resulting in loss of chlorophyll and necrosis. The frosted crown buds and young roots are first water-soaked, and later they show necrosis and browning. Frequently these frosted areas furnish avenues of entrance for parasitic fungi, especially of the crown rot type, such as *Helminthosporium sativum* Pamm., King, & Bakke. Frost injury during the late summer and autumn is manifest by the shriveling and the green color of the pericarp. This type of injury affects not only the yield of grain, but also the quality of the crop. Fungi, especially *Alternaria* and *Penicillium* spp., invade the frosted tissues to cause further damage. Frosted grain is graded as damaged, and it is discounted on the markets. Frosted kernels are lower in germination and seedling vigor. Geddes *et al.* (1932), Johnson and Whitcomb (1927), Newton and McCalla (1934, 1935), and others have discussed the types of damage and the effect on composition and baking quality.

Winter killing consists of two kinds of injury, each associated with different environmental complexes. (1) The freezing and desiccation of the seedling tissues during periods of low temperature and low relative humidity in the absence of snow covering is the more common type of winter wheat injury. (2) The depletion of reserves by high physiological activity of the dormant plant tissues and the invasion of the weakened tissues by fungi occur in periods of wet, cloudy warm weather or under heavy snow covering when the temperature remains above freezing for long periods. Wheat varieties respond differently to these two types of winter conditions, as shown by Akerman (1927), Salmon (1933), Tumanov (1931), and others. Winter wheats developed for winter survival under conditions of very low temperatures, dry air, and limited snow covering are generally very susceptible to winter injury of the second type when the temperatures are around freezing and the snow covering is heavy for long periods. The relation of storage reserves, bound water, and other physiological and chemical factors to cell killing, tissue rupturing, and tissue desiccation are discussed by Harvey (1930), Levitt (1941), Maximov (1929), Newton and associates (1924, 1931), and others.

Winter killing is due to a complex of factors that vary in different regions. The prevention of winter killing is associated with proper soil conditions and adapted varieties. The development of winter-resistant varieties is the best means of controlling this loss, as discussed by Clark (1936), Quisenberry (1931), Quisenberry and Clark (1929), Salmon (1933), and others.

2. Gray Speck, Nonparasitic. The gray speck or yellow-gray leaf spotting of wheat due to soil mineral deficiencies in alkaline organic soils is common in both winter and spring wheats. The disease is manifest frequently on wheat and oats in the United States, although it is of minor importance (Fig. 28). The symptoms and control are discussed in Chap. 6.

A leaf necrosis with similar symptoms was described by Straib (1935) as due to gene mutations in the Kolben wheat lines and in a number of the tetraploid wheats. Environment influenced the expression and severity of the necrosis.

3. Virus Diseases. Several virus diseases, chiefly of the mosaic type, are distributed widely on wheat and other cereals. The mosaic-rosette, incited by a soil-borne virus, was reported first, followed by the wheat streak mosaic and yellow dwarf. Several other cereal and grass virus diseases of the mosaic group occur on wheat. A strain of the corn streak virus, a yellows type, transmitted by leaf hoppers, occurs in Asia and

Africa. The wheat striate virus, similar to a wheat virus occurring on wheat in Europe and Asia, occurs in Central North America. The detailed information on many of these virus diseases is still meager. Fortunately, a relatively large number of wheat varieties are resistant to one or more of the virus diseases.

Mosaic-rosette, Frequently Called Eastern Wheat Mosaic Virus, Soil-borne and Transmitted Mechanically. The disease was observed first in the United States in 1919 in southern Illinois (Koehler *et al.*, 1952, McKinney, 1923) and later found in a number of locations in the eastern soft red winter wheat area of the United States (McKinney, 1937). A similar soil-borne mosaic occurs in the eastern portion of the hard red winter wheat area and probably represents a strain of the same virus. A similar virus of wheat occurs in Japan, Egypt, and probably other areas (Ikata and Kawai, 1937, Melchers, 1931, Wada and Fukano, 1937). The symptoms of the disease vary with wheat variety and strain of the virus. The symptoms of yellow mosaic are light-yellowish-green mottling and striping with dwarfing and excessive tillering in some varieties. The green mosaic is distinguished by the dark-bluish-green color with indistinct white mottling and striping and rosetting in susceptible varieties. Intracellular bodies are present in the cells of the mosaic tissue (McKinney *et al.*, 1923, Wada and Fukano, 1934, 1937). The disease is similar in etiology to the soil-borne oat mosaic (Chap. 6).

The method of transmission of the virus in nature is uncertain. Mechanical transfer is possible, through leaf or crown tissues. Transmission through infested soil occurs even when soil and plants are maintained in insect-proof cages. Soil, treated with heat or chemical disinfectants, does not induce the disease. Nematodes and other soil microflora investigated give negative results, although mites and other small soil-inhabiting insects are possible vectors (McKinney 1923, 1930, Webb, 1927, 1928). The environmental conditions are important in the expression of the disease; low temperatures and short days favor development. Control of the soil-borne mosaic in the United States is accomplished by the use of resistant varieties (McKinney *et al.*, 1925, Webb *et al.*, 1923). Most wheat varieties derived from Turkey are resistant to moderately resistant. The disease develops aggressively only on certain soil types and in areas of moderate winters.

Wheat Streak Mosaic, Virus Transmitted Mechanically and by Eriophyd Mites. The disease is distributed extensively through the western hard red winter and spring areas of the United States and Canada. The host range of the virus is wide, including many weed grasses (Slykhuis, 1952). Losses are severe in seasons favorable for the development of the disease. The symptoms frequently are similar to those of the soil-borne mosaic. Greenish, yellow, narrow linear, intermittent stripes and streaks located between the vascular bundles of the leaves are characteristic. Entire leaves frequently show chlorosis followed by necrosis. Plants frequently are stunted, and size of head and yield of grain are reduced. The absence of leaf mottle and excessive rosetting differentiates this mosaic from the mosaic-rosette type. The disease symptoms are expressed at high temperatures rather than at low as in the mosaic-rosette (McKinney, 1937, Sill, 1953, Slykhuis, 1952, 1953).

The virus is transmitted readily mechanically and by *Aceria (Acarina) Tulipae* (Keifer) (Slykhuis, 1952, 1953, 1955). Further investigations with this group of vectors common on the cereals and grasses may relate them to the transmission of other viruses. Heavy populations of mites on wheat cause chlorosis and stunting in addition to streak mosaic symptoms. The commercial wheats are susceptible to streak mosaic although they vary in tolerance. Some Agropyron-Triticum and rye-wheat crosses are resistant (Sill, 1953). Late sowing of winter wheat and early sowing of spring wheat and plowing under volunteer wheat and grass hosts reduce infection.

Yellow Dwarf, Virus Transmitted by Aphids. The disease is generally prevalent on barley and oats and causes some damage on wheat. See Chaps. 3 and 6.

Stripe Mosaic, Transmitted Mechanically and Seed-borne. See Chap. 3.

Several other grass mosaics have been transmitted to wheat although little is known regarding their natural occurrence on wheat.

Striate, Virus Transmitted by Leaf Hopper. The striate virus (striate mosaic, Slykhuis, 1953) represents a yellows disease based on classification by vector. Apparently the disease or similar diseases are distributed widely although somewhat localized in occurrence. The fine, intermittent stripes or streaks appear along the veins. They are first chlorotic, followed by necrosis and browning in some varieties. Plants are stunted, and yields are reduced.

The virus is transmitted by the leaf hopper *Endria inimica* (Say) with an incubation period in the vector. *Deltocephalus striatus* (Linn.) is a vector in Europe. The virus persists over winter in winter wheat.

Corn Streak, Virus Transmitted by Leaf Hoppers. See Chap. 4.

Strain A of this virus incites fine, linear, chlorotic streaks and shortened, curled leaves; excessive tillering, shortened culms and spikes; and sterility in wheat, oats, and barley. The disease is reported in Africa and India, where wheat is grown in association with corn. *Cicadulina* spp. migrating from corn are the vectors. Many of the wheat varieties are resistant. Marquis (C.I. 3641), Renown (C.I. 11709), Reward (C.I. 8182), Regent (C.I. 11869) are highly resistant (Gorter, 1947). Delaying seeding date in wheat helps control the disease.

4. Black Chaff, *Xanthomonas translucens f. sp. undulosa* (E. F. Sm., L. R. Jones, & Reddy) Hagb. The disease is distributed widely on wheat. Damage is usually very light, as the disease rarely occurs in epiphytotics over extensive areas. Black chaff is frequently a complex of diseases when it occurs in association with pseudo-black chaff and brown necrosis on Hope and H-44 and derivatives from hybrids with these wheats. Hart and Zaleski (1935), McFadden (1939), and others have shown that the brown necrosis or melanistic reaction of these latter wheats and hybrid selections from these wheats carrying the mature plant factor for stem rust resistance is in some instances a type of reaction to stem rust. Melanism is induced in high temperature and humidity in the Hope and H-44 wheats, as shown by Broadfoot and Robertson (1933), Hagborg (1936), and Johnson and Hagborg (1944). Apparently this type of necrosis as well as black chaff was considered by Goulden and Neatby (1929), Hayes *et al.* (1934), and perhaps some of the reports from other countries, as the Hope and H-44 wheats are used widely in breeding for resistance to stem rust.

The bacterial black chaff occurs on the floral bracts, culms, and leaves of wheat, rye, and some grasses (Hagborg, 1942, Wallin, 1946). On the lemma, including the awn, the lesions appear as small, linear to striated, brown to black spots frequently coalescing to blacken the lemma. On the rachis and culm the lesions are longer and striated after they coalesce. The leaf lesions are light brown, translucent in the centers, irregularly linear, and frequently coalesce to form blotches. The young lesions on all tissues are water-soaked, and as they mature, pigmentation occurs and droplets or scales of exudate appear on the surface of the lesions, as described by Fang *et al.* (1950), Smith (1917), Smith *et al.* (1939), and Wallin (1946).

THE BACTERIUM

Xanthomonas translucens f. sp. undulosa (E. F. Sm., L. R. Jones, & Reddy) Hagb.

[*Phytomonas translucens* var. *undulosa* (E. F. Sm., L. R. Jones, & Reddy) Hagb.]

(*Bacterium translucens* var. *undulosum* E. F. Sm., L. R. Jones, & Reddy)

The short rods usually with one polar flagellum are similar in morphology to the species described on barley. The physiology of the variety differs somewhat, and the variety shows specialization on wheat.

The etiology and control are similar to the bacterial blight of barley (Chap. 3).

5. Basal Glume or Spikelet Rot, *Pseudomonas atrofaciens* (McCull.) Stapp. The disease occurs in most of the wheat-growing sections of the world, especially where moisture is prevalent during the heading of the crop. The disease is of minor importance, as reported by McCulloch (1920) and Noble (1933). The light-brown bacterial rot at the base of the spikelet is not conspicuous unless examined closely. When the disease is severe, the rot extends into the rachis and into the base of the kernels. The lesions appear depressed and the tissues partly eroded with an inconspicuous gray bacterial exudate present in the sunken areas. These symptoms are distinctly different from those of "black point" caused by *Helminthosporium sativum* and other fungi, although the latter frequently occur in association with the bacteria to produce atypical symptoms.

THE BACTERIUM

Pseudomonas atrofaciens (McCull.) Stapp
[*Phytomonas atrofaciens* (McCull.) Bergey *et al.*]
(*Bacterium atrofaciens* McCull.)

The colonies are white in culture and consist of short cylindrical rods with one to four polar flagella.

The etiology and control of the disease are similar to that of the bacterial blight of barley (Chap. 3).

6. Bacterial Spike Blight, *Corynebacterium tritici* (Hutch.) Burkholder. A spike blight with yellow exudate conspicuous on the diseased inflorescence occurs in India, Egypt, China, and Western Australia, in association with the gall nematode (Vasudeva and Hingorani, 1952). A similar disease occurs on several grasses in Europe and North America, frequently without the nematode. See grasses, Chap. 12. The gall-forming nematode, *Anguillulina* (*Tylenchus*) *tritici* (Steinb.) Gervais & von Beneden, occurs in the Southeastern United States on wheat, but the bacterial spike blight has not been reported in this area. Control of the nematode by removal of the invaded kernels from seed, the use of rotation and sanitation, proper tillage management, and resistant varieties prevent the bacterial disease.

7. Pythium Root Rot, *Pythium* spp. The *Pythium* root rots are distributed throughout the world on wheat, the other cereals, and grasses and especially on the fine prairie and loess soil types of North America. Under moist soil conditions and continuous cereal and grass culture, these browning root rots cause considerable damage (Ho and Melhus, 1941, Sprague, 1944, Subramanian, 1928, Vanterpool and associates, 1930, 1932, 1935, 1938, 1940, 1942). The characteristic symptom of the disease is the light-brown soft rot of the rootlets and roots and the pale-green stunted growth of the tillers. The sporangia and oospores are abundant in the freshly rotted roots. A severe development of the disease results in a soft rot of the leaf sheaths and cortical tissues of the crown below the soil surface and the browning of the leaves.

THE FUNGI. A number of morphologically similar species of *Pythium* occur as parasites on the Gramineae. The more prevalent are the following:

> *Pythium arrhenomanes* Drechsl.
> *Pythium tardicrescens* Vanterpool
> *Pythium graminicola* Subrm.
> *Pythium aristosporium* Vanterpool
> *Pythium volutum* Vanterpool & Truscott

These closely related species are differentiated by Drechsler (1936), Middleton (1943), and Vanterpool (1938) by minor differences in morphology. The general morphology is given in Chap. 4.

ETIOLOGY. These *Pythium* spp. develop in the soil in association with roots and crop residues of the cereals and grasses. The aggressive parasitism of these fungi apparently is conditioned in part by their predominance in the soil microflora, or conversely by the absence of microorganisms that function dominantly in utilizing the nutritive constituents in the crop residues or that produce antibiotic substances inhibiting the development of the *Pythium* spp. The physiological condition of the cereal and grass plants especially as it affects root development and composition is important in determining both root invasion and the extent of root regeneration. The development of root rot is increased by a number of environmental factors, such as tightly compacted fine soils, high nitrogen in relation to phosphate, continuous cropping of cereals, or wheat and summer fallow. Garrett (1944), Simmonds (1941), and others have reviewed the literature and discussed the physiology and pathology of this type of parasitism. Further investigations are necessary on the etiology of this group of fungi and their control.

CONTROL. The methods of control of the *Pythium* root rot are inadequate at present. Wheat varieties show differences in reaction to the disease, but resistance under conditions favorable for the disease has not been demonstrated. Legumes in the crop-rotation system help reduce the damage. Balanced soil fertility is important in the control of the disease.

8. Downy Mildew, *Sclerospora macrospora* Sacc. or *Sclerophthora macrospora* (Sacc.) Thir., Shaw, & Naras. This downy mildew occurs in locally restricted outbreaks on the cereals and grasses throughout the world. While it exists as a potential menace, especially to wheat and oats, the restricted and sporadic development of the disease in the Southern, Central, and Eastern United States probably indicates a continued minor relationship in cereal and grass production.   The destructive potentialities of the disease under favorable environmental conditions are sufficiently great, however, to warrant careful watch for its occurrence and spread.

SYMPTOMS.   The infected plants are erect, yellowish green, somewhat dwarfed, and they tiller excessively.   The leaves are thickened, remain erect, and develop in a close whorl around the culm because of reduced internodal elongation and the stiff thickened conditions of the leaf blade. Many of the infected tillers turn brown and die.   Disintegration of the parenchymatous tissue and shredding of the leaves are uncommon.   The large brown oospores held rather permanently in the mesophyll tissue between the veins of the leaf blade and sheath constitute the important diagnostic symptom.   The inflorescences are proliferated into leaf-like structures (Fig. 26).

THE FUNGUS

Sclerospora macrospora Sacc. or
Sclerophthora macrospora (Sacc.) Thir., Shaw, & Naras.

The sporangial stage of the fungus is evanescent and difficult to find. The oospores are embedded firmly in the mesophyll and parenchyma of leaf and culm.   The large oospores, light brown in mass, are light yellow under the microscope, large, about 60 microns in diameter, globose, and smooth-walled.   Peglion (1930) described the germination of the oospores by the formation of a large papillate sporangium containing zoospores. The free oospores as well as those embedded in the leaf tissues apparently germinate in the same manner.   The *Phytophthora*-like sporangia of the imperfect stage are the basis for the new genus.

ETIOLOGY.   The oospores remain viable for long periods within the dead tissue, and they are capable of causing infection under favorable conditions.   The localized sporadic occurrence of the disease and the type of oospore germination indicate the high moisture requirements of the fungus in relation to infection.   The differences between the general distribution of *Sclerospora graminicola* in which the oospores form a germ tube and the limited occurrence of *S. macrospora* in which the oospores produce motile zoospores are explained in part by this difference in germination and the requirement of abundant free moisture for zoospore distribution and infection in the latter parasite.   The pathogen occurs commonly in the Southern United States and Mexico on *Phalaris tuberosa* and a semiprostrate *Phalaris* sp.   Infection on wheat seedlings was obtained from the pathogen on these grasses.   Good surface drainage and soil preparation, control of weed hosts, and crop rotation are the presently known methods of control.   The pathogen occurs on corn and sugarcane as well as wheat, oats, and barley.

9. Powdery Mildew, *Erysiphe graminis tritici* E. Marchal. Wheat and barley are the chief cereals damaged appreciably by powdery mildew. See Chap. 3 for the complete discussion. The disease is prevalent on winter and spring wheats during periods of cool, cloudy weather. Heavy mildew infection, especially during the period of tillering and internodal elongation of the plant, reduces the size of kernels and yield of grain. Allen and Goddard (1938), Allen (1942), and the Treleases (1929) have studied the influence of powdery mildew on the metabolism of the wheat plant. Apparently a low ratio of carbohydrate supply to nitrogenous compounds in the wheat tissues increases mildew development and damage. The respiration activity of infected tissues is increased, which further depletes the total available carbohydrates in the plant, although there is a localized accumulation of carbohydrates in the infected cells. Root development and size of grain are reduced, owing largely to the constant deficit of available carbohydrates. Cherewick (1944) and Futrell and Dickson (1954) have shown that the effect of light and temperature upon the physiological condition of the wheat plant influences the initial stages of fungus infection more than disease development after infection.

Physiologic specialization of the parasite is developed highly on the cereals and grasses (Graf-Marin, 1934, Mains, 1933, Marchal, 1902, 1903, Vallega and Cenoz, 1941). Two physiologic races of *Erysiphe graminis tritici* occur commonly on wheat in the United States and one race in Canada. Norka (C.I. 4377) and Axminister (C.I. 1839) are resistant to race 1 and susceptible to race 2. Vallega and Cenoz (1941) described three races in Argentina differing from those in the United States. Chul (C.I. 2277) is susceptible to two, and Dixon (C.I. 6295) is susceptible to all three races. Schlichtling (1939) listed six races in Germany. Chul (C.I. 2277), Dixon (C.I. 6295), Hope (C.I. 8178), Huron (C.I. 3315), Sonora (C.I. 4293) are resistant to the two races in the United States. The newer soft red winter wheat varieties produced in the United States are mostly resistant to intermediate in reaction to powdery mildew. The recent spring wheat varieties are largely resistant to intermediate, although Thatcher (C.I. 10003), Mida (C.I. 12008), Rival (C.I. 11703), and Henry (C.I. 12265) are susceptible.

Varieties of *Triticum monococcum* and *T. timopheevi* are resistant to the races occurring in the United States and Argentina. Mains (1934) has shown that at least two independent factor pairs are associated with powdery mildew resistance in wheat. The disease is discussed in detail in Chap. 3.

10. Ergot, *Claviceps purpurea* (Fr.) Tul. Light infections are common in wheat in most humid and semihumid sections. Some varieties of *Triticum vulgare* and a larger number of *T. durum* are susceptible, and occasionally these are damaged by ergot, especially in the durum wheat areas of North America. The detailed discussion of the disease is given in Chap. 8.

11. Gibberella and Fusarium Blight or Scab, *Gibberella* and *Fusarium* spp. The disease occurs on the cereals and grasses in the temperate

Fig. 55. Head blight, or scab, of wheat caused by *Gibberella zeae* showing the blighted spikes (*A*), seedlings (*B*), and scabbed kernels (*C*).

humid and semihumid areas of the world. *Fusarium* head blight causes severe damage to wheat, especially where temperature and relative humidity are high during the heading and blossoming periods. Stand, yield, and quality are affected by the disease. The scabbed wheat kernels are removed by modern cleaning equipment; therefore the diseased kernels do not occur in large quantities in commercially milled wheat products. The scabbed kernels separated out are used chiefly for poultry feeds. The growth of the *Gibberella* and *Fusarium* spp. in the developing kernel results in the formation of compounds that act as strong emetics in man, pigs, dogs, and animals with similar digestive systems. The exact chemical nature of the substances is not known, although the specific biological action on the nerve center controlling the stomach muscles is ascertained. The physiological effects are manifest by violent nausea, dizziness, and temporary irritation and soreness of the stomach and intestinal membranes. Continued intake results in loss of appetite and general digestive disturbances. Pigs and dogs fed 10 per cent or more of badly scabbed grain vomit and then refuse the grain mixture. In Russia where the infected grain was used extensively for bread, the reaction was known as "intoxicating bread" (Agronomoff, 1934, Gabrilovitch, 1906, Naumov, 1916, Shapovalov, 1917, and others). Cattle, sheep, and poultry, with the exception of the pigeon, do not react to the infected grain.

SYMPTOMS AND EFFECT. The disease occurs as a seedling blight, foot rot, and head blight. The blighted seedlings are characterized by a light-brown to reddish-brown water-soaked cortical rot and blight either before or after emergence (Fig. 55). The crown and culm rot phase of the disease occurs as the plants approach maturity. The head blight is conspicuous before the spikes mature. The infected spikelets first appear water-soaked, followed by the loss of chlorophyll, and a final bleached straw color (Fig. 55). During warm, humid weather, conidial development is abundant and the infected spikelets show a pink or salmon-pink

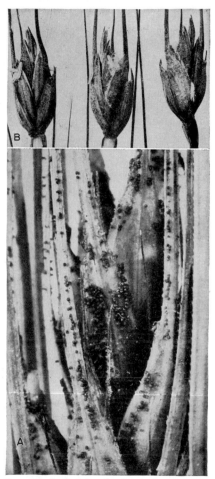

FIG. 56. Spikelets of wheat showing the initial contact infection from the anther tissue followed by mycelial and conidial development on the lemma and palea (B); perithecia on the same structures of Turkey winter wheat (A).

cast, especially at the base and in the crease of the kernel. The infection frequently spreads to adjacent spikelets or through the entire spike. Perithecia develop on the infected floral bracts in some varieties of wheat under conditions of continued warm, wet weather (Fig. 56). The infected kernels are more or less shriveled, with a scabby appearance due to the tufty mycelial outgrowths from the pericarp, and they range in color from white, pink, to light brown, depending upon the time of infection and environmental conditions during disease development.

THE FUNGI. Several species or one specie of *Fusarium* and *Gibberella* are associated with this disease on cereals and grasses depending upon the system of classification followed. Snyder and Hansen (1945) combine all species in the sections, *Roseum, Arthrosporiella, Gibbosum*, and *Discolor*, under *Fusarium roseum* (Lk.) on the basis that the variation in morphology extends across former species delimitations. The group pathogenic on the cereals is designated *Fusarium roseum f. cerealis* (Cke.) Snyder & Hansen. Varieties are used to designate known pathogens. These varieties correspond morphologically approximately with the species listed as such in the following pages. The designation as suggested by Snyder and Hansen for the imperfect stage becomes:

> *Fusarium roseum f. cerealis* (Cke.) Snyder & Hansen. All inclusive
> *Fusarium roseum f. cerealis* "Graminearum"
> *Fusarium roseum f. cerealis* "Culmorum"
> *Fusarium roseum f. cerealis* "Avenaceum" etc.

For the perfect stage:

> *Gibberella roseum f. cerealis* (Cke.) Snyder & Hansen
> [*Gibberella saubinetii* (Mont.) Sacc.]
> [*Gibberella zeae* (Schw.) Petch.]

Only the one variety is known under this trinomial, namely, the perfect stage of "Graminearum."

The morphological description of the species is amended to include the characters of the group (Snyder and Hansen, 1945, 1954). Both the Wollenweber and the revised nomenclature are included, since both systems of classification are in use at present.

1. *Gibberella zeae* (Schw.) Petch.
   Fusarium graminearum Schw. Conidial stage
   [*Gibberella saubinetii* (Mont.) Sacc.]
   (*Pionnotes flavicans* Sacc. & Sacc.)
   (*Fusarium graminearum* Schw.)
   (*Fusarium roseum* Lk.)
   [*Fusarium bufonicola* (Speg.) Sacc. & Trott.]
   (*Fusarium rostratum* App. & Wr.)
   (*Fusarium discolor* var. *majus* Wr.)
   (*Fusarium funicolum* Tassi.)
   [*Fusarium insidiosum* (Berk.) Sacc.]

(*Sphaeria zea* Schw.)
[*Botryosphaeria saubinetii* (Mont.) Niessl.]
(*Botryosphaeria dispersa* Ntrs.)
[*Gibbera pulicaris* (Fr.) Sacc.]
(*Gibberella tritici* Henn.)
(*Gibbera saubinetii* Mont.)

The mycelium is white to pink within and on the tissues. The conidia are borne in sporodochia or pionnotes, sickle-shaped, gradually tapering toward the apex, not constricted appreciably toward the base, have thin hyaline walls and septations, are generally 5-septate, and measure 41–60 by 4.3–5.5 microns. Thickened vegetative resting spores (chlamydospores) are absent. The perithecia are scattered on the surface, somewhat embedded in the mycelium, smooth at the base with protuberant projections near the apex, purplish black to dark blue, and ovoid to subconical, varying in size and shape. Asci are numerous, cylindrical, tapering to the base and hyaline. Ascospores are regularly eight, borne in one or two irregular rows, fusiform, slightly curved, largely three-celled, and measure 20–30 by 3.4–5.0 microns.

2. *Fusarium culmorum* (W. G. Sm.) Sacc.
   (*Fusisporium culmorum* W. G. Sm.)
   (*Fusarium culmorum* var. *leteius* Sher.)
   (*Fusarium culmorum* var. *majus* Wr.)
   (*Fusarium heidelbergense* Sacc.)
   (*Fusarium mucronatum* Faut.)
   (*Fusarium neglectum* Jacz.)
   (*Fusarium roseum* var. *rhei* Karst.)
   (*Fusarium rubiginosum* App. & Wr.)
   (*Fusoma ochraceum* Corda.)
   (*Fusarium sambucinum* Fuckl.)
   (*Fusarium schribauxii* Del.)
   (*Fusoma tenue* Grove)
   (*Fusarium versicolor* Sacc.)

The conidia are borne the same as in the previous species. The conidia are somewhat wider and longer than in the previous species, slightly constricted toward the base, 5-septate when mature, and measure 30–50 by 4.8 7.5 microns. Thick-walled vegetative resting spores (chlamydospores) are common. No ascigerous stage is known.

3. *Fusarium culmorum* var. *cereale* (Cke.) Wr.
   (*Fusarium cereale* Cke.)
   [*Fusarium cerealis* (Cke.) Sacc.]
   (*Fusarium equiseti* var. *crassum* Wr.)

The variety differs from the species by longer conidia, frequently with 7 to 9 septations.

4. *Fusarium avenaceum* (Fr.) Sacc.

The synonymy of this species is lengthy and complicated, as reported by Wollenweber and Reinking (1935).

The conidia are borne similarly to the previous two species. The conidia are much narrower and more tapering toward the apex than in the previous species, usually 3- to 5-septate, and measure 45–66 by 3.1–4.1 microns. Vegetative resting spores (chlamydospores) are uncommon. See Fig. 6.

ETIOLOGY. *Gibberella zeae* or *G. roseum f. cerealis* is the common species on wheat in the corn belt of North America, with the other species or varieties more prevalent northward into Canada. A similar distribution occurs in Europe. In Asia, *G. zeae* is common in the Pacific coastal area and the other species in the northern interior.

These fungi are associated closely with crop residues of the cereals. The mycelium develops abundantly on the residues following maturity of the crop until the tissues are disintegrated the following season. Conidial development is profuse when conditions are favorable. Perithecial development in the corn belt occurs sparingly in the autumn but is abundant on the corn and small-grain straw and stubble the following late spring and summer. In warmer climates, the development of perithecia is abundant during the autumn and early spring. Seedling infection is from seed-borne inoculum and from the crop residue. Crown tissues are invaded largely from the mycelium in the crop residue. Head blight occurs during moist, warm seasons from conidia and ascospores produced on the crop residues, especially when such refuse is on the soil surface. The initial invasions of the developing kernels frequently are from contact infection of the mycelium growing saprophytically on dehisced anthers (Fig. 56). The spores are carried considerable distances by wind and air currents. Secondary infection is from conidia and mycelium, as reported by Dickson (1941), Koehler *et al.* (1924), Pugh *et al.* (1933), and others. Warm, moist conditions are important in relation to perithecial development and in influencing seedling- and head-blight development (Bayles, 1936, Dickson, 1923, Garrett, 1944).

CONTROL. Crop rotation, sanitation, soil preparation, and seed treatment are important control measures. Covering the crop residue completely when plowing and treating the seed with the organic mercury compounds aid in the control of head blight and seedling blight, respectively. Wheat varieties vary greatly in susceptibility; however, no highly resistant wheats are known. Christensen *et al.* (1929), Hanson *et al.* (1950), Scott (1927), and others have listed the scab reaction of spring and winter wheats.

12. Snow Mold, Foot Rot, and Head Blight, *Fusarium* spp. and *Calonectria* spp. The disease occurs uncommonly in the Northern United

States and Canada; it is common in northern Europe and Asia where it causes considerable damage. Winter wheat and rye and the grasses are damaged in areas where the snow covering is heavy and the soil temperatures are mild.

SYMPTOMS AND EFFECTS. The fungus is conspicuous on the leaf and crown tissues of the winter cereals and grasses as the snow is melting in the spring. The white superficial mycelium is abundant under moist conditions. Penetration and killing of the leaf and bud tissues result in bleaching and drying out of these plant parts. Frequently local areas or spots show the killing with healthy plants adjoining. Conidia of the *Fusarium* stage of the fungus are present on the dead tissues. Several other fungi cause similar symptoms but are differentiated by types of mycelium, conidia, or sclerotia. The crown rot stage of the disease is inconspicuous and frequently associated with other fungi. The head blight also is less noticeable, as frequently individual kernels are infected without extensive blighting of the floral bracts. The kernels are shriveled and light brown in color.

THE FUNGI. Here again the two classifications are in use. Snyder and Hansen (1945) emended the species as follows:

Fusarium nivale (Fr.) Snyder & Hansen. All inclusive
Fusarium nivale f. graminicola (Berk. & Br.) Snyder and Hansen, conidial stage, with varieties "Major", etc.
Calonectria nivale (Fr.) Snyder & Hansen
Calonectria nivale f. graminicola (Berk. & Br.) Snyder & Hansen, the perithecial stage.

The Wollenweber classification follows:

1. *Calonectria graminicola* (Berk. & Br.) Wr.
   Fusarium nivale (Fr.) Ces. Conidial stage
   (*Nectria graminicola* Berk. & Br.)
   (*Calonectria nivalis* Schaf.)
   [*Fusarium nivale* (Fr.) Ces.]
   (*Lanosa nivalis* Fr.)
   (*Fusarium hibernans* Lind.)
   (*Fusarium minimum* Fuckl.)
   (*Fusarium loliaceum* Duc.)
   (*Fusarium miniatulum* Sacc.)
   (*Fusarium miniatum* Prill. & Del.)
   (*Fusarium oxysporum* Klot.)
   (*Fusarium secalis* Jacz.)
   [*Fusarium tritici* (Lieb.) Eriks.]
   (*Fusarium ustiliginis* Rostr.)

The mycelium is abundant, white to gray in superficial mass. The conidia are borne in sporodochia or pionnotes, light-salmon color, thickly sickle-shaped, tapering at the apex, usually 3-septate, and average about 23 by 3 microns. Perithecia are scattered on the surface or within the mycelial mat, frequently associated with clumps of sterile mycelium, round to oval, nearly smooth, and dark red or reddish brown in color. Asci are numerous, cylindrical, tapering gradually to the base, and hyaline. The eight ascospores are borne in one or two irregular rows in the ascus. They are spindle-shaped, 1- to 3-septate, and average about 15 by 3 microns. Vegetative resting spores and sclerotia are absent.

2. *Calonectria graminicola* var. *neglecta* Krampe.
(*Fusarium nivale* var. *majus* Wr.)

The conidia are larger and thicker than in the species, without the pronounced tapering at the apex. The asci are much shorter than in the species.

Under the same environmental conditions, other species of *Fusarium* frequently produce similar symptoms, as described by Wollenweber and Reinking (1935).

The etiology is similar to the other *Fusarium* diseases of the cereal crops. Mycelium, conidia, and perithecia develop on crop residues. These fungi are aggressive as parasites only on weakened plants or, more specifically, plants depleted in cellular reserves and inactive in vegetative development, a condition frequently associated with excessive respiration rates and low light intensity. The fungus develops parasitically under somewhat lower temperature conditions than the other *Fusarium* spp. occurring on the cereals, with the possible exception of *Fusarium culmorum*.

Control measures are similar to the other *Fusarium* diseases. Soil drainage is especially important. In turf grasses, the disease is held in check by the use of fungicides. Mercurous chloride is used extensively as well as the organic mercury compounds (Dahl, 1934, Sampson, 1931, Schaffnit and Meyer, 1930, Sprague, 1950, and others).

13. Take-all, *Ophiobolus graminis* Sacc. The disease is common on the cereals and grasses in rather specific areas throughout the world (Kirby, 1925). Take-all occurs more commonly in the drier winter wheat sections of Southwestern and Northwestern North America. The greatest losses from the disease occur in the porous alkaline soils where winter wheat culture is continuous or associated with the culture of grasses and with the breaking of the native grass sod. This is one of a complex of diseases attacking the roots, crown, and basal culm tissues of the cereals and grasses. The plant parts invaded, the geographical distribution, the environmental conditions under which the diseases develop, and the fungi concerned are somewhat different for each disease. The terms "take-all," "Pietin," and "Fusskrankheit" through common usage are associated more generally with the disease caused by *Ophio-*

bolus graminis and similar species of this genus, although as shown by Garrett (1944) this is not indicated too clearly in the literature.

SYMPTOMS AND EFFECTS. The symptoms vary greatly under different environmental conditions. Under relatively moist conditions, the conspicuous symptoms of take-all appear about the time wheat is heading. Localized areas occur in which growth is checked, the green color fades, and rapid bleaching of the leaves, culms, and heads follows. This sequence of symptoms is characteristic for the disease under very favorable conditions, although not entirely specific for take-all. The main roots, crown, and basal culm tissues show a dry rot accompanied by a dark-brown to black surface mat of thick-walled coarse mycelium. This fungal mat is conspicuous, especially on the culm base under the leaf sheaths (Fig. 57).

Under drier conditions the symptoms are less conspicuous. The plant tillering is reduced, the plants are short, few plants show the dead bleached condition, and the mycelial mat is less pronounced. The presence of the fungus on the diseased tissues is the only sure criterion (Carne and Campbell, 1924, Fellows, 1928, 1938, Hynes, 1937, McKinney, 1925, Samuel, 1937).

THE FUNGUS

Ophiobolus graminis Sacc.
[*Ophiobolus cariceti* (Berk. & Br.) Sacc.]

The mycelium comprises a limited growth of fine hyphae, grayish in color, and an abundant development of coarse, thick-walled, brown to black, irregular hyphae. The perithecia are formed in or beneath the leaf sheath with strands of mycelium associated with the base and the cylindrical curved beaks extending through the sheath tissues. They are round to oblong, black, and about 400 microns in diameter (Fig. 57). The asci are numerous, elongate, clavate, straight or curved, and numerous thread-like paraphyses are present in the young perithecium. The mature asci are ejected from the perithecium during periods of abundant moisture. The ascus wall disintegrates in contact with free water to liberate the eight spores. The ascospores are hyaline, slender, tapering toward the ends, measure 3 by 70–80 microns, and are 5- to 7-septate when mature. The production of minute falcate conidia in the germination of the ascospores occurs under some conditions. *Ophiobolus graminis* var. *avenae* Turner is reported on oats.

Wojinowicia graminis (McAlp.) Sacc. & Sacc., weakly parasitic on wheat, occurs commonly in association with the take-all disease. The smooth-walled pycnidia of *W. graminis* frequently are confused with the perithecia of *Ophiobolus graminis* unless examined carefully. The conidia are long, slender, slightly curved, tapering at the ends, many-septate, and are borne singly on short cells in the base of the pycnidium.

ETIOLOGY. *Ophiobolus graminis* is soil-borne in rather direct association with straw and roots of the cereals and grasses. Garrett (1944) and

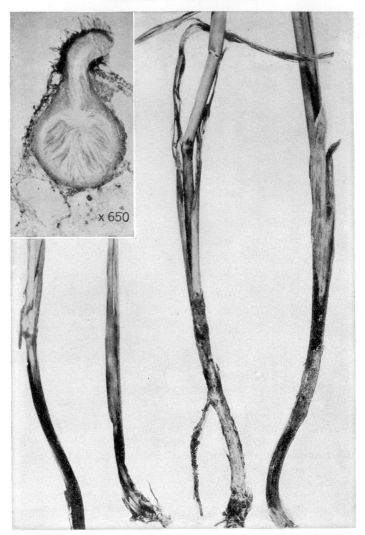

×650

Fig. 57. The base of wheat culms showing the mycelial mat of *Ophiobolus graminis* beneath the leaf sheath.   A section through the perithecium of *O. graminis*, highly magnified, is shown in the insert.

Sprague (1950) have reviewed the literature on the influence of environmental conditions on the development of the disease as well as the antagonistic phenomena of the microflora on the parasite.   Undecomposed cereal straw and roots are necessary for survival and parasitic activity of *O. graminis*.   Infection occurs from the active mycelium in the crop residue penetrating the root, crown, and culm tissues.   Frequently localized areas of diseased wheat plants are associated with the presence of the pathogen on weed grasses, especially *Agropyron repens.* The damage is dependent largely upon the presence or absence of soil micro-organisms that suppress the activity of *O. graminis;* alkaline sandy soils, low phosphate and potash levels, and the absence of residues of crops other than wheat apparently suppress the soil microflora that inhibit the development and parasitic potentiality of *O. graminis.*

CONTROL. Crop rotations, involving legumes, other dicotyledonous crops, oats, or corn are important in reducing damage. A balanced fertility with a good supply of available phosphate and potash reduces losses. Wheat varieties show only small differences in susceptibility to the disease that is associated apparently with adaptability of the varieties rather than resistance to the disease.

14. Foot Rot or Culm Rot, *Cercosporella herpotrichoides* Fron. The disease caused by this fungus is reported in limited areas in northern Europe and the Northwestern United States and western Canada. In the earlier literature *Leptosphaeria herpotrichoides* deNot. was considered a pathogen in the disease complex, but Sprague (1934, 1950), and Sprague and Fellows (1934), have shown it to be a saprophyte. Other *Cercosporella* spp. cause leaf spots and scald on the grasses.

The disease is conspicuous near the end of the growing season by the lodging of the diseased plants. The lesions are evident first on the leaf sheath as elliptical to ovate spots with light-straw-colored centers and brown margins. Similar spots occur on the culm beneath the lesions in the sheath. Necrosis also occurs around the roots in the upper crown nodes. Under moist conditions the lesions enlarge, and a black stroma-like mycelium develops over the surface of the crown and base of the culms, giving the tissues a charred appearance. When infection occurs early, individual culms and weaker plants are killed before the grain is mature.

THE FUNGUS

Cercosporella herpotrichoides Fron.

The conidiophores on the lesions are simple or branched, short, erect, and originate from the stromal cells of the macrohyphae. The large ends of the obclavate conidia are attached terminally or subterminally. The conidia are curved slightly, chiefly 5- to 7-septate, and measure 30–80 by 1.5–3.5 microns. No perithecial stage is known.

The pathogen persists in crop residue, invades the crown tissues of fall-sown wheat, especially when sown early, and develops in the crown and basal culm tissue during the following growing season. Conidia of *Cercosporella* are abundant on the lesions in the spring and early summer, and they apparently are responsible as well as the mycelium for secondary spread.

Crop rotation, especially legumes, the use of spring grains, and delayed fall sowing aid in control of the disease. The wheats and grasses show differences in susceptibility, according to Sprague (1934, 1936).

15. Crown Rot and Root Rot, *Helminthosporium sativum* Pamm., King, & Bakke and Other Species. The disease complex caused by this group of fungi is distributed widely on wheat. Frequently, the *Helminthosporium* is associated with species of *Fusarium*, especially *F. culmorum*, further to complicate the relationship of this group of fungi to the root rot and foot rot of wheat. Henry (1924), Hynes (1935, 1937), Simmonds (1941), and Stakman (1920) have described the malady and reviewed the literature. *H. sativum*, under favorable conditions, causes severe damage to wheat, barley, and grasses, especially as the plants approach maturity. According to Sallans (1940), wheat recovers from the malady when growing conditions are unfavorable for the continued development of the disease. The disease complex on wheat was discussed by Bolley (1909, 1913) who associated the increase in prevalence and severity of the malady to continuous wheat culture. During the 40 years since this warning, the disease has been studied in most of the wheat-producing sections of the world. The kernel infection caused by *Helminthosporium* spp. and *Alternaria* spp. is severe on barley and the durum wheats, and certain of the stem rust-resistant wheats are relatively susceptible, notably Apex (C.I. 11636) and Thatcher (C.I. 10003), according to Brentzel (1944) and Greaney and Wallace (1943).

The symptoms and development of the disease are similar on the several cereals. They are discussed specifically for *Helminthosporium sativum* in Chap. 3.

THE FUNGI. The species of *Helminthosporium* associated with the disease in wheat and barley apparently are diverse. Much confusion exists in the earlier literature regarding the species described as occurring on wheat, as reviewed by Drechsler (1923).

1. *Helminthosporium sativum* Pamm., King, & Bakke

This predominant species is widely variable. The morphology is discussed in Chap. 3. Dastur (1942) described *Cochliobolus tritici* Dast. on wheat culms, which likely is synonymous with *C. sativus* (Ito & Kuribay.) Dast. although differing somewhat in morphology. The other large conidial types of the genus with dark conidial walls and germ tubes developing from the apical cells of the conidia that have been described on wheat probably differ sufficiently from *H. sativum* to be retained as species.

2. *Helminthosporium* spp.

Other species of this genus and related have been described as inciting crown rot on wheat, other cereals, and grasses. Many of these pathogens attack a wide range of hosts as in *H. sativum* (Boedijn, 1933, Drechsler, 1923, Henry, 1924, Hynes, 1937, Sprague, 1950). *Helminthosporium bicolor* Mitra, *H. halodes* Drechsl., *H. halodes* var. *tritici* Mitra, *H. tetramera* McKinney or *Curvularia specifera* (Bainer) Boed., *C. geniculata* (Tr. & Er.) Boed. or *C. ramosa* (Bainer) Boed., *Brachycladium* spp., and others are discussed by Garrett (1944), Simmonds (1941), Sprague (1950), and others.

16. Leaf Spot, *Helminthosporium tritici-vulgaris* Nisikado. The disease was first reported from Japan, later from Eastern United States, and at present is distributed through the soft and hard red winter wheat and adjacent spring wheat areas of North America. Fusiform spots on leaf blades and sheaths are yellowish brown to dark brown, frequently zonate. Later the spots spread with indefinite areas of necrotic tissue. Perithecia are produced in abundance on the old straw. The disease is not severe on most wheat varieties.

THE FUNGUS

Helminthosporium tritici-vulgaris Nisikado or
Pyrenophora tritici-vulgaris Dickson.

Conidiophores form on infected tissues singly or in groups of 2 to 3; dark olive with swollen basal cell. Conidia are yellowish brown, cylindrical, straight with slight curvature to one side, basal cell enlarged, 118 by 17 microns, usually 8- to 10-septate, and forming germ tubes from polar and middle cells. Perithecia are abundant on leaf sheath and culm of straw and stubble during the spring from Ohio to Kansas and northward to Wisconsin. Perithecia are black, conical to elongate, beak not well defined, and numerous conidiophores over surface and short black setae around the apex. Asci are irregularly clavate with thickened wall extending to near apex. Ascospores, usually eight, are arranged distichously, brown, characteristically 3-septate with 1 to 2 longitudinal septa in center cell, size variable 40–70 by 16–24 microns. Germination occurs from several cells, and ascospores produce typical leaf spot on wheat. Morphologically this fungus is similar to *H. tritici-repentis* Died., *P. tritici-repentis* (Died.) Drechsl.; Wehmeyer (1949) combines these and other species under *Pleospora trichostoma* (Fr.) Ces. & deNot. The morphologically similar group of species on cereals and grasses apparently need more study. See barley, Chap. 3.

Both conidia and ascospores serve as primary inoculum for spread of the leaf spot during the spring and summer. The disease occurs on spring wheats in the North Central area. Plowing under stubble and rotation help control early spring spread. Varieties differ in susceptibility, especially in size and spread of the spots.

17. Anthracnose, *Colletotrichum graminicolum* (Ces.) G. W. Wils. The disease occurs on wheat in light soils of low fertility throughout the world. Wheat is damaged less than most of the other cereals.

Gloeosporium bolleyi Sprague frequently occurs as a root rot alone or in association with *C. graminicolum* in North Central North America. This pathogen contributes to the unhealthy condition of wheat in poor soils in continuous wheat and grass culture (Sprague, 1950). Anthracnose is discussed in detail in Chap. 8.

18. Septoria Leaf Blotch and Glume Blotch, *Septoria tritici* Rob. and *S. nodorum* Berk. Two *Septoria* blotches occur on wheat throughout the world. The leaf blotch is generally the more important disease of the two as it is distributed over a wider area in both the hard red winter and soft red winter wheat sections of the United States and occurs more consistently than glume blotch. Epiphytotics of the glume blotch, however, cause severe shriveling of the kernels in occasional seasons in the soft red winter wheat section. In Argentina especially, the diseases cause kernel shriveling and reduced yields in susceptible varieties. According to Sprague (1934, 1938, 1944, 1950) and Weber (1922), physiologic races of these two species occur on some other cereals and grasses.

SYMPTOMS AND EFFECTS. The leaf blotch appears first as light-green to yellow spots between the veins of the leaves. The lesions spread rapidly to form light-brown irregular blotches with a speckled appearance as the pycnidia develop. The small submerged brown pycnidia in the blotches are the final diagnostic symptom of the disease. Under favorable conditions, especially in the late autumn and early spring, defoliation and invasion of the crown tissues occur, resulting in weakened or dead plants (Fig. 58). Lesions on the culms, floral bracts, and pericarp of the kernels are less conspicuous and much smaller, with sparse pycnidial development.

The glume blotch caused by *Septoria nodorum* occurs more generally on the floral bracts and nodal tissues of the culm. The lesions are small, linear to oblong, light brown to dark brown in color, and the submerged pycnidia less conspicuous because of the darker color of the blotch than in the leaf blotch. Examination of the conidia is the only sure means of distinguishing between the two diseases.

THE FUNGI

1. *Septoria tritici* Rob.
   (*Septoria graminum* var. *tritici* Desm.)
   (*Septoria cerealis* f. *tritici-vulgaris* Thuem)
   (*Septoria tritici* Thuem)
   (*Septoria triticina* Unam.)
   (*Septoria tritici* Desm.)

According to Sprague (1944), Roberge, to whom Desmazières credits the description, is the correct authority for the binomial. Two types of mycelium develop: a thin hyaline intercellular mycelium and a coarser olivaceous sterile superficial type. The pycnidia are subepidermal, usually in the stomatal cavity and subglobose, smooth-walled, brown to black with slightly raised ostioles. Two types of spores are produced, macrospores, which predominate, and microspores. The macroconidia are slender, cylindrical with rounded ends, usually straight, hyaline, 3- to 7-septate, and

Fig. 58. Leaf blotch (*A*) and glume blotch (*B*) infection of winter wheat caused by *Septoria tritici* and *S. nodorum,* respectively.

vary considerably in size, averaging 50 by 2.2 microns for the spores produced in summer and slightly longer during the winter and spring. The microconidia are curved, aseptate, hyaline, and measure 5–9 by 1.0–1.3 microns. These occur in association with the macroconidia or alone in periods of low temperature.

2. *Leptosphaeria   nodorum*   Müller
    Septoria nodorum (Berk.) Berk.   Conidial stage
    (*Septoria glumarum* Pass.)

(*Phoma hennebergii* Kühn)
(*Macrophoma hennebergii* Ber. & Vog.)

The mycelium is branched, irregular, hyaline at first to dark olivaceous when mature. Pycnidia are subepidermal, usually in the stomatal cavity, irregular in shape with the ostioles stomatal, brown to black, and larger than in the former species. Spores are oblong to cylindrical with rounded ends, hyaline, usually 3-septate, and measure on the average 26 by 3 microns. Perithecia of *Leptosphaeria* were described by Weber (1922) as associated with the fungus. The relationship of this ascigerous stage was established by Müller (1952).

ETIOLOGY. Spores within the pycnidia and mycelium within the tissues persist for long periods of unfavorable conditions. The mycelium in infected kernels produces seedling infection, as reported by Machacek (1945). The spores are produced, germinate, and cause infection under a wide range of conditions. Infection by *Septoria tritici* is general in the autumn on the leaves of winter wheat, and the mycelium remains active even at temperatures near freezing. Early-spring spread of the disease is rather general. Infection in the spring wheat occurs from spores within pycnidia formed the previous season, spores produced on dead-wheat refuse during the spring, and from infected seedlings. The etiology of the two species is similar. Evidence indicates that *S. nodorum* is less resistant to severe winter conditions and develops at higher temperatures than *S. tritici*. In North America, *S. nodorum* is prevalent in the more southern winter wheat regions.

CONTROL. Crop rotation, sanitation, and plowing under volunteer wheat plants in the fall are important control measures. Mackie (1929) and others have reported many of the commercial wheat varieties of the United States as moderately resistant. Many of the important commercial varieties of Argentina are susceptible. Seed treatment with the organic mercury compounds kills the fungus borne in or on the grain.

19. Rhizoctonia Blight, *Rhizoctonia solani* Kuehn and Other Species. The disease occurs on wheat, oats, and many grasses, and it is distributed widely throughout the world. While the malady is important in local areas, as reported by Goeffrey and Garrett (1932), Hynes (1937), and Subramaniam (1928), the disease is of minor importance on the wheat crop as a whole.

SYMPTOMS. The disease appears in patches in which the plants are stunted, and the leaves in many varieties show a purple cast. Plants are weakened and killed or more generally recover, in which case maturity is delayed and yield is low. The tan-colored cortical rot of the root system and tan zonate lesions on the basal leaf sheaths are the conspicuous symptoms. The characteristic mycelium of the fungus is present in the rotted tissues and root stubs near the crown.

THE FUNGUS

Rhizoctonia solani Kuehn

The *Corticium* or *Pellicularia* stage is not common on the cereals and grasses.

The mycelium is white and brown intermixed. The young mycelium branches characteristically at acute angles with a constriction of the mycelium at the union and a septum in the mycelium at the constriction. The dark-gray to black irregularly shaped sclerotia are uncommon on the diseased roots, but they occur more frequently on the sheath tissues. They usually germinate to form mycelium. The *Corticium* stage consists of the club-shaped basidia with four apical sterigmata bearing sporidia that are hyaline and oval with tapering base. Rogers (1943) suggested the binomial *Pellicularia filamentosa* (Pat.) Rogers for the basidial stage of *Rhizoctonia solani*.

Peltier (1916) has described the fungus on numerous plants. Dickinson (1930), Monteith (1926), and others have described the disease and its control on the grasses used for golf greens. As in most other crops, the disease develops best at low temperatures, although strains of the fungus occur that incite disease at high temperatures (Sprague, 1950).

The disease is controlled by good soil drainage, balanced fertility, and proper rate of seeding. The use of chemicals such as mercurous chloride and the organic mercury compounds controls the disease.

20. Typhula Blight, *Typhula itoana* Imai and *T. idahoensis* Remsberg. This disease develops under rather special environmental conditions and, therefore, is restricted in its occurrence. The blight occurs on the winter cereals and grasses during periods of heavy snow covering or cloudy weather with temperatures remaining slightly above freezing. In this respect the disease is similar to the snow mold caused by *Fusarium* spp., but apparently the two diseases are not found commonly in the same areas. The *Typhula* disease on wheat, especially, occurs in the Intermountain valleys at relatively high altitudes. The disease is distributed more generally on the grasses and is world-wide in occurrence.

SYMPTOMS. The *Typhula* blight is conspicuous as the snow disappears as a felty white mycelial mat over the plants and adjacent soil. The presence of numerous light- to dark-brown spherical sclerotia as the mycelium matures is the characteristic symptom. The plants overrun by the mycelium gradually lose the deep-green color, wither, and turn brown. The disease occurs in spots varying in size and shape with less definite symptoms toward the margins of the areas. The killing of the plants ranges from dead leaf tissue to invasion and rotting of the culm, crown, and root tissues. The disease disappears as the temperature rises, moisture decreases, and sunlight increases.

THE FUNGI

1. *Typhula itoana* Imai
   (*Typhula graminum* Karst.)

(*Typhula elegantula* Karst.)
(*Sclerotium fulvum* Fr.)

The sclerotia are tawny to hazel brown, spherical to slightly flattened, and superficial or embedded in the diseased plant tissues.   The sclerotia commonly germinate to form mycelium or thickened branching sterile structures with clamp connections apparent. The sclerotia, in soil or sand and in the presence of light high in ultraviolet rays, produce one to four thickened basidia, sometimes branched.   The basidia are clavate, flesh-colored, nonseptate with four apical sterigmata bearing sporidia.   The sporidia are hyaline, ovate, slightly curved, and average 11.1 by 6.0 microns in size.   This is probably the more common and widely distributed species on the cereals and grasses, according to Imai (1936), Remsberg (1940), Tasugi (1929, 1930, 1935), and Volk (1937).

2. *Typhula idahoensis* Remsberg

This species is similar in general morphology to the former.   The sclerotia are chestnut brown, and the basidia are fawn to wood brown in color.   This species on wheat is restricted apparently to the Intermountain areas of the United States and Canada, although it is distributed more widely on the grasses.   It occurs extensively in Europe and Asia.

3. *Typhula graminum* Karst. is retained as a species by Remsberg (1940), but its pathogenicity on the cereals and grasses is questioned.

Cormack has found still another, at present unnamed, Basidiomycete developing at very low temperatures in Alberta, Canada.   The mycelium develops parasitically on wheat among other crops at temperatures slightly above freezing.

This latter pathogen and some isolates of the former produce HCN which under snow cover and at low temperatures causes extensive killing of plant tissues (LeBeau and Dickson, 1953).   The fall application of organic mercury compounds is effective in controlling these diseases, and the treatment appears economically practical (Holton and Sprague, 1949).

Sclerotium rolfsii Sacc. occurs occasionally on wheat and other cereals and grasses in the warmer areas of the United States and other countries, according to Godfrey (1918), Tisdale (1921), and others.   The small spherical brown sclerotia occur on the rotted culm, crown, and root tissues.

21. Loose Smut, *Ustilago tritici* (Pers.) Rostr. or *U. nuda* (Jens.) Rostr. The loose smut of wheat is distributed generally with the crop in the humid and semihumid wheat-producing areas.   The disease is severe in the soft red winter and hard red spring wheat areas of Central and Eastern North America.   The loose spore mass replaces the floral bracts and ovaries and is conspicuous from the heading to blossom period of the crop. The symptoms and etiology are similar to the loose smut of barley discussed in Chap. 3.

THE FUNGUS

Ustilago tritici (Pers.) Rostr. or
Ustilago nuda (Jens.) Rostr.
See Fischer (1953) for synonymy.

The morphology is the same as *Ustilago nuda,* causing the loose smut of barley
(Chap. 3). This is another case where long usage of a binomial appears to warrant
retaining it, although the two species, *U. tritici* and *U. nuda,* cannot be differentiated
morphologically. Fischer (1953) and others have suggested the combination of the
two as specialized races of the same parasite.

CONTROL. The loose smut infection within the kernel tissues is controlled by the modified hot-water treatment, water steep, and use of
resistant varieties. The hot-water treatment is expensive and difficult
to use except on a limited amount of foundation seed. The commercial
varieties and selections show a wide range in loose smut infection. The
club wheats and spelt are relatively susceptible. The durum wheats show
more resistance than most of the common break wheats. Some of the
strains of *Triticum dicoccum* and *T. timopheevi* are highly resistant to the
known physiological races.

The more resistant spring wheats are Preston (C.I. 3328), Hope (C.I.
8178), and hybrid selections and *T. timopheevi* and hybrid selections
(Allard and Shands, 1954, Tingey and Tolman, 1934). The latter two
have been crossed with winter wheats, especially Kanred (C.I. 5146)
and Turkey (C.I. 1558). Kawvale (C.I. 8180), Currell (C.I. 3326), Leap
(C.I. 4283), Thorne (C.I. 11856), Trumbull (C.I. 5657) of American
winter varieties, Grüne Dame and Jubile of the European winter wheats
are reported resistant to most races of the pathogen (Alkins *et al.,* 1947,
Oort, 1947, Roemer *et al.,* 1938).

Resistance is conditioned by one or more factor pairs both dominant
and recessive, and in some varieties the maternal tissues influence the
expression of resistance, as in barley, Chap. 3 (Heyne and Hansing,
1955, Tingey and Tolman, 1934). Oort (1947) has investigated the
relation of pathological histology to the expression of resistance.

Physiological specialization occurs in the pathogen. However, race
differentiation in the several countries is based on the reaction of varieties in which little is known regarding the factors conditioning resistance
(Bever, 1947, 1953, Grevel, 1930, Hanna, 1937, Oort, 1947, Radulescu,
1935, Stakman *et al.,* 1935).

22. Bunt or Stinking Smut, *Tilletia caries* (DC.) Tul., *T. foetida* (Wallr.)
Liro, and *T. controversa* Kühn. The stinking smut or bunt of wheat is
associated closely with the historical development of plant pathology.
It was among the first smuts to receive attention. The early literature
on bunt was descriptive. Tillet in 1755 differentiated between la Carie

(bunt) and le Charbon (loose smut) of wheat and established the infective principle of bunt dust; later, Tessier in 1783 and others considered it a degeneration of the grain. Perhaps the symptoms and fishy odor of this smut functioned unduly in the formulation of the concept that disease manifestations were morbid eruptions of vegetable matter, a concept that dominated the thinking of European mycologists for a time and reached its climax in Unger's book "Exantheme der Pflanzen" in 1833. The bunt disease also, during this same period, was the basis for the formulation of the experimental concept of parasitism in fungi by Prevost in 1807. The noted mycologists of the period, Bulliard, DeCandolle, Link, Tulasnes, and Léveillé, were concerned with the disease and the two fungi. DeBary in 1853 reconfirmed phases of the parasitic nature of the bunt fungus, and Kühn in 1858 summarized the information on practical smut control. The disease has received increasing attention through the years, especially since the detailed experiments of Brefeld in 1883.

The bunt of wheat is world-wide in distribution. Whereas loose smut occurs in the more humid wheat sections, bunt is more prevalent in the drier sections as well as occurring in the areas with high summer moistures.

Two of the three *Tilletia* spp. on wheat occasionally infect rye and several grasses. *T. foetida* apparently occurs on only wheat and occasionally rye. Many additional species of *Tilletia* occur on grass hosts only (see Chap. 12). Dwarf bunt, described as incited by a variety of *T. caries* when found in the Western United States, and later found to be the species described much earlier from Central Europe as *T. controversa* Kühn, is widely distributed on wheat and a few grasses in North America and Europe and probably other areas where infested seed has been introduced (Fischer, 1952, 1953, Wagner, 1950). The bunt or stinking smut has been discussed in detail, including important literature citations by Holton and Heald (1941).

The geographic distribution of the three species differs somewhat. *T. foetida*, the smooth-spored species, is more common in the central and eastern humid areas of North America than *T. caries*, while both species occur in the western, drier areas, although the latter species is more prevalent (Rodenhiser and Holton, 1945). A similar distribution is apparent in other areas (Churchward, 1932). The dwarf bunt pathogen, first associated with the western area in North America and central Europe, has been found more recently widely distributed in many local areas in eastern North America. Apparently seed-borne spores, resistant to the ordinary seed treatments used for bunt control, account for the general distribution. The long duration of spores in infested soils is responsible for the continuation of the infestation once established.

The disease causes losses in yield, produces difficulties in threshing, and lowers quality. Prior to the general use of seed treatments and resistant varieties, bunt caused large reductions in yield of wheat. The presence of the spores in quantities caused explosions in separators and fire losses. The spores adhere to the threshed grain, and they are removed only by washing before the wheat is milled (Bates *et al.*, 1929). The Federal grain standards designate wheat that has an unmistakable odor of smut or that contains smut balls, portions of balls, or spores of smut in excess of a quantity equal to 14 balls of average size in 250 grams of wheat as "light smutty" and wheat with smut balls and spores in excess of 30 balls in 250 grams as "smutty." These descriptive terms are included with the regular grade designation. Smutty wheat is discounted, frequently more than the cost of cleaning the grain.

SYMPTOMS. The symptoms of bunt usually are not apparent until the wheat is headed. However, some varieties of wheat infected with certain races of the parasites show dwarfing of the plants, small light-colored spots on the leaves, and a grayish cast of the foliage and culms during the period of tillering and internodal elongation (Angell, 1934, Churchward, 1934, Holton and Rodenhiser, 1942). According to Helyet *et al.* (1938), root development in smutted plants is reduced from shortly after heading to maturity. The smutted plants of many varieties appear bluish green to grayish green in color, and the heads frequently show characters unlike the healthy spikes of the variety (Fig. 59). The odor of trimethylamine is characteristically present from the period of spore formation to maturity and in the threshed grain. The smut balls that replace the kernels frequently are conspicuous in the smutted spike. The smut balls, somewhat the shape of the wheat kernels, are grayish green changing to brown as the grain ripens. The presence of the smut balls and the dark spores adhering to the kernels are evident in the threshed grain. The effect of bunt on the morphology and physiology of the wheat plant is reviewed by Holton and Heald (1941). In the dwarf bunt, the symptoms are evident by the reduced internodal elongation of the infected plants, and the spore balls are nearly round instead of the shape of the kernels.

THE FUNGI

1. *Tilletia caries* (DC.) Tul.
   (*Uredo caries* DC.)
   (*Lycoperdon tritici* Bjerk.)
   (*Tilletia secalis* (Cda.) Kühn)
   [*Tilletia tritici* (Bjerk.) Winter]

The latter synonym (*T. tritici*) was used extensively as the binomial for this fungus until recently.

The sori are formed in the ovaries, the pericarp of the caryopsis persists, and the sori are more or less the same shape as the kernel. The chlamydospores are brownish

FIG. 59. Bunt, or stinking smut, of wheat showing healthy (*left*) and infected (*right*) spikes of two wheat varieties (*A*) and (*B*) and the modification of the spikes by the smut. Healthy kernels and smut balls are shown (*C*). The chlamydospores of *Tilletia caries* (*a*) and *T. foetida* (*b*) are shown, highly magnified, in the insert.

black, globose to subglobose, light to dark brown, reticulate with reticulations ranging from minute shallow meshes to deep indentations, and 15 to 23 microns in diameter (Fig. 59). The sterile cells are globose to subglobose, thin-walled, hyaline, smooth to faintly reticulate, and 12 to 18 microns in diameter. The spores germinate to form a promycelium and an apical crown of long filiform sporidia, variable in number, usually 8 to 16, which fuse in pairs while still attached.

2. *Tilletia foetida* (Wallr.) Liro
   (*Erysibe foetida* Wallr.)
   (*Ustilago foetens* Berk. & Curt.)
   [*Tilletia foetens* (Berk. & Curt.) Schr.]
   (*Tilletia laevis* Kühn)

Tilletia foetens was used extensively for this fungus until recently.

The sori are formed as in the previous species. The chlamydospores are globose to ovoid or elongate, light grayish brown to olivaceous brown, smooth, usually 17 to 20 microns in diameter or 17-20 by 18-22 microns in size (Fig. 59). Sterile cells are hyaline, thin-walled, smaller than the spores, mostly about 14 microns in diameter. Germination is similar to the former species. Holton (1944) described the morphology of hybrids between these two species.

3. *Tilletia controversa* Kühn

Tilletia controversa was designated as a variety of (*T. tritici*) *T. caries* and later described as a new species, *T. brevifaciens* Fischer, before early European specimens were found.

Sori are formed as in the previous species, characteristically globose; host plants are considerably stunted one-third to one-fourth normal height. Spores are hard, subglobose to globose, yellowish to grayish brown, surface exposure with large polygonal reticulations and encased in a hyaline, almost imperceptible, gelatinous sheath, diameter 17 to 22 microns. Sterile cells few and usually appear hyaline or pale green, globose with smooth wall occasionally enclosed in gelatinous sheath. The gelatinous wall of the spore is very impermeable to water and chemicals and requires a long period of soaking at low temperatures before germination (Holton et al., 1949, Lowther, 1948). Germination is similar to the other species.

4. *Tilletia indica* Mitra
   Neovossia indica (Mitra) Mundkur

Based on Fischer's (1953) limitations of the genus *Neovossia:* "Spores simple, produced singly within the end cells of special sporogenous hyphae which invest the spores and remain attached as an investing sheath and a long pedicel-like appendage on the spore," this species is placed in *Tilletia* as in the similar kernel smut of rice (Chap. 7).

Sori are formed in ovaries frequently replacing only part of the tissues of the mature caryopsis. Pericarp and kernel shape are not modified greatly by the sorus. Spores are mostly globose, brown with a light-brown tuberculate exposure embedded in a pale-yellow to hyaline sheath often ending at one side in a short apiculus, diameter average 35 microns (Mitra, 1931). Germination is by the formation of an unbranched basidium and apical whorl of numerous sporidia. Fusion of attached sporidia is rare, as most primary sporidia are binucleate (Holton, 1949).

ETIOLOGY. The etiology of the 3 species is similar. Seedling infection occurs, resulting in a systemic invasion of the seedling primordium, and spores develop in the ovaries. The inoculum is from seed-borne chlamydospores or spores in the soil in the drier winter wheat areas and in all areas with dwarf bunt. Churchward (1940) has investigated penetration and establishment of the fungus in the wheat seedling. Penetration and the early phases of systemic infection apparently occur in the resistant wheats. Environmental conditions influence the infection and establishment of the parasite in the wheat seedlings, as reviewed by Holton and Heald (1941).

Apparently in *T. indica* aerial sporidia enter flowers after pollination and infection of the ovary and spore development occur during the period of kernel development. The durum wheats tested were resistant and the common wheats (*vulgare*) susceptible (Bedi *et al.*, 1949).

CONTROL. Seed treatment and the use of bunt-resistant varieties are the combined methods of control. The use of seed treatment even with bunt-resistant varieties is recommended to prevent the propagation and distribution of races of the pathogen to which the resistant variety is susceptible. This recommendation is based on field experience with resistant varieties and on investigations of the occurrence of new physiologic races in nature or by experimental hybridization (Holton, 1947, 1953, Rodenhiser and Holton, 1953). Seed treatment for bunt control is associated closely with the history of plant pathology (Holton and Heald, 1941). The earlier use of copper compounds and formaldehyde is superseded by the volatile organic mercury compounds, although copper carbonate dust is used in some areas. The dwarf bunt is not amenable to control by these treatments as generally used (Bamberg, 1941, Fischer and Hirschorn, 1945). Many wheat varieties resistant to groups of races of the pathogen are known. Certain of these, such as Hohenheimer (C.I. 11458), Hussar (C.I. 4843), Martin (C.I. 4463), Oro (C.I. 8220), Rio (C.I. 10061), and Turkey (C.I. 1558 and 10016) winter wheats and Florence (C.I. 4170) and Hope (C.I. 8178) spring wheats are resistant to many races of the three pathogens. Varieties resulting from hybrids between these resistant wheats are used successfully in controlling bunt (Briggs and Holton, 1950, Holton and Jackson, 1951, Rodenhiser and Holton, 1945).

The bunt-resistant varieties used and the inheritance of resistance are summarized by Briggs and Holton (1950), Holton and Heald (1941), and Roemer *et al.* (1938). Briggs and coworkers (1933, 1940, 1945, 1950) have explained the intermediate percentages of bunt in resistant lines on the basis of multiple factors, with certain factor pairs functioning as weak factors in conditioning resistance to specific groups of races of the pathogens. Their genetic analyses have defined five major gene pairs and two weak: Martin MM and M_2M_2, Hussar HH, Rio RR, Turkey TT, and the weak genes $XXYY$. The R and T genes are linked closely, and a variety carrying both would give a reaction similar to the *T* gene acting alone. The Martin and either the Turkey or Rio gene together give resistance to 25 races of *T. caries* and *T. foetida* and several races of *T. controversa*.

Specialization in the three species is clearly defined and apparently fairly stable (Holton and Rodenhiser, 1942, Rodenhiser and Holton, 1945, 1953). Lines homozygous and heterozygous for pathogenicity have been demonstrated by inbreeding (Holton, 1953). Hybrids between races have been produced, and they have been demonstrated as occurring in nature. Certain races produce specific morphological or physiological responses on particular wheat varieties. The reaction of physiological races of the three species to the differential wheat varieties, as summarized by the authors cited above, is given in Table 12.

Briggs and Holton (1950) arrange 25 races of *T. caries* (T-) and *T. foetida* (L-), as determined on the standard set of wheat differentials, into six groups based on their control by the respective gene pairs conditioning resistance. Stated conversely, the races are regrouped on the basis of having genes capable of inciting bunt in wheats with the specific genes or gene combinations. Group I (T-1, T-2, T-9, T-10, L-1, and L-2) are controlled by either the M or H gene pairs. Group II (T-3, T-11, T-16, L-3, L-8, and L-10) are controlled by the M, but not by the H gene. Group III (T-4, T-6, T-12 and L-4) are controlled by the H, but not by the M gene. Group IV (T-5, T-7, T-8, T-15, L-5, and L-7) are not controlled by either the M or H genes. Group V (T-13 and L-9) differ from those in Group IV only by the resistance of Martin and Odessa which carry a second gene M_2. Group VI (T-14) controlled by either the M_2 or H gene pairs.

23. Flag Smut, *Urocystis tritici* Koern or *Urocystis agropyri* (Preuss.) Schröt. The disease is distributed widely throughout the world on wheat, but apparently it is of major importance only in a limited number of areas. Flag smut on wheat was found first in the United States in 1919 (Humphrey and Johnson, 1919), and it has appeared since in scattered localized areas in the South Central winter wheat region and in the Pacific Northwest. The disease was eliminated partly in these local areas by a shift in crops and by the use of resistant varieties of wheat in the adjacent territory. Flag smut is of major importance only in the winter wheat sections of mild winter climates and with the general use of susceptible varieties and continuous wheat culture. According to Miller and Millikan (1934), the flag smut of wheat in South Australia in 1931 was damaging 3 to 4 per cent of the crop, owing to the general use of susceptible varieties and continuous wheat culture. The smut damage started decreasing with the wider use of the resistant variety Nabawa. Noble (1937) reported for New South Wales as follows: "Flag smut has

TABLE 12. REACTION OF WHEAT VARIETIES TO PHYSIOLOGIC RACES OF
Tilletia caries AND *T. foetida*

Physiologic race number	Ridit (C.I. 6703)	Oro (C.I. 8220)	Hohenheimer (C.I. 11458)	Hussar (C.I. 4843)	Albit (C.I. 8275)	Martin (C.I. 4463)	White Odessa (C.I. 4655)	Ulka (C.I. 11478)	Marquis (C.I. 3641)	Canus (C.I. 11637)	Mindum (C.I. 5296)
T. caries											
T- 1	R	R	R	R	R	R	R	S	I	R	R
2	R	R	R	R	R	R	R	S	R	R	S
3	R	R	R	R	R	R	R	S	S	S	I
4	R	R	R	R	I	S	S	S	S	R	I
5	R	R	R	R	I	S	S	S	S	S	I
6	R	R	R	R	S	S	S	S	S	R	I
7	R	R	R	I	S	S	S	S	S	I	I
8	R	R	R	S	S	S	S	S	S	S	I
9	R	R	I	R	R	R	R	S	I	R	I
10	R	R	S	R	R	R	R	R	I	R	R
11	S	R	R	R	R	R	R	I	S	S	I
12	R	R	S	R	S	I	S	S	R	R	
13	S	R	R	S	S	R	S	—	I	I	
14	R	R	R	R	S	R	S	S	R	R	
15	R	R	S	S	S	S	S	S	S	S	
16	R	S	S	R	R	R	R	S	I	I	
17	R	S	R	R	R	R	R				
T. foetida											
L- 1	R	R	R	R	R	R	R	S	I	R	I
2	R	R	R	R	R	R	R	S	S	R	I
3	R	R	R	R	R	R	R	S	S	S	I
4	R	R	R	R	S	S	S	S	I	R	I
5	R	R	R	R	S	S	S	S	S	S	I
6	R	R	R	I	S	S	S	S	S	S	I
7	R	R	R	I	S	S	S	S	S	S	I
8	R	S	R	R	R	R	R	S	S	S	I
9	S	R	R	I	S	R	S	S	I	S	
10	I	R	R	R	R	R	R	S	I	S	
11	R	R	R	R	R	R	R	S	I	I	
12	R	R	R	R	S	R	R	S	I	R	
13	R	R	S	R	R	R	R	S	S	R	
14	R	R	R	R	S	I	S	S	S	R	
15	R	R	R	S	S	S	S	S	S	I	

T-1, etc., race numbers for *T. caries*, L-1, etc., race numbers for *T. foetida*.

R—resistant (0 to 10 per cent infection); I—intermediate (11 to 40 per cent); and S—susceptible (41 to 100 per cent).

ceased to be the serious disease it was some years ago before resistant varieties were grown. . . ."

SYMPTOMS AND EFFECTS. The disease is evident from the late seedling stage until maturity of the crop. The early symptoms of the smut are the gray to grayish-black linear sori in the older leaf blades and leaf sheaths (Fig. 60). The sori form in the mesophyll tissue between the veins, and they are covered by the epidermis of the leaf during the earlier stages of development. Later the epidermis ruptures, releasing the black spore mass, and finally the leaf tissue frays along the linear sori. The symptoms appear on the leaves as they unfold and in the culm tissues as the culms elongate. In most susceptible varieties, the plants are dwarfed by reduced internodal elongation. Usually spike development is stopped prior to its emergence from the leaf whorl. The disease in wheat and most grasses is characteristically a leaf smut in contrast to the stalk smut produced by the species on rye.

THE FUNGUS

Urocystis tritici Koern. or *Urocystis agropyri* (Preuss.) Schröt. See Fischer (1953) for synonyms.

Fischer (1943) suggested the combination of the morphologically similar species *Urocystis tritici*, *U. occulta*, and *U. agropyri* (Preuss.) Schroet. on wheat, rye, and grasses, respectively, under *U. agropyri*, recognizing the specialized varieties of the fungus on the different cereals and grasses. Fischer (1953) includes *U. tritici* under *U. agropyri* and retains *U. occulta* on rye largely on the basis of smaller spore balls, incomplete investment of spore ball by sterile cells, and the location of sori in the culm and leaf sheath rather than in the leaves as in *U. agropyri*. Long usage argues against this change as in the case of the generic name *Urocystis* (Zundel *et al.*, 1940).

The sori are linear, black, at first covered by the epidermis, chiefly in the leaves and upper culm tissues and frequently stunting plant development and aborting the spike (Fig. 60). The spore balls are globose to oblong, measure 18–35 by 35–40 microns, are composed of one to four dark chlamydospores, and are hyaline to brown with somewhat smaller sterile cells mostly completely surrounding the fertile cells. The chlamydospores are angular to globose, dark reddish brown, smooth, and 14 to 20 microns in diameter. The spores germinate in place to form a short promycelium, with or without septations, and three to four hyaline cylindrical sporidia are borne near the apex.

ETIOLOGY. Seedling infection from soil or seed-borne chlamydospores is followed by the systemic invasion of the primordia and the development of spores in the unfolding leaf tissues. The spores are in part soil-borne in the regions where the disease is severe. Environmental conditions influence both the infection and the complete establishment of the mycelium in the primordial tissues, as reported by Angell (1934), Angell *et al.* (1937, 1938), Faris (1933), Jones and Seif-El-Nasr (1940), Noble (1923, 1924, 1934), and others.

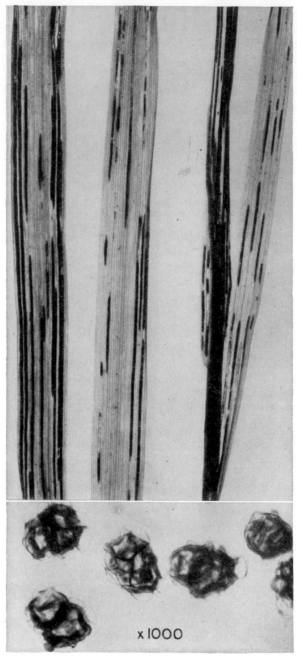

FIG. 60. Flag smut of wheat caused by *Urocystis tritici* or *U. agropyri* and chlamydospores, highly magnified.

CONTROL. Seed treatment, crop rotation, and the use of resistant varieties are important control measures. Seed treatment is partly effective under conditions of soil infestation. Resistant varieties, as cited earlier, offer the most practical means of control. The lists of the important resistant wheats and the genetics of inheritance of resistance are reported by Carne and Limbourn (1927), Pridham and Dwyer (1930), Shelton (1924), Shen (1934), Tisdale et al. (1923, 1927), Wu (1949), Yu et al. (1931, 1934, 1936, 1945), and others.

Physiologic specialization occurs in *Urocystis tritici*. According to Yu et al. (1936, 1945), 12 races differentiated on five wheats occur in China. However, Verwoerd (1929), in Africa, found no evidence of specialization in collections he tested. Holton and Johnson (1943) have differentiated two races in the United States that are separated widely geographically: race 1 from Kansas and probably representative of the Illinois, Missouri, Kansas area and race 2 from Washington. Oro × Federation–38 and –40 are resistant to race 1 and susceptible to race 2. Oro (C.I. 8220) apparently gives the same differential reaction as the two hybrid selections. Hafiz (1951), comparing the reaction of the pathogen from several major world areas differentiated: the races 1 (Kansas) and 2 (Washington) from the United States with the Italian and Australian collections similar to race 2, the distinct race 4 from Pakistan with some similarity to the Italian collections, and a questionable group 3 from China and Cyprus with no infection on the wheat varieties used. The varietal reaction of the pathogen from Azerbaijan indicates a possible fifth group although similar to 2 (Gorlenko, 1946). Holton and Jackson (1951) list Golden (C.I. 10063), Rio (C.I. 10061), a selection from a Crimean wheat resistant to many races of the bunt pathogens but susceptible to dwarf bunt, and Rex (C.I. 11689), possibly Rex Ml, as resistant to both flag smut and dwarf bunt.

24. Stem Rust, *Puccinia graminis tritici* Eriks. & Henn. Stem rust is distributed generally with the wheat crop. In the drier areas, the disease develops in epiphytotic form only in moist seasons. This disease probably has caused greater and more spectacular damage than any other disease of the wheat crop. Losses are usually higher in the spring wheat sections of North America than in either the soft or hard winter wheat areas, as summarized by Craigie (1944, 1945). This is due apparently to two main factors: (1) the relatively high summer precipitation in the spring wheat areas and (2) the plant growth occurring over a longer period of favorable summer conditions. Greaney et al. (1941) have reviewed the earlier literature on the effect of stem rust on composition and have shown that nitrogen is decreased and ash constituents are modified. Yield of grain is reduced, and composition is modified, varying considerably with the stage of plant growth when the rust develops. In addition to wheats and other small grains, a large number of grasses are susceptible to this and other physiologic varieties of the pathogen.

SYMPTOMS AND EFFECTS. Two stages of the rust occur on the wheats and grasses. The red rust or uredial stage is evident on the leaves and culms at any stage of plant growth. The uredia are reddish brown in color, usually oblong in shape, and the epidermis of the leaves and culms is ruptured and pushed back around the pustule. The uredia of this rust are distinguished readily from those of the leaf rust, except perhaps in the very early spring or late autumn when the uredia of the two rusts appear somewhat similar. The differences in morphology of the uredio-spores of the two are always distinguishing characters. In resistant varieties, development of the uredia is reduced or prevented, and yellow flecking or brown necrosis are the characteristic symptoms (Stakman et al., 1944). The black rust or telial stage develops more abundantly on the leaf sheaths and culms of the rusted plants, especially during and just prior to maturation of the wheat tissues. The telia are oblong to linear, dark brown to black, and the teliospores are exposed. Severe stem rust development results in numerous uredia and telia on the leaves, culms, and spikes and the drying out and early maturity of the wheat plant.

The aecial stage of this rust occurs on *Berberis* and *Mahonia* spp. early in the spring. The orange-yellow lesions are common on the leaves, petioles, and blossoms of several *Berberis* spp. The pycnial (sperma-gonial) lesions occur first, usually on the upper surface of the leaf. The lesion is slightly elevated, orange yellow, and produces an exudate when mature that attracts insects. The aecia develop on the undersurface of the leaf immediately below the spermagonial lesions or surrounding them where the infections develop on the petioles or blossoms. The elongated aecial cups with serrate marginal peridia are conspicuous over the surface of the lesion (Fig. 61).

THE FUNGUS

Puccinia graminis tritici Eriks. & Henn.
(*Uredo linearis* Pers. and others)
(*Uredo graminis* Eriks. & Henn.)
(*Uredo frumenti* Mart.)
(*Puccinia cerealis* Mart.)
(*Puccinia linearis* Roehl.)
(*Puccinia agropyri* Otth.)
(*Lycoperdon lineare* Schrk. and others.)
(*Erysibe linearis* Wallr.)
(*Caeoma lineare* Schlecht)
(*Caeoma berberidis* Schlecht)

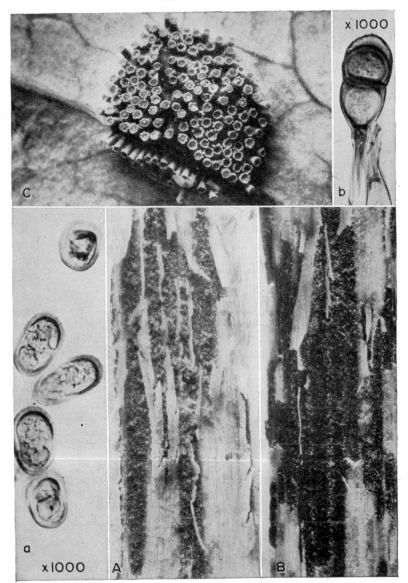

Fig. 61. *Puccinia graminis.* The aecium on the leaf of *Berberis vulgaris* (*C*) and the uredia (*A*) and telia (*B*) on wheat are shown, × 20. The urediospores (*a*) and teliospores (*b*) are shown magnified, approximately × 1,000.

(*Aecidium elongatum* Lk.)
(*Aecidium lineare* Gmel.)
(*Aecidium berberidis* Gmel.)

The synonomy and history of this fungus is given in more detail by Sydow.[1] Excellent color plates of the spore forms and symptoms are included with a detailed discussion of the disease by Eriksson and Henning.[2]

The spermagonia (pycnia) and aecia develop on some of the *Berberis* and *Mahonia* spp., chiefly *B. vulgaris* L. Prior to aecial development the flask-shaped pycnia of the upper surface of the leaf bear numerous hyaline thread-like spermatia and flexuous paraphyses-like hyphae in a slightly viscous exudate. Allen (1930, 1933), Buller (1938), Craigie (1927, 1928, 1931), and others have demonstrated the role of the spermatia and flexuous hyphae in the initiation of the binucleate or dicaryotic stage of the fungus. The aecial stage develops on the undersurface of the leaf below the spermagonium in which the fusion of spermatium and flexuous hypha occurs. The aecial lesion is elevated, orange yellow, and numerous shallow tube-shaped aecial cups develop on the surface of the lesion (Fig. 61). The length of the peridial tubes varies with environmental conditions. The margins of the peridia are turned out and serrated. Aeciospores are borne in chains within the cups. Aeciospores are subglobose to hexagonal, light orange-yellow, smooth, and germinate by the formation of a hyaline germ tube.

The uredia develop on the cereals and grasses. The uredia are oblong to circular with the epidermis ruptured and turned back around the margin of the pustule, orange red to chestnut brown in color, and dusty in appearance because of the presence of the numerous urediospores (Fig. 61). The urediospores are elliptical or pyriform with four conspicuous germ pores arranged around the equator of the spores, dark orange-yellow, echinulate, and free from the short pedicels when released (Fig. 61).

The telia form in the uredia and independent of the uredia as the cereal and grass plants approach physiological maturity. The telia are naked, oblong to linear, and dark brown to black, with the ruptured epidermis usually conspicuous around the margins. The teliospores are borne on persistent pedicels. The spores are fusiform to clavate, two-celled, slightly constricted at the septum with the apex thickened and rounded or slightly pointed, smooth, and chestnut brown in color. The teliospores germinate in place after a rest period to form a four-celled promycelium (basidium) and laterally borne sporidia on minute pointed sterigmata.

Specialized varieties of the fungus are restricted to the larger groups of the cereals and grasses. It is a common practice to indicate these varieties by the use of trinomials (Eriksson, 1894, Eriksson and Henning, 1896). Numerous physiologic races occur within these varieties, as summarized by Stakman *et al.* (1944).

ETIOLOGY. The parasite is characteristically a heteroecious long-cycle rust fungus. The complete cycle occurs in the cooler climates where the barberry, cereals, and grasses occur in close proximity. The teliospores are formed as the infected cereal and grass tissues approach maturity. The teliospores remain dormant over winter and usually require low temperatures such as freezing and thawing to germinate. Nuclear

[1] Monographia Uredinearum, vol. 1, pp. 692–698.   1904.
[2] Die Getreideroste.   1896.

fusion and the initiation of the true diploid phase of the parasite occur during the teliospore maturation or prior to germination.   The teliospores germinate to produce basidia with characteristically four haploid cells and laterally borne sporidia.   The sporidia are wind-borne and infect the susceptible *Berberis* spp. by direct penetration through the epidermis. The spermatia and flexuous hyphae form in the initial (spermagonial or pycnial) lesions on the barberry.   The spermatia are transferred largely by insects or meteoric water, and fusions between the spermatia and flexuous hyphae apparently initiate the binucleate or dicaryotic phase of the fungus.   The fungus is predominantly heterothallic in nature. Inbreeding has been reported by Johnson and Newton (1938) and Newton and Johnson (1937) as well as the abnormal production of uredia and telia on the barberry.   The literature on hybridization between varieties and races of the pathogen and on inbreeding of races is summarized by Johnson (1953).   The aecia develop following the sexual fusion, and aeciospores are produced in great abundance early in the spring.   The aeciospores are wind-borne to the gramineous plants, where infection occurs through the stomata.   The mycelium in the cereal and grass tissues produces the urediospores which cause numerous secondary infections on the susceptible cereals and grasses.   The urediospores frequently furnish the inoculum for the progressive spread of the rust over an extensive area, as in the spread northward in North America and Europe (Craigie, 1945, Lehmann, 1937).   Later as these plant tissues approach physiological maturity, the mycelium produces teliospores. As in most of the heteroecious rust fungi, two widely different suscepts support the aecial and the uredial and telial stages in the complete life cycle of *Puccinia graminis*.

The barberry supports the sexual stage of this economically important fungus.   The discovery of the role of the spermatia and the flexuous hyphae borne in the spermagonia on the barberry reemphasized the importance of the aecial host in inducing genetic variability in the parasite.   The barberry species are susceptible to infection by sporidia of many of the different specialized varieties and numerous physiologic races of the parasite.   Therefore cross fertilization between these numerous races is possible and affords the mechanism for the production of new potentially dangerous races.   Investigations by Johnson *et al.* (1932, 1934), Newton *et al.* (1930, 1931, 1932), and Stakman *et al.* (1930, 1934) and summarized by Johnson (1953) have demonstrated experimentally the role of the barberry in the production of hybrids among the various specialized varieties and physiologic races.   The significance of the barberry, therefore, becomes twofold: (1) the completion of the life cycle of the parasite in the cooler temperate zones, the local propagation of newly introduced races of the pathogen, and early spring spread of the rust to the cereals and (2) the production of new potentially dangerous races of the parasite through hybridization occurring in the aecial stage on the barberry.

Environmental conditions influence greatly the etiology of stem rust. Infection and development of the rust on the cereals and grasses and the barberry are determined in large part by temperature and moisture conditions. In nature, low temperatures or freezing and thawing are necessary before the teliospores will germinate. The reaction of many resistant varieties is influenced also by high temperatures. The persistence of the parasite in the uredial stage either as perennial mycelium or freshly produced viable urediospores is determined by mild winters and moist summers. The spread of urediospores from regions of mild climate to those of less favorable environment is important in the production of stem rust epiphytotics. Craigie (1945) has reviewed the mass of literature on the epidemiology and the control of stem rust.

CONTROL. The control of stem rust is largely by means of barberry eradication and by the use of resistant varieties. Barberry eradication is important in the temperate zones where the aecial infection occurs naturally. The breaking of the cycle of the parasite, in the regions where the uredial stage does not survive the winter, prevents the early spring spread of the rust to the cereals and grasses. In areas where the barberry is abundant and the climate favorable for the rust, as in the Transcaucasian mountains, small-grain production is impractical and corn is the chief grain produced. The elimination of the susceptible barberries prevents further genetic variation of the parasite through hybridization and recombination of characters for broader parasitic potentialities. Many countries have passed laws preventing the sale, distribution, and growing of the susceptible barberry species. Early-maturing varieties of the grains and early planting aid in escaping stem rust damage.

Varieties resistant to stem rust offer the most practical means of control. However, the history of stem rust control by the use of resistant varieties illustrates the complexity of the problem and the inadequacy of basic information on the genetics of the hosts and pathogen, on the physiology of the fungus, and on the nature of resistance. Information available indicates at least two types of resistance to stem rust, morphological and physiological. Differentiation as mature-plant resistance and physiologic resistance frequently is suggested. Probably most resistant wheat varieties now known express in some degree both types of resistance. The present knowledge that some varieties are resistant in the seedling stage to a given race or group of races of the pathogen and susceptible in the adult plant stage, or the reverse, susceptible in the seedling stage and resistant in the adult stage (Campos *et al.*, 1953), does not define the type of resistance but rather illustrates the complex nature of resistance.

Morphologic resistance to stem rust probably involves more specifically the extent of damage incited by the pathogen on the graminicolous hosts. The rate of penetration and number of haustoria entering the host cells possibly are involved in cell-wall developmental morphology. The rate and extent of development of sclerenchyma and other thick-wall cells delimiting the parenchymatous or mesophyll regions determine the spread of mycelium, the size of sori, and the area of ruptured epidermis (Allen, 1923, Hart, 1931, Peterson, 1931, Stakman and Hart, 1943). The development of wheats with semibrachytic culm type of growth increases the ratio of grain yield to straw and reduces lodging. In such wheats and barleys the culm tissues are enclosed in the leaf sheaths until maturity. The culms of such varieties are protected from stem rust, and total rust damage is reduced materially. The combination of morphologic resistance with physiologic resistance is desirable.

Physiologic resistance involves the degree of compatibility between the cell protoplasts of host and pathogen. The rust pathogen is an obligate parasite and, therefore, dependent upon the host cells for its nutrition. The physiology of this type of parasitism is either a direct nutritive relationship in which the host supplies the synthesized metabolites and specific compounds necessary for fungus development or an indirect one in which the host supplies the missing link in one or more enzyme systems or the specific hormones essential for fungus metabolism and development. A compatible reaction between the cells of host and pathogen establishes this physiological relationship, and rust develops. An incompatible reaction blocks the development of the pathogen. All degrees in the interaction called compatibility are evident in various combinations of host and pathogen genotypes. The reaction types 0, 1, 2, 3, 4 for size of uredia and chlorosis or necrosis are used to express degree of compatibility (Fig. 62). The flecking type of reaction and other types of local necrosis are external manifestations of this incompatibility. Stakman (1915) described the reaction as "hypersensitivity" and suggested the nutritional isolation of the obligate parasite by the collapse of the host cells surrounding the advancing hyphae. The interaction between some varieties of wheat and specific races of the pathogen as in *Triticum timopheevi* and race 56 of *Puccinia graminis tritici* frequently is manifest by collapse of the appressorium over the stoma or of the substomatal vesicle as it forms in the substomatal cavity. In contrast race 15B establishes a compatible relationship in this host.

The nature of the physiological type of resistance is not simple. Compounds synthesized or stored in the fungus regulate germination, type of

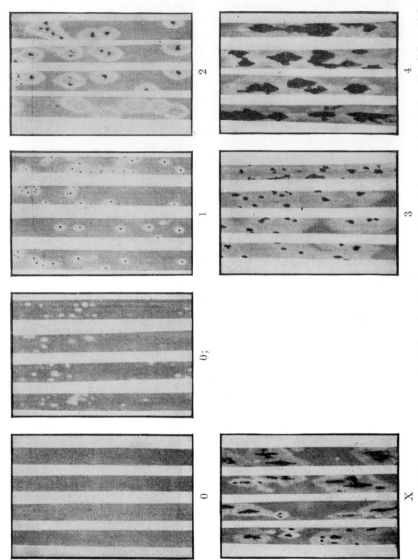

FIG. 62. Uredial infection types produced by physiologic races of *Puccinia graminis tritici* on the differential varieties of *Triticum* spp. 0, 0;, 1, and 2 are resistant types of reaction; X is mesothetic (resistant and susceptible on the same plant); and 3, 4 are susceptible reactions.   (*Courtesy of Conference for the Prevention of Grain Rust.*)

mycelial growth, and other physiological processes and morphological development (Allen, 1955). Compounds present in the host change permeability of the fungus membranes, supply general as well as specific metabolites, and apparently act specifically on the protoplasm of the pathogen. Certain of these processes and their interactions are modified by environment, especially light and temperature. Only a relatively few stem rust-resistant wheats express physiological resistance in either the seedling or adult stage at 29°C. (84°F.) and reduced light.

Physiologic resistance is specific for specific pathogenicity factors in the pathogen. There is a specific interaction between the rust-conditioning factors or genes of the host and the factors or genes for virulence in the fungus. The combination of pathogenicity factors in the presently designated physiologic races are not classified against the factors for physiologic resistance in the various wheats and related genera. The factors conditioning pathogenicity or virulence in the fungus have been demonstrated, but little is known about their number or inheritance (Johnson, 1953). Several factor pairs in the wheats studied individually condition resistance to a group of these races (Aamodt, 1923, Johnson, 1953, Neatby, 1931). However, no wheat practical for use in commercial production is resistant to all races of *Puccinia graminis*.

Present practice in breeding for resistance to stem rust is based largely on concentrating in one wheat variety factors governing resistance to all the races present in the region. In the period 1920 to 1950 few factor pairs at most were involved in protecting commercially grown wheats from stem rust. When a new race appears capable of attacking this variety, the source of resistance necessary to control this race is added to the variety. In field practice the extensive use of such varieties results in the screening out and increase of the race capable of attacking these varieties almost as rapidly as the new sources of resistance can be added. This is well illustrated in Ceres susceptible to race 56, the varieties derived from Hope and the durums susceptible to race 15B, and Eureka susceptible to 126B (Borlaug, 1954, Craigie, 1945, De Urries, 1951, Johnson, 1953, Stakman *et al.*, 1943, 1954, Vallega, 1955, Watson and Waterhouse, 1949).

An alternative plan involves the introduction into a commercially acceptable, ecologically adapted variety of several distinct factors or factor groups for resistance followed by a selection process resulting in several agronomically similar varieties possessing these different sources of stem rust resistance. The several types of resistance give better protection against the increase and wide distribution of races capable of attacking any one of these sources of resistance. Wheat production is stabilized in so far as stem rust losses are involved by having adapted resistant varieties in reserve to replace the susceptible when these shifts in race populations occur (Athwal and Watson, 1954, Borlaug, 1954, Watson, 1949, Watson and Waterhouse, 1949).

Inheritance of resistance to stem rust has received increasing attention since first placed on a scientific basis by Biffen (1905) and Nilsson-Ehle (1909). Resistance to stem rust is controlled by single dominant factor pairs or at most a few factor pairs. However, the total number of factors interacting is large when all of the races of the pathogen are involved. Minor factors apparently modify the type and extent of the reaction in some instances (Clark, 1936).

The earlier studies on inheritance involved the transfer of stem rust resistance from the emmer ($n = 14$) wheats to the vulgare ($n = 21$) wheats. Some of the more important wheats resulting from these interspecific crosses in which inheritance has been studied are Marquillo (C.I. 6887) and lines (Hayes et al., 1920, 1925, 1955); Hope (C.I. 8178) and H-44 (McFadden 1930, 1949) involving emmers; and Timstein (C.I. 12347)[1] and unnamed selections involving T. timopheevi (Allard and Shands, 1954, Ko and Ausemus, 1951). Kostoff (1936, 1938) produced T. timococcum reported with high resistance in crosses involving T. timopheevi with T. monococcum and common wheats although no inheritance studies are reported. Probably some of the stem rust-resistant wheats from the Mediterranean area and South America contain emmer parentage from natural crossing (Love, 1951). The inheritance of resistance to stem rust in this group appears to be conditioned by two, usually dominant, factor pairs, each factor functioning independently for separate groups of races unless the two factors are linked as reported by Allard and Shands (1954), Ausemus (1943), Hayes et al. (1955), and Johnson (1953).

The inheritance of stem rust resistance in the common wheats, notably Kanred (C.I. 5146), McMurachy (C.I. 11876), and the Kenya varieties, again, is relatively simple. Aamodt (1923) showed a single dominant factor in Kanred conditioned resistance to 16 races. Watson (1941, 1943) found resistance conditioned by single factors with the factor carried by McMurachy allelomorphic to one in the Kenya. Later investigations by Watson and associates (Watson and Waterhouse, 1949, Watson and Singh, 1952, Athwal and Watson, 1954) indicate a series of independently inherited, mostly dominant factors in the Kenya selections. Recent unpublished investigations of Borlaug and associates in Mexico and of several investigators in the United States and Kenya indicate that at least three factor pairs conditioning resistance to a large number of races under favorable conditions are combined in Kenya Farmer (C.I. 12880). Other sources of resistance for further study and use are being

[1] Recent information indicates this accession number is not Timstein derived from T. timopheevi.

catalogued by means of the International Wheat Nursery established by the late Dr. B. B. Bayles and given official status following the International Rust Conference in 1953. The wheats investigated and their rust reaction are included in annual reports issued by the Cereal Section, Field Crops Research Branch, Agricultural Research Service, U.S. Department of Agriculture, Beltsville, Md.

Physiologic specialization is another expression of the rather specific compatibility between the rust pathogen and its suscept. According to Stakman *et al.* (1944), biotypes form the basic concept for physiologic races. By using the uredial infection types produced on 12 varieties and species of *Triticum* (Fig. 62), more than 189 physiologic races of *Puccinia graminis tritici* are differentiated. On a gene basis for virulence, *i.e.*; the establishment of the uredial stage of the pathogen in the tissues of a host containing a specific gene conditioning resistance, the above classification is inadequate. The identification and transfer of genes or gene groups at a single locus conditioning resistance into suitable tester hosts, preferably into a common genotype, would result in a more useful, cumulative catalogue of races (Johnson, 1953, Vallega, 1955).

Control of stem rust by means of protectant and eradicant fungicides has been investigated, but with less emphasis than breeding for resistance. Sulfur has been recognized as a good rust fungicide even before resistance to rust was reported. The problem of keeping sulfur on the young host tissues as they are exposed to infection has been economically impractical. With the introduction and use of organic sulfur compounds and chemicals, functioning systemically in plants, investigations on chemical control have increased greatly. The carbamates control rust, but require several applications to ensure adequate protection. Some of the sulfa drugs are effective as fungicides, but again presently known compounds are not economically practical. Some of the sulfamates control rust; however, those effective at present affect adversely both germination and flour quality. Some other chemicals acting as systemics are either highly phytotoxic or ineffective under field conditions. Present expansion of the investigations with newly synthesized compounds and modern spraying and dusting equipment probably will make chemical control practical as an additional method of defense (Hassebrauk, 1951, Hotson, 1953, Livingston, 1953).

25. Stripe Rust, *Puccinia glumarum* (Schm.) Eriks. & Henn. The stripe rust is restricted in distribution in contrast to both the stem rust and leaf rust of wheat. This rust is found along the Pacific Coast and Intermountain areas of North America. It extends east of the Rocky Mountains in the northern, cooler sections of Canada. Essentially the same distribution occurs in South America with the rust spreading out eastward over the cooler plains area of southern Argentina. Again in

Europe and Asia, the stripe rust is prevalent in the northern and eastern cooler regions and in the mountainous areas of Central Europe and Asia. Apparently this rust is limited in its distribution to the areas of relatively cool summer temperatures and humid winters.   The stripe rust is common on wheat, barley, rye, and many grasses.   In areas where the disease is severe, it causes reduction in yield of the grains and grasses, as shown by Bever (1937).   The rust is known as glume rust and yellow rust as well as stripe rust.

SYMPTOMS.   The stripe rust appears early in the spring before the other rusts are abundant.   This is especially evident in sections of mild winters.   The linear, narrow, citron-yellow uredia appear on the autumn foliage and on new growth early in the spring.   The uredia coalesce to produce long stripes between the veins on the leaf blade and sheath and small linear lesions on the floral bracts (Fig. 63).   The pustules frequently break through the epidermis on the inner surface of the leaf sheath and lemma and palea.   Under favorable conditions, uredia form on the culms, especially inside the leaf sheaths.   The telia develop as narrow linear pustules covered permanently by the epidermis.

THE FUNGUS

Puccinia glumarum (Schm.) Eriks. & Henn.
(*Uredo glumarum* Schm.)
(*Trichobasis glumarum* Lév.)
(*Puccinia tritici* Oerst.)
(*Puccinia neglecta* West.)
(*Puccinia rubigo-vera* Aut.)

The aecial stage of this rust is unknown.   The uredia are linear, citron to orange yellow, usually narrow, with a marked tendency to form stripes on the leaves and culms. Spatulate paraphyses sometimes occur around the margin of the uredium.   The urediospores are round to ovate, finely echinulate, with three or four germ pores.   The telia are linear and covered by the epidermis.   Brown paraphyses border the telia and are intermingled with the teliospores along the margin of the telium.   The teliospores are oblong to cuneiform, smooth, slightly constricted at the septum, with the apex less thickened and pointed than in *Puccinia graminis*.   The spore morphology is similar to *P. rubigo-vera* (DC.) Wint. or *P. recondita* Rob.

ETIOLOGY.   The mycelium and urediospores develop abundantly in the late autumn on the cereals and grasses, and they are important in the persistance of the fungus.   The mycelium and, less frequently, the spores remain viable over winter and develop early the following spring.   The late-autumn and early-spring development of this rust is characteristic. Secondary infection and aggressive development of the parasite occur during the early growth period of the susceptible cereals and grasses. In the cooler areas, especially where light intensity is low, the rust develops aggressively until maturity of the crop.   In less favorable areas, the rust development is checked in late spring, and summer survival becomes the critical period in the life cycle of the parasite. The summer environment apparently functions more specifically than the winter period in restricting the distribution of this rust.

Fig. 63. The uredial stage of stripe rust of wheat caused by *Puccinia glumarum*.

CONTROL. Many of the commercial varieties of wheat and the other cereals are resistant to this rust. Resistant strains are found in many of the grasses common in the areas where this rust is prevalent, as shown by Hassebrauk (1932), Hungerford and Owen (1923), Newton and Johnson

(1936), Sanford and Broadfoot (1933), Vallega (1955), and others. The inheritance of resistance to this rust was among the first studied, as reported by Armstrong (1922), Biffen (1905), and Nilsson-Ehle (1908, 1909, 1911) and summarized by Roemer *et al.* (1938). A single factor or a few factor pairs function in the inheritance of resistance.

Physiologic specialization occurs, although the differentiation of races is complicated by the response of the parasite to environmental conditions. Bever (1934) and Newton and Johnson (1936) differentiated three races in the United States and Canada. Eriksson (1894) included this parasite in the early studies on specialization. Extensive investigations on specialization of this parasite are reported from Europe, with considerable confusion in the number of races and their response on different varieties, as reported by Gassner and Straib (1932, 1934), Roemer *et al.* (1938), Rudorf (1929), Straib (1937), and others. The Straib differentials generally used are unsatisfactory in North and South America. Chino 166 is resistant to a group of races identified by Straib (1937) and those found later in Argentina, but it is susceptible to races isolated in Chile (Vallega, 1955).

26. Leaf Rust, *Puccinia rubigo-vera tritici* (Eriks.) Carleton or *P. recondita* Rob. The leaf rust of wheat is generally distributed through the humid and semihumid wheat-producing areas of the world. The leaf rust is distributed more uniformly and occurs more regularly than either of the other rusts of wheat. Until recently, leaf rust has caused damage in the Central and Eastern United States on the winter wheats. In recent years damage has been severe in the Central United States and Canada on the hard red spring wheats. This rust also occurs on barley and a number of grasses, as shown by Johnston (1936, 1940) and Mains (1933). Damage is severe especially when the rust infection occurs early and continues during the growing season, as reported by Caldwell *et al.* (1934), Johnston and Miller (1934), Waldron (1937), and others in the United States; Greaney *et al.* (1941) and Peterson and Newton (1939) in Canada; Neill (1931) in New Zealand; Phipps (1938) in Australia; and Butler and Hayman (1906) in India. Total yield of grain is reduced appreciably. Kernel volume is lowered without appreciable shriveling of the grain, and nitrogen content of the grain is decreased.

SYMPTOMS. The uredial stage of the rust is evident on the leaves from the seedling stage to maturity. In the early spring, new uredia frequently form in a circle around uredia of the previous autumn. The uredia are round to slightly oblong, orange yellow, and generally the ruptured epidermis is inconspicuous around the uredia. The covered telia form adjacent to the uredia and in new locations, especially on the leaf sheaths (Fig. 64).

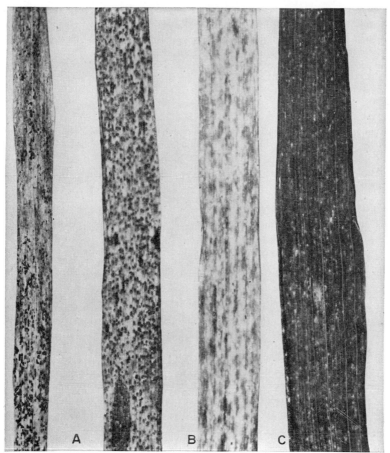

Fig. 64. Reaction of wheat varieties to *Puccinia rubigo-vera tritici*. (*A*) Two leaves fully susceptible showing telial and uredial development, (*B*) chlorosis and necrosis following infection, and (*C*) highly resistant reaction.

The Fungus

Puccinia recondita Rob. ex Desm.
Puccinia rubigo-vera tritici (Eriks.) Carleton
(*Uredo rubigo-vera* DC.)
[*Puccinia rubigo-vera* (DC.) Wint.]
(*Puccinia dispersa sp.f. tritici* Eriks. & Henn.)
(*Puccinia triticina* Eriks.)

Puccinia triticina Eriks. was used generally following Eriksson's (1894) division of the morphological species *P. rubigo-vera* into the several species based on specialization. Mains (1933) investigated the aecial and uredial suscepts and suggested the use of the trinomial, which has been generally accepted.   Cummins and Caldwell (1956) report *P. recondita* Rob. ex Desm. antedates *P. rubigo-vera* Wint.

The aecial stage occurs infrequently on several species of *Thalictrum* as citron-yellow swollen lesions.   The spermagonia form on the upper surface of the leaflet, followed by the small aecia on the undersurface.   The uredia on wheat and some grasses are round to slightly oblong, orange yellow, and occur on the leaves and less commonly on the floral bracts.   The urediospores are round to ovate, echinulate, with three to four germ pores distributed over the surface.   Both the uredia and urediospores differ morphologically from the comparable structures in *Puccinia graminis*.   The telia are small, oval to oblong, black, and covered by the epidermis.   The teliospores are surrounded by a thin layer of brown paraphyses.   The teliospores are oblong to cuneiform, smooth, brown, slightly constricted at the septum, and rounded at the apex without pronounced apical thickening of the wall.   Germination of the different spore forms is similar to that of *P. graminis*.

ETIOLOGY.   This rust persists largely in the uredial stage.   Infection of the more common aecial host *Thalictrum polygamum* Muhl. in North America occurs occasionally.   Mains (1933) has listed 12 species of *Thalictrum* as susceptible to this rust.   D'Oliveira (1951) reports *T. speciosiasimum* infected naturally in the Mediterranean area and adds two species of *Isopyrum* as aecial hosts.   The rust collections from several geographic areas react differentially on the aecial hosts.   In North and South America, Europe, and Asia the aecial stage is not important in the etiology of this rust pathogen although local infections occur infrequently. The uredial mycelium and urediospores, to a lesser extent, survive the winter on winter wheat plants as far north as winter wheat survives. New uredia develop early in the spring from this mycelium, and infection spreads from the new urediospores.   There is some spring spread from urediospores produced the previous autumn.   In the areas of moderate to plentiful moisture, the urediospores on the mature crop reinfect the volunteer wheat seedlings and the fall-sown wheat.   In such areas the grasses are not important in the production or continuance of the fungus. It is questionable whether the grasses play an important part in the maintenance or distribution of this rust, although Johnston (1940), Mains (1933), Vallega (1947) and others have shown that a number of grasses including perennials are susceptible.   The urediospores spread northward similar to those of *Puccinia graminis*, as discussed by Craigie (1945) and Vallega (1955).

CONTROL.   Leaf rust-resistant varieties offer the chief means of control of this disease.   Leaf rust-resistant wheats are essential to economical wheat production, especially in the humid regions where winter and spring varieties overlap, as in the North Central United States.   Most of the commercial winter varieties of this area are moderately to highly resistant to the races commonly occurring in this region.   Many of the newer spring and winter wheats are resistant to the common races of both *Puccinia graminis tritici* and *P. rubigo-vera tritici* or *P. recondita*.

According to Guard (1938), Johnston (1940), Vallega (1942, 1944), and others, more varieties resistant to leaf rust occur in the wheat species with lower chromosome numbers. *Haynaldia villosa* (L.) Schur. and *Triticum monococcum* L. are highly resistant. Nearly all the durums are resistant, whereas the emmers range from susceptible to resistant. *T. timopheevi* Zhuk. is resistant in both North and South America. Resistance to relatively large groups of races of the parasite are found in the *T. vulgare* Vill. group of varieties. Hope is intermediate in reaction to some few races and resistant to most. Resistance is controlled by several single factor pairs in the different sources of resistance, as discussed by Mains *et al.* (1926), Roemer *et al.* (1938), and others.

Specialization is well developed in this parasite as over 128 physiologic races are differentiated on eight wheat varieties, as reported by Hassebrauk (1937), Johnston and Levine (1954), Johnston and Mains (1932), Newton and Johnson (1941), Vallega (1942, 1944), and others. Many races are localized geographically, while others are widely distributed, especially when urediospore dissemination occurs over wide areas.

In the leaf rust as in stem rust, new virulent races of the pathogen soon increase on the resistant varieties and race populations change with the varieties used. The cumulative catalogue of races based on factors for resistance in the hosts is needed for this pathogen (Johnson, 1953, Vallega, 1955).

REFERENCES

AAMODT, O. S. Breeding wheat for resistance to physiologic forms of stem rust. *Jour. Am. Soc. Agron.* **19**:206–218. 1927.

AASE, H. C. Cytology of Triticum, Secale, and Aegilops hybrids with reference to phylogeny. *Res. Studies State Col. Wash.* **2**:1–60. 1930.

AGRONOMOFF, E. A., *et al.* Biochemistry and microbiology of stored wheat grain infected by Fusarium. *Food Supply Tech. Pub. Leningrad.* 1934.

AKERMANN, A. Studien über den Kalteted und die Kalteresistenz der Pflanzen nebst Untersuchungen über die Winterfestigkeit des Weizens. Berlingska Boktryckeriet. Lund. 1927.

ALLARD, R. W., and R. G. SHANDS. Inheritance of resistance to stem rust and powdery mildew in cytologically stable spring wheats derived from *Triticum timopheevi*. *Phytopath.* **44**:266–274. 1954.

ALLEN, P. J. Changes in the metabolism of wheat leaves induced by infection with powdery mildew. *Am. Jour. Bot.* **29**:425–435. 1942.

———. The role of a self-inhibitor in the germination of rust uredospores. *Phytopath.* **45**:259–266. 1955.

——— and D. R. GODDARD. A respiratory study of powdery mildew of wheat. *Am. Jour. Bot.* **25**:613–621. 1938.

ALLEN, R. F. A cytological study of heterothallism in *Puccinia graminis*. *Jour. Agr. Research* **40**:585–614. 1930.

———. Further cytological studies of heterothallism in *Puccinia graminis*. *Jour. Agr. Research* **47**:1–16. 1933.

ANGELL, H. R. A preliminary note on the recognition of flag smut or bunt infection based on the deformation of seedlings. *Jour. Counc. Sci. Ind. Research Aust.* **7**:110–112. 1934.

—— *et al.* The effect of *Urocystis tritici* Koern. on the extent of development of the roots and aerial parts of the wheat plant. I and II. *Jour. Counc. Sci. Ind. Research Aust.* **10**:136–142. **11**:256–257. 1938.

ARMSTRONG, S. F. The mendelian inheritance of susceptibility and resistance to yellow rust (*Puccinia glumarum* E. and H.) in wheat. *Jour. Agr. Sci.* **12**:57–96. 1922.

ATHWAL, D. S., and I. A. WATSON. Inheritance and the genetic relationship of resistance possessed by two Kenya wheats to races of *Puccinia graminis tritici.* *Proc. Linn. Soc. N.S. Wales* **79**:1–14. 1954.

ATKINS, I. M., E. D. HANSING, and W. M. BEVER. Reaction of varieties and strains of winter wheat to loose smut. *Jour. Am. Soc. Agron.* **39**:363–377. 1947.

AUSEMUS, E. R. Breeding for disease resistance in wheat, oats, barley, and flax. *Bot. Rev.* **9**:207–260. 1943.

BAMBERG, R. H. Fall-sown spring wheat susceptible to dwarf bunt. *Phytopath.* **31**:951–952. 1941.

BATES, E. N., *et al.* Removing smut from Pacific Northwest wheat by washing. *U.S. Dept. Agr. Cir.* 81. 1929.

BAYLES, B. B. Influence of environment during maturation on the disease reaction and yield of wheat and barley. *Jour. Agr. Research* **53**:717–748. 1936.

BEDI, K. S., M. R. SIKKA, and B. B. MUNDKUR. Transmission of wheat bunt due to *Neovossia indica* (Mitra) Mundkur. *Indian Phytopath.* **2**:20–26. 1949.

BEVER, W. M. Physiologic specialization of *Puccinia glumarum* in the United States. *Phytopath.* **24**:686–688. 1934.

——. Influence of stripe rust on growth, water economy and yield of wheat and barley. *Jour. Agr. Research* **54**:375–385. 1937.

——. Physiologic races of *Ustilago tritici* in the eastern soft wheat region of the United States. *Phytopath.* **37**:889–895. 1947.

——. Further studies on physiologic races of *Ustilago tritici.* *Phytopath.* **43**:681–683. 1953.

BIFFEN, R. H. Mendel's laws of inheritance and wheat breeding. *Jour. Agr. Sci.* **1**:4–48. 1905.

BOEDIJN, K. B. Ueber einige Phragmosporen Dematiazeen. *Bul. Jardin Bot. Buitenzorg* **13**:120–134. 1933.

BOLLEY, H. L. Deterioration in wheat fields due to root rots and blight producing diseases. *No. Dak. Agr. Exp. Sta. Press. Bul.* 33. 1909.

——. Wheat: soil troubles and seed deterioration. *No. Dak. Agr. Exp. Sta. Bul.* 107. 1913.

BORLAUG, N. E. Mexican wheat production and its role in the epidemiology of stem rust in North America. *Phytopath.* **44**:398–404. 1954.

BRENTZEL, W. E. The black point disease of wheat. *No. Dak. Agr. Exp. Sta. Bul.* 330. 1944.

BRIGGS, F. N. A third genetic factor for resistance to bunt, *Tilletia tritici*, in wheat. *Jour. Gen.* **27**:435–441. 1933.

——. Linkage between the Martin and Turkey factors for resistance to bunt, *Tilletia tritici*, in wheat. *Jour. Am. Soc. Agron.* **32**:539–541. 1940. (Also Khishen and Briggs.)

—— and C. S. HOLTON. Reaction of wheat varieties with known genes for resistance to races of bunt, *Tilletia caries* and *T. foetida.* *Agron. Jour.* **42**:483–486. 1950.

BROADFOOT, W. C., and H. T. ROBERTSON. Pseudo-black chaff of Reward wheat. *Sci. Agr.* **13**:512–514. 1933.

BULLER, A. H. R. Fusions between flexuous hyphae and pycnidiospores in *Puccinia graminis*. *Nature* **141**:33. 1938.

BUTLER, E. J., and J. M. HAYMAN. Indian wheat rusts. *Mem. Dept. Agr. India* 1. 1906.

CALDWELL, R. M., *et al.* Effect of leaf rust (*Puccinia triticina*) on yield, physical characters, and composition of winter wheats. *Jour. Agr. Research* **48**:1049–1071. 1934.

CAMPOS, T. A., J. W. GIBLER, and N. E. BORLAUG. Correlation of seedling and adult plant reaction to stem rust of wheat. *Phytopath.* **43**:468. 1953.

CARLETON, M. A. The small grains. The Macmillan Company. New York. 1916.

CARNE, W. M., and J. G. C. CAMPBELL. Take-all of wheat and similar diseases of cereals. *West. Aust. Dept. Agr. Bul.* 119. 1924.

—— and E. J. LIMBOURN. Flag smut of wheat. Varietal resistance test. *Jour. Dept. Agr. Aust.* (2) **4**:4–7. 1927.

CHRISTENSEN, J. J., *et al.* Susceptibility of wheat varieties and hybrids to Fusarial head blight in Minnesota. *Minn. Agr. Exp. Sta. Tech. Bul.* 59. 1929.

CHURCHWARD, J. G. The geographic distribution of *Tilletia* spp. on wheat in Australia in 1931. *Proc. Linnean Soc. N.S. Wales* **62**:403–408. 1932.

——. A note on the occurrence of seedling lesions caused by cereal smuts. *Proc. Linnean Soc. N.S. Wales* **59**:197–199. 1934.

——. The initiation of infection by bunt of wheat (*Tilletia caries*). *Ann. Appl. Biol.* **27**:58–64. 1940.

CLARK, J. A. Improvement in wheat. *U.S. Dept. Agr. Yearbook*, pp. 207–302. 1936.

—— and B. B. BAYLES. Classification of wheat varieties grown in the United States. *U.S. Dept. Agr. Tech. Bul.* 459. 1935. *Tech. Bul.* 1083. 1954.

CRAIGIE, J. H. Discovery of the function of the pycnia of the rust fungi. *Nature* **120**:765–767. 1927.

——. On the occurrence of pycnia and aecia in certain rust fungi. *Phytopath.* **18**:1005–1015. 1928.

——. An experimental investigation of sex in rust fungi. *Phytopath.* **21**:1001–1040. 1931.

——. Increase in production and value of the wheat crop in Manitoba and Eastern Saskatchewan as a result of the introduction of rust resistant wheat varieties. *Sci. Agr.* **25**:51–64. 1944.

——. Epidemiology of stem rust in Western Canada. *Sci. Agr.* **25**:285–401. 1945. (Excellent review and literature list.)

CUMMINS, G. B. and R. M. CALDWELL. The validity of binomials in the leaf rust complex of cereals and grasses. Phytopath. **46**:81–82. 1956.

DAHL, A. S. Snow mold of turf grasses as caused by *Fusarium nivale*. *Phytopath.* **24**:197–214. 1934.

DASTUR, J. F. Notes on some fungi isolated from black point affected wheat kernels in the central provinces. *Indian Jour. Agr. Sci.* **12**:731–742. 1942.

DE URRIES, M. J. Razas fisiológicas de *Puccinia graminis tritici* en España. *Pub. Cons. Sup. Invest. Cient. Madr.* 1951.

DICKINSON, L. S. The effect of air temperature on the pathogenicity of *Rhizoctonia solani* parasitizing grasses on putting green turf. *Phytopath.* **20**:597–608. 1930.

DICKSON, J. G. The influence of soil temperature and moisture on the development

of seedling blight of wheat and corn caused by *Gibberalla saubinetti* (Mont.)
Sacc. *Jour. Agr. Research* **23** :837–870. 1923.

————. The nature of immunity in plants. *5th Pacific Sci. Cong. Proc.* (1933) **4**:
3211–3219. 1934.

————. Scab of wheat and barley and its control. *U.S. Dept. Agr. Farmer's Bul.*
1599. 1942.

D'OLIVEIRA, B.   The centers of origin of cereals and the study of their rusts. *Agron.
Lusit.* **13** :221–226.   1951.

DRECHSLER, C.   Some graminicolous species of Helminthosporium. I. *Jour. Agr.
Research* **24** :641–739.   1923.

————. *Pythium graminicolum* and *P. arrhenomanes.* *Phytopath.* **26** :676–684.
1936.

ERIKSSON, J.   Ueber die Specialisierung des Parasitismus bei den Getreiderostpilzen.
Ber. deut. bot. Gesell. **12** :292–331.   1894.

———— and E. HENNING.   Die Getreideroste, ihre Geschichte, Natur, sowie Mass-
regeln gegen Dieselben.   Norstadt and Soner.   Stockholm.   1896.

FANG, C. T., O. N. ALLEN, A. J. RIKER, and J. G. DICKSON.   The pathogenic, physiological, and serological reactions of the form species of *Xanthomonas translucens.*
Phytopath. **40** :44–64.   1950.

FARIS, J. A.   Influence of soil moisture and soil temperature on infection of wheat by
Urocystis tritici. *Phytopath.* **23** :10–11.   1933.

FELLOWS, H.   Some chemical and morphological phenomena attending infection of
the wheat plant by *Ophiobolus graminis* Sacc.   *Jour. Agr. Research* **38** :647–661.
1928.

FISCHER, G. W.   Some evident synonymous relationships in certain graminicolous
smut fungi. *Mycologia* **35** :610–619.   1943.

————. *Tilletia brevifaciens sp. nov.* causing dwarf bunt of wheat and certain grasses.
Wash. State Col. Res. Studies **20** :11–14.   1952.

———— and E. HIRSCHHORN.   The Ustilaginales or "smuts" of Washington. *Wash.
Agr. Exp. Sta. Bul.* 459.   1945.

FUTRELL, M. C., and J. G. DICKSON.   The influence of temperature on the development of powdery mildew on spring wheats. *Phytopath.* **44** :247–251.   1954.

GABRILOVITCH, O. E.   Intoxicating bread. *Diss. Med. Acad. St. Petersburg, Russia.*
1904.   (In Russian.)

GARRETT, S. D.   Root disease fungi.   Chronica Botanica Co.   Waltham, Mass.
1944.

GASSNER, G., and W. STRAIB.   Die Bestimmung der biologischen Rassen des Weizen-
gelbrostes [*Puccinia glumarum* f. sp. *tritici* (Schmidt) Erikss. u. Henn.] *Arb.
biol. Reichs. Land- u. Forstw. Berlin-Dahlem* **21** :141–163.   1932.

————. Weiterer Untersuchungen über biologische Rassen und über die Spezial-
isierungsverhältnisse des Gelbrostes, *Puccinia glumarum* (Schm.) Erikss. und
Henn. *Arb. biol. Reichs. Land- u. Forstw. Berlin-Dahlem* **21** :121–145.   1934.

GEDDES, W. F., *et al.*   The milling and baking quality of frosted wheat of the 1928
crop. *Can. Jour. Research* **C 6** :119–155.   1932.

GODFREY, G. H. *Sclerotium rolfsii* on wheat. *Phytopath.* **8** :64–66.   1918.

GOEFFREY, S., and S. D. GARRETT.   *Rhizoctonia solani* on cereals in South Australia.
Phytopath. **22** :827–836.   1932.

GORLENKO, M. V.   Twenty-five years of cereal diseases in United Socialistic Soviet
Republics. *Jour. Bot. U.S.S.R.* **31** :3–17.   1936.

GORTER, G. J. M. A.   Wheat stunt—a new cereal disease. *Fmg. So. Africa* **22** :29–32.
1947.

GOULDEN, C. H., and K. W. NEATBY.   A study of disease resistance and other varietal characters of wheat—application of the analysis of variance and correlations. *Sci. Agr.* **9**:575–586. 1929.

GRAF-MARIAN, A. Studies on powdery mildews of cereals. *N.Y. (Cornell) Agr. Exp. Sta. Mem.* 157. 1934.

GREANEY, F. J., *et al.* The effect of stem rust on the yield, quality, chemical composition, and milling and baking properties of Marquis wheat. *Sci. Agr.* **22**:40–60. 1941.

——— and H. A. H. WALLACE. Varietal susceptibility to kernel smudge in wheat. *Sci. Agr.* **24**:126–134. 1943.

GREVEL, F. K. Untersuchungen über das Vorhandensein biologischer Rassen des Flugbrandes des Weizens (*Ustilago tritici*). *Phytopath. Zeitschr.* **2**:209–234. 1930.

GUARD, A. T. Studies on cytology and resistance to leaf rust of some interspecific and intergeneric hybrids of wheat. *Am. Jour. Bot.* **25**:478–480. 1938.

HAFIZ, A. Physiologic specialization in *Urocystis tritici* Koern. *Phytopath.* **41**: 809–812. 1951.

HAGBORG, W. A. F. Black chaff a composite disease. *Can. Jour. Research* **C 14**: 347–359. 1936.

———. Classification revision in *Xanthomonas translucens*. *Can. Jour. Research* **C 20**:312–326. 1942.

HANNA, W. F. Physiologic forms of loose smut of wheat. *Can. Jour. Research* **C 15**:141–153. 1937.

HANSON, E. W., E. R. AUSEMUS, and E. C. STAKMAN. Varietal resistance of spring wheats to Fusarial head blight. *Phytopath.* **40**:902–914. 1950.

———, H. E. MILLIRON, and J. J. CHRISTENSEN. The relation of the Bluegrass billbug, *Calendra parvula* (Gyllenhal), to the development of basal stem rot and root rot of cereals and grasses in North-Central United States. *Phytopath.* **40**: 527–543. 1950.

HART, H. Morphologic and physiologic studies on stem rust resistance in cereals. *U.S. Dept. Agr. Tech. Bul.* 266. 1931.

———. Stem rust of new wheat varieties and hybrids. *Phytopath.* **34**:884–899. 1944.

——— and K. ZALESKI. The effect of light intensity and temperature on infection of Hope wheat by *Puccinia graminis tritici*. *Phytopath.* **25**:1041–1066. 1935.

HARVEY, R. B. An annotated bibliography of the low temperature relations of plants. Burgess Publishing Co. Minneapolis. 1935.

HASSEBRAUK, K. Gräserinfektionen mit Getreiderosten. *Arb. Biol. Reichs. Land- u. Forstw. Berlin-Dahlem* **20**:165–182. 1932.

———. Untersuchungen über die physiologische Spezialisierung von *Puccinia triticina* Eriks. in Deutschland und einigen anderen europäischen Staaten während der Jahre 1934 und 1935. *Arb. biol. Reich. Land- u. Forstw. Berlin-Dahlem* **22**:71–89. 1937.

———. Untersuchen über die Einwirkung von Sulfonamiden und Sulfonen auf Getreideroste. I. Die Beeinflussung des Fruktifikationsvermögens. *Phytopath. Zeitschr.* **17**:384–400. 1951.

HAYES, H. K., *et al.* Correlated inheritance of reaction to stem rust, leaf rust, bunt, and black chaff in spring wheat. *Jour. Agr. Research* **48**:59–66. 1934.

———, F. R. IMMER, and D. C. SMITH. Methods of plant breeding. McGraw-Hill Company, Inc. New York. 1955.

HELY, F. W., *et al.* Bunt infection and root development in wheat. *Jour. Counc. Sci. Ind. Research Aust.* **11**:254–255. 1938.

HENRY, A. W. Root-rots of wheat. *Minn. Agr. Exp. Sta. Tech. Bul.* 22. 1924.

HEYNE, E. G., and E. D. HANSING. Inheritance to loose smut of wheat in the crosses of Kawvale × Clarkan. *Phytopath.* **45**:8–10. 1955.

HO, WEN-CHUN, C. H. MEREDITH, and I. E. MELHUS. *Pythium graminicola* Subr. on barley. *Iowa Agr. Exp. Sta. Res. Bul.* 287. 1941.

HOLTON, C. S. Chlamydospore germination in the fungus causing dwarf bunt of wheat. *Phytopath.* **33**:732–735. 1943.

———. Inheritance of chlamydospore and sorus characters in species and race hybrids of *Tilletia caries* and *T. foetida. Phytopath.* **34**:586–592. 1944.

———. Host selectivity as a factor in the establishment of physiologic races of *Tilletia caries* and *T. foetida* produced by hybridization. *Phytopath.* **37**:817–821. 1947.

———. Observations on *Neovossia indica. Indian Phytopath.* **2**:1–5. 1949.

———. Pathogenicity studies with inbred lines of physiologic races of *Tilletia caries. Phytopath.* **43**:398–400. 1953.

——— R. H. BAMBERG, and R. H. WOODWARD. Progress in the study of dwarf bunt of wheat in the Pacific Northwest. *Phytopath.* **39**:986–1000. 1949.

——— and F. D. HEALD. Bunt or stinking smut of wheat. Burgess Publishing Co. Minneapolis. 1941. (Good review of literature.)

——— and T. L. JACKSON. Varietal reaction to dwarf bunt and flag smut, and the occurrence of both in the same wheat plant. *Phytopath.* **41**:1035–1037. 1951.

——— and A. G. JOHNSON. Physiologic races in *Urocystis tritici. Phytopath.* **33**:169–171. 1943.

——— and H. A. RODENHISER. New physiologic races of *Tilletia tritici* and *T. livis. Phytopath.* **32**:117–129. 1942.

——— and R. SPRAGUE. Studies on the control of snow mold of winter wheat in Washington. *Phytopath.* **39**:860. 1949.

HOTSON, H. H. Some chemotherapeutic agents for wheat stem rust. *Phytopath.* **43**:659–662. 1953.

HUMPHREY, H. B., and A. G. JOHNSON. Take-all and flag smut, two wheat diseases new to the United States. *U.S. Dept. Agr. Farmers' Bul.* 1063. 1919.

HUNGERFORD, C. W., and C. E. OWENS. Specialized varieties of *Puccinia glumarum* and hosts of variety *tritici. Jour. Agr. Research* **25**:363–401. 1923.

HYNES, H. J. Studies on Helminthosporium root-rot of wheat and other cereals. I and II. *Dept. Agr. N.S. Wales Sci. Bul.* 47. 1935.

———. Studies on Rhizoctonia root-rot of wheat and oats. *Dept. Agr. N.S. Wales Sci. Bul.* 58. 1937.

———. Species of *Helminthosporium* and *Curvularia* associated with root-rot of wheat and other graminaceous plants. *Jour. Proc. Roy. Soc. N.S. Wales* **70**: 378–391. 1937.

———. Studies on take-all of wheat. I. *Jour. Aust. Inst. Agr. Sci.* **3**:43–48. 1937.

———. Studies on Helminthosporium root-rot of wheat and other cereals. III and IV. *Dept. Agr. N.S. Wales Sci. Bul.* 61. 1938.

IKATA, S., and I. KAWAI. Some experiments concerning the development of yellow mosaic disease (white streak) of wheat: relation between the development of yellow mosaic disease of wheat and soil temperature. *Jour. Plant. Prot.* **24**:491–501, 847–854. 1937.

IMAI, S. On the causal fungus of the Typhyla-blight of graminaceous plants. *Japanese Jour. Bot.* **8**:5–18. 1936.

ITO, S., and K. KURIBAYASHI. The ascigerous forms of some graminicolous species of Helminthosporium in Japan. *Jour. Fac. Agr. Hokkaido Imp. Univ. Sapporo* **29**:85–125. 1931.

JOHNSON, A. H., and W. O. WHITCOMB. A comparison of some properties of normal and frosted wheats. *Mont. Agr. Exp. Sta. Bul.* 204. 1927.

JOHNSON, T., Variation in the rusts of cereals. *Biol. Revs.* **28**:105–157. 1953.

———. Selfing studies with physiologic races of wheat stem rust, *Puccinia graminis* var. *tritici. Can. Jour. Bot.* **32**:506–522. 1954.

———, *et al.* Hybridization of *Puccinia graminis tritici* with *Puccinia graminis secalis* and *Puccinia graminis agrostidis. Sci. Agr.* **13**:141–153. 1932.

———. Further studies of the inheritance of spore color and pathogenicity in crosses between physiologic forms of *Puccinia graminis tritici. Sci. Agr.* **14**:360–373. 1934.

——— and W. A. F. HAGBORG. Melanism in wheat induced by high temperature and humidity. *Can. Jour. Research* C **22**:7–10. 1944.

——— and M. NEWTON. The origin of abnormal rust characteristics through the inbreeding of physiologic races of *Puccinia graminis tritici. Can. Jour. Research* C **16**:38–52. 1938.

JOHNSTON, C. O. Reaction of certain varieties and species of the genus Hordeum to leaf rust of wheat, *Puccinia triticina. Phytopath.* **26**:235–245. 1936.

———. Some species of Triticum and related grasses as hosts for the leaf rust of wheat, *Puccinia triticina* Eriks. *Kans. Acad. Sci. Trans.* **43**:121–132. 1940.

——— and M. N. LEVINE. Physiologic races of *Puccinia rubigo-vera tritici* in the United States. *Plant Dis. Reptr.* **38**:647–648. 1954.

——— and E. B. MAINS. Studies on physiologic specialization in *Puccinia triticina. U.S. Dept. Agr. Tech. Bul.* 313. 1932.

——— and E. C. MILLER. Relation of leaf rust infection to yield, growth, and water economy of two varieties of wheat. *Jour. Agr. Research* **49**:955–981. 1934.

JONES, G. H., and ABD EL GHANI SEIF-EL-NASR. The influence of sowing depth and moisture on smut diseases, and the prospects of a new method of control. *Ann. Appl. Biol.* **27**:35–57. 1940.

KHISHEN, El A. A., and F. N. BRIGGS. Inheritance of resistance to bunt (*Tilletia caries*) in hybrids with Turkey wheat selections C.I. 10015 and 10016. *Jour. Agr. Research* **71**:403–413. 1945.

KIHARA, H. Genomanalyse bei Triticum und Aegilops. VII. *Mem. Col. Agr. Kyoto Imp. Col.* 41. 1937. Also previous series in *Cytologia* 1 to 6. 1930–1935.

KIRBY, R. S. The take-all disease of cereals and grasses caused by *Ophiobolus cariceti* (Berkeley and Broome) Saccardo. *N.Y. (Cornell) Agr. Exp. Sta. Mem.* 88. 1925.

KLAGES, K. H. W. Ecological crop geography. The Macmillan Company. New York. 1942.

KO, K. S., and E. R. AUSEMUS. Inheritance of reaction to stem rust in crosses of Timstein with Thatcher, Newthatch and Mida. *Agron. Jour.* **43**:194–201. 1951.

KOEHLER, B., W. M. BEVER, and O. T. BONNETT. Soil-borne wheat mosaic. *Ill. Agr. Exp. Sta. Bul.* 556, pp. 567–599. 1952.

KOSTOFF, D. Studies on the polyploid plants. XI. Amphidiploid *Triticum timopheevi* Zhuk. × *Triticum monococcum* L. *Zeitschr. Zücht.* A **21**:41–45. 1936.

———. *Triticum timococcum*, the most immune wheat experimentally produced. *Chron. Bot.* **4**:213–214. 1938.

LEBEAU, J. B., and J. G. DICKSON. Preliminary report on production of hydrogen cyanide by a snow-mold pathogen. *Phytopath.* **43**:581–582. 1953.

LEHMANN, E., H. KUMMER, and H. DANNENMANN. Der Schwarzrost seine Geschichte, seine Biologie und seine Bekämpfung in Verbindung mit der Berberitzenfrage. J. F. Lehmanns Verlag. Munchen-Berlin. 1937.

LEVITT, J.   Frost killing and hardiness of plants.   Burgess Publishing Co.   Minneapolis. 1941.

LIVINGSTON, J. E.   The control of leaf and stem rust of wheat with chemotherapeutants. *Phytopath.* **43**:496–499. 1953.

LOVE, R. M.   Varietal differences in meiotic chromosome behavior in Brazilian wheats. *Agron. Jour.* **43**:72–76. 1951.

LOWTHER, C. V.   Low temperature as a factor in the germination of dwarf bunt chlamydospores. *Phytopath.* **38**:309–310. 1948.

MAINS, E. B.   Host specialization of *Erysiphe graminis tritici.   Proc. Nat. Acad. Sci.* **19**:49–53. 1933.

———.   Host specialization in the leaf rust of the grasses, *Puccinia rubigo-vera. Mich. Acad. Sci., Arts, Letters* **17**:289–394. 1933.

———.   Inheritance of resistance to powdery mildew, *Erysiphe graminis tritici* in wheat. *Phytopath.* **24**:1257–1261. 1934.

——— et al.   Inheritance of resistance to leaf rust, *Puccinia triticina* Eriks. in crosses of common wheat, *Triticum vulgare* Vill.   *Jour. Agr. Research* **32**:931–972. 1926.

MARCHAL, E.   De la specialisation du parasitisme chez l'*Erysiphe graminis* DC. *Compt. rend. Acad. Sci. Paris* **135**:210–212. 1902. **136**:1280–1281. 1903.

MARTINEZ, L. M., and E. R. AUSEMUS.   Inheritance of reaction to leaf rust, *Puccinia rubigo-vera tritici* (Erikss.) Carleton, and of certain other characters in a wheat cross. *Minn. Agr. Exp. Sta. Tech. Bul.* 205. 1953.

MAXIMOV, N. A.   Internal factors of frost and drought resistance in plants. *Protoplasma* **7**:259–291. 1929.

McCULLOCH, L.   Basal glume rot of wheat. *Jour. Agr. Research* **18**:543–551. 1920.

McFADDEN, E. S.   A successful transfer of emmer characters to vulgare wheat. *Jour. Am. Soc. Agron.* **22**:1020–1034. 1930.

———.   Brown necrosis, a discoloration associated with rust infection in certain rust-resistant wheats. *Jour. Agr. Research* **58**: 805–819. 1939.

———.   New sources of resistance to stem rust and leaf rust in foreign varieties of common wheat. *U.S. Dept. Agr. Cir.* 814, pp. 1–16. 1949.

——— and E. R. SEARS.   The origin of *Triticum spelta* and its free-threshing hexaploid relatives. *Jour. Heredity* **37**:81–116. 1946.

McKINNEY, H. H.   Investigations of the rosette disease of wheat and its control. *Jour. Agr. Research* **23**:771–800. 1923.

———.   Foot-rot diseases of wheat in America.   *U.S. Dept. Agr. Bul.* 1347. 1925.

———.   A mosaic of wheat transmissible to all cereal species in the tribe Hordeae. *Jour. Agr. Research* **40**:547–556. 1930.

———.   Mosaic disease of wheat and related cereals.   *U.S. Dept. Agr. Cir.* 442. 1937.

——— et al.   The intracellular bodies associated with the rosette disease and a mosaic-like leaf mottling of wheat. *Jour. Agr. Research* **26**:605–608. 1923.

———.   Wheat rosette and its control.   *Ill. Agr. Exp. Sta. Bul.* 264. 1925.

MELCHERS, L. E.   Wheat mosaic in Egypt.   *Science* (N.S.) **73**:95–96. 1931.

MIDDLETON, J. T.   The taxonomy, host range and geographic distribution of the genus Pythium. *Mem. Torrey Bot. Club* **20**:1–171. 1943.

MILLER, W. R., and C. R. MILLIKAN.   Investigation on flag smut of wheat caused by *Urocystis tritici* Koern. *Jour. Dept. Agr. Victoria* **32**:365–380, 418–432. 1934.

MITRA, M.   A comparative study of species and strains of *Helminthosporium* on certain Indian cultivated crops. *Trans. Brit. Myc. Soc.* **15**:254–293. 1930.

———.   A new bunt of wheat in India. *Ann. Appl. Biol.* **18**:178–179. 1931.

———.   Saltation in the genus Helminthosporium. *Trans. Brit. Myc. Soc.* **16**:115–127. 1931.

MONTEITH, J. The brown patch disease of turf; its nature and control. *U.S. Golf Assn. Greens Soc. Bul.* **6**:127–142. 1926

MÜLLER, E. Pilzliche Erreger der Getreideblattdüre. *Phytopath. Zeit.* **19**:403–416. 1952.

NAUMOV, N. A. Intoxicating bread. *Min. Zeml. Trudy Biuro. Mikol. Fitopat. Uchen. Kom.* 12. 1916. (In Russian.)

NEILL, J. C. Effects of rusts and mildews on yield and quality of wheat. *New Zealand Jour. Agr.* **43**:44–45. 1931.

NEWTON, M., *et al.* A preliminary study of the hybridization of physiologic forms of *Puccinia graminis tritici*. *Sci. Agr.* **10**:721–731. 1930.

————. Hybridization between *Puccinia graminis tritici* and *Puccinia graminis secalis*. *Phytopath.* **21**:106–107. 1931.

————. Specialization and hybridization of wheat stem rust, *Puccinia graminis tritici* in Canada. *Can. Dept. Agr. Bul.* 160. 1932.

———— and T. JOHNSON. Stripe rust, *Puccinia glumarum*, in Canada. *Can. Jour. Research* **C 14**:89–108. 1936.

———— and ————. Production of uredia and telia of *Puccinia graminis* on *Berberis vulgaris*. *Nature* **139**:800–801. 1937.

———— and ————. Environmental reaction of physiologic races of *Puccinia triticina* and their distribution in Canada. *Can. Jour. Research* **C 19**:121–133. 1941.

NEWTON, R. A. The nature and practical measurement of frost resistance in winter wheat. *Alberta Col. Agr. Res. Bul.* 1. 1924.

———— and J. A. ANDERSON. Respiration of winter wheat plants at low temperatures. *Can. Jour. Research* **C 5**:87–110. 1931.

———— and A. G. McGALLA. Effect of frost on wheat at progressive stages of maturity. I, III. *Can. Jour. Research* **C 10**:414–429. 1934. **13**:263–282. 1935.

NILSSON-EHLE, H. Einige Ergebnisse von Kreizungen bei Hafer und Weizen. *Bot. Notiser.* 1908. Kreuzungsuntersuchungen an Hafer und Weizen. *Lunds Univ. Arrsskrift N.F.* 2 **5**, 2:1–122. 1909.

————. Kreuzungsuntersuchungen an Hafer und Weizen. *Lunds Univ. Arrsskrift N.F.* 2 **7**, 6:1–82. 1911.

NISHIKADO, Y. Preliminary notes on yellow spot disease of wheat caused by *Helminthosporium tritici-vulgaris* Nishikado. *Ber. Ōhara Inst. Landw. Forsch.* **4**:103–109. 1929.

NOBLE, R. J. Studies on *Urocystis tritici* Koern., the organism causing flag smut of wheat. *Phytopath.* **13**:127–139. 1923.

————. Studies on the parasitism of *Urocystis tritici* Koern., the organism causing flag smut of wheat. *Jour. Agr. Research* **27**:451–489. 1924.

————. Basal glume rot. *Agr. Gaz. N.S. Wales* **44**:107–109. 1933.

————. Note on the longevity of spores of the fungus *Urocystis tritici* Koern. *Jour. Proc. Roy. Soc. N.S. Wales* **67**:403–410. 1934.

————. The occurrence of plant diseases in New South Wales. *N.S. Wales Dept. Agr. Sci. Bul.* 57. 1937.

OORT, A. J. P. Stuifbrand specialisatie, een Probleem voor den Kweker. Onderzoekingen over Stuifbrand. III. *Tijdschr. Plantenzeikt.* **53**:25–43. 1937.

PEGLION, V. La formazione dei conidi e la germinazione delle oospore delle *Sclerospora macrospora* Sacc. *Bol. R. Staz. Pat. Veg.* (N.S.) **10**:153–164. 1930.

PELTIER, G. L. Parasitic Rhizoctonia in America. *Ill. Agr. Exp. Sta. Bul.* 189. 1916.

PERCIVAL, J. The wheat plant. E. P. Dutton & Company, Inc. New York. 1921.

PETERSON, R. F. Stomatal behavior in relation to the breeding of wheat for resistance to stem rust. *Sci. Agr.* **12**:155–173. 1931.

PETURSON, B., and M. NEWTON. The effect of leaf rust on the yield and quality of Thatcher and Renown wheat in 1938. *Can. Jour. Research* **C 17**:380–387. 1939.

PHIPPS, I. F. The effect of leaf-rust on yield and baking quality of wheat. *Jour. Aust. Inst. Agr. Sci.* **4**:148–151. 1938.

PRIDHAM, J. T., and R. E. DWYER. Reaction of wheat varieties to flag smut. *Agr. Gaz. N.S. Wales* **41**:413–415. 1930.

PUGH, G. W., *et al.* Factors affecting infection of wheat heads by *Gibberella saubinetii*. *Jour. Agr. Research* **46**:771–797. 1933.

PUGSLEY, A. T. The resistance of Oro and Orfed wheats to *Tilletia foetida*. *Aust. Jour. Agr. Res.* **1**:391–400. 1950.

QUISENBERRY, K. S. Inheritance of winter hardiness, growth habit, and stem rust reaction in crosses between Minhardi winter and H-44 spring wheats. *U.S. Dept. Agr. Tech. Bul.* 218. 1931.

—— and J. A. CLARK. Breeding hard red winter wheats for winter hardiness and high yields. *U.S. Dept. Agr. Tech. Bul.* 136. 1929.

RADULESCU, E. Untersuchungen über die physiologische Spezialisierung bei Flugbrand des Weizens, *Ustilago tritici* (Pers.) Jens. *Phytopath. Zeitschr.* **8**:253–258. 1935.

REMSBERG, R. E. Studies in the genus Typhula Fries. *Mycologia* **32**:52–96. 1940. *Phytopath.* **30**:178–180. 1940.

RODENHISER, H. A., and C. S. HOLTON. Distribution of races of *Tilletia caries* and *Tilletia foetida* and their relative virulence on certain varieties and selections of wheat. *Phytopath.* **35**:955–969. 1945.

—— and ——. Differential survival and natural hybridization in mixed spore populations of *Tilletia caries* and *T. foetida*. *Phytopath.* **43**:558–560. 1953.

ROEMER, T., *et al.* Die Züchtung resistenter Rassen der Kulturpflanzen. Paul Parey. Berlin. 1938. Also *Kühn-Archiv.* **45**:1–427. 1938.

RUDORF, W. Beiträge zur Immunitätszüchtung gegen *Puccinia glumarum tritici*. *Phytopath. Zeitschr.* **1**:465–525. 1929.

SALLANS, B. The use of water by wheat plants when inoculated with *Helminthosporium sativum*. *Can. Jour. Research* **C 18**:178–198. 1940.

SALMON, S. C. Resistance of varieties of winter wheat and rye to low temperature in relation to winter hardiness and adaptation. *Kans. Agr. Exp. Sta. Tech. Bul.* 35. 1933.

SAMPSON, K. The occurrence of snow mold on golf greens in Britain. *Jour. Board Green Keeping Research* **2**:116–118. 1931.

SAMUEL, G. White heads or take-all in wheat. *Jour. Min. Agr. Gr. Brit.* **44**:231–241. 1937.

SANDO, W. T. Intergeneric hybrids of Triticum and Secale with *Haynaldia villosa*. *Jour. Agr. Research* **51**:759–800. 1935.

SANFORD, G. B., and W. C. BROADFOOT. The relative susceptibility of cultivated and native hosts in Alberta to stripe rust. *Sci. Agr.* **13**:714–721. 1933.

SCHAFFNIT, E., and K. MEYER-HERMANN. Ueber den Einfluss der Bodenreaktion auf die Lebensweise von Pilzparasiten und das Verhalten ihrer Wirtspflanzen. *Phytopath. Zeitschr.* **2**:99–166. 1930.

SCHLICHTLING, I. Untersuchungen über die physiologische Spezialisierung des Weizenmeltaus, *Erysiphe graminis tritici* (DC.) in Deutschland. *Kühn Arch.* **48**:52–55. 1939.

SCOTT, I. T. Varietal resistance and susceptibility to wheat scab. *Mo. Agr. Exp. Sta. Res. Bul.* 111. 1927.

SHANDS, R. G. Disease resistance of *Triticum timopheevi* transferred to common wheat. *Jour. Am. Soc. Agron.* **33**:709–712. 1941.

SHAPOVALOV, M. Intoxicating bread. *Phytopath.* **7**:384–386. 1917. (Review of Russian papers.)

SHELTON, J. P. Breeding wheats resistant to flag smut. *Agr. Gaz. N.S. Wales* **35**: 336–338. 1924.

SHEN, T. H. The inheritance of resistance to flag smut (*Urocystis tritici* Koern.) in ten wheat crosses. *Col. Agr. For. Nanking Bul.* 17. 1934.

SILL, W. H. Some characteristics of the wheat streak-mosaic virus and disease. *Trans. Kans. Acad.* **56**:411–424. 1953.

SIMMONDS, P. M. Root rots of cereals. *Bot. Rev.* **7**:308–332. 1941.

SLYKHUIS, J. T. Virus diseases of cereal crops in South Dakota. *So. Dak. Agr. Exp. Sta. Tech. Bul.* 11, pp. 1–29. 1952.

―――. Striate mosaic, a new disease of wheat in South Dakota. *Phytopath.* **43**:537–450. 1953.

―――. Wheat streak mosaic in Alberta and factors related to its spread. *Can. Jour. Agr. Sci.* **33**:195–197. 1953.

―――. *Aceria tulipae* Keifer (Acarina: Eriophyidae) in relation to the spread of wheat streak mosaic. *Phytopath.* **45**:116–128. 1955.

SMITH, E. F. A new disease of wheat. *Jour. Agr. Research* **10**:51–54. 1917.

―――― *et al.* The black chaff of wheat. *Science* (N.S.) **50**:48. 1919.

SNYDER, W. C., and H. N. HANSEN. The species concept of Fusarium with reference to Discolor and other sections. *Am. Jour. Bot.* **32**:657–666. 1945.

―――― and ――――. Variation and speciation in the genus Fusarium. *Annals N.Y. Acad. Sci.* **60**:16–23. 1954.

SPRAGUE, R. A physiologic form of *Septoria tritici* on oats. *Phytopath.* **24**:133–143. 1934.

―――. The relative importance of *Cercosporella herpotrichoides* and *Leptosphaeria herpotrichoides* as parasites on winter cereals. *Phytopath.* **24**:167–168. 1934.

―――. The association of *Cercosporella herpotrichoides* with the Festuca consociation. *Phytopath.* **24**:669–676, 946–948. 1934.

―――. Relative susceptibility of certain species of Gramineae to *Cercosporella herpotrichoides*. *Jour. Agr. Research* **53**:659–670. 1936.

―――. The status of *Septoria graminum*. *Mycologia* **30**:672–678. 1938.

―――. Septoria disease of Gramineae in Western United States. *Ore. State Monogr.* 6. 1944.

―――. Diseases of cereals and grasses in North America. The Ronald Press Company. New York. 1950.

―――― and H. FELLOWS. Cercosporella foot rot of winter cereals. *U.S. Dept. Agr. Tech. Bul.* 428. 1934.

STAKMAN, E. C. Reaction between *Puccinia graminis* and plants highly resistant to its attack. *Jour. Agr. Research* **4**:193–200. 1915.

―――. Recent studies of wheat stem rust in relation to breeding resistant varieties. *Phytopath.* **44**:346–351. 1954.

―――― *et al.* Origin of physiologic forms of *Puccinia graminis* through hybridization and mutation. *Sci. Agr.* **10**:707–720. 1930.

―――. Relation of the barberry to the origin and persistence of physiologic forms of *Puccinia graminis*. *Jour. Agr. Research* **48**:953–969. 1934.

―――. Die Bestimmung physiologischer Rasses pflanzenpathogener Pilze. *Nova Acta Leopoldina* **3**:281–336. 1935.

―――. Observations of stem rust epidemiology in Mexico. *Am. Jour. Bot.* **27**: 90–99. 1940.

―――. Identification of physiologic races of *Puccinia graminis tritici*. *U.S. Dept. Agr. Bur. Ent. Plant Quarantine* E-617. 1944.

STAKMAN, L. J.   A Helminthosporium disease of wheat and rye.   *Minn. Agr. Exp. Sta. Bul.* 191.   1920.

STRAIB, W.   Untersuchungen über erbliche Blattnekrosen des Weizens.   *Phytopath. Zeitschr.* **8**:541–587.   1935.

———.   Die Untersuchungsergebnisse zur Frage der biologischen Spezialisierung des Gelbrostes (*P. glumarum*) und ihre Bedentung für die Pflanzenzüchtung.   *Züchter* **9**:118–129.   1937.

SUBRAMANIAM, L. S.   Root rot and sclerotial diseases of wheat.   *Agr. Res. Inst. Pusa Bul.* 177.   1928.

SUKHOV, K. S., and P. T. PETLYUK.   *Delphax striatella* Fallen as the vector of the virus disease (Zakuklivanie) in grains.   *Compt. rend. Acad. Sci. U.S.S.R.* (N.S.) **26**:483–486; **26**:479–482.   1940.   From *Rev. Appl. Myc.* **20**:155–157.   1941.

TASUGI, H.   On the snow-rot fungus, *Typhula graminum* Karst. of graminaceous plants.   *Jour. Imp. Agr. Exp. Sta. Nishigahara, Tokyo* **1**:41–56.   1929.

———.   On the pathogenicity of *Typhula graminum* Karst.   *Jour. Imp. Agr. Exp. Sta. Nishigahara, Tokyo* **1**:183–198.   1930.

———.   On the physiology of *Typhula graminum* Karst.   *Jour. Imp. Agr. Exp. Sta. Nishigahara, Tokyo* **2**:443–458.   1935.   (In Japanese; English summaries.)

THOMPSON, W. P.   The causes of the cytological results obtained in species crosses in wheat.   *Can. Jour. Research* **C 10**:190–198.   1934.

TINGEY, D. C., and B. TOLMAN.   Inheritance of resistance to loose smut in certain wheat crosses.   *Jour. Agr. Research* **48**:631–635.   1934.

TISDALE, W. H.   Two sclerotium diseases of rice.   *Jour. Agr. Research* **21**:649–657.   1921.

——— *et al.*   Flag smut of wheat with special reference to varietal resistance.   *Ill. Agr. Exp. Sta. Bul.* 242.   1923.

———.   Further studies on flag smut of wheat.   *U.S. Dept. Agr. Cir.* 424.   1927.

TRELEASE, S. F., and H. M. TRELEASE.   Susceptibility of wheat to mildew as influenced by carbohydrate supply.   *Bul. Torrey Bot. Club* **56**:65–92.   1929.   **55**:41–67.   1928.

TUMANOV, I. I.   The hardening of winter plants to low temperature.   *Bul. Appl. Bot. Genet. and Plant Breed.* **25**:69–109.   1931.

VALLEGA, J.   Razas fisiologicas de *Puccinia triticina y P. graminis tritici* comunes en Chile.   *Min. Agr. Chile Dept. Gen. Fitot. Tech. Bol.* 3.   1942.

———.   Reacción de algunos Trigos con respecto a las razas fisiologicas de "*Puccinia rubigo-vera tritici*," communes en Argentina.   *Revista Fac. Agron. Vet. LaPlata Univ.* **11**:1–27.   1944.

———.   Reacción de algunas especies espontáneas de "Hordeum" con respecto a las ragas que afectan el Trigo.   *Rev. Invest. Agr. Buenos Aires* **1**:52–62.   1947.

———.   Wheat rust races in South America.   *Phytopath.* **45**:242–246.   1955.

——— and H. CENOZ.   Reacción de algunos Trigos a las razas fisiologicas de *Erysiphe graminis tritici* communes en Argentina.   *Inst. Fitotecn. Santa Catalina Pub.* 30.   1941.

VANTERPOOL, T. C.   Studies on browning root rot of cereals.   III. Phosphorus-nitrogen relations of infested fields.   IV. Effects of fertilizer amendments.   V. Preliminary plant analyses.   *Can. Jour. Research* **C 13**:220–250.   1935.

———.   Some species of Pythium parasitic on wheat in Canada and England.   *Ann. Appl. Biol.* **25**:528–543.   1938.

———.   Studies.   VI. Further contributions on the effects of various soil amendments on the incidence of the disease in wheat.   *Can. Jour. Research* **C 18**:240–257.   1940.

—— and G. A. LEDINGHAM. Studies. I. The association of *Lagena radiciola* n. gen. n. sp. with root injury of wheat. *Can. Jour. Research* **C 2**:171–194. 1930.

—— and R. SPRAGUE. *Pythium arrhenomanes* on cereals and grasses in the northern great plains. *Phytopath.* **32**:327–328. 1942.

—— and J. H. L. TRUSCOTT. Studies. II. Some parasitic species of Pythium and their relation to the disease. *Can. Jour. Research* **C 6**:68–93. 1932.

VASUDEVA, R. S., and M. K. HINGORANI. Bacterial disease of wheat caused by *Corynebacterium tritici* (Hutchinson) Bergey *et al. Phytopath.* **42**:291–293. 1952.

VAVILOV, N. I. Immunity to fungous diseases as a physiological test in genetics and systematics, exemplified in cereals. *Jour. Gen.* **4**:49–65. 1914.

VOLK, A. Untersuchungen über *Typhula graminum* Karst. *Zeitschr. Pflanzenkrankh.* **47**:338–365. 1937.

WADA, E., and H. FUKANO. On the differences of X-bodies in green and yellow mosaic of wheat. *Agr. Hort. Japan* **9**:1778–1790. 1934. (Abstract) *Japan Jour. Bot.* **7**:3–4. 1935.

—— and ——. On the difference and discrimination of wheat mosaics in Japan. *Jour. Imp. Agr. Exp. Sta.* **3**:93–128. 1937.

WAGNER, F. Auftreten, Sporenkeimung und Infektion des Zwergsteinbrandes an Weizen. *Zeitschr. Pflanzenbaum u Pflanzenschutz* **1**:1–13. 1950.

WALDRON, L. R. The effect of leaf rust accompanied by heat upon yield, kernel weight, bushel weight, and protein content of hard red spring wheat. *Jour. Agr. Research* **53**:399–414. 1937.

WALLIN, J. R. Parasitism of *Xanthomonas translucens* (J. J. and R.) Dowson on grasses and cereals. *Iowa State Col. Jour. Sci.* **20**:171–193. 1946.

WATKINS, A. E. The wheat species: a critique. *Jour. Gen.* **23**:173–263. 1930.

WATSON, I. A. Inheritance of resistance to stem rust in crosses with Kenya varieties of *Triticum vulgare* Vill. *Phytopath.* **31**:558–560. 1941.

——. Inheritance studies with Kenya varieties of *Triticum vulgare* Vill. *Proc. Linn. Soc. N.S. Wales* **68**:72–90. 1943.

—— and E. P. BAKER. Linkage of resistance to *Erysiphe graminis tritici* and *Puccinia triticina* in certain varieties of *Triticum vulgare. Proc. Linn. Soc. N.S. Wales* **68**:150–152. 1943.

—— and D. SINGH. The future of rust resistant wheat in Australia. *Jour. Aust. Inst. Agr. Sci.* **18**:190–197. 1952.

—— and W. L. WATERHOUSE. Australian rust studies. VII. Some recent observations on wheat stem rust in Australia. *Proc. Linn. Soc. N.S. Wales* **74**:113–131. 1949.

WEBB, R. W. Soil factors influencing the development of the mosaic disease in winter wheat. *Jour. Agr. Research* **35**:587–614. 1927.

——. Further studies on the soil relationships of the mosaic disease of winter wheat. *Jour. Agr. Research* **36**:53–75. 1928.

—— *et al.* Varietal resistance in winter wheat to the rosette disease. *Jour. Agr. Research* **26**:261–270. 1923.

WEBER, G. F. Septoria diseases of wheat. *Phytopath.* **12**:537–583. 1922. (Good summary of early literature.)

WOLLENWEBER, H. W., and O. A. REINKING. Die Fusarien. Paul Parey. Berlin. 1935.

WU, Y. S. Temperature and cultural studies on *Urocystis tritici* Koern. *Can. Jour. Res.* **C 27**:67–72. 1949.

—— and E. R. AUSEMUS. Inheritance of leaf rust reaction and other characters in a spring wheat cross. *Agron. Jour.* **45**:43–38. 1953.

Yu, T. F., *et al.*　Varietal resistance and susceptibility of wheats to flag smut.　II. *Agr. Sinica* **1**:79–81.　1934.

————.　Varietal resistance and susceptibility of wheats to flag smut.　III. Physiologic specialization in *Urocystis tritici* Koern.　*Bul. Chinese Bot. Soc.* **2**:111–113. 1936.

————.　Varietal resistance and susceptibility of wheats to flag smut.　IV. Further studies on physiologic specialization in *Urocystis tritici* Koern.　*Phytopath.* **35**: 332–338.　1945.

———— and H. K. Chen.　A Chinese wheat resistant to flag smut.　*Phytopath.* **21**: 1202–1203.　1931.

Zazhurilo, V. K., and G. M. Sitnikova.　Mosaic of winter wheat.　*Compt. rend. Acad. Sci. U.S.S.R.* (N.S.) **25**:798–801.　1939.　From *Rev. Appl. Myc.* **19**:268. 1940.

———— and ————.　Mosaic of spring cereals in the Voronezh district.　*Compt. rend. Acad. Sci. U.S.S.R.* (N.S.) **26**:474–478.　1940.　From *Rev. Appl. Myc.* **20**:157–158.　1941.

Zundel, G. L., *et al.*　A note on the status of the generic name Urocystis.　*Phytopath.* **30**:453–454.　1940.

CHAPTER 12

GRASS DISEASES

The grasses are receiving increasing attention as farm crops in many agricultural areas of the world. The more intensive cultivation of selections obtained from the indigenous or introduced species and the re-establishment and maintenance of the more desirable natural ones are changing the concepts of grassland agriculture. Grasses are becoming specific crop plants in the rotation system on the farm, in the soil-conservation program, in pastures, and on the range. As such, they are evaluated not as grass associations, but rather as species, varieties, or strains which have distinct and widely varied propensities. The multiplication of individual strains, by seed or clonal propagation, results in large populations similar or alike. The comparisons of large numbers of strains usually demonstrate differences, including disease reaction, between those single strain units within a species. Under these conditions diseases become a more evident factor in the development and economy of certain selected strains. Fortunately, the variability within a species generally affords ample scope for selecting disease-resistant lines when a sufficient array of clonal or selfed strains are investigated under conditions favorable for disease development. Furthermore, the planting of an area with a single strain of grass without first evaluating it for disease reaction increases the chances for a disease normally of minor importance in a mixed-grass population to become one of major significance. Therefore the investigation of the diseases of the grasses coincident with the breeding procedure is an essential phase of grass improvement.

The importance of the disease and the etiology of the parasite vary with the production of the grass in pure stands, in nursery rows, or in seed fields and its use in pastures or on the range. Frequently a disease is severe in nursery rows or seed fields and unimportant in the pasture mixtures. Control measures often differ greatly under these varied conditions of growth. Much of the information available concerning the diseases of grasses is based largely on observation under pasture conditions or where the plants are growing in their natural habitats. The chapter on grass diseases must deal largely with symptoms of the diseases, the morphology of the parasites, and only the more general phases of etiology and control. Information on the geographical distribution of the diseases frequently is inadequate, as the diseases have been studied intensively in some regions and in others not at all. The wide variety of grass species involved and the still wider range of parasites make the task more difficult in the space allotted. Reference should be made to the diseases on the cereal crops for more detailed discussions of many of the diseases common to the Gramineae and also to Sprague (1950).[1]

1. Nonparasitic Maladies. The grasses, like the cereal crops, are subject to external manifestations of unfavorable environmental conditions. The response to soil mineral deficiencies is similar in manifestation to that in the different cereal groups. Chlorosis, leaf flecking, and spotting are the common symptoms.

Pigmented spots or blotches commonly occur on the leaves of many species, such as those of *Andropogon*, *Bromus*, *Dactylis*, *Panicum*, and *Sorghum*, especially when inbreeding is employed in the breeding program.

Bends and proliferation reported in many genera of the forage and pasture grasses are listed in this group, chiefly since no cause for either condition is known (Fischer, 1941, Nielsen, 1941). The upper portion of the culm or the rachis bends downward, usually in a hairpin-like turn. The inflorescences develop into proliferated leaf-like structures similar to crazy top in corn. Both conditions frequently occur on the same plant.

2. Mosaics and Yellows, Viruses. Two viruses are relatively specific to certain genera of the grasses. Probably others occur that have not been investigated.

Agropyron Mosaic, Virus Transmitted Mechanically at Low Temperatures, Possibly by Aphids. This mosaic is distributed widely on *Agropyron repens* (L.) Beauv., *A. elongatum* (Host.) Beauv., *A. inerme* (Scrib. & Smith) Rydb., *A. intermedium* (Host.) Beauv., and *Festuca rubra* L. Leaf mottling and striping (both yellow and green), necrofis, and some stunting are symptoms. Virus persists in crown and rhizomes, not seedborne (McKinney, 1934, Slykhuis, 1952).

Brome Mosaic, Virus Transmitted Mechanically, Not Seed-borne. The disease is distributed widely on *Bromus inermis* Leyss. The virus has a wide host range based on mechanical inoculation including other grasses, cereals, tobacco, beets, and lamb's-quarters. The symptoms are chlorotic mottling and striping, brown necrotic stripes and blotches, shortened culms, frequently with sterile inflorescences and occasionally

---

[1] General reference to the *U.S. Department of Agriculture*, *Plant Disease Reporter and Supplements*, for the compilations and notes contained therein; those by G. W. Fischer, J. R. Hardison, C. O. Johnston, and R. Sprague were especially valuable.

excessive tillering. The symptoms are similar to those of stripe mosaic of barley, although the particles are round in the brome virus. The virus persists in the plant crowns. This virus is lethal in the growing point of Golden Giant sweet corn seedlings which are used for biological assay (McKinney *et al.*, 1942).

Other Mosaic Diseases on Economic Grasses

1. Sugarcane mosaics (Chap. 10) occur on tropical grasses of economic importance: *Saccharum* spp., *Sorghum vulgare* Pers. and varieties, *Setaria* spp., *Pennisetum glaucum* (L.) R.Br., *Eleusine indica* (L.) Gaertn., *Miscanthus sinensis* Anderss., *Tripsacum laxum* Nash.

2. Sugarcane chlorotic streak on *Miscanthus* spp.

3. Yellow dwarf mosaic, on cereals and grasses. The virus is distributed widely in barley, oats, and many grasses. The grass hosts appear to be the source of virus inoculum as the aphid vectors move to the cereals. The grasses show a range of symptoms: red or yellow discoloration of areas or entire leaf blade, reduced culm elongation, and poorly developed inflorescence. The economic grasses affected are *Bromus catharticus* Vahl., *B. inermis* Leyss., and several annual species of *Bromus* and *Festuca*, *Hordeum*, *Avena*, *Phalaris paradoxa* L., and *Andropogon barbinodis* Lag. Among several grasses reported as carrying the virus but showing no symptoms in California are *Dactylis glomerata* L., *Festuca elatior arundinacea* Schreb., *Poa pratensis* L., *Agropyron trachycaulum* (Link) Malte, *Phalaris tuberosa* L., *Chloris gayana* Kunth., *Cynodon dactylon* (L.) Pers., and *Sorghum vulgare* var. *sudanense* (Piper) Hitch. *Phleum pratense* L. was among 19 species reported immune (Oswald and Houston, 1953). In Wisconsin some clones of timothy and Kentucky bluegrass show characteristic yellow dwarf symptoms. Aphid transmission to Blackhulless (C.I. 666) barley is a method of indexing for the virus.

4. Barley stripe mosaic. The virus is transmitted mechanically and with pollen in some hosts. It is seed-borne in barley and wheat and probably in some of the more compatible grass hosts. The disease is distributed widely in North America. The symptoms range from chlorosis to necrotic brown stripes on all leaves in the systemic distribution to local narrow to broad stripes in the grasses showing local lesion response (Hagborg, 1954, Slykhuis, 1952). Weed grasses in the genera *Bromus*, *Echinochloa*, *Eragrostis*, *Panicum*, and *Setaria* are susceptible, and the millet grasses *Panicum miliaceum* L. and *Setaria italica* (L.) Beauv. Local lesions occur on *Agropyron* spp., *Bromus inermis* Leyss., *Festuca rubra* L., *Phalaris arundinacea* L., and *Phleum pratense* L. Chevron (C.I. 1111) barley is used as an assay host for this virus.

5. Wheat streak mosaic, virus transmitted mechanically and by the grass mite, *Aceria tulipae*. The virus is distributed in west central North

America on weed grasses *Bromus* and *Setaria* spp. and occurs in mild form on some economic grasses (McKinney and Fellows, 1951, Slykhuis, 1952).

6. Striate wheat virus, transmitted by leaf hoppers. The disease occurs in limited areas in the north central prairie area of North America and a similar area in Russia. The virus occurs naturally in *Eragrostis cilianensis* (All.) Lutati and *Panicum capillare* L. The range of grass hosts is not known. Fine chlorotic streaks over the veins of the leaves are the characteristic symptom (Slykhuis, 1953).

7. Rice yellow dwarf virus, transmitted by leaf hoppers. The disease occurs in Japan and possibly Formosa and the Philippines. Stunted, chlorotic plants of *Alopecurus fulvus* L., *Echinochloa crus-galli* Beauv., *Panicum miliaceum* L., and *Poa pratensis* L. were produced by the virus from rice (Padwick, 1950).

8. Corn streak virus, transmitted by leaf hoppers, *Cicadulina* spp. The disease is restricted to Africa and India. Susceptible grasses are *Digitaria* spp., *Eleusine* spp., and *Sorghum arundinaceum* Stapf. (McClean, 1947, Storey, 1929).

3. Bacterial Blights, *Pseudomonas coronafaciens* var. *atropurpureum* (Reddy, & Godkin) Stapp, *Xanthomonas translucens* (L. R. Jones, A. G. Johnson, & Reddy) Dows., and Other Species. Specialized varieties of the two common cereal bacterial pathogens occur on the grasses. The bacterial blight caused by *Pseudomonas coronafaciens* var. *atropurpurea* (Reddy & Godkin) Stapp is distributed widely on *Bromus inermis* Leyss and *Agropyron repens* (L.) Beauv. Tessi (1953) has shown *P. coronafaciens* grown on glucose media capable of producing the halo lesion on *Agropyron, Briza, Bromus, Danthonia, Elymus, Hordeum, Lolium, Phalaris* spp.; in contrast *P. striafaciens* on the same media incites lesions on only *Briza maxima* L., *Bromus catharticus* Vahl., *Bromus hordeaceus* L., *Bromus rubens* L. The halo blight occurs naturally on some species in the genera listed and *Phleum* (Allison and Chamberlain, 1946, Godkin, 1923). The lesions are first circular to elliptical, water-soaked, later browning and coalescing to form purplish-brown areas on the leaf blade and sheath (Fig. 65), usually with a halo around the lesion. Exudate on the surface of the lesion is absent. Spots on the pedicels and panicles are smaller and restricted. Several varieties of *X. translucens* incite stripe blight on *Agropyron, Bromus, Hordeum, Phleum,* and *Secale* spp. The lesions are usually long stripes or interrupted linear to irregular fine stripes, water-soaked at first, later showing brown pigmentation, translucent areas, and exudate on the surface (Fig. 65). Frequently the apical leaf whorl and inflorescence are blighted with exudate conspicuous on the diseased tissues. Later infections on the floral bracts are fine linear to elongate dark-brown spots with sparse exudate (Fang *et al.*, 1950, Wallin, 1946, Wallin and Reddy, 1945).

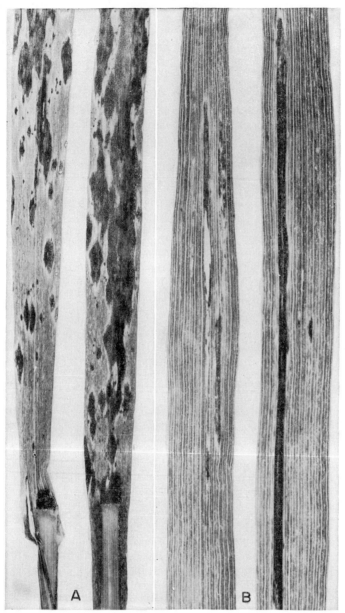

FIG. 65. Bacterial blights caused by *Pseudomonas coronafaciens* var. *atropurpureum* (*A*) and *Xanthomonas translucens f. sp. cerealis* (*B*) on *Bromus inermis*.

The head blight caused by *Corynebacterium agropyri* (O'Gara) Burk. [*Bacterium agropyri* (O'Gara) Stapp], [*Phytomonas agropyri* (O'Gara) Bergey *et al.*], (*Aplanobacter agropyri* O'Gara), described by O'Gara (1916), occurs on *Agropyron* spp., *Sitanion jubatum* J. G. Smith, *Sporobolus airoides* (Torr.) Torr., and other grasses. The disease is similar to the European bacterial blight (Sampson and Western, 1941) of *Dactylis glomerata* L. caused by *Corynebacterium rathayi* (E. F. Sm.) Dows., [*Phytomonas rathayi* (E. F. Sm.) Bergey *et al.*], (*Aplanobacter rathayi* E. F. Sm.), which apparently occurs occasionally on this grass in North America.

The head blight caused by the former species is prevalent in northern North America, where it causes some damage in seed production. The lesions occur sparsely on the leaves; they involve mainly the inflorescences. Severe infection results in erect leaves and distorted spikes, or panicles partly enclosed in the leaves, with an abundant yellow exudate over the diseased surfaces. Seed transmission of these two diseases is common. Exclusion of seed lots from areas where the disease occurs and seed treatment with solutions of formaldehyde or organic mercury compounds are recommended.

The sorghum and sugarcane bacterial blights occur on the related grasses. The more prevalent of this group are the bacterial stripe and streak of Johnson grass [*Sorghum halepense* (L.) Pers.] and Sudan grass [*S. vulgare* var. *sudanense* (Piper) Hitchc.] (see Chap. 9).

The bacterial wilts or gumosis diseases occur on some tropical grasses in addition to sugarcane, sorghum, and corn. The bacteria invade the xylem of the host causing chlorotic leaf striping and shortened culms, frequently with leaves erect and final necrosis and release of the gummy bacterial mass. The yellow mass of bacteria in the xylem vessels are conspicuous in culms and leaves. *Axonopus, Miscanthus, Saccharum,* and *Sorghum* spp. are among the grasses damaged. *Xanthomonas vasculorum* (Cobb) Dows., *X. rubrilineans* (Lee *et al.*) Starr & Burk., *X. axonoperis* Starr & Garces, similar in character, are the incitants of the disease (Garces, 1947, Starr and Garces, 1950). Resistant varieties and elimination of sources of inoculum are suggested as control measures. As this group of bacterial pathogens is transmitted in cuttings and frequently on seed, forage-plant stocks from areas where the diseases occur should be avoided (see Chap. 10).

4. Pythium Seed Rot, Seedling Blight and Root Rot, *Pythium* spp. The *Pythium* spp. common on the cereal crops cause severe damage to the grasses, as reviewed by Sprague (1944, 1950). This disease complex apparently is more severe in the heavy or fine soil areas of west central North America, but it is general in occurrence in all areas. Preemergence seed rotting, seedling blight, and damping-off of young seedlings more generally are caused by *P. debaryanum* Hesse. *P. graminicola* Subrm. causes a seedling blight and root rot of grasses. *P. arrhenomanes* Drechsl., similar to the above, is distributed widely on the grasses, especially in association with the root browning and root rotting of seedlings and growing plants. The plants are weakened and yellowed or killed by the root rot, depending upon the environmental conditions. Differences in the tolerance of lines of the grasses are evident, and selection for resistance to this root-disease complex is important in grass breeding. Crop sequence, soil fertility, soil preparation, and time of seeding are important in control. Other species of *Pythium* are associated with the complex, as discussed in Chap. 11.

5. Downy Mildews, *Sclerospora* spp. The downy mildews are not serious diseases on the major forage and pasture grasses. *Sclerospora graminicola* (Sacc.) Schroet. occurs especially on species of *Pennisetum* and *Setaria*, but the grasses of economic importance in this group are grown largely in tropical or in the Asiatic countries according to Hiura (1935). *S. macrospora* Sacc. or *Sclerophthora macrospora* (Sacc.) Thir., Shaw, & Naras. occurs in localized areas on a number of economic grasses, as well as the cereals, in the more temperate zones. Apparently the disease is restricted in its spread, although its occurrence in increasing frequency in the Central United States warrants further consideration, especially for forage grasses (see Chaps. 5 and 11).

6. Powdery Mildew, *Erysiphe graminis* DC. Powdery mildew is distributed widely on many genera of the grasses. However, the disease is of direct economic significance at present on a relatively few grasses, notably *Agropyron*, *Agrostis*, *Avena*, *Festuca*, and *Poa* spp. and strains. Very resistant and susceptible types occur in the nursery investigations with many of these grasses. Powdery mildew is potentially more dangerous in nurseries and seed fields than in pastures. The disease appears during cool, somewhat cloudy seasons, especially on the leaves. The powdery superficial, white, gray, or buff mycelium and conidia develop in blotches or spread uniformly over the leaf surface. The leaves frequently brown and dry out, gradually reducing the leaf area and forage. The symptoms are less conspicuous where the grasses are grazed or cut closely.

Erysiphe graminis consists of many specialized varieties and races on the cereals and grasses. According to Hardison (1944, 1945) and others, specialization is less restricted in the grasses than in the cereals. Certain of the wild grasses furnish inoculum for the cereal crops as well as for some commercial grasses. Resistant varieties or strains offer the best means of control of the disease (see Chaps. 3 and 11).

7. Ergot, *Claviceps* spp. Several species of *Claviceps* somewhat similar morphologically occur on the cereals and grasses. According to

Langdon (1941, 1950), *Claviceps pusilla* Ces. is the species causing ergot on 11 genera of grasses of Australia and the South Pacific. One genus of this group of grasses, *Digitaria*, is common in the United States, and several others are introduced. Apparently the disease is not reported on these grasses in the United States. *C. paspali* F. L. Stevens & Hall is severe on Dallis grass, *Paspalum dilatatum* Poir, and other species in the warmer areas (Lefebvre, 1939). *C. cinerea* Griff. occurs sporadically on *Hilaria* spp. in Southwestern United States and adjacent Mexico. Sclerotia of this species as in others of the warmer climates germinate after a short period of dormancy. *C. maximensis* Theis occurs extensively on *Panicum maximum* Jacq. and probably other species in the Caribbean Islands (Theis, 1952). The author found it prevalent on this grass in Florida. Ergot develops on *Pennisetum* and *Cenchrus* spp. in India; alkaloid content is low (Romakrishnan, 1952). The common species, widely distributed on the grasses, is *C. purpurea* (Fr.) Tul. This species occurs on the small grains and grasses common in the temperate zones of the world. The disease is particularly severe on *Agropyron, Agrostis, Bromus, Calamogrostis, Dactylis, Elymus, Festuca, Glyceria, Lolium, Phleum,* and *Poa* spp., or most of the economic forage grasses of the temperate zones. Specialized races occur in this species (McFarland, 1921; Mastenbroek and Oort, 1941, Stäger, 1903, 1905, 1922). One race of the pathogen occurs on the cereals and most of the more common grasses. A race is specialized to *Lolium* and related grasses. The race on *Glyceria* apparently is restricted to this genus. Petch (1938) and others have shown that *C. microcephala* (Wallr.) Tul. is not morphologically distinct from *C. purpurea*. Other species have been described or listed on various hosts on the basis of host specialization rather than distinct morphological characters.

The fungi infest the young ovaries of the grass flowers by direct penetration, produce the sphacelial stroma, and finally the sclerotia that replace the seed in the infected flowers. In some tropical areas only the conidial (sphacelial) or "honeydew" stage develops (Langdon, 1954, Theis, 1953). The sclerotia develop below the conidial stromata and grow into elongated brownish-black or purplish-black fungus bodies somewhat the shape of the grass caryopsis (Fig. 66). When mature, the sclerotia usually extend beyond the flora bracts. The sclerotia remain attached within the grass flower until the host is fully mature. Some of the sclerotia are harvested with the hay or seed. Others fall on the ground where they germinate, following a period of dormancy, to form stalked stromatic heads in which the perithecia develop. These contain ascospores which are wind-borne to reinfect the grass flowers. The production of ascospores usually is synchronized with the blossoming of the grasses (see Chap. 8).

The sclerotia usually contain alkaloids and other compounds adversely affecting animal physiology. Fully matured sclerotia free from fungus infection contain compounds causing principally constriction of the capillary blood vessels. Ingestion of large numbers of sclerotia causes abortion. Continuous ingestion of relatively small quantities of sclerotia results in reduced milk flow and frequently tissue necrosis in animal extremities, such as hoof, tail, and ear. Ergot sclerotia parasitized by *Fusarium* spp. during their development contain compounds affecting the nervous system and result in partial or complete paralysis. This latter type of ergot poisoning apparently is associated with ergot on Dallas grass in the South Central humid section of the United States (Brown, 1916, Simpson and West, 1952). *Cerebella andropogonis* Ces. and *Cladosporium* spp. as well as *Fusarium* develop in the exudate of the conidial stage and apparently function in reducing sclerotial formation (Theis, 1953).

FIG. 66. Ergot sclerotia on *Bromus inermis.*

Ergot damage to livestock is controlled by close grazing prior to heading of grasses and by mowing pastures soon after heading. Apparently resistant varieties are uncommon in most grasses, especially in cross-pollinated species. A resistant selection similar to Dallas grass from the cross *Paspalum dilatatum* × *P. mallacophyllum*, the latter not infected by *Claviceps paspali*, resulted in one immune and six resistant lines more vigorous than Dallas grass and better seed production (*Miss. Agr. Exp. Sta. Ann. Repts.* 1952, p. 14; 1954, pp. 16–17).

8. Epichloe Head Blight or Choke, *Epichloë typhina* (Fr.) Tul. The disease is of minor importance in North America. It occurs sparingly on a large number of grasses but is sometimes abundant on *Calamagrostis,*

Glyceria, Koeleria, and *Poa* spp. in north central North America. In Europe and Asia, however, the disease is a serious factor in seed production, especially in the cool, humid northern regions, as summarized by Sampson and Western (1941). The disease is restricted apparently to limited areas in North America where the seasons are cool and the winters are mild or where the plants are protected by snow. The fungus *Epichloë typhina* (Fr.) Tul. produces a perennial mycelium in the crown buds of infected plants and a systemic infection of the primordial tissues. In summer, the mycelium forms a white felt over the surface of late tillers or more characteristically a white stroma with minute conidia over the inflorescence as it emerges. The stroma usually encloses all the spike or part of the inflorescence in the case of the more open panicle type of grass species. The color of the stroma changes to orange as the plants approach maturity, and numerous perithecia with papillate ostioles develop submerged in the stroma. The asci contain eight hyaline filiform many-celled ascospores, which are forcibly discharged and wind-borne. The stroma disintegrates after the plants are ripened fully. Seed transmission occurs in *Festuca rubra* L. and probably in other grasses in which diseased plants produce seed.

9. Tar Spot, *Phyllachora* spp. Tar spots are distributed widely on a large number of grasses; however, they are localized and rarely become general in nursery, field, or pasture. The black, sunken glossy spots on the leaves are conspicuous when the plants are infected heavily. *Phyllachora graminis* (Fr.) Fckl. is one of the more common species, although many others are described on the grasses. The perithecia are immersed in the black stroma with the ostioles opening on both surfaces of the leaves. The asci are cylindrical with short pedicels, intermixed with filiform paraphyses, and contain eight ovoid hyaline unicellular ascospores. Orton (1944) has summarized the literature concerning the species on the Gramineae. Over 40 species were described, the differentiation being based on the clypei and the size and shape of the ascospores.

10. Snow Mold and Foot Rot, *Calonectria graminicola* (Berk. & Br.) Wr., *Fusarium nivale* (Fr.) Ces. This disease of the grasses is similar to that occurring on winter wheat. Damage is more severe on turf species, especially where they are grown under conditions of heavy nitrogen application to force late summer vegetative development (Fig. 67). The disease and its control is discussed by Broadfoot (1938), Bennett (1933), Dahl (1934), Wollenweber and Reinking (1935), and Sampson and Western (1941). The treatment of turf of golf greens and lawns with fungicides, especially the mercury compounds, is practical. This also controls the snow scald and snow mold caused by *Typhula, Sclerotium,* and *Rhizoctonia* spp. (see Chap. 11).

11. Seedling Blight and Crown Rot, *Gibberella* and *Fusarium* spp. This type of seedling blight and crown rot, so common on certain of the cereal crops, is of minor importance on most of the grasses. *Fusarium culmorum* (W. G. Sm.) Sacc. causes some damage in the more northern sections of North America and Europe. The many species associated

Fig. 67. Snow mold damage on a golf green, photographed early in the spring.

with decaying roots and crowns of the grasses apparently develop as secondary or saprophytic organisms on the weakened or dead tissues, according to Sprague (1944) and others. Spikelet and spike blight caused by *Gibberella zeae* (Schw.) Petch or *G. roseum* f. *cerealis* (Cke.) Snyder & Hansen occurs sparingly on some grasses in the humid, warm sections of the corn belt and other humid regions.

12. Take-all, *Ophiobolus graminis* Sacc. Take-all occurs on native grasses, and it is world-wide in distribution. Evidence indicates that it is indigenous on the grasses in most regions and spreads to wheat from the grasses. Infrequently the disease reduces vigor and stand of pasture grasses such as *Agrostis alba* L., *Bromus inermis* Leyss, and *Festuca* *elatior* var. *arundinacea* (Schreb.) Wimm. The common weed grass, *Agropyron repens* (L.) Beauv., frequently is a source of inoculum to wheat (see Chap. 11 and Sprague, 1950).

13. Blind Seed Disease, *Phialea temulenta* Prill. & Del. The disease is prevalent in the grass-seed-producing area in Oregon in the United States, in New Zealand and Tasmania, and in northern Europe (De Tempe, 1950, Hardison, 1948, Osborne, 1947, Wade, 1949). Although several grass hosts are reported, the disease is important economically chiefly on *Lolium* spp. The fungus invades the pericarp and adjacent tissues of the caryopsis soon after pollination. Unicellular, cylindrical, hyaline macroconidia are produced in abundance, forming a pink slime on the surface of the pericarp. Nongerminable, unicellular, ovoid, hyaline microconidia are produced in sporodochia later on the seed in the soil. Apothecia develop on the dead or diseased seed in the soil to produce ascospores at blossoming time of host (Sprague, 1950). The pathogen reduces the yield of germinable seed.

The elimination of badly diseased fields following inspection, crop rotation, cutting grass early for hay when infection is heavy, and early seed harvest are suggested as control. Some evidence indicates that varieties of rye grass vary in susceptibility.

14. Helminthosporium Foot Rots and Leaf Blights. Some 30 species of *Helminthosporium* occur on the grasses in various parts of the world, as reported by Drechsler (1923, 1929, 1930, 1935), Henry (1924), Hynes (1937), Nishikado (1929), and Sprague (1950). Some, such as *Helminthosporium sativuum* Pamm., King, & Bakke, are of major importance in the economy of grass production; others are more restricted in distribution and in the grasses attacked and are potential hazards in the production of susceptible lines of the grasses only.

Helminthosporium sativum Pamm., King, & Bakke probably ranks as the species of first importance on the grasses. The general symptoms, morphology, and etiology of diseases caused by this cosmopolitan species are discussed in Chaps. 3 and 11. The parasite causes seedling blight and root rot on a wide range of grass species, especially in the central prairie area of North America. According to Andrews (1943), Christensen (1922), and others, it is probably the most important seedling and root parasite on the grasses in the humid North Central area. This fungus is exceeded in importance only by *Pythium* spp. in the drier prairie areas of the United States and Canada, as summarized by Sprague (1944). The disease also occurs on grasses in other countries. Both *Pythium* and *Helminthosporium* root rots are important in reducing the stands in many of the economic grasses. Species of Poa are resistant to *H. sativum*, but stands in *Poa pratensis* L. are damaged, especially in the more southern range of this grass, by another species, *H. vagans* Drechsl.

Helminthosporium vagans Drechsl. causes a leaf spot and culm and crown rot of *Poa pratensis* and some other species of this genus. The organism is widely distributed in North America, and it is common in

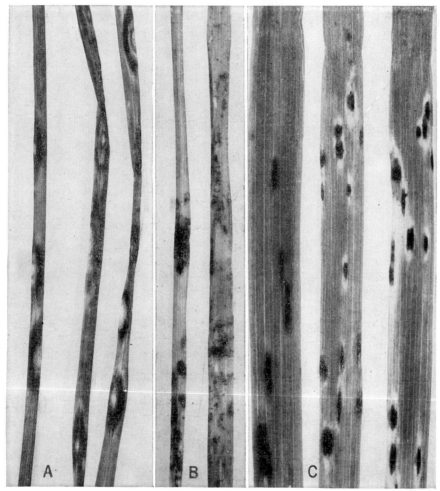

Fig. 68. (*A*) Leaf spots caused by *Helminthosporium vagans* on *Poa pratensis*. (*B*) Leaf blotch caused by *H. dictyoides* on *Festuca elatior*. (*C*) Brown leaf spot caused by *H. bromi*, or *Pyrenophora bromi* on *Bromus inermis*.

Europe. The bluish-brown leaf spots are numerous on the leaves. The diffuse brown lesions on the base of the culm and crown are abundant from mid-summer to autumn (Fig. 68). Young plants, especially seedlings in old sod, are killed, and older plants are weakened. The conidia are cylindrical to slightly tapering toward the apex, rounded at the ends, measure 8-10 by 50-280 microns, and are dark olivaceous in color. They germinate by the formation of germ tubes from the end and middle segments.   Blue grass strains show differences in susceptibility.   *H. poae* Baudys occurs on *Poa secunda* Presl. and *P. trivialis* L. in North America.

Pyrenophora bromi (Died.) Drechsl. or *Helminthosporium bromi* Drechsl. is common on *Bromus inermis* Leyss and some other species of the genus.   The disease appears as small dark-brown oblong spots on the first leaves to develop in the spring and continues until mid-summer (Fig. 68).   The older lesions are surrounded by a yellow margin.   Severely infected leaves turn yellow and dry out.   Conidia form sparsely on the lesions and withered leaves.   Perithecia initials form in the lesions during the summer, and mature ascospores develop the following spring.   The conidia are cylindrical with rounded ends, measure 14-26 by 45-265 microns, and 1- to 10-septate without pronounced constriction at the septations, and germination occurs from the middle and end cells.   The ascospores are light brown, measure 20-30 by 45-72 microns, are uniformly divided by 3 transverse septa, and longitudinal septa occur in one or both of the middle cells (Chamberlain and Allison, 1945, Drechsler, 1923). The disease is generally more abundant in nursery or seed fields than in meadows.   Variation occurs in the reaction of *B. inermis* strains to the organism.

Pyrenophora tritici-repentis (Died.) Drechsl., *Helminthosporium tritici-repentis* Died., is widely distributed on some *Agropyron* and *Elymus* spp. The disease is found more commonly in North America on *Agropyron repens* (L.) Beauv.   The leaf spots are indefinite brown blotches with conidia on the older areas.   The brown lesions around the base of the culms are not defined clearly.   The conidia are cylindrical, straight, the basal cell tapering sharply to the hilum, and measure 7-9 by 80-220 microns.   The ascospores are light brown, with 3 transverse and 2 longitudinal septa, and are 18-28 by 45-70 microns in size.

Three species of *Helminthosporium* occur on the creeping bent grasses and redtop (*Agrotis* spp.).   *Helminthosporium stenacrum* Drechsl. causes a mild leaf blighting without definite pigmented spots.   The conidia are hyaline to yellow, cylindrical with rounded ends, and occasionally produce narrowed distal prolongations.   They are 15-23 by 53-135 microns in size and germinate from several or all segments.   *H. erythrospilum* Drechsl. is distributed widely in North America on *Agrostis* spp. on which it produces small reddish-brown or russet spots followed by the killing of the leaves.   The conidia are yellow to light olivaceous in color, straight cylindrical with abruptly rounded ends, measure 8-16 by 25-105 microns, and germinate from any or all cells.   *H. triseptatum* Drechsl. causes straw-colored leaf spots and some wilting of leaves on the *Agrostis* spp.   The disease occurs sparingly on velvet grass, *Holcus lanatus* L.;

orchard grass, *Dactylis glomerata* L.; redtop, *Agrostis alba* L.; and timothy, *Phleum pratense* L. The conidia are dark olivaceous in color, ellipsoidal or short cylindrical, sometimes tapering toward the base, with hemispherical end cells, 2- to 3-septate, measure 15-21 by 35-50 microns, and germinate from the basal cell.

Helminthosporium giganteum Heald & Wolf produces a zonate eye spot with brown margins and tan centers on the foliage and inflorescence of a large number of grasses, especially in the warmer climates of North America. The huge conidia are pale brown, cylindrical, slightly tapering to both ends, with rounded apical cells, 5-septate, measure 15-21 by 300-315 microns, and germinate from the middle and end cells.

Helminthosporium siccans Drechsl. occurs on several cultivated species of *Lolium* in North America and Europe. The dark-brown elongated spots coalesce, forming mottled discolored areas and causing drying of the leaf blades and finally of the sheaths. The inflorescences are spotted and reduced in development. The conidia are yellow to light brown, straight, slightly tapering toward the apex, with rounded end cells, measure 14-20 by 35-130 microns, and germinate from middle and end cells.

Helminthosporium cynodontis Marig. occurs abundantly on the same grasses as *H. giganteum* and causes leaf withering and bleaching. The conidia taper toward both ends, are generally curved and 3- to 9-septate, measure 11-14 by 27-80 microns, and germinate from the end cells only.

Helminthosporium dictyoides Drechsl. produces a brown net blotch on the leaves of *Festuca elatior* L. Graham (1955) reports a variety of this species on timothy. The disease is widely distributed in North America, and when severe it causes spotting, browning, and wilting of the leaves (Fig. 68). The conidia are yellow, straight, tapering gradually to the apex, typically 3- to 5-septate, 14 to 17 microns wide at the base by 23 to 115 microns long, and germinate from the end cells.

Helminthosporium turcicum Pass. causes leaf blight and defoliation of Sudan grass and Johnson grass. The races of the parasite on these grass sorghums do not infect corn readily. The morphology of the fungus is given in Chap. 4.

Helminthosporium sacchari (Breda de Haan) Butl. produces a severe eyespot disease on Napier grass, *Pennisetum purpureum* Schumach., and on lemon grass, *Cymbopogon citratus* (DC.) Stapf. (see Chap. 10).

Helminthosporium ravenelii Curt. & Berk causes sooty spike on several species of *Eragrostis* and *Sporobolus* and probably on *Panicum*. The disease is widely distributed in the warmer, humid climates. The conidia are straight or sigmoid curved, tapering somewhat more at the base, rounded at the ends, usually 3- to 4-septate, measure 12-19 by 22-78 microns, and germinate from the apical cells. Nishikado (1929) listed *H. miyakei* Nishikado as an additional species causing sooty spike in Asia on *Eragrostis pilosa* (L.) Beauv.

A few of the small-spored types (*Curvularia*) occur commonly on the grasses in addition to those described on wheat (Chap. 11). *Curvularia geniculata* (Tracy & Earle) Boed. is common on the dead tissues of the grasses. Under favorable conditions, root and crown tissues are invaded. The conidia form in dense clusters on the conidiophores. They are curved more on one side than on the other, taper toward both ends, generally 4-septate with the center cell larger and darker-colored, and measure 7-14 by 19-45 microns.

The *Helminthosporium* spp. on the grasses are differentiated into two groups, as discussed by Drechsler (1934), Ito and Kuribayashi (1931), and others. In the species with the light-colored epispore wall, generally cylindrical conidia, and germinating from the middle and end cells of the conidia, the ascigerous stages are in the genus *Pyrenophora* (*Pleospora*). In contrast, the species forming conidia with darker olivaceous epispore wall, conidia tapering toward the ends, and germination from the apical cells only, the perithecial stages are in *Cochliobolus* (*Ophiobolus*). A few species are associated with *Leptosphaeria*. See Sprague (1950) for other species.

15. Anthracnose, *Colletotrichum graminicolum* (Ces.) G. W. Wils. The disease on the grasses occurs primarily as a root, crown, and culm rot with the exception of Sudan grass on which it causes a zonate leaf spot. Considerable damage results on old sod in areas of light soils and depleted fertility.

Gloeosporium bolleyi Sprague incites a severe seed rot and root necrosis on grasses (Sprague, 1950).

Anthracnose is discussed in Chaps. 8 and 9.

16. Mastigosporium Leaf Fleck. Leaf fleck occurs in spring and autumn on a large group of grasses. Numerous purplish-brown flecks occur on the leaves. The lesions elongate and coalesce to form irregular blotches. The center of the mature lesion is light-colored, and groups of glistening white to gray conidia are conspicuous on these areas. In Europe, *Mastigosporium album* Riess is the predominating species. The oblong conidia are 3- to 5-septate, with several long bristle-like appendages arising from the apical cell, measure 14-18 by 40-62 microns, and are borne on short thick conidiophores. The variety of this species on orchard grass in Europe forms smaller conidia without appendages (Sampson and Western, 1941). According to Sprague (1938, 1940, 1950), the muticate type without appendages, commonly occurring on *Agrostis*, *Calamagrostis*, and *Dactylis* spp. in North America, is referred to another species, *M. rubricosum* (Dearn. & Barth.) Nannf. *M. album* var.

calvum Ell. & Davis, *M. album* var. *muticum* Sacc., *M. calvum* (Ell. & Davis) Sprague, *Fusoma rubicosa* Dearn. & Barth., and *Amastigosporium graminicola* Bond-Mart. are synonyms. The conidia are usually 3-septate without appendages. A third species, *M. cylindricum* Sprague,

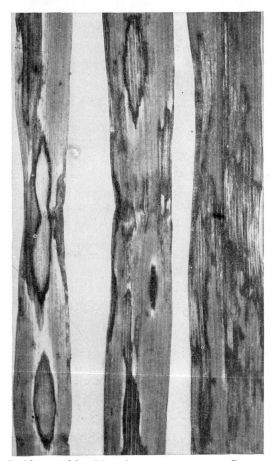

FIG. 69. Scald caused by *Rhynchosporium secalis* on *Bromus inermis.*

on *Bromus vulgaris* (Hook.) Shear occurs in North America. The conidia of this species are cylindrical with blunt rounded ends and are much smaller than in the others. The leaf fleck diseases, when severe, defoliate susceptible lines of the grasses. Bollard (1950) gives host range and reaction.

17. Leaf Scald, *Rhynchosporium secalis* (Oud.) J. J. Davis and *R. orthosporium* Caldwell. Scald is widely distributed on several genera of the grasses, as well as on barley and rye. Races of *Rhynchosporium* *secalis* (Oud.) J. J. Davis, the species causing scald on the cereals, are of major importance on *Bromus, Agropyron,* and *Elymus* in Northwest and North Central North America, Europe, and Asia. Specialized races occur on these grasses and cereals, as discussed by Caldwell (1937) (see Chap. 3). The leaf scald on *Bromus inermis* Leyss is severe during the early spring and late autumn (Fig. 69). Lines vary greatly in their reaction to the disease. *R. orthosporium* Caldwell is differentiated from the former species by the longer conidia and the absence of the beak on the apical cell. This species occurs in localized areas on *Alopecurus, Calamagrostis,* and *Dactylis* spp.

18. Streak or Brown Leaf Blight, *Scolecotrichum graminis* Fckl. Many grasses are susceptible to this disease, which is distributed widely in the temperate zones. *Bromus inermis* Leyss is among the few economic grasses resistant to the disease. The grayish-brown to dark-brown linear lesions occur on the leaf blade and extend into the leaf sheath. Defoliation results from the lateral spread of the infection, especially in the leaf sheath. In the mature lesions, the dark-gray masses of conidiophores, arranged in rows as they emerge through the stomata on the upper surface of the leaf, are the distinguishing symptom of this leaf blight (Fig. 70). The conidiophores are fasciculate, olive gray, unbranched, irregular in shape, and occur in dense clusters in the stomatal openings. The conidia are elongate with brown to olive-brown slightly tapering ends and are 1-septate. Several species of *Scolecotrichum* have been described on the grasses, although most authors recognize the one species *S. graminis* Fckl. Von Hohnel transferred this species to the genus *Passalora* on the basis that the type species of the genus *Scolecotrichum* was founded on a misconception. Horsfall (1929), on the basis of the description of *Cercospora graminicola* Tracy & Earle (1895), included the species under this binomial (Sprague, 1950).

19. Gray Spot or Blast, *Piricularia grisea* (Cke.) Sacc. The gray spot followed by leaf necrosis is common on many grasses. The disease is distributed widely, but it is of importance on economic grasses, both turf and pasture, chiefly in the warmer climates. The spots are elliptical or round, gray with brown or purple borders, and frequently kill the leaf blade. Gray to brown indefinite areas occur on the culm and inflorescence. The gray-brown conidiophores and conidia borne in coiled cymes readily identify the pathogen. Conidia are ovate, 2-septate with apical cell cone-shaped and basal hilum-like terminal where it breaks from the stalk. See rice diseases, Chap. 7.

20. Ascochyta Leaf Spot. Several species of *Ascochyta* occur on the grasses (Grove, 1935, 1937, Sprague, 1943, 1950). The indefinite spots of variable color appear on the older leaves and culms. The light-brown lens-shaped or globose pycnidia with the pores opening to the surface

FIG. 70. Streak, or brown leaf spot, caused by *Scolecotrichum graminis* on *Phleum pratense* (*A*), *Dactylis glomerata* (*B*), and *Festuca elátior* (*C*).

form in groups within the tissues on these spots. The pycnidial wall is usually thin and frequently poorly defined at the base. The spores are hyaline to yellow, 1-septate, cylindrical to fusoid, and relatively short. Several species occur on the grasses. *Ascochyta sorghi* Sacc. (*A. graminicola* Sacc.) and *A. agropyrina* (Fairm.) Trott. are the more common species.

This group of leaf spot diseases occurs chiefly on the mature tissues of many grasses, and most of them are of minor importance (Sprague, 1950).

21. Stagonospora Leaf Spot. The pycnidia form on the older leaves and culms with or without a conspicuous lesion. The pycnidia are globose to flattened, dark-colored, and immersed or partly projecting through the epidermis, often with a pore at the apex. The spores are hyaline, oblong, fusoid, or ellipsoid with two or more septa. *Stagonospora arenaria* Sacc. is distributed widely on the grasses.

22. Septoria Leaf Blotch, *Septoria* spp. Numerous species of *Septoria* occur on the grasses and cereals, especially in the temperate and subtropical zones. These leaf blotches occasionally cause defoliation and reduce seed yield. The irregular blotches are straw-colored to brown with dark-brown to black pycnidia on the older portions of the lesions. In many instances, the blotches are similar to those produced by *Stagonospora* or *Ascochyta*. Likewise the morphological distinctions between the three genera on the grasses are not clearly defined, as pointed out by Sprague (1944, 1950). *Septoria* spp. generally are more aggressive as parasites on the cereals and grasses than those of closely related genera.

In the genus *Septoria*, the pycnidia are subepidermal, slightly erumpent, and are formed in the older portions of the flecks or spots on leaf, culm, and inflorescence of the Gramineae (Frandsen, 1943, Grove, 1935, 1937, and Sprague, 1944). The pycnidia are globose to lens-shaped, brown to black, ostiolate, parenchymatous, comprising the outer several layers of brown polygonal cells and the inner layers of subhyaline to hyaline flattened or bulbous cells differentiating conidiophores. The conidia (pycnospores) are hyaline to chlorinous, nonseptate to multiseptate, predominantly at least ten times as long as broad, straight to curved, and fusiform, filiform, or scolecosporous. Some of the species occurring on the grasses are listed in addition to those given under the respective cereal crops. Sprague (1944, 1950) gives a more detailed description of these and other species. He (1943) has differentiated the wider multiseptate yellowish-brown to light-brown spore color of the Septoria-like types on the grasses and transferred these to the genus *Phaeoseptoria*.

Septoria tritici Rob. on *Triticum* spp. and the varieties *S. tritici* f. *avenae* (Desm.) Sprague on oats, *S. tritici* var. *lolicola* Sprague & A. G. Johnson on *Lolium* spp., and *S. tritici* f. *holci* Sprague on *Holcus lanatus* L. are long-spored types (see Chap. 11). *S. passerinii* Sacc. occurs on *Hordeum* spp. and *Sitanion hystrix* (Nutt.) J. G. Smith (see Chap. 3). *S. secalis* var. *stipae* Sprague occurs on *Stipa* and *Agrostis* spp. *S. bromi* Sacc. is distributed widely on *Bromus* spp., and it occurs early in the spring, decreases in abundance during the hot summer, and reappears again in the autumn. The conidia are slender, whip-like, usually 2-septate, and stouter in the late autumn. This leaf blotch is not so abundant as

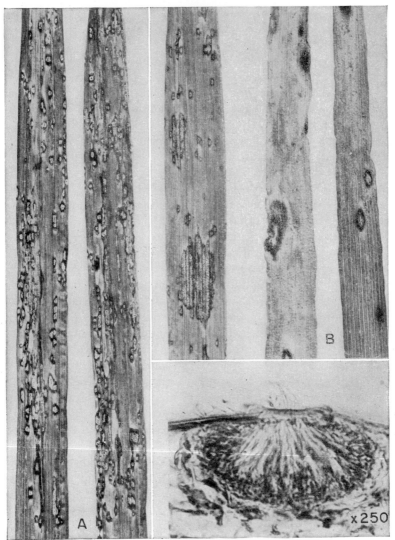

FIG. 71. Typical leaf lesions of *Selenophoma bromigena* (A) and *Septoria bromi* (B) on *Bromus inermis* and pycnidium of *S. bromi*.

Selenophoma bromigena (Sacc.) Sprague & A. G. Johnson on *Bromus inermis* Leyss in Central North America (Fig. 71). *S. jaculella* Sprague with straight 2- to 5-septate conidia occurs on *Bromus* spp. in the Western United States. *S. macropoda* Pass. and varieties are distributed widely on *Poa* spp. The pycnidia are flattened to globose, light brown, and produce small 1- to 3-septate filiform conidia. A second species, *S. oudemansii* Sacc., is common on *Poa* spp., and it is distinguished from the former by the lighter golden color and large closely packed cells of the pycnidia. The conidia are mostly cylindrical, wider, and shorter than in the former species and usually 1-septate. *S. elymi* Ell. & Ev. (*S. agropyri* Ell. & Ev.) is common on *Elymus* and *Agropyron* spp. The pycnidia are flattened, thin-walled, and brown. The conidia are cylindrical to filiform and usually 3-septate. *S. infuscans* (Ell. & Ev.) Sprague (*Cylindrosporium infuscans* Ell. & Ev.) occurs on *Elymus* spp. in western North America and possibly elsewhere; a long-spored species *Septoria pacifica* Sprague on *Elymus mollis* Trin. occurs in the western region. Other species occur on the grasses (Davis, 1942, Frandsen, 1943, Grove, 1935, 1937, Sprague, 1944, 1950).

The etiology and control are similar to that given for the cereals. Strains of the cultivated grasses show differences in susceptibility to the specific fungi causing these diseases.

23. Selenophoma Leaf Blotch. Sprague and Johnson (1940, 1945) transferred the species with falcate nonseptate spores to the amended genus *Selenophoma*. *Selenophoma bromigena* (Sacc.) Sprague & A. G. Johnson on *Bromus inermis* Leyss is one of the more common species of this group. The initial lesions appear as small brown flecks on the leaves early in the spring. The spots enlarge to form irregular blotches with translucent older portions in which the black pycnidia develop (Fig. 71). The mature pycnidia frequently drop out of the tissue, leaving small holes in the lesions. The spores are hyaline, narrow, tapering at the ends, slightly sickle-shaped, and nonseptate. Septa are formed during germination on nutrient media. Lines of smooth brome grass show some differences in reaction to the disease (Allison, 1945). *S. donocis* (Pass.) Sprague & A. G. Johnson and related forms occur on many species of grasses in North America.

24. Rhizoctonia Root and Crown Rot. Root rot, crown rot, and eyespot lesions occur on the lower leaf sheath tissues and culms of many grasses. The disease appears in localized spots early in the growing season. Brown patch of turf grasses in which the crown and leaf tissues are killed in local spots is a severe manifestation of the disease. The snow mold type of injury also occurs occasionally. *Rhizoctonia solani* Kuehn and other species of this genus are associated with the disease (Monteith, 1926, Sprague, 1944, 1950). Control of the disease on turf grasses is obtained by fungicides and the use of resistant strains in the creeping bents (Broadfoot, 1936, Monteith and Dahl, 1932). The disease is discussed in more detail in Chap. 11.

25. Typhula Snow Mold or Snow Scald. The snow mold caused by several species of *Typhula* is common on the grasses early in the spring, especially where the grass has been under a heavy snow covering (Fig. 67). Remsberg (1940) has summarized the literature on the genus and given the morphology of a number of species, including those associated with the disease on grasses. *Typhula itoana* Imai is apparently the most prevalent and widely distributed species on the cereals and grasses. Wernham (1941) reported a difference in the reaction of bent strains to this disease. Several resistant to *Rhizoctonia* and *Fusarium* were susceptible to *Typhula*. Andrews (1944) and Broadfoot and Cormack (1941) reported high- and low-temperature Basidiomycetes, respectively, different from those above, which cause severe damage to grass crowns of *Agropyron cristatum* (L.) Beauv. in Minnesota and westward during the summer, in the former, and to certain legumes and grasses early in the spring in western Canada, in the latter.

26. Leaf Smuts, *Ustilago and Urocystis* spp. Numerous smuts occur on the grasses (Clinton 1906, Fischer, 1953, Fischer and Hirschhorn, 1945, Liro, 1924, 1938, Zundel, 1939, 1946). Like those present on the cereals, the spore-bearing sori are more or less specific to certain morphological parts of the grasses, such as the leaves, culms, inflorescences, and caryopses; in some species the sori are formed in gall-like structures on the young tissues of any part of the plant. As an aid in recognition of the smuts and on the same basis used in the cereals, the grass smuts are grouped, primarily on the basis of symptoms, as presented by Fischer and Hirschhorn (1945).

The leaf smuts of the grasses appear chiefly in the leaf blades and leaf sheaths. These smuts are subdivided on the basis of symptoms into stripe smut, flag smut, and spot smut. While sori frequently develop in the inflorescences of the infected plants, the latter structures generally fail to develop and emerge from the leaf whorl, and therefore they are inconspicuous on the smutted plants. The sori are linear, forming as long or short stripes between the veins of the leaves, depending upon the species involved. After the leaf epidermis is ruptured and the spores are discharged, the leaf tissues frequently present a brown shredded appearance. Chlamydospores in the shredded tissues aid in the identification of the fungus species. The sori in the leaf spot smuts are covered by the epidermis and are more permanent. The etiology of this group of smut fungi on the grasses is similar. Seedling or young crown-bud infection is followed by the systemic infection of the primordia. Spore formation occurs in the leaves as the latter become fully developed. In the perennial grasses the mycelium usually persists in the crown tissues and dormant buds over several years. In some strains of stripe smut, infection of bud primordia occurs in established perennial plants. Seed infection occurs in some grasses (Fischer, 1940). Two leaf smuts are distributed generally on a large number of the important pasture and turf grasses.

Stripe smut caused by *Ustilago striiformis* (West.) Niessl occurs commonly on redtop (*Agrostis alba* L. and var. *gigantea* Roth), the creeping bents (*A. palustris* Huds. and *A. tenuis* Sibth.), orchard grass (*Dactylis glomerata* L., timothy (*Phleum pratense* L.), and blue grass (*Poa pratensis* L.). It is less widely distributed on certain other *Agropyron, Agrostis, Beckmannia, Elymus, Festuca, Holcus, Lolium, Poa,* and *Sitanion* spp. and some other grasses. Physiologic varieties and races are distinguished on certain of the grasses (Davis, 1930, Fischer, 1940, Thirumalachar and Dickson, 1953). Resistant strains occur in some of the grasses. The sori form as long, narrow, almost black stripes in the leaves and leaf sheaths (Fig. 72). The leaves are shredded after the dispersal of the spores. Internodal elongation and the development of the inflorescence are restricted in many grasses. The spores are globose to ellipsoid, dark olive brown, prominently echinulate, and chiefly 9 to 12 microns in diameter. Spore germination occurs directly or after a resting period with the formation of a branched promycelium and less frequently sporidia (Davis, 1924, Fischer, 1940, 1953, Kreitlow, 1943, 1944).

Flag smut caused by *Urocystis agropyri* (Preuss) Schröt. is manifest especially in the upper leaves of a somewhat similar group of grasses as in the former species. This smut is widely distributed but occurs less abundantly than the former (Fig. 72). The spore balls are globose to elongate and are composed of one to four reddish-brown smooth fertile spores surrounded by smaller hyaline to light-brown sterile cells. The spore germination in this species is controlled by an inhibitor, but is probably similar to that of the species occurring on wheat and rye. Fischer (1953) includes the morphologically similar *U. tritici* Koern. under *U. agropyri.*

Brown stripe smut caused by *Ustilago longissima* (Schlecht.) Meyen is generally distributed on species of *Glyceria*. The long numerous sori in the leaves are light brown in contrast to the darker lesions of the former smuts. Spores are globose to elongate, golden yellow to light olivaceous, minutely echinulate, and 7 to 9 microns in diameter. *U. macrospora* Desm. [*U. aculeata* (Ule) Liro] occurs occasionally on *Agropyron* and *Elymus* spp. and cannot be distinguished macroscopically from the common stripe smut. The spores are dark brown, ovoid to globose, 12 to 19 microns in diameter, coarsely verrucose, and germinate with a branched promycelium. *U. macrospora* Desm. (*U. echinata* Schroet.), occurring sparingly on Reed canary grass (*Phalaris arundinaceae* L.), is similar to the large-spored type previously described (Fischer, 1953).

Leaf spot smuts caused by *Entyloma* spp. occur on many of the grasses throughout the world. The sori are formed in the leaves and less frequently in the floral bracts, and they appear as tar-like angular to oblong spots or blisters covered by the epidermis of the suspect. This group of *Entyloma* spp. is relatively similar in morphology, and confusion exists in

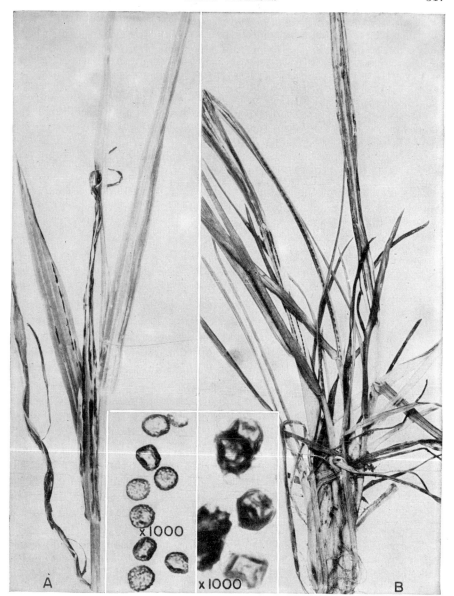

FIG. 72. Stripe smut on *Phleum pratense* (*A*) and flag smut on *Poa pratensis* (*B*) caused by *Ustilago striiformus* and *Urocystis agropyri*, respectively. The chlamydospores of the two species are shown, highly magnified, in the inserts.

the specific names used. The chlamydospores are angular to globose, 6 to 14 microns in diameter, closely packed, usually smooth-walled, light brown to reddish brown in color, and germinate, frequently in place, by the formation of a promycelium with terminal sporidia (Chap. 7). *Entyloma lineatum* (Cke.) J. J. Davis is common on *Zizania aquatica* L. and rice in North America and Europe. *E. irregulare* Johans on *Poa* spp. is distributed widely in Europe and North America, especially in the Northwest. *E. crastophilum* Sacc., similar and perhaps synonymous with *E. irregulare*, occurs over the same area as the former on *Phleum* and *Agrostis* spp. and some other grasses. Fischer (1953) includes the latter two species under *E. dactylidis* (Pass.) Cif.

27. Culm Smuts. The culm smuts are found chiefly in western North America and in somewhat similar habitats in South America, Europe, North Africa, and Asia. Four species are differentiated on the grasses by Fischer and Hirschhorn (1945).

Ustilago spegazzinii Hirschh. (*U. hypodytes* Aucht.) and the variety *U. spegazzinii* var. *agrestis* (Syd.) G. W. Fisch. & Hirschh. occur on *Agropyron*, *Elymus*, *Poa*, and some other genera. The sori are chiefly superficial on the internodes of the culms and sometimes on the aborted inflorescences. The naked linear dark-brown to black sori are covered by the leaf sheaths (Fig. 73). The spores are yellowish brown to olivaceous, globose to angular, with bipolar subhyaline crests of epispore echinulations; the remainder of the epispore wall is minutely echinulate. The chlamydospores are 4 to 6 microns in diameter and germinate by the formation of a branched promycelium and aerial sporidia. According to Bond (1940) and Fischer (1945), the mycelium persists in the crown and stolon tissues of the perennial grasses. Infection occurs in crown-bud primordia, and sori appear from two to three seasons following infection. This may be associated with the period required for the grass primordia to develop into culms, as suggested by Bond (1940). The increase of smut in permanent stands of grasses makes this an important disease in areas where inoculum is present. Some selections of crested wheat grass [*Agropyron cristatum* (L.) Beauv.] apparently are resistant.

Ustilago williamsii (Griff.) Lavrov, formerly included under *U. hypodytes*, occurs on *Oryzopsis* and *Stipa* spp. in western North America and Argentina. The symptoms are similar to those caused by *U. spegazzinii*. Spores are globose to subglobose, dark olivaceous brown, and the epispore is smooth but deeply cracked and bearing two bipolar cap-like appendages. Germination occurs readily with the formation of a promycelium and sporidia. Culm smut caused by this species appears earlier after infection than in the former species.

Ustilago halophila Speg. produces a culm smut on *Distichlis* spp. in the drier sections of North and South America and Australia. The plants are dwarfed, and the sori are covered early by the epidermis of the culm. The spores are globose to irregular, yellow brown to olive brown, smooth, without bipolar areas or appendages, and 5 to 7 microns in diameter.

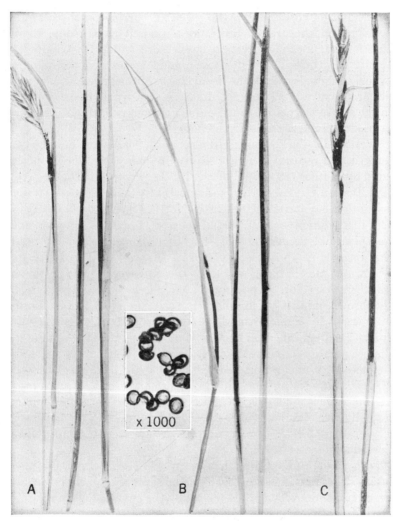

FIG. 73. Culm smut caused by *Ustilago spegazzinii* on (*A*) *Agropyron cristatum*, (*B*) *A. trichophorum*, and (*C*) *A. elongatum*. The chlamydospores are shown, highly magnified, in the insert.

Spore germination is typically by the formation of a promycelium and sporidia.

Ustilago hypodytes (Schlecht.) Fries occurs chiefly on *Melica, Oryzopsis,* and *Stipa* spp. in western and Intermountain North America,

Argentina, and northern Europe. Sori are naked on the internodes and into the inflorescences. Spores are smaller than the former species and germinate by means of a branched promycelium without sporidia.

28. Head Smuts. The smuts forming sori in grass inflorescences are numerous and include some of the species occurring on the cereal crops. These smuts on the grasses are more abundant as a group in western North America, although certain species are widely distributed. The head smuts, as grouped, include those producing spores in the inflorescence as a whole or those in which the floral bracts of the spikelets as well as the ovaries are involved. These are differentiated from the kernel smuts in which the sori usually are formed only in the ovaries. Only the smuts of major economic importance are included.

Ustilago bullata Berk., causing the common head smut on a large number of grasses, results in major losses in many economic species, as reviewed by Fischer (1940). This smut is less common in the North Central and Eastern United States, and apparently it is restricted to the drier areas in other countries as well as in North America. The sori formed in the spikelets involve all or part of the floral bracts. They are enclosed in the epidermal membranes of the floral structures, which persist in varying degrees in the different grasses (Fig. 74). The spore mass is dark brown to purple black and loose to semicovered in type, depending upon the grass species. The spores are globose to irregular, dark brown to olive brown, minutely echinulate to verrucose, 5 to 14 but more often 7 to 9 microns in diameter, and germinate to form a promycelium and sporidia. Seedling infection occurs, and in some of the perennial grasses the disease persists for several years. Fischer (1940) differentiated eight physiologic races and reported resistant lines in some grasses. Parker (1942) reported resistant lines in *Bromus catharticus* Vahl. (*B. unioloides* H.B.K.). This head smut is attributed to three species, similar morphologically, by Zundel (1939, 1946) and others: *U. bromivora* (Tul.) Fisch. von Waldh., *U. bullata*, and *U. lorentziana* Thuem. These are considered synonyms of *U. bullata*.

Ustilago mulfordiana Ell. & Ev. occurs on *Festuca* spp. in the Intermountain states and in western North America. Sori form in the aborted inflorescences and are concealed partly by the enveloping leaves. Spores are dark olivaceous brown, globose to ovoid, coarsely echinulate, and 12 to 19 microns in diameter.

Ustilago trebouxii H. & P. Syd. (*U. sitanii* G. W. Fisch.) produces a brown loose smut of the inflorescences and linear sori in the upper leaves of *Elymus, Hordeum, Poa,* and *Sitanion* spp. Spores are globose to ovate, light brown, minutely echinulate, 4 to 5 microns in diameter, and germinate to form a promycelium frequently without sporidia.

Several head smuts incited by *Sorosporium* spp. occur on specific

groups of important grasses (Fischer, 1953). *Sorosporium cenchri* Henn.
[*S. syntherisme* (Peck) Farl.] is distributed widely on *Cenchrus* and
Panicum spp. The sori involve the entire inflorescence, leaving only

FIG. 74. Head smut of grasses caused by *Ustilago bullata* showing the range in symptoms on (*A*) *Bromus carinatus,* (*B*) *Agropyron pauciflorum,* (*C*) *Bromus hordeaceus,* (*D*) *B. polyanthus,* (*E*) *B. catharticus,* and (*F*) *Elymus canadensis.* The chlamydospores, highly magnified, are shown in the insert.

shredded mass of vascular tissue after spores are discharged. Sorus is
early covered by periderm of fungus, plate-like cells that shed and rarely
appear with spores when examined. Spores are arranged in balls at
first, later separating. Spores are irregular-shaped, minutely echinulate,
and 9 to 13 microns. *S. confusum* Jacks. with sori enclosed in floral
bracts occurs on *Aristida* spp.; *S. ellisii* Wint. and *S. everhartii* Ell. & Gall.

with sorus involving all of the inflorescence occur on *Andropogon* spp.; and other less important smuts of this genus are listed by Fischer (1953).

A few species of *Sphacelotheca* incite head smuts on the economic grasses. The peridium covers the sorus, and at maturity these cells occur as individual or groups of empty "sterile" cells mixed with spores. The spores are formed around a central columella. *S. andropogonis* (Opiz) Bubak occurring on *Andropogon* spp., *S. destruens* (Schlect.) Stev. & Johns. on *Panicum* spp., and *S. reiliana* (Kühn) Clint. on *Sorghum* spp. are important head smuts in this genus, often confused with *Sorosporium*.

29. Kernel Smuts. The more important kernel smuts of the grasses are those caused by species of *Tilletia*. These smuts occur on the grasses of the western and Intermountain areas of North America and similar areas in other countries (Fischer, 1953, Fischer and Hirschhorn, 1945). They are of economic importance on some of the better range and forage grasses. The sori form in the ovaries and assume the general shape and color of the caryopses of the species. The *Tilletia* spp. on some of the more important grasses are described briefly. Some species of *Sphacelotheca* cause kernel smuts on the grasses.

Sphacelotheca sorghi (Lk). Clint. and *S. cruenta* (Kühn) Potter are distributed widely on Sudan grass and other sorghum grasses (see Chap. 9).

Tilletia guyotiana Hariot on *Bromus* spp. Spores are yellow to chestnut, deeply reticulate, and 21 to 26 microns in diameter. The sterile cells are hyaline, smooth, and 14 to 18 microns in diameter.

Tilletia fusca Ell. & Ev. on *Festuca* spp. Spores are dark, reddish brown, deeply and prominently reticulate, and 21 to 27 microns in diameter. Sterile cells are grayish yellow, smooth, and 12 to 20 microns in diameter.

Tilletia cerebrina Ell. & Ev. on *Deschampsia* spp. Spores are grayish brown, reticulate with the reticulations varying in size and shape, and 21 to 25 microns in diameter. The sterile cells are hyaline to greenish yellow and 12 to 17 microns in diameter.

Tilletia elymi Diet. & Holw. on *Elymus glaucus* Buckl., *E. canadensis* L. Spores are light olive brown to dark olivaceous brown, globose to subglobose, and deeply reticulate. Reticulations are variable in size and shape, and the spores are 21 to 28 microns in diameter. Sterile cells are hyaline to greenish yellow, smooth, thin-walled, and 17 to 21 microns in diameter.

Tilletia holci (West.) De Toni on *Holcus lanatus* L. is similar in spore morphology and size to *T. cerebrina*.

Tilletia asperifolia Ell. & Ev. on *Muhlenbergia* spp. Spores are yellowish to dark brown, 17 to 21 microns in diameter, and are enveloped by a

thin hyaline membrane, distinctly reticulate to cerebriform.   The sterile
cells, larger than the spores, are hyaline, and the walls are laminated.

    T. caries (DC.) Tul. and *T. controversa* Kühn occur on the wheat grasses
(see Chap. 11).

    Many other smuts occur less generally on the cultivated and wild
grasses (Clinton, 1906, Fischer, 1953, Fischer and Hirschhorn, 1945,
Zundel, 1946) (see also Chaps, 3, 5 to 9, and 11).

    30. Stem Rust, *Puccinia graminis* Pers.   Stem rust is common on
many of the grasses.   The uredia develop on the leaves and culms, and
the telia generally form on the leaf sheaths and culms.   The symptoms
and etiology are the same as on the cereal crops (see Chaps. 6 and 11).
The specialized varieties and physiologic races occurring on the cereals
as well as *Puccinia graminis agrostidis* Eriks., *P. graminis phlei-pratensis*
(Eriks. & Henn.) Stakman & Piemeisel, *P. graminis poae* Eriks. & Henn.,
P. graminis lolii Waterh. (Waterhouse, 1951) and other varieties cause
damage on the grasses.   Both hay and seed production in many of the
economic grasses are reduced by stem rust.   Some species, *e.g.*, *Bromus
inermis* Leyss, are resistant to stem rust.   The reaction of certain of the
grasses to *P. graminis*, *P. rubigo-vera* (DC.) Wint., *P. glumarum* (Schm.)
Eriks. & Henn., and *P. coronata* Cda. is given in Table 13.

    31. Leaf Rusts.   The leaf rusts on the grasses are caused by several
species, the most important being *Puccinia rubigo-vera* (DC.) Wint.
Mains (1932) combined a large number of morphologically similar leaf
rust species under this binomial.   On the basis of the reaction of the
grasses to the rusts and in most instances including the reaction of the
plants known to be the aecial hosts of these rust parasites, Mains grouped
them into specialized varieties of *P. rubigo-vera* using the trinomial system.
He listed 56 trinomials for the varieties with uredia and telia on the cereals
and grasses and with aecia on species or groups of species of *Thalictrum,
Clematis, Anemone* and *Hepatica; Aquilegia, Delphinium, Ranunculus,
Actaea, Aconitum, Anchusa, Onosmodium* and *Macrocalyx; Symphytum*
and *Pulmonaria; Lithospermum* and *Myosotis; Phacelia, Hydrophyllum,*
and *Impatiens*, as well as some few with aecia unknown.   These specialized varieties of *P. rubigo-vera* cause the more important leaf rusts on the
grasses.   *P. recondita* Rob. recently has been suggested as the correct
binomial.   The urediospores are ovate to globose with germ pores distributed in contrast to the elliptical or pyriform urediospores with equatorial
germ pores in *P. graminis*.   The telia of *P. rubigo-vera* are covered by
the epidermis in contrast to the naked telia in *P. graminis*.   See Chap. 11
for the detailed morphology.   The reactions of some of the grasses to this
species are given in the following table.

    Crown rust caused by *Puccinia coronata* Cda. is distributed widely and

TABLE 13. RANGE IN REACTION OF SOME GENERA AND SPECIES OF THE GRASSES TO *Puccinia graminis* AND SPECIALIZED VARIETIES, *P. rubigo-vera*, *P. glumarum*, AND *P. coronata* IN THE UNITED STATES AND CANADA*

Genera of grasses or species	Varieties of *P. graminis*						*P. graminis*†	*P. rubigo-vera*	*P. glumarum*	*P. coronata*
	agrostidis	*avenae*	*phlei-pratensis*	*poae*	*secalis*	*tritici*				
Aegilops (5 species)............							S	R-S	S	
Agropyron (33 species)...........		R	R		R-S	R-S	R-S	R-S	R-S	R-S
Agrostis (19 species)...........	S	R-S			R	R	R-S	R-S		R-S
Alopecurus (6 species)..........	S	R-S			R	R	R-S	R-S		R-S
Ammophila arenaria (L.) Link....		S					S			R-S
Andropogon (2 species)..........		R					R			R
Anthoxanthum (2 species)........	R	R-S			R	R	R-S		R	R-S
Arctagrostis latifolia (R. Br.) Griseb...............										S
Arrhenatherum elatius (L.) Mert & Koch..................		R-S	R-S		R	R	R-S	R	R	R-S
Avena (11 species)...............		S	S				S	R-S		R-S
Beckmannia (2 species)..........		R	R		R	R	R-S		R	R-S
Bouteloua (2 species)..........		R			R	R	R-S			R
Brachypodium distachyon (L.) Beauv...................		R					R			R
Briza (3 species)...............							S	R		R
Bromus (39 species)............	R-S	R-S	R		R-S	R-S	R-S	R-S	R-S	R-S
Buchloë dactyloides (Nut.) Engelm.							S			
Calamagrostis (9 species)........	R	S				R	R-S			R-S
Calamovilfa longifolia (Hook.) Scrib....................							S			R-S
Catabrosa aquatica (L.) Beauv....							S			
Chloris (2 species).............									S	R
Cinna (2 species)...............							S	S		R-S
Corynephorus canescens (L.) Beauv....................							S			
Cynodon dactylon (L.) Pers.......		R				R	R-S			R
Cynosurus cristatus L...........		R				R	R			R
Dactylis glomerata L............	S	S	R-S	R	R	R-S	R-S	R	R-S	R-S
Dactyloctenium aegyptiacum (L.) Richt.................		R					R			R
Danthonia (5 species)...........		R					R			R
Deschampsia (6 species).........		R-S				S	R-S	R		R-S
Distichlis spicata (L.) Greene....							S			R
Echinochloa crusgalli (L.) Beauv..		R					R-S			R
Elymus (26 species).............		R	R		R-S	R-S	R-S	R-S	R-S	R-S
Eragrostis (2 species)...........		R	R		R	R	R			R
Festuca (25 species)............		R-S	R-S		R	R-S	R-S	R-S	R-S	R-S
Glyceria (6 species).............		R-S					R-S	R-S		R-S

TABLE 13. RANGE IN REACTION OF SOME GENERA AND SPECIES OF THE GRASSES
TO *Puccinia graminis* AND SPECIALIZED VARIETIES, *P. rubigo-vera*,
P. glumarum, AND *P. coronata* IN THE UNITED STATES
AND CANADA* (*Continued*)

Genera of grasses or species	Varieties of *P. graminis*						*P. graminis*†	*P. rubigo-vera*	*P. glumarum*	*P. coronata*
	agrostidis	*avenae*	*phlei-pratensis*	*poae*	*secalis*	*tritici*				
Hierochloë (2 species)	....	S	....	....	....	....	R-S	....	....	R-S
Holcus (2 species)	S	R-S	R-S	....	R	R	R-S	R-S	....	R-S
Hordeum (10 species)	R	R	R	....	R-S	R-S	R-S	R-S	R-S	R-S
Hystrix (2 species)	....	R	....	....	S	S	R-S	R-S	S	R-S
Koeleria cristata (L.) Pers	S	R-S	S	....	R	R-S	R-S	R-S	S	R
Lagurus ovatus L	....	....	....	....	....	....	S	....	....	R-S
Lamarckia aurea (L.) Moench	....	....	....	....	....	....	S	....	....	R-S
Limnodea arkansana (Hutt.) Dewey	....	S	....	....	....	....	S	....	....	S
Lolium (4 species)	....	R	R-S	....	R	R	R-S	R	R	R-S
Melica (6 species)	....	....	....	....	....	....	R	R-S	R-S	R
Milium effusum L	....	S	....	....	....	....	S	....	....	
Molinia caerulea (L.) Moench	....	....	....	....	....	....	S	....	....	R
Muhlenbergia (3 species)	....	R	R	....	R	R	R-S	....	....	R
Panicum virgatum L	....	....	....	....	....	....	S	S	....	R
Paspalum setaceum Mich	....	....	....	....	....	....	....	....	....	S
Phalaris (6 species)	R	R-S	S	....	R	R	R-S	....	S	R-S
Phleum (4 species)	....	R-S	S	R	R	R	R-S	....	....	R-S
Phragmites communis Trin	....	....	....	....	....	....	S	....	....	R
Poa (36 species)	....	R-S	R-S	R-S	....	....	R-S	R-S	....	R-S
Polypogon monspeliensis (L.) Desf.	....	S	....	....	....	....	S	....	....	R-S
Puccinellia (6 species)	....	....	....	....	....	....	S	R-S	....	R
Schedonnardus paniculatus (Nut.) Trel	....	R	....	....	....	....	R	....	....	R
Secale montanum Gaus	....	....	....	....	....	....	....	R-S	S	
Sitanion (3 species)	....	R	....	....	....	S	R-S	R-S	S	
Sorghastrum (2 species)	....	....	....	....	....	....	R-S	....	....	R
Sphenopholis (3 species)	....	R-S	....	....	....	....	R-S	R	....	R
Sporobolus (3 species)	....	R	....	....	S	R-S	R-S	R	....	R
Stipa (5 species)	....	R	R	....	R	R-S	S	....	....	R
Triodia flava (L.) Smyth	....	R	....	....	....	....	R-S	....	....	R
Trisetum (6 species)	....	S	....	....	....	....	R-S	R-S	S	R-S

* G. W. Fischer, and M. N. Levine. Summary of the recorded data on the reaction of wild and cultivated grasses to stem rust (*Puccinia graminis*), leaf rust (*P. rubigo-vera*), stripe rust (*P. glumarum*), and crown rust (*P. coronata*) in the United States and Canada. *U.S. Dept. Agr. Plant Dis. Reptr. Suppl.* 130. 1941. (Mimeographed.)

† *Puccinia graminis* without designation of race or including all of the physiologic races.

R—resistant or failed to infect; S—susceptible, based on observation or inoculation.

is important economically on oats and many grasses (see Table 13). Several specialized varieties of the fungus occur on the grasses as well as physiologic races. The aecial stages of these varieties are specialized somewhat on *Rhamnus* spp. and on *Berchemia scandens* (Hill) Trel., *Elaeagnus commutata* Bernh., and *Shepherdia canadensis* (L.) Nutt. The morphology and etiology of this species are given in Chap. 6. The digitate projections, forming a crown on the apex of the teliospore, differentiate the fungus from those causing other leaf rusts.

Puccinia poae-sudeticae (West.) Jørstad is distributed widely on many species of *Poa* and a few other grasses. It is prevalent on *Poa pratensis* L. in the humid temperate zones. The aecial stage of this fungus is unknown. The uredia are epiphyllous, orange-yellow, with numerous peripheral paraphyses. The paraphyses around the urediospores differentiate this species from *P. rubigo-vera*. The urediospores are similar in the two species. Telia are covered rather permanently by the epidermis, and numerous subepidermal paraphyses are present, especially in the periphery of the telium. The teliospores are oblong or clavate, dark brown, and short-pedicled.

Several leaf rusts are common on the tropical grasses of economic importance (Theis, 1953). The yellow leaf rust, *Puccinia polysora* Underw., occurs on *Erianthus* and *Tripsacum* spp. Species of especially *Uromyces* and *Puccinia* are numerous on the grasses in both the tropical and temperate zones (Arthur, 1934).

32. Stripe Rusts. The brown stripe rust caused by *Puccinia montanensis* Ell. occurs on the leaves of *Agropyron* and *Elymus* spp., *Melica imperfecta* Trin., and some few other grasses. The aecia occur infrequently on *Berberis fendleri* Gray. This rust is distributed chiefly through western North and South America. The uredia on the leaves form narrow long lines, dark reddish brown in color. Numerous clavate paraphyses surround the urediospores. The telia are oblong, grayish brown, and often form narrow lines below the leaf epidermis. This rust is differentiated from the yellow stripe rust by the differences in color, shape, and time of appearance of the uredia and the shape and abundance of paraphyses in the uredia.

The yellow stripe rust caused by *Puccinia glumarum* (Schm.) Eriks. & Henn. is common on many grasses in the Intermountain and Pacific Coast area of North and South America and northern Europe and Asia. The geographic distribution of this rust is limited to the areas of relatively cool climates. Many of the economically important grasses are damaged in early spring by this rust. (See Table 13 for the reaction of grasses.) The uredia form orange-yellow stripes on the leaves and floral bracts, especially during cool cloudy weather. The uredia are orange-yellow, with occasional hyphoid paraphyses around the outer edges. Telia form less abundantly than the uredia, and they appear as fine dark lines below the epidermis. The morphology, etiology, and control are discussed in Chap. 11.

Many other rusts occur on the grasses, as listed by Arthur (1934) and others.

REFERENCES

ALLISON, J. L. *Selenophoma bromigena* leaf spot on *Bromus inermis*. *Phytopath.* **35**:233–240. 1945.

——— and D. W. CHAMBERLAIN. Distinguishing characters of some forage grass diseases prevalent in the Northcentral states. *U.S. Dept. Agr. Cir.* 747. 1946.

ANDREWS, E. A. Seedling blight and root rot of grasses in Minnesota. *Phytopath.* **33**:234–239. 1943.

———. The pathogenicity of a nonsporulating Basidiomycete on grasses in Minnesota. *Phytopath.* **34**:352–353. 1944.

ARTHUR, J. C. Manual of the rusts in the United States and Canada. Purdue Research Foundation. Lafayette, Ind. 1934.

BENNETT, F. T. Fusarium patch disease of bowling and golf greens. *Jour. Board Greenkeep. Res.* **3**:79–86. 1933.

BOLLARD, E. G. Studies on the genus Mastigosporium. I. General account of the species and their host range. II. Parasitism. *Trans. Brit. Myc. Soc.* **33**:250–264, 265–275. 1950.

BOND, T. E. T. Observations on the disease of sea lyme-grass (*Elymus arenarius* L.) caused by *Ustilago hypodytes* (Schlecht.) Fries. *Ann. Appl. Bot.* **27**:330–337. 1940.

BROADFOOT, W. C. Experiments on the control of snow-mould of turf in Alberta. *Sci. Agr.* **16**:615–618. 1936.

———. Snow-mould of turf in Alberta. *Jour. Board Greenkeep. Res.* **5**:182–183. 1938.

——— and M. W. CORMACK. A low-temperature Basidiomycete causing early spring killing of grasses and legumes in Alberta. *Phytopath.* **31**:1058–1059. 1941.

BROWN, H. B. Life history and poisonous properties of *Claviceps paspali*. *Jour. Agr. Research* **7**:401–406. 1916.

CALDWELL, R. M. Rhynchosporium scald of barley, rye and other grasses. *Jour. Agr. Research* **55**:175–198. 1937.

CHAMBERLAIN, D. W., and J. L. ALLISON. The brown leaf spot on *Bromus inermis* caused by *Pyrenophora bromi*. *Phytopath.* **35**:241–248. 1945.

CHRISTENSEN, J. J. Studies on the parasitism of *Helminthosporium sativum*. *Minn. Agr. Exp. Sta. Bul.* 11. 1922.

CLINTON, G. P. Ustilaginales; Ultilaginaceae, Tilletiaceae. *North American Flora* **7**:1–82. 1906.

DAHL, A. S. Snow-mold of turf grasses as caused by *Fusarium nivale*. *Phytopath.* **24**:197–214. 1934.

DAVIS, J. J. Parasitic fungi of Wisconsin. Privately printed. Madison, Wis. 1942.

DAVIS, W. H. Spore germination of *Ustilago striaeformis*. *Phytopath.* **14**:251–267. 1924.

———. Two physiologic forms of *Ustilago striaeformis*. *Phytopath.* **20**:65–74. 1930.

De Tempe, J. De Phialea-ziekte van raaigras in Nederland. *Tijdschr. Planten-ziekt.* **56**:164–169. 1950.

Drechsler, C. Some graminicolous species of Helminthosporium. I. *Jour. Agr. Research* **24**:641–739. 1923.

——. Occurrence of the zonate-eyespot fungus *Helminthosporium giganteum* on some additional grasses. *Jour. Agr. Research* **39**:129–136. 1929.

——. Leaf spot and foot rot of Kentucky bluegrass caused by *Helminthosporium vagans. Jour. Agr. Research* **40**:447–561. 1930.

——. Phytopathological and taxonomic aspects of Ophiobolus, Pyrenophora, Helminthosporium, and a new genus, Cochliobolus. *Phytopath.* **24**:953–983. 1934.

——. A leaf spot of bent grasses caused by *Helminthosporium erthrospilum*, n. sp. *Phytopath.* **25**:344–361. 1935.

Fang, C. T., O. N. Allen, A. J. Riker, and J. G. Dickson. The pathogenic, physiological and serological reactions of the form species of *Xanthomonas translucens. Phytopath.* **40**:44–64. 1950.

Fischer, G. W. Fundamental studies of the stripe smut of grasses (*Ustilago striaeformis*) in the Pacific Northwest. *Phytopath.* **30**:93–118. 1940.

——. Host specialization in the head smut of grasses, *Ustilago bullata. Phytopath.* **30**:991–1017. 1940.

——. "Bends" a new disease of cereals and grasses. *Phytopath.* **31**:674–676. 1941.

——. The mode of infection and the incubation period in the stem smut of grasses. *Ustilago spegazzinii* (*U. hypodytes*). *Phytopath.* **35**:525–532. 1945.

——. Manual of the North American smut fungi. The Ronald Press Company. New York. 1953.

—— and E. Hirschhorn. A critical study of some species of Ustilago causing stem smut on various grasses. *Mycologia* **37**:236–266. 1945.

—— and ——. The Ustilaginales or "smuts" of Washington. *Wash. Agr. Exp. Sta. Bul.* 459. 1945.

Frandsen, N. O. Septoria-arten des Getreides und anderer Gräser in Dänemark. *Meddel. Plantepatol. afd. Kgl. Vet. Landb. København.* 26. 1943.

Garces, C. O. Informe preliminar sobre la gomosis de los pastos Micay e imperial o Gramalote en Columbia. *Revis. Fac. Nal. Agron.* **7**:1–23. 1947.

Gieger, M., and B. F. Barrentine. Isolation of the active principle in *Claviceps paspali*—a progress report. *Jour. Am. Chem. Soc.* **61**:966–967. 1939.

Graham, J. H. Helminthosporium leaf streak of timothy. *Phytopath.* **45**:227–228. 1955.

Grove, W. B. British stem- and leaf-fungi. Cambridge University Press. London, England. 2 vols., 1935, 1937.

Hagborg, W. A. F. Dwarfing of wheat and barley by the barley stripe-mosaic (false stripe) virus. *Can. Jour. Bot.* **32**:24–37. 1954.

Hardison, J. R. Specialization of pathogenicity in *Erysiphe graminis* on wild and cultivated grasses. *Phytopath.* **34**:1–20. 1944.

——. Specialization in *Erysiphe graminis* for pathogenicity on wild and cultivated grasses outside the tribe Hordeae. *Phytopath.* **35**:394–405. 1945.

——. Field control of blind seed disease of perennial rye grass in Oregon. *Phytopath.* **38**:404–419. 1948.

Henry, A. W. Root rots of wheat. *Minn. Agr. Exp. Sta. Tech. Bul.* 22. 1924.

Hiura, M. Mycological and pathological studies on the downy mildew of Italian millet. *Gifu Imp. Col. Agr. Res. Bul.* 35. 1935. (Good literature review.)

Höhnel, F. Von. Studien über Hyphomyzeten. *Zentrlb. Bakt.* (2) **60**:1–26. 1924.

HORSFALL, J. G. A study of meadow-crop diseases in New York. *N.Y.* (*Cornell*) *Agr. Exp. Sta. Mem.* 130. 1930.

HYNES, H. J. Species of Helminthosporium and Curvularia associated with root-rot of wheat and other graminaceous plants. *Jour. Proc. Roy. Soc. N.S. Wales* **70**:378–391. 1937.

ITO, S., and K. KURIBAYASHI. The ascigerous forms of some graminicolous species of Helminthosporium in Japan. *Jour. Fac. Agr. Hokkaido Univ. Sapporo* **29**:85–125. 1931.

KREITLOW, K. W. *Ustilago striaeformis.* 1. Germination of chlamydospores and culture of *forma agrostidis* on artificial media. *Phytopath.* **33**:707–712. 1943.

———— and W. M. MYERS. Prevalence and distribution of stripe smut of *Poa pratensis* in some pastures of Pennsylvania. *Phytopath.* **34**:411–415. 1944.

LANGDON, R. F. N. Occurrence of ergot in Queensland with special reference to *Claviceps pusilla* Cesati. *Jour. Aust. Inst. Agr. Sci.* **7**:85–87. 1941.

————. Studies on Australian ergots. I. *Claviceps pusilla* Cesati. *Rept. Dept. Biol. Univ. Queensland* **2**:1–12. 1950.

————. Ergot of *Paspalum dilatatum. Jour. Aust. Inst. Agr. Sci.* **20**:52–53. 1954.

LEFEBVRE, C. L. Ergot of Paspalum. *Phytopath.* **29**:365–367. 1939.

LIRO, J. I. Die Ustilagineen Finnlands. I and II. *Ann. Acid. Sci. Fenn. Helsinki* A **17**:1–636, 1924. **42**:1–720. 1938.

MAINS, E. B. Host specialization in the leaf rust of grasses, *Puccinia rubigo-vera. Mich. Acad. Sci., Arts, Letters* **17**:289–394. 1933.

MASTENBROEK, C., and A. J. P. OORT. Het voorkomen van moederkoren (Claviceps) op granen en grassen en de specialisatie van de moederkorenschimmel. *Tijdschr. Plantenziekt.* **47**:165–182. 1941.

McCLEAN, A. P. D. Some forms of streak virus occurring in maize, sugarcane, and wild grasses. *Dept. Agr. So. Africa Sci. Bul.* 265. 1947.

McFARLAND, F. T. Infection experiments with Claviceps. *Phytopath.* **11**:41. 1921. Thesis. University Wisconsin. Madison, Wis.

McKINNEY, H. H. Mosaic diseases of wheat and related cereals. *U.S. Dept. Agr. Cir.* 442, pp. 1–22. 1937.

———— and H. FELLOWS. Wild and forage grasses found to be susceptible to the wheat streak-mosaic virus. *Plant Dis. Reptr.* **35**:441–442. 1951.

————, ————, and C. O. JOHNSTON. Mosaic of *Bromus inermis. Phytopath.* **32**:331. 1942.

MONTEITH, J. The brown-patch disease of turf; its nature and control. *U.S. Golf Assn. Greens Sec. Bul.* **6**:127–142. 1926.

————. Experiments on brown-patch control. *U.S. Golf Assn. Greens Sec. Bul.* **7**:210–216. 1927.

———— and A. S. DAHL. Turf diseases and their control. *U.S. Golf Assn. Greens Sec. Bul.* **12**:140–144. 1932.

NIELSEN, E. L. Grass studies. 5. Observations on proliferation. *Bot. Gaz.* **103**:177–181. 1941.

NISHIKADO, Y. Studies on the Helminthosporium diseases of Gramineae in Japan. *Ber. Ōhara Inst. Landw. Forsch.* **4**:111–126. 1929.

O'GARA, P. J. A bacterial disease of western wheat-grass, *Agropyron Smithii.* Occurrence of a new type of bacterial disease in America. *Phytopath.* **6**:341–350. 1916.

ORTON, C. R. Graminicolous species of *Phyllachora* in North America. *Mycologia* **36**:18–53. 1944.

OSBORNE, W. L. Importance of blind seed disease in rye grass. *New Zealand Jour. Agr.* **75**:595–602. 1947.

OSWALD, J. W., and B. R. HOUSTON. Host range and epiphytology of the cereal yellow dwarf disease. *Phytopath.* **43**:309–313. 1953.

PADWICK, G. W. Manual of rice diseases. Commonwealth Mycological Institute. Kew, Surrey. 1950.

PARKER, D. L. A note on perennial prairie grass. *Jour. Aust. Inst. Agr. Sci.* **7**:29–30. 1942.

PETCH, T. British Hypocreales. *Trans. Brit. Myc. Soc.* **21**:243–305. 1938.

RAMAKRISHNAN, T. S. Observations on ergots on Pennisetum and other grasses. *Proc. Indian Acad. Sci.* B **36**:97–101. 1952.

REDDY, C. S., and J. GODKIN. A bacterial disease of brome grass. *Phytopath.* **13**:75–86. 1923.

SAMPSON, K., and J. H. WESTERN. Diseases of British grasses and herbage legumes. Cambridge University Press. London. 1941.

SIMPSON, C. F., and E. WEST. Ergot poisoning in cattle. *Fla. Agr. Exp. Sta. Cir.* 543. 1952.

SLYKHUIS, J. T. Virus diseases of cereal crops in South Dakota. *So. Dak. Agr. Exp. Sta. Tech. Bul.* 11, 1–29. 1952.

―――. Striate mosaic, a new disease of wheat in South Dakota. *Phytopath.* **43**: 537–540. 1953.

SPRAGUE, R. Two Mastigosporium leaf spots of Gramineae. *Jour. Agr. Research* **57**:287–299. 1938.

―――. A third species of Mastigosporium on Gramineae. *Mycologia* **32**:43–45. 1940.

―――. Some leaf spot fungi on western Gramineae. *Mycologia* **33**:655–665. 1941.

―――. The genus Phaeoseptoria on grasses in the Western hemisphere. *Mycologia* **35**:483–491. 1943.

―――. Root rots of cereals and grasses in North Dakota. *No. Dak. Agr. Exp. Sta. Bul.* 332. 1944.

―――. Septoria diseases of Gramineae in Western United States. *Ore. State Monogr. State Col. Press* 6. 1944.

―――. Diseases of cereals and grasses in North America. The Ronald Press Company. New York. 1950.

――― and A. G. JOHNSON. Selenophoma on grasses, I and II. *Mycologia* **32**:415. 1940. **37**:638–639. 1945.

STÄGER, R. Infektionsversuche mit Gramineen bewohnenden *Claviceps*-arten. *Bot. Zeitschr.* **61**:58–111. 1903.

―――. Weiterer Beiträge zur Biologie der Mutterkorns. *Zentrlb. Bakt.* (2) **14**: 25–32. 1905.

―――. Beiträge zur Verbreitungsbiologie der Claviceps-sklerotien. *Zentrlb. Bakt.* (2) **54**:329–339. 1922.

STARR, M. P., and C. O. GARCES. El agente causante de la gomosis bacterial del pasto imperial en Colombia. *Revis. Fac. Nal. Agron.* **12**:73–83. 1950.

STOREY, H. H. A mosaic virus of grasses not virulent to sugarcane. *Ann. Appl. Biol.* **16**:525–532. 1929.

TESSI, J. L. Estudio comparativo de dos bacterios patogenos en Avena y determinación de una toxina que origina sus diferencias. *Rev. Invest. Agrico.* **7**:131–145. 1953.

THEIS, T. An undescribed species of ergot on *Panicum maximum* Jacq. var. common guinea. *Mycologia* **44**:789–794. 1952.

―――. Some diseases of Puerto Rican forage crops. *Fed. Exp. Sta. U.S. Dept. Agr. Mayaguez, P.R., Bul.* 51, pp. 1–31. 1953.

THIRUMALACHAR, M. J., and J. G. DICKSON. Spore germination, cultural characters and cytology of varieties of *Ustilago striiformis* and the reaction of hosts. *Phytopath.* **43**:527–535. 1953.

TRACY, S. M., and F. S. EARLE. New species of parasitic fungi. *Torrey Bot. Club Bul.* **22**:174–179. 1895.

WADE, G. C. Blind disease of rye grass. *Tasm. Jour. Agr.* **20**:226–227. 1949.

WALLIN, J. R. Parasitism of *Xanthomonas translucens* (J. J. and R.) Dowson on grasses and cereals. *Iowa State Col. Jour. Sci.* **20**:171–193. 1946.

———— and C. S. REDDY. A bacterial streak disease of *Phleum pratense* L. *Phytopath.* **35**:937–939. 1945.

WATERHOUSE, W. L. Australian rust studies. VIII. *Puccinia graminis lolii* an undescribed rust of *Lolium* spp. and other grasses in Australia. *Proc. Linn. Soc. N.S. Wales* **76**:57–64. 1951.

WERNHAM, C. C. New facts about eastern snow mold. *Phytopath.* **31**:940–943. 1941.

WOLLENWEBER, H. W., and O. A. REINKING. Die Fusarien. Paul Parey. Berlin. 1935.

ZUNDEL, G. L. I. (Ustilaginales) Additions and corrections. *North American Flora.* **7**:971–1030. 1939

DISEASES OF LEGUMES

CHAPTER 13

ALFALFA AND SWEETCLOVER DISEASES

The common alfalfa or lucerne, *Medicago sativa* L., is grown extensively for hay and pasture. Varieties of this species constitute one of the most valuable hay and seed crops of the forage legumes. Some few additional species, notably *M. falcata* L., are used as forage crops in some areas (Tysdal and Westover, 1937). Many species, both perennial and annual, occur wild in central and western Asia, southern Europe, and northern Africa. Grossheim (1930) lists 25 species of *Medicago* occurring in the Transcaucasian area. Some of the annual species, introduced from Europe, occur wild in pastures and waste places in North and South America. Some of the diseases described occur on the annual as well as the perennial species.

The genus comprises a polyploid series with a basic chromosome number of 8 pairs (Wipf, 1939). One annual species, *Medicago hispide* Gaertn., is apparently the exception with 7 chromosome pairs. *M. sativa* and *M. falcata* are tetraploid species with 16 chromosome pairs. Polyploids of the former species are reported.

Alfalfa is one of the ancient perennial forage legumes adapted especially to semiarid regions. In this widely diverse cross-pollinated species, strains or varieties occur that are adapted to cold, relatively high summer humidity and other variations in climate. The generally high self-sterility of the plants makes the isolation and stabilization of the desirable characters for plant growth, disease resistance, and seed production difficult. Selfing, sib-pollination, and clonal propagation are important in the breeding program and especially in the comparisons of disease reaction.

The perennial species develop a crown of stem branches, including axillary and secondary buds and a taproot. The stems and roots develop secondary thickenings. The annual stems develop from the crown buds. The crown and root tissues constitute the important storage tissues.

Several species of *Melilotus* are used for forage and soil improvement. Varieties of white sweetclover, *Melilotus alba* Desv., yellow sweetclover, *M. officinalis* (L.) Lam., and *M. indica* All. are used more commonly. The species studied have eight chromosome pairs. The adaptation of the biennial and annual varieties is relatively wide.   The plant structure and general anatomy are similar to alfalfa, and many of the diseases are similar in the two crops.

1. Cold and Winter Injury. Winter injury is common in alfalfa, especially in the northern range of its cultivation.   The crown and root tissues of alfalfa and clover plants are damaged by low temperatures and desiccation, smothering by ice sheets and tissue injury, and heaving by

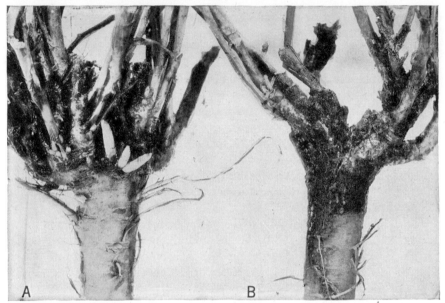

Fig. 75. Winter injury in alfalfa root and crown.   (A) Sound; (B) injured.   (*Courtesy of F. R. Jones.*)

freezing and thawing.   Frequently one or a combination of these types of injury results in winterkilling of extensive acreages of these crops. Plants with less severe injury are damaged further by the entrance of fungi and bacteria into the tissues before the injured tissues recover or are repaired by cellular activity (Jones, 1928, Peltier and Tysdal, 1931, Weimer, 1929, 1930) (Fig. 75).

Damage in 1-year old plants is located usually in the phloem and phloem rays or is restricted to the large cells exterior to the phloem fibers. Injury in the exterior of the phloem results in a sheath of damaged tissues surrounding the root, which kills the phellogen or cork cambium exterior to it.   Injury in the phloem usually is accompanied by necrosis of the large cells in the xylem rays in the center of the root.   In more severe winter damage, portions of the root near the crown are killed completely or disorganized sufficiently for fungi to enter and rot the tissues, including those of the base of the crown.   Under such conditions, the plants fail to recover the following spring.

Damage in plants the second season or later is largely in the parenchymatous cells of the phloem and phloem rays.   The location of the injured cells is different than in the first year, however, as they are located inside rather than outside the last group of fibers of the phloem.   This injury usually extends inward through the cambium and xylem rays of the previous season's growth and results in breaks in the cambium cylinder, but usually the injury does not extend inward to the center of the root. Damage in the sheath of cells immediately beneath the phellogen occurs independently or in conjunction with phloem injury (Fig. 76).

Injury of the crown stems is similar to that in the roots.   The large cells beneath the phellogen show damage first.   The ray cells are damaged, and the bundles become separated by these dead cells.   Phloem parenchyma is injured as in the root.   Crown buds are damaged or killed in some instances.   Jones (1945), using clonal populations from a large number of plants, has shown differences in type and degree of injury in plant lines and in stem growth of injured plants.

The nature of the injury appears to be due in part to the separation of the cells along the middle lamellae.   When extensive, this results in the physiological isolation and death of the tissue.   The cells are killed and sometimes ruptured by the freezing.   These injured cells and those adjacent respond by biochemical changes and the deposition of brown amorphous substances in and between the injured cells.   The injured cells are isolated ultimately by the meristematic activity of the surrounding cells.

The response to injury is conditioned by the character of the cells adjoining the injured tissues.   If these tissues are capable of rapid meristematic activity, repair is rapid and complete before decay of the injured tissues progresses far beyond the damaged cells.   Under less favorable situations, fungal or bacterial development in the tissues is extensive. According to Jones (1928), Jones and Weimer (1928), and Wiant and Starr (1936), bacterial wilt of alfalfa occurs in greater abundance in plants damaged by winter injury.   The initial tissue damage and recovery in injured plants differ with varieties, the storage of reserves in the previous season, the environmental conditions and other factors (Albert, 1927, Graber et al., 1927, Nelson, 1925, Steinmetz, 1926).

Control of winter injury depends largely upon the use of adapted varieties, resistant lines, and management of the crop.   The varieties adapted to the more northern sections appear to go into dormancy in better physiological condition to withstand winter injury than those that continue vigorous vegetative growth until late autumn.   Hardening and freezing tests on young plants under artificial conditions are not always

FIG. 76. Transections of alfalfa roots showing the tissues damaged by low winter temperatures. (*Courtesy F. R. Jones.*)

reliable means of measuring varietal response to this complex as it functions in the field.   Selection of surviving plant lines growing under field conditions, as suggested by Jones (1945), and combining those lines in synthetic varieties are the most reliable methods of control.   Crop management designed to ensure adequate storage of reserves in the crown and root is essential to winter survival and recovery in the spring.   Continuous pasturing and frequent cutting especially late into the autumn deplete the reserves essential to plant development and winter survival. The importance of growth and decay of transient or noncambial roots in these crops has been suggested by Jones (1943).

2. **Yellowing, Leaf Discoloration, and Dwarfing,** Nonparasitic.   Several types of yellowing and other discoloration of foliage and shortened stem growth are common in alfalfa and sweetclover.   Frequently, the symptoms and plant response are similar for a number of unfavorable environmental conditions.   These causal complexes cannot be determined without careful study of past and present environmental conditions. Deficiency of mineral nutrients is cumulative in effect, and frequently production is reduced materially before symptoms are expressed fully.

Boron deficiency in the soils of many agricultural areas results in dwarfed plants and yellow to bronze foliage, especially in alfalfa.   Bauer *et al.* (1941), Colwell and Lincoln (1942), McLarty *et al.* (1937), Piland and Ireland (1941), and others have reviewed the symptoms and effects of boron deficiency in alfalfa.   The symptoms are short terminal internodes, death of terminal buds, and yellow or red foliage.   These symptoms differ from the marginal yellowing and white spotting of the leaflets due to potash and phosphate deficiencies and are somewhat less distinctly differentiated from leaf hopper injury.   White spots on the leaflets are characteristic of some plants at certain stages of growth, as shown by Jones (1945, 1955).

Heavy potato leaf hopper infestation causes yellowing and dwarfing of plants.   The leaflets yellow and frequently brown along the margins. Internodal elongation, plant vigor, crown bud development, and transient root survival are direct or indirect effects of the injury.   The terminal buds and flower primordia appear grayish green and greatly retarded in development, or they dry out and fail to develop.   The primary tissues of the stem show punctures and necrotic spots (Smith and Poos, 1931); however, the secondary growth from the cambium gives better evidence on the type of injury (Jones, 1945).   The necrosis of cambium cells around the punctures and excess phloem production in adjacent portions of the cambium result in irregular thickening of the stunted stems.   The presence of the leaf hoppers is further evidence of the cause (Granovsky, 1928, Hollowell *et al.*, 1927, Johnson, 1934, 1936, 1938, Jones, 1945).

Yellowing and stunting of the plants occur as a response to other unfavorable environmental conditions, such as heat and drought. The abnormal appearance of the plants is localized or general in the field or area. The floral buds are retarded and under severe conditions yellow and dry out (Jones, 1937).

Mineral deficiencies are manifest by various types of leaf discoloration and spotting. The use of plant symptoms for diagnosis of specific compounds is risky, as frequently several agencies produce similar symptoms. Boron deficiency in alfalfa affects the development of the whole plant. Grayish-white irregular marginal lesions on the leaflets, reduced stem elongation, and blossom fall are characteristic symptoms. Zinc, copper, and molybdenum deficiencies cause somewhat similar symptoms. Low potash is manifest by leaflet yellowing and yellow blotches, frequently similar to other agents causing yellowing. Low phosphorus is evident by purple discoloration of especially new growth (Berger, 1949, Brown and Viets, 1952, Kipps, 1947).

3. Alfalfa Mosaic, Witches' Broom, and Dwarf, Viruses Transmitted by Aphids and Leaf Hoppers. Several mosaics occur on alfalfa and sweetclover. These plants and the clovers apparently harbor a large number of viruses occurring on annual legumes and other crop plants. The virus complex on these plants needs further investigation, especially as the perennial species of this group not only function as sources of virus infection, but also as winter habitat for the aphid vectors. The viruses occurring on alfalfa and sweetclover are numerous, and many of them are studied inadequately (Black and Price, 1940, Holmes, 1939, Pierce, 1934, Price, 1940, Snyder and Rich, 1942, Zaumeyer, 1938, Zaumeyer and Wade, 1935). Weiss[1] summarizes the viruses described, primarily on the leguminous crops.

Alfalfa Mosaic, Virus Transmitted Mechanically and by *Aphis gossypii* (Glover), *Macrosiphum pisi* (Kltb.) and Others. This mosaic is distributed widely in North America, although it is more common west of the Rocky Mountains and possibly occurs in Europe and Asia. The virus or strains occur on a wide range of crop plants, and it is important naturally on especially bean, pea, potato, celery, and paprika pepper.

The first symptoms on alfalfa are small greenish-yellow spots followed by more extensive and diffuse chlorosis. The leaflets become crinkled, irregular in shape, brittle, and there is no necrosis of tissues (Weimer, 1934). The symptoms are masked by high temperatures.

The virus and strains are transmitted mechanically to a wide range of plants in 27 families in addition to the Leguminosae. The pea aphid, *Macrosiphum pisi* (Kltb.), is the common vector in nature.

Several other aphids are vectors of some strains: *Aphis gossypii* (Glover), *A. fabae* (Scop.), *A. medicaginis* (Koch), *Macrosiphum solani-*

[1] *U.S. Dept. Agr. Plant Dis. Reptr., Suppl.* 154, 1945.

folii (Ash.), *A. persicae* (Sulz.) (Oswald, 1950, Swenson, 1952). Aphid transmission to alfalfa apparently results in higher infection than mechanical with several viruses studied. The virus persists in the crown tissues of infected plants. Laufer and Ross (1940) and Ross (1941) have studied the virus protein.

Several other legume viruses occur on the alfalfa, and in some instances this perennial crop may function in carrying the viruses through the winter.

Witches' broom, Virus Transmitted by Grafting, Dodder, and Leaf Hoppers *Scaphytopius acutus* (Sag.), *S. dubius* (Van Duzee), Considered by Some as a Variety of *S. acutus*, and *Orosius argentatus* (Evans). Witches' broom occurs in western North America and Australia. The disease is serious on alfalfa in local areas. Insect transmission to one or more species of *Medicago*, *Melilotus*, *Trifolium*, *Lotus*, *Astragalus*, *Lathyrus*, and *Hedysarum* with typical symptoms is reported (Helson, 1951, Klostermeyer and Menzies, 1951). Transmission to a wider group of hosts was accomplished by use of dodder (Kunkel, 1952). The virus incites the development of numerous crown buds that form fine stems shorter than on the healthy plants. Leaves are small, and leaflets exhibit marginal chlorosis and crinkling. Flower buds develop poorly. Infected plants show the disease each season, although they are relatively short-lived (Edwards, 1936, Menzies, 1946). By means of dodder the virus was transmitted from alfalfa to carrot, tomato, and periwinkle, but not to alfalfa (Kunkel, 1952). By means of grafting the virus was transmitted in alfalfa, tomato, potato, and periwinkle (Edwards, 1936, Kunkel, 1952). The disease is distinct from aster yellows and potato witches' broom (Kunkel, 1952). Because of the long incubation period associated with leaf hopper transmission (74 days), annual legumes apparently are unlikely hosts.

Alfalfa Dwarf, Virus Transmitted by Amblycephalinae Leaf Hoppers. Alfalfa dwarf occurs in the Southwestern and Southeastern United States and possibly in New South Wales. The disease is severe in limited areas, especially where the virus disease, Pierce's disease of grapes, is prevalent. The plants are dark green, dwarfed, and gradually decline in vigor until they die after several seasons. The roots show the deposition of brown gum-like material in the xylem bundles (Stoner, 1953, Weimer, 1936, 1937). The disease is prevalent under high moisture conditions and where alfalfa and vineyards are in close proximity.

The host range of the virus is wide and includes grasses, herbs, and shrubs as well as common legumes (Freitag, 1951, Freitag and Frazier, 1954). The virus is transmitted by grafting (Hewitt *et al.*, 1946). Vectors of the virus include leaf hoppers of several genera and spittle insects of three genera (Freitag *et al.*, 1952, Hewitt *et al.*, 1946). Three vectors, the green sharpshooter, *Draeculacephale minerva* (Ball), red-headed sharpshooter, *Carneocephala fulgida* (Nott.), and blue-green sharpshooter, *Hordnia circellata* (Baker), appear to be the most important naturally (Freitag and Frazier, 1954).

Aster Yellows Virus on Alfalfa, Transmitted by Leaf Hoppers.   Alfalfa infected with aster yellows virus occurs in local areas infrequently.   The plants show various degrees of stunting and excess branching.   The flowers show chlorosis, malformation, and sterility, generally remaining on the plant rather than shedding as in witches' broom (Kunkel, 1926).

Wound tumor Disease of Legumes, Virus Transmitted by Leaf Hoppers.   The disease earlier called "clover big vein" is distributed widely, but it is economically of minor importance.   Several legumes, *Melilotus*, *Trigonella*, and *Trifolium* spp. and other plants, react by the formation of swellings or tumor-like proliferations principally of phloem at the site of puncture by viruliferous leaf hoppers of *Agallia constricta* (Van Duzee), *A. quadripunctata* (Prov.), and *Agalliopsis novella* (Say), (Black, 1945, 1950, 1951, 1955).   Investigations on the biology of virus development in the insect vectors have been associated with this and other viruses (Black, 1955).

4. Sweetclover Mosaic, Ring Spot, and Streak, Viruses.   *Mosaic* on sweetclover apparently may be caused by any one of the following viruses: alfalfa mosaic, bean yellow mosaic, pea common mosaic, pea mottle, pea streak, pea wilt, alsike clover mosaic, subclover mosaic, broadbean mild mosaic, and red clover vein mosaic.   The symptoms produced by these various viruses are not differentiated on sweetclover.   Leaf mottling, chlorosis, and other mild-type symptoms usually develop.

Tobacco ring spot virus produces light-yellow irregular spots and mottling of the leaflets in nature.   A strain of the virus produces veinal chlorosis, pronounced puckering of the leaflets, and dwarfing.

Tobacco streak apparently may be harbored on the common white sweetclover on which it produces general chlorosis and chlorotic ring and line patterns.

5. Bacterial Wilt, *Corynebacterium insidiosum* (McCull.) H. L. Jens. The bacterial wilt of alfalfa is distributed widely in North America, and it is apparently less common in South America, Europe, and Asia.   The disease is probably the most important malady of the crop in the United States.   During the past 20 years it has spread over most of the important alfalfa-producing areas of the country.   Alfalfa plants are killed so rapidly that fields are unprofitable after 3 or 4 years.

SYMPTOMS.   The plants are reduced in vigor, the leaves yellow and bleach, and the plants die in the late summer.   The leaflets on infected plants are smaller and thicker prior to the loss of the green color.   The stems are smaller and more numerous in the earlier stages of disease development. The taproot shows a pale-brown discoloration of the outer woody tissue. This is evident when the outer bark is peeled off or when the stem is sectioned (Fig. 77). The bacteria are present in the xylem

FIG. 77. (A) Alfalfa plant, lower center, showing the symptoms of wilt caused by *Corynebacterium insidiosum*. (B) Sections through the root showing the brown discoloration of the outer woody tissues. (*Courtesy of F. R. Jones.*)

bundles of the new growth and spread tangentially into the parenchymatous tissue and adjacent bundles, resulting in distribution around the circumference of the root. The initially invaded areas are associated with wounds in the phloem rays caused especially by winter injury. The wilted plants occur first in the lower portions of the fields, either as scattered plants or more frequently in groups of plants.

THE BACTERIUM

Corynebacterium insidiosum (McCull.) II. L. Jens.
[*Phytomonas insidiosa* (McCull.) Bergey *et al.*]
(*Aplanobacter insidiosum* McCull.)

The colonies are pale yellow and consist of short rods with rounded ends without flagella.

ETIOLOGY.   The infection of the plants occurs during the spring and early summer.   The bacteria in the older diseased plants are released by the breakdown of the infected tissues, and they are distributed in the soil water.   Entrance into the plants occurs commonly through rifts in the root tissues caused by winter injury during the previous winter (Jones, 1928, Jones and Weimer, 1928).   The rapid spread of the bacteria in the root tissues occurs in the xylem tissue formed during the spring and early summer.   The tissues later in the summer are resistant to invasion.   The infected plants usually die during the late summer of the second year.

CONTROL.   The control of the disease is difficult, as indicated by the general spread in the alfalfa-producing areas.   Proper management of the crop to prevent injury of the root tissues helps reduce the spread of the disease.   Resistant varieties offer the only satisfactory means of controlling the disease.   Resistance to bacterial wilt must be associated with resistance to winter injury to secure long life in plants in wilt-infested areas (Jones, 1945).   Some lines or composite varieties are resistant to winter injury and capable of survival over long periods (Brink *et al.*, 1934, Jones, 1934, 1940, Peltier, 1933, Peltier and Schroeder, 1932, Peltier and Tysdal, 1934).   The Turkestan types and varieties contain more wilt tolerance than the common alfalfa.

6. Bacterial Stem Blight, *Pseudomonas medicaginis* Sackett.   The disease occurs locally in the western part of North America, on alfalfa principally.   The disease is of minor importance.   The lesions occur on the stems and foliage of the younger plants and extend into the crown and roots in some older plants.   The dark to light-brown spots are linear, with droplets or scales of bacterial exudate.   The lesions commonly occur on the stems in association with frost cracks.

THE BACTERIUM

Pseudomonas medicaginis Sackett
[*Phytomonas medicaginis* (Sackett) Bergey *et al.*]
[*Bacterium medicaginis* (Sackett) E. F. Sm.]

Light-yellow colonies of short rods that are motile by means of 1 to 4 polar flagella develop on media.

The disease occurs in the early spring, especially when low temperatures cause injury or cracking of the young stem epidermis, and therefore it is common only in the spring growth.   Varieties react differently to the spread of the disease in the stem tissues (Richards, 1934, 1936–1937, Sackett, 1910).

7. Crown Wart, *Urophlyctis alfalfae* (Lage.) Magn. or *Physoderma alfalfae* (Lage.) Karling. The crown wart on alfalfa occurs in the warmer areas of North and South America and Europe. The similar disease on white clover, common in Europe, occurs in the Gulf area of the United

Fig. 78. Crown wart of alfalfa caused by *Urophlyctis alfalfae*. (*Courtesy of F. R. Jones.*)

States. The disease is confined to local wet areas where some damage occurs.

The galls are swollen and modified stem bud primordia (Fig. 78). The primordial scales, leaves, and stipules in the infected buds thicken to form the scale-like galls around the central axis of the undeveloped stem. The galls are usually near the soil surface. Small leaf galls are formed occasionally. The galls are white when young, and gray to brown as they decay and dry out in mid-summer. The swollen cells of the gall tissue contain the resting spores of the fungus in various stages of development.

THE FUNGUS

Urophlyctis alfalfae (Lage.) Magn. or
Physoderma alfalfae (Lage.) Karling

(*Cladochytrium alfalfae* Lage.)

(*Oedomyces alfalfae* Lage.)

The branched haustorial processes form in a well-defined zone near the apex of the vegetative cell and the resting spore. The resting spores are initiated from the apex of the vegetative cell by the enlargement of the axial haustorial element. The mature resting spores are thick-walled, light brown to golden yellow, with the zone of scars of the haustorial elements where the wall is thinner (Jones and Drechsler, 1920, Wilson, 1920). Karling (1950) transfers this and other species to *Physoderma, Physoderma alfalfae* (Lage.) Karling.

Urophlyctis trifolii (Pass.) Magn., similar to the above species in morphology, occurs in Europe and Asia and is reported on white clover in the South Central United States. *Olpidium trifolii* (Pass.) Schroet., causing leaf curl on white clover and possibly the same as *U. trifolii*, occurs in the United States as well as in Europe. *U. trifolii* is listed as synonymous with *O. trifolii* by Chilton *et al.* (1943), but distinguished morphologically by Atkinson (1940).

8. Pythium Seed Rot, Damping-off, and Root Rot, *Pythium debaryanum* Hesse and Other Species. The *Pythium* spp. associated with the legumes occur throughout the world, especially in the finer texture acid soils (Buchholtz, 1942). The seed-rotting, preemergence blighting, and post-emergence damping-off reduce the stands in the small-seeded legumes, especially alfalfa and sweetclover. The disease is primarily associated with the seed rot and young seedlings, although rootlet rot and stem rot occur in established plants, especially in wet locations. The soft rot of the tissues and presence of the sporangia and oospores in the freshly rotted tissues are the most certain symptoms. Several species of *Pythium* are associated with the disease on the small-seeded legumes (Halpin *et al.*, 1954). These species also occur on a wide range of cultivated and wild plants.

Middleton (1943) described this group of morphologically similar species with somewhat spherical nonproliferous sporangia. They differ primarily in the number and origin of the antheridia attached to the oogonia. *Pythium debaryanum* Hesse is distinguished by the terminal or intercalary sporangia and oogonia and by one to six monoclinous and declinous antheridia attached to each oogonium. The sporangia germinate by either zoospores or germ tubes. *P. ultimum* Trow has chiefly terminal sporangia and oogonia and usually one monoclinous antheridium attached to each oogonium. The sporangia germinate by germ tubes only. *P. splendens* Braun, *P. vexans* DBy., and several others of this group occur on these crops. The species more common on the grasses, *P. arrhenomanes* and *P. graminicola*, are relatively nonpathogenic on the legumes.

The etiology of the group is similar. The seed and seedling attack from the soil-borne mycelium is associated with the maturity and general condition of the seed and the soil environment in which the seed germinates. Fully matured seed, well-prepared seed bed, a balanced fertility, including liming of acid soils, and seed treatment under some conditions (Allison and Torrie, 1945) are the best control measures.

9. Phytophthora Root Rot of Sweetclover, *Phytophthora* spp. A soft rot of the root and crown of sweetclover plants occurs in the spring in the North Central United States and Canada. The tops bleach and wither due to the spongy soft rot of the root and lower portion of the crown. The disease is of minor importance. *Phytophthora cactorum* (Leb. & Cohn) Schroet. is associated with the disease (Benedict, 1954). Resistant strains of sweetclover are the best means of control where the disease becomes severe (Jones, 1939). *P. megasperma* Drechsl. occurs on sweetclover but is less pathogenic (Cormack, 1940). A root rot of alfalfa incited by *P. cryptogea* Peth. & Laff. occurs in California (Erwin, 1954).

10. Downy Mildew, *Peronospora trifoliorum* DBy. The downy mildew is distributed widely on alfalfa in the temperate zones of the world. In the Northern United States the disease causes considerable damage to young plants and reduces stand of plants in the second year. The disease occurs rarely on sweetclover and not on the clovers in the United States, although the downy mildew is severe on clover in some sections of Europe. The characteristic symptoms are the light-green leaves especially at the apex of the stem and the grayish-white mycelium on the surface of the leaves. Internodal elongation is reduced, the stems are smaller, and the leaflets are twisted and rolled in severe infections. The conidiophores and violet conidia are conspicuous on the undersurface of the leaflets. Where the infection is systemic, the stems are swollen and the foliage chlorotic, with conidial production abundant just prior to the collapse of the leaf tissue (Fig. 79). Oospores are formed in the leaf tissue.

THE FUNGUS. *Peronospora trifoliorum* DBy.

The extensive synonymy is given by Chilton *et al.* (1943).

The mycelium is abundant in the leaf and stem tissues, and the grayish surface mycelium is extensive under humid conditions. The conidiophores are slender, dichotomously branched at acute angles, with the secondary branches curving downward in the older conidiophores. The conidia are globose to broadly elliptical, violet-colored, measure 15–20 by 18–36 microns, and germinate by the formation of germ tubes. The oospores are globose, smooth-walled, light brown, and 24 to 30 microns in diameter.

ETIOLOGY. The oospores in the dead tissues and the perennial mycelium in the crown buds enable the parasite to persist when once established. The mycelium is probably carried on the seed in some areas. Secondary spread occurs from the conidia whenever environmental con-

FIG. 79. Downy mildew of alfalfa caused by *Peronospora trifoliorum* showing (*A*) malformation, chlorosis, and necrosis of tissues of the systemically infected plant and (*B*) downy mass of conidiophores on the leaflets.

ditions are favorable. Clonal lines of alfalfa show differences in susceptibility. Some varieties are damaged more than others. Specialization
apparently occurs on the clovers and alfalfa.

FIG. 80. The spring black stem disease of (A) alfalfa incited by *Ascochyta imperfecta*,
(B) sweetclover (left) incited by *A. meliloti* and (right) stem canker caused by *A.
caulicola*, and (C) red clover incited by *Phoma trifolii*.

11. Spring Black Stem, *Ascochyta imperfecta* Pk., *Ascochyta meliloti*
Trus. or *Mycosphaerella lethalis* Stone, and *Phoma trifolii* E. M. Johnson &
Valleau. This complex of similar diseases on alfalfa, sweetclover, and
the clovers is common in North and South America and Europe. The
disease on alfalfa causes appreciably more damage than the others.
Disease development is severe during the cold wet conditions of early
spring, and in cold seasons damage extends into the early summer. The
infections spread extensively late in the autumn, but stem blackening is

less pronounced, although pycnidial development is abundant. Crown buds frequently are invaded and damaged. Seed production is reduced, and seed quality is lowered (Cormack, 1945, Kernkamp and Hemerick, 1954).

SYMPTOMS. On alfalfa and sweetclover the disease appears as dark-brown to black lesions on the stems and petioles. When the disease is severe, young shoots are blackened and killed and stems are girdled by the lesions. The brown spots on the leaves are small, irregular in shape, and coalesce to form the blackened areas (Fig. 80). The infected leaves turn yellow and wither before they drop. The brown spots appear on the pods under cool growing conditions. Pycnidia are not common on the lesioned stems during the growing season. Pycnidia are numerous, how-ever, on the old stems produced in the previous autumn and interspersed with the new spring growth of stems. Black stem symptoms occur on other legumes, but they are caused by other species of *Ascochyta*. Although the fungi on alfalfa and sweetclover infect the other crops, each occurs predominantly on the one crop (Cormack, 1945, Johnson and Valleau, 1933, Rosemblit, 1950, Sprague, 1929, Toovey *et al.*, 1936).

THE FUNGI

1. *Ascochyta imperfecta* Pk.
    (*Phoma medicaginis* Malbr. & Roum.)
    (*Diplodina medicaginis* Oud.)

Schenck and Gerdemann (1956) classify the pathogen on alfalfa as *Phoma herbarum* var. *medicaginis* West ex Rab. on the basis of the presence of the one-celled *Phoma* conidia and early reports.

The pycnidia are globose, ostiolate, without a beak, and light to dark brown in color. The spores are hyaline, oval or cylindrical with rounded ends, straight or slightly curved, and uniseptate when mature. Spore septation and size are variable. No perfect stage is known. This fungus causes the spring black stem on alfalfa. Remsberg and Hungerford (1936) wrongly described *Pleospora rehmina* Staritz as the per-fect stage of *Phoma medicaginis*. The fungus is seed-borne as well as persisting on crop residue (Cormack, 1945).

2. *Phoma trifolii* E. M. Johnson & Valleau

The species differs in morphology from that above by the nonseptate oval spores and is the cause of black stem on the clovers.

3. *Mycosphaerella lethalis* Stone or *Ascochyta meliloti* Trus.
    (*Ascochyta lethalis* Ell. & Barth.)

The conidial stage is *Ascochyta meliloti* (Trel.) J. J. Davis, as described by Jones (1944), although *A. meliloti* Trusova has priority. The other synonyms are given by Chilton *et al.* (1943).

The perithecia are submerged, globose with the ostioles elongated into a beak. The asci are cylindrical to clavate with an outer and inner wall. The ascospores are hyaline, ellipsoidal, two-celled, and slightly constricted at the septum. The pycnidia are globose, ostiolate, and brown. The spores are hyaline oblong, slightly curved, 1-septate when mature, and measure 5–6 by 13–20 microns. This fungus causes spring black stem on sweetclover.

4. *Ascochyta caulicola* Laub.

Causes a stem canker and hypertrophy of the stems of sweet clover without stem blackening as a characteristic symptom. This species is morphologically similar to the pycnidial stage of the former fungus (Jones, 1938).

ETIOLOGY. The development of the disease caused by the several parasites is similar. Pycnidia develop abundantly on the stems in late autumn and the following early spring. The fungi persist in the crop refuse on the surface of the soil. They are seed-borne in areas where the environment is favorable for pod infection. The infection and development of the disease is favored by cool, wet weather (Cormack, 1945, Jones, 1939, Peterson and Melchers, 1942). The disease appears the year of seeding and thereafter. Local cortical infection occurs through natural openings or injuries. Crown and root infections followed by rotting occur in alfalfa.

The control of the disease is accomplished by crop management and adapted resistant varieties. The fungi apparently persist in crop residues for only one season; therefore, crop rotation is practical, especially in the biennials. Burning the old stems early in the spring before the plants start growth reduces the inoculum. Commercial varieties show differences in reaction to the disease, and resistant plants occur in alfalfa (Koepper, 1942, Reitz, 1948). Selfed lines of sweetclover are resistant to both spring and summer black stem diseases (Jones, 1944). Seed treatment is advisable, especially when infected seed is used in areas free from the disease.

12. Summer Black Stem and Leaf Spot, *Cercospora zebrina* Pass. and *Cercospora davissii* Ell. & Ev. or *Mycosphaerella davisii* F. R. Jones. The summer black stem and leaf spot diseases develop on alfalfa, clover, and sweetclover during warm, moist weather. This disease complex frequently occurs in association with the spring black stem during early summer and again in the early autumn. The disease is distributed extensively on alfalfa, clover, sweetclover, and several wild legumes in central and eastern North America and in Europe. Damage to white clover pastures is severe during late summer in the Southern United States.

The symptoms vary somewhat on the different legumes. The leaf spots range from light-brown linear lesions on red clover to large circular ash-gray to tawny sunken spots on sweetclover. The center of the lesions is gray to black when conidial production is abundant. The lesions on the stems, petioles, and inflorescences are reddish brown to dark brown, depending upon the tissues. The lesions are sunken, and necrosis of tissues is extensive. The lesions are conspicuous when the plants are in the blossom stage of development. Infected seed is shriveled and discolored with mycelium on the surface. The fungi are restricted closely in the legumes attacked (Hopkins, 1921, Horsfall, 1929, Jones, 1944, Nagel, 1934).

THE FUNGI

1. *Cercospora zebrina* Pass.
   (*Cercospora medicaginis* Ell. & Ev.)

The complete synonymy is given in Chilton *et al.* (1943) in which they include *C. davisii* Ell. & Ev.

The conidiophores are hyaline to brown, nonseptate, and rather long (35 to 45 microns). The conidia are cylindrical fusoid, hyaline to light yellow, 3- to 6-septate, and average 3 by 50 microns in size. No spermagonial or perithecial stage is known. This species occurs on alfalfa and the clovers.

2. *Mycosphaerella davisii* F. R. Jones

The conidial stage is *Cercospora davissi* Ell. & Ev., not *C. meliloti* Oud. according to Jones (1944), and he retains *C. davisii* distinct from *C. zebrina* on alfalfa and the clovers.

The perithecia are inconspicuous, often few and scattered on dead overwintered sweetclover stems. They develop beneath the epidermis through which the ostiole opens, and they are spherical and dark-colored. Asci are cylindrical to clavate, grouped at the base of perithecia, and develop in succession throughout the summer. Ascospores are irregularly biseriate, hyaline, straight or slightly curved along one side, bluntly pointed, and measure 4–5 by 12–20 microns. Spermagonia develop at low temperatures and are thickly scattered, black, subepidermal, erumpent, often approaching an acervulus in form, and bear rod-shaped spermatia. Conidiophores are amphigenous, tufted, straight or subflexuous, pale brown, continuous or 1- to 2-septate. Conidia are first cylindrical, later acicular, hyaline to greenish yellow, 1- to 13-septate, and 2.2 to 4.5 microns wide at the base by 20 to 140 long. This fungus occurs on the several species of *Melilotus*, or sweetclovers.

ETIOLOGY. The fungi persist in the old stems of the legumes and produce conidia in abundance under warm, moist conditions. They are seed-borne in areas where moisture is plentiful during the period of seed development. Secondary spread from conidia occurs during wet, warm weather.

CONTROL. Crop management and use of resistant or adapted varieties helps reduce the damage, especially on sweetclover. Burning or removal of crop residue aids in reducing the summer inoculum. The disease is more severe in second year's growth that is cut or grazed than in stands that make complete growth without retarded development. Resistant lines of sweetclover suggest the use of resistant varieties (Jones, 1944).

13. Leaf Spot and Root Rot, *Leptosphaeria pratensis* Sacc. & Briard or *Stagonopora meliloti* (Lasch.) Petr. The disease occurs on alfalfa, sweetclover, and some other legumes, and it is distributed widely in humid

FIG. 81. Leaf spots on alfalfa (*A*) and sweetclover (*B*) caused by *Stagonospora* stage of *Leptosphaeria pratensis*.

areas. Albrecht (1942) reported the disease causing severe damage during the spring in white clover pastures in the Southeastern United States.

On alfalfa the disease is primarily a root rot, although the leaf spot is common. The leaf spot is circular to angular, pale buff with light-brown margins, and with numerous pycnidia in the central portion of the lesion (Fig. 81). The ashy-gray lesions on the annual stems are uncommon.

The brown to black root and crown rot is the common symptom on alfalfa. Necrosis and dry rot frequently involve the upper portion of the taproot and crown. No new buds form above the lesion, and the plant eventually dies. Secondary organisms enter the rotted areas, especially under conditions of high moisture, to confuse the symptoms. On sweetclover and white clover the circular tan leaf spots are numerous in spring and late autumn with brown pycnidia abundant (Fig. 81). The inconspicuous stem lesions are prevalent during the growing season and spread rapidly as the plants mature. In the late autumn the stems are brown internally and golden brown on the surface as the pycnidia of the *Phoma* stage develop. The fungus develops three stages: *Stagonospora*, largely on the leaf spots; *Phoma*, on the stems in late fall; and the perfect stage on the stems in the late autumn and spring (Jones and Weimer, 1938).

THE FUNGUS

Leptosphaeria pratensis Sacc. & Briard.
Stagonospora meliloti (Lasch.) Petr. Conidial stage
Phoma meliloti Allesch.

The extensive synonomy is given by Chilton *et al.* (1943) and Jones and Weimer (1938).

The perithecia, formed in the old stems, are globose, dark brown, and the ostiole is usually papillate. The ascospores are oblong to fusoid, yellow, usually with 3 septa, and 25 to 30 microns long. The *Stagonospora* pycnidia are submerged in the tissues, with the neck or rostrum extending through the epidermis. The rostrum forms a central canal narrower at the base than at the apex. This funnel-shaped rostrum is a distinguishing morphological character and occurs also in the deeply submerged *Phoma* pycnidia. The spores vary greatly in size and the presence or absence of the septum.

ETIOLOGY. Both the conidia and ascospores serve as primary inoculum for the early-spring leaf infection. Secondary spread from conidia occurs during wet-weather conditions. The stem and root infections apparently are associated with injuries of various types. The influence of environment, especially temperature, on the development of the different stages of the fungus is pronounced; the *Stagonospora* stage develops at summer temperature, and the *Phoma* stage during low autumn temperatures. The disease is prevalent in the second and following years.

Control of the disease is largely by crop management. The removal or burning of old stems and crop residue reduces the inoculum for spring crown and leaf infection. Differences in susceptibility occur in both alfalfa and sweetclover. The latter crop is relatively more susceptible, according to Jones *et al.* (1941).

14. Common Leaf Spot, *Pseudopeziza medicaginis* (Lib.) Sacc. and *P. meliloti* Syd. This leaf spot is distributed rather generally on alfalfa throughout the world. It is probably one of the most common diseases on alfalfa and yet of minor importance in most areas, as the infection is generally light. Severe infection causes defoliation (Jones, 1919).

FIG. 82. The common leaf spot of alfalfa incited by *Pseudopeziza medicaginis* (*A*), the spots with apothecia shown slightly magnified in the insert, the leaf spot of red clover incited by *P. trifolii* (*B*), and the yellow leaf blotch of alfalfa incited by *P. jonesii* (*C*).

SYMPTOMS. The circular small brown spots occur on the leaflets. The spots are restricted in size, usually do not coalesce, and generally do not cause discoloration of the surrounding leaf tissues. The small dark-brown to black raised disk (apothecium) in the center of the mature spot is a distinguishing characteristic (Fig. 82). Small elliptical spots occur on the succulent stems but rarely form apothecia. Heavy infections cause defoliation, especially of the lower leaves. On sweetclover the spots are less pronounced.

THE FUNGI

1. *Pseudopeziza medicaginis* (Lib.) Sacc.

The apothecia arise in a stroma beneath the epidermis, solitary or clustered, the epidermis is ruptured, and the hymenia are raised above the surface of the leaf. The ascus-bearing surface of the hymenial layer is first covered by a thin layer of small rounded fungus cells and partly covered by the rupture epidermis. The asci are borne among numerous, nonseptate paraphyses slightly longer than the asci. Under conditions of suitable environment, the ascospores are discharged by the rupture of the end of the ascus, and spores are expelled with some force. Asci are clavate with eight ascospores arranged in two rows. The spores are hyaline, ovate to oblong, without septations, and 8 to 14 microns long.

2. *Pseudopeziza meliloti* Syd.

This species is similar in morphology to the previous species and occurs less abundantly on *Melilotus alba*. *Pseudopeziza trifolii* (Biv.-Bern.) Fckl. occurs sparingly on the clovers. The synonymy of the species is given by Chilton *et al.* (1943).

ETIOLOGY. The fungus persists in the leaf tissues until they are decomposed. Apothecia overwinter on the fallen leaflets, and new ones are formed on the undecomposed leaves in the spring. Ascosporic inoculum is abundant during the entire growing season. The disease develops whenever environmental conditions are favorable.

Varieties differ in their reaction to the disease. The varieties Du Puits from France and Caliverdi from California contain a fair percentage of resistant plants. Resistant plants showing near immunity in the young leaflets and only showing development of the mycelium between the pallisade cells in older leaves have been selected (Jones, 1953, Jones *et al.*, 1941).

15. Yellow Leaf Blotch, *Pseudopeziza jonesii* Nannf. The disease apparently is distributed less generally than the common leaf spot although world-wide in occurrence. Although leaf blotch occurs less frequently than the spot disease, it causes severe defoliation when present in any abundance. Alfalfa is the only legume damaged by the disease.

SYMPTOMS. The young lesions appear as yellow stripes and blotches elongated parallel to the leaf veins. The lesions enlarge, and the color changes to orange-yellow or brown, shading to yellow at the margins. Small orange to brown pycnidia develop, especially on the upper surface of the blotch (Fig. 82). Apothecia are scarce on the undersurface of the blotch until the leaflets drop. The stem lesions are elongate yellow blotches which soon turn dark brown; frequently apothecia form on the surface. The stem lesions generally are not abundant (Jones, 1918, Nannfeldt, 1932).

THE FUNGUS

Pseudopeziza jonesii Nannf.
(*Pyrenopeziza medicaginis* Fckl.)

Pseudopeziza jonesii Nannf. is used in the current literature, especially the European, although *Pyrenopeziza medicaginis* Fckl. is preferred by some pathologists. The synonyms are given in Chilton *et al.* (1943).

Apothecia develop on the dead leaf tissue and are partly closed by mycelial strands until mature. The asci are borne among numerous nonseptate paraphyses. Asci are cylindrical to clavate with eight ascospores usually in two rows. The spores are ovoid, hyaline, nonseptate, and 8 to 11 microns long. The pycnidia are formed subepidermally and rupture the epidermis in irregularly lobed cavities. The bottle-shaped conidiophores are closely packed on the interior of the pycnidial cavities. The conidia vary greatly; usually they are cylindrical with rounded ends.

ETIOLOGY. Ascospores are produced in the late spring from over-wintered apothecia. Primary infection occurs from ascospores that are produced in decreasing numbers as the growing season advances, and they are produced again late in the autumn. The conidia apparently do not cause infection.

Cutting the crop before leaf drop is desirable in both leaf spot diseases. The leaves are saved for hay, and ascosporic inoculum is reduced. Plants in waste areas also supply a source of inoculum. Burning the leaves and stubble in the early spring also reduces the spring inoculum. Apparently all the alfalfa varieties are moderately susceptible to the disease; however, resistant plants occur in most varieties (Jones, 1953).

16. Root Rots and Wilt, *Fusarium, Cylindrocarpon* spp., and Other Fungi. The root rots included in this complex of diseases are common on the perennial and biennial leguminous crops. Apparently they are associated in many instances with plants weakened by unfavorable winter or summer conditions, such as winter injury of various types and summer dormancy due to drought. Usually the plants are attacked before they become active vegetatively after winter or summer dormancy (Cormack, 1934, 1937, Scott, 1926, Weimer, 1928). The fungi associated with the rots vary greatly in areas and seasons. The root rots of this type are prevalent more generally in the colder and drier climates. The wilt type of disease occurs more generally in the warmer climates. Damage, especially to older stands of the perennial crops, is severe when environmental conditions and other diseases are weakening the plants.

SYMPTOMS. These root rots are conspicuous in the early spring. The mycelium is found around the roots and crowns and in ruptured and injured tissues. The lesions vary from irregular brown rotted areas to the complete disintegration of the root and lower crown. Such rots progress slowly in their development, with final weakening and death of the plants during the spring and summer.

The wilt of alfalfa caused by *Fusarium oxysporum* f. *medicaginis* (Weimer) Snyder & Hansen shows relatively little cortical rot during the early stages of disease development. The stems yellow, wilt, and dry out; the symptoms vary considerably depending upon environment. The vascular system shows the characteristic browning due to fungus invasion and root-tissue response.

THE FUNGI. *Fusarium avenaceum* (Fr.) Sacc., *F.* *arthrosporoides* Sherb., *F. culmorum* (W. F. Sm.) Sacc., *F. poae* (Pk.) Wr. *F. scirpi* var. *acuminatum* (Ell. & Ev.) Wr., *Cylindrocarpon ehrenbergi* Wr., *C. obtusis-porum* (Cke. & Hark.) Wr., *Plenodomus meliloti* Mark.-Let., and other species are reported associated with the root rot. *F. oxysporum* f. *medicaginis* (Weimer) Snyder & Hansen causes a wilt disease of alfalfa. The morphology of the wilt fungus is given in Chap. 16.

17. Rhizoctonia Root Rots. Two types of root rots caused by species of *Rhizoctonia* occur on the legumes. (1) The violet root rot caused by *Rhizoctonia violacea* Tul. or *Helicobasidium purpureum* (Tul.) Pat. (*R. crocorum* Fr.) is common in many European areas (Sampson and Western, 1941), and it occurs rarely in North America. The rotted areas are brown with mats of mycelium attached in the early stages. Later the roots are rotted and shredded with a brown to dark-violet discoloration. The rotted portion extends from the crown to as far as 6 inches below the soil line. The disease is associated with low areas subject to flooding and frequently follows root injuries. (2) Smith (1943, 1946) described a root cankering due to *R. solani* Kuehn in limited areas in the Southwestern United States. The dark sunken areas occur on the taproot and branches. The root is not rotted completely as in the former disease. This fungus is active on alfalfa during the high summer temperatures and relatively inactive in the cooler periods. Strains of the same species cause a seedling cortical rot and root rot of older sweetclover plants in the more northern areas (Benedict, 1954, Erwin, 1954).

Cormack (1941, 1948) reported damage to legumes in early spring by a Basidiomycete that killed the plants in a relatively short period. The disease apparently is northern in distribution in North America. The fungus growing on legume residue produces HCN in quantities lethal to the plant tissues under the snow covering and at the low temperatures. The dead tissues are invaded, and the mycelium grows over the plants and adjacent soil surface (LeBeau and Dickson, 1953). *Medicago falcata* appears more resistant than common alfalfa.

The *Sclerotium* rots occur on alfalfa but are more common on the clovers. Sanford and Cormack (1933) reported indications of resistance in alfalfa. Alfalfa stands are damaged by *Phymatotrichum omnivorum* (Shear) Dugg. in the Southwestern United States (see Chap. 15).

18. Smuts of Sweetclover and Legumes, *Thecaphora deformans* Dur. &
Mont. and *Entyloma meliloti* McAlp.   Both the seed smut and leaf smut
occur on legumes.   Both smuts are distributed widely, although damage
is minor on the economic legumes.   The seed smut usually is inconspicuous, as the sori are formed in the ovules and remain enclosed in the pods.
As the infected seed pods mature and dehisce, the dusty spore mass is
evident.   Fischer (1953) combines the several species of *Thecaphora* on
the Leguminosae under *T. deformans* Dur. & Mont.   The smut occurs
commonly on *Astragalus, Desmodium, Lathyrus, Lotus, Lupinus,* and
Vicia spp.   Occasionally *Trifolium* spp. are smutted; *T. tridentatum*
Lindl. in California.   The author collected this smut on *T. pratense* L. in
Wisconsin.   Gutner (1941, p. 185) reports this smut on *Medicago* and
Trifolium spp. in Europe and Asia.

Thecaphora deformans Dur. & Mont.   See Fischer (1953, p. 156) for
synonomy.

Sori are formed in the ovules enclosed in swollen pods, spores not
evident until pods open.   Sorus is enclosed by seed coat of enlarged seed.
Spore balls are globose or elongated, brown, coarsely verrucose, 24–60
by 30–7 microns, and composed of 5 to 20 firmly united spores.

The leaf smut occurs on *Melilotus* and *Trigonella* spp., mostly in warm
climate areas.

Entyloma meloloti Mc.Alp. (*E. trigonellae* Stevenson).   Small circular to
confluent pustular areas occur on the leaves, gray to brown with lighter-
colored center and covered by epidermis.   Spores are globose to ovoid or
irregular, closely packed, 12 to 15 microns, frequently with hyaline sheath.

In both smut pathogens sporidia are the source of local infections in
flowers and leaves.   Information on varietal reaction in the various
legumes is observational.

19. Rust, *Uromyces striatus medicaginis* (Pass.) Arth.   The rust is
general on alfalfa in North America and Europe, and it is probably world-
wide in the humid temperate zones.   The uredia and telia occur on
alfalfa.   In Europe the aecial stage occurs sparingly on *Euphorbia
cyparissias* L.   The reddish-brown uredia and telia develop on the leaves
and stems late in the season.   The telia form in the same lesions and in
independent sori.

THE FUNGUS

Uromyces striatus medicaginis (Pass.) Arth. or
U. striatus Schroet.

The synonyms are given in Arthur (1934) and Chilton *et al.* (1943).
The aecia are scattered somewhat elevated, and the infected *Euphorbia* plants are
dwarfed.   Peridia are white with irregular edges; spores are orange-colored and finely
verrucose.   The uredia on the alfalfa are small, round, and mostly on the leaves.   The
urediospores are globose to elliptical and light brown.   The telia, formed later in the

uredia and independently, are naked, round to oblong, and dark brown. The telio-spores are ovate to pyriform and dark brown in color.

In Europe the rust is heteroecious, with the aecial stage forming a perennial mycelium in the *Euphorbia* sp. In North America the rust persists in leaf tissue in the milder climates and apparently spreads northward during the summer. The rust occurs on several species of *Medicago*. Glasscock (1946) was unable to obtain infection of alfalfa with urediospores from *Medicago lupulina* L. Koepper (1942) reported *M. ruthenica* Trantr., with eight chromosome pairs, immune; *M. falcata* L. variable in reaction; and varieties of *M. sativa* L. ranging from suscep-tible to resistant. Ladak is the most resistant commercial variety. The Turkestan types in general are susceptible.

REFERENCES

AITKEN, Y., and B. J. GRIEVE. A mosaic virus of subterranean clover. *Jour. Aust. Inst. Agr. Sci.* **9**:81–82. 1943.

ALBERT, W. B. Studies on the growth of alfalfa and some perennial grasses. *Jour. Am. Soc. Agron.* **19**:624–654. 1927.

ALBRECHT, H. R. Effect of diseases upon survival of white clover, *Trifolium repens* L., in Alabama. *Jour. Am. Soc. Agron.* **34**:725–730. 1942.

ALLISON, J. L., and J. H. TORRIE. Effect of several seed protectants on germination and stands of various forage legumes. *Phytopath.* **34**:799–804. 1944.

ARTHUR, J. C. Manual of the rusts in United States and Canada. Purdue Research Foundation. Lafayette, Ind. 1934.

ATKINSON, R. E. Studies of *Olpidium trifolii* and *Urophlyctis trifolii* on white clover in Louisiana. *Phytopath.* **30**:2. 1940.

BAUR, K., *et al.* Boron deficiency of alfalfa in Western Washington. *Wash. Agr. Exp. Sta. Bul.* 296. 1941.

BENEDICT, W. G. Studies on sweetclover failures in Southwestern Ontario. *Can. Jour. Bot.* **32**:82–94. 1954.

———. Stunt of clovers caused by *Rhizoctonia solani. Can. Jour. Bot.* **32**:215–220. 1954.

BERGER, K. C. Boron in soils and crops. *Adv. Agron.* **1**:321–351. 1949.

BLACK, L. M. A virus tumor disease of plants. *Am. Jour. Bot.* **32**:408–415. 1945.

———. A plant virus that multiplies in its insect vector. *Nature, London* **166**:852–853. 1950.

———. Hereditary variations in the reaction of sweetclover to the wound-tumor virus. *Am. Jour. Bot.* **38**:256–267. 1951.

———. Concepts and problems concerning purification of labile insect-transmitted plant viruses. *Phytopath.* **45**:208–216. 1955.

——— and W. C. PRICE. The relationship between viruses of potato calico and alfalfa mosaic. *Phytopath.* **30**:444–447. 1940.

BRINK, R. A., *et al.* Genetics of resistance to bacterial wilt in alfalfa. *Jour. Agr. Research* **49**:635–642. 1934.

BROWN, L. C., and F. G. VIETS. Zinc deficiency of alfalfa in Washington. *Agron. Jour.* **44**:276. 1952.

BUCHHOLTZ, W. F. Influence of cultural factors on alfalfa seedling infection by *Pythium debaryanum* Hesse. *Iowa Agr. Exp. Sta. Res. Bul.* 296. 1942.

CHILTON, S. J. P., *et al.* Fungi reported on species of Medicago, Melilotus, and Trifolium. *U.S. Dept. Agr. Misc. Pub.* 499. 1943.

COLWELL, W. E., and C. LINCOLN. A comparison of boron deficiency symptoms and potato leafhopper injury in alfalfa. *Jour. Am. Soc. Agron.* **34**:495–498. 1942.

CORMACK, M. W. On the invasion of roots of Medicago and Melilotus by *Sclerotinia* sp. and *Plendomus meliloti* D. and S. *Can. Jour. Research* **C 11**:474–480. 1934.

———. *Cylindrocarpon ehrenbergi* Wr. and other species as root parasites of alfalfa and sweet clover in Alberta. *Can. Jour. Research* **C 15**:403–424. 1937.

———. *Fusarium* spp. as root parasites of alfalfa and sweet clover in Alberta. *Can. Jour. Research* **C 15**:493–510. 1937.

———. *Phytophthora cactorum* as a cause of root rot in sweet clover. *Phytopath.* **30**:700–701. 1940.

———. A low-temperature Basidiomycete causing early spring killing of grasses and legumes in Alberta. *Phytopath.* **31**:1058–1059. 1941.

———. Studies on *Ascochyta imperfecta*, a seed- and soil-borne parasite of alfalfa. *Phytopath.* **35**:838–855. 1945.

———. Winter crown rot or snow mold of alfalfa, clovers, and grasses in Alberta. I. Occurrence, parasitism, and spread of the pathogen. *Can. Jour. Research* **C 26**:71–85. 1948.

EDWARDS, E. T. The witches' broom disease of lucerne. *Dept. Agr. N.S. Wales Sci. Bul.* 52. 1936.

ERWIN, D. C. Relation of *Stagonospora, Rhizoctonia* and associated fungi to crown rot of alfalfa. *Phytopath.* **44**:137–144. 1954.

———. Root rot of alfalfa caused by *Phyphthora cryptogea*. *Phytopath.* **44**:700–704. 1954.

FISCHER, G. W. Manual of the North American smut fungi. The Ronald Press Company. New York.

FREITAG, J. H. Host range of the Pierce's disease virus of grapes as determined by insect transmission. *Phytopath.* **41**:920–934. 1951.

——— and N. W. FRAZIER. Natural infectivity of leafhopper vectors of Pierce's disease virus of grape in California. *Phytopath.* **44**:7–11. 1954.

———, ———, and R. A. FLOCK. Six new leafhopper vectors of Pierce's disease virus. *Phytopath.* **42**:533–534. 1952.

GLASSCOCK, H. H., and W. M. WARE. *Uromyces striatus* Schroet. on *Medicago lupulina* L. and other host plants in Britain. *Trans. Brit. Mycol. Soc.* **29**:167–169. 1946.

GORZ, H. J. Inheritance of reaction to *Ascochyta caulicola* in sweet clover (*Melilotus alba*). *Agron. Jour.* **47**:379–383. 1955.

GRABER, L. R., *et al.* Organic food reserves in relation to the growth of alfalfa and other perennial herbaceous plants. *Wis. Agr. Exp. Sta. Res. Bul.* 80. 1927.

GRANOVSKY, A. A. Alfalfa yellow-top and leafhoppers. *Jour. Econ. Ent.* **21**:261–266. 1928.

GROSSHEIM, A. A. Flora of Kavkazia. State Printer. Tiflis and Erivan. 1930.

GUTNER, L. S. The smut fungi of U.S.S.R. State Printing Office. Moscow-Leningrad. 1941. (In Russian.)

HALPIN, J. E., E. W. HANSON, and J. G. DICKSON. Studies on the pathogenicity of seven species of *Pythium* on alfalfa, sweetclover and Ladino clover seedlings. *Phytopath.* **44**:572–574. 1954.

HELSON, G. A. H. The transmission of witches' broom virus disease of lucerne by the common brown leafhopper, *Orosius argentatus* (Evans). *Aust. Jour. Sci. Res.* **B 4**:115–124. 1951.

HEWITT, W. B., *et al.* Leafhopper transmission of the virus causing Pierce's disease of grape and dwarf of alfalfa. *Phytopath.* **36**:117–218. 1946.

HOLLOWELL, E. A., *et al.* Leafhopper injury in clover. *Phytopath.* **17**:399–404. 1927.

HOPKINS, E. F. Studies on the Cercospora leaf spot of bur clover. *Phytopath.* **11**:311–318. 1921.

HORSFALL, J. G. Species of *Cercospora* on Trifolium, Medicago, and Melilotus. *Mycologia* **21**:304–312. 1929.

JOHNSON, E. M., and W. D. VALLEAU. Black-stem of alfalfa, red clover, and sweet clover. *Ky. Agr. Exp. Sta. Bul.* 339. 1933.

JOHNSON, F. The complex nature of white-clover mosaic. *Phytopath.* **32**:103–116. 1942.

JOHNSON, H. W. Nature of injury of forage legumes by the potato leafhopper. *Jour. Agr. Research* **49**:379–406. 1934.

————. Effect of leafhopper yellowing on the carotene content of alfalfa. *Phyto-Path.* **26**:1061–1063. 1936.

————. Further determinations of the carbohydrate-nitrogen relationship and carotene in leafhopper-yellowed and green alfalfa. *Phytopath.* **28**:273–277. 1938.

JONES, F. R. Yellow-leafblotch of alfalfa caused by the fungus *Pyrenopeziza medicaginis. Jour. Agr. Research* **13**:307–329. 1918.

————. The leaf-spot diseases of alfalfa and red clover caused by the fungi *Pseudopeziza medicaginis* and *P. trifolii*, respectively. *U.S. Dept. Agr. Bul.* 759. 1919.

————. Development of the bacteria causing wilt in the alfalfa plant as influenced by growth and winter injury. *Jour. Agr. Research* **37**:545–569. 1928.

————. Winter injury of alfalfa. *Jour. Agr. Research* **37**:189–211. 1928.

————. Testing alfalfa for resistance to bacterial wilt. *Jour. Agr. Research* **48**:1085–1098. 1934.

————. A foliage yellowing and floral injury of alfalfa associated with heat and drought. *Phytopath.* **27**:729–730. 1937.

————. A seed-borne disease of sweet clover. *Phytopath.* **28**:661–662. 1938.

————. Evidence of resistance in sweet clover to a Phytopthora root rot. *Phytopath.* **29**:909–911. 1939.

————. Four fungus parasites of sweet clover infecting seed. *Phytopath.* **29**:912–913. 1939.

————. Bacterial wilt of alfalfa and its control. *U.S. Dept. Agr. Cir.* 573. 1940.

————. Growth and decay of the transient (noncambial) roots of alfalfa. *Jour. Am. Soc. Agron.* **35**:628–634. 1943.

————. *Ascochyta meliloti* (Trel.) Davis as the conidial stage of *Mycosphaerella lethalis* Stone. *Trans. Wis. Acad. Sci., Arts, Letters* **35**:137–138. 1944.

————. Life history of *Cercospora* on sweet clover. *Mycologia* **36**:518–525. 1944.

————. Winter injury and longevity in unselected clones from four wilt-resistant varieties of alfalfa. *Jour. Am. Soc. Agron.* **37**:828–838. 1945.

————. Measurement of resistance in alfalfa to common leaf spot. *Phytopath.* **43**:651–654. 1953.

————. Testing for disease resistance. *Proc. 6th Intern. Grassland Cong.*, pp. 1567–1572. 1953.

————. White spot of alfalfa as a genetic character. *Phytopath.* **45**:289–290. 1955.

———— and C. DRECHSLER. Crown wart of alfalfa caused by *Urophlyctis alfalfae. Jour. Agr. Research* **20**:293–324. 1920.

———— *et al.* Evidence of resistance in alfalfa, red clover, and sweet clover to certain fungus parasites. *Phytopath.* **31**:765–766. 1941.

———— and O. F. SMITH. Sources of healthier alfalfa. *U.S. Dept. Agr. Yearbook,* pp. 228–237. 1953.

———— and J. L. WEIMER. Bacterial wilt and winter injury of alfalfa. *U.S. Dept. Agr. Cir.* 39. 1928.

———— and ————. Stagonospora leaf spot and root rot of forage legumes. *Jour. Agr. Research* **57**:791–812. 1938.

KARLING, J. S. The genus *Physoderma* (Chytridiales). *Lloydia* **13**:29–71. 1950.

KERNKAMP, M. F., and G. A. HEMERICK. The relation of *Ascochyta imperfecta* to alfalfa seed production in Minnesota. *Phytopath.* **43**:378–383. 1953.

KIPPS, E. H. The calcium/manganese ratio in relation to growth of lucerne in Canberra, A.C.T. *Jour. Comm. Sci. Ind. Res. Aust.* **20**:176–189. 1947.

KLOSTERMEYER, E. C., and J. D. MENZIES. Insect transmission of alfalfa witches'-broom virus to other legumes. *Phytopath.* **41**:456–458. 1951.

KOEPPER, J. M. Relative resistance of alfalfa species and varieties to rust caused by *Uromyces straitus.* *Phytopath.* **32**:1048–1057. 1942.

KUNKEL, L. O. Studies on aster yellows. *Am. Jour. Bot.* **13**:646–705. 1926.

————. Transmission of alfalfa witches broom to nonleguminous plants by dodder and cure in periwinkle by heat. *Phytopath.* **42**:27–31. 1952.

LAUFFER, M. A., and A. F. ROSS. Physical properties of alfalfa mosaic virus. *Jour. Am. Chem. Soc.* **62**:3296–3300. 1940.

LEBEAU, J. B., and J. G. DICKSON. Preliminary report on production of hydrogen cyanide by a snow-mold pathogen. *Phytopath.* **43**:581–582. 1953.

McLARTY, H. R., *et al.* A yellowing of alfalfa due to boron deficiency. *Sci. Agr.* **17**:515–517. 1937.

MENZIES, J. D. Witches' broom in alfalfa in North America. *Phytopath.* **36**:762–774. 1946.

————. Methods of reducing the spread of the witches' broom disease in alfalfa. *Phytopath.* **43**:649–650. 1953.

MIDDLETON, J. T. The taxonomy, host range and geographic distribution of the genus Pythium. *Mem. Torrey Bot. Club* **21**:1–171. 1943.

MURPHY, D. M., and W. H. PIERCE. Common mosaic of the garden pea, *Pisum sativum.* *Phytopath.* **27**:710–721. 1937.

NAGEL, C. M. Conidial production in species of *Cercospora* in pure culture. *Phytopath.* **24**:1101–1110. 1934.

NANNFELDT, J. A. Studien über die Morphologie und Systematik der nichtlichenisierten Inoperculaten Discomyceten. *Nova Acta Regiae Soc. Scient. Upsaliensis* (4) **8**:5–368. 1932.

NELSON, N. T. The effect of frequent cutting on the production, root reserves, and behavior of alfalfa. *Jour. Am. Soc. Agron.* **17**:100–113. 1925.

OSBORNE, H. T. Vein-mosaic of red clover. *Phytopath.* **27**:1051–1058. 1937.

OSWALD, J. W. A strain of the alfalfa-mosaic virus causing vine and tuber necrosis in potato. *Phytopath.* **40**:973–991. 1950.

PELTIER, G. L. The relative susceptibility of alfalfas to wilt. *Nebr. Agr. Exp. Sta. Res. Bul.* 66. 1933.

———— and F. R. SCHROEDER. The nature of resistance in alfalfa wilt. *Nebr. Agr. Exp. Sta. Res. Bul.* 63. 1932.

———— and H. M. TYSDAL. Hardiness studies with two-year-old alfalfa plants. *Jour. Agr. Research* **43**:931–955. 1931.

———— and ————. Wilt and cold resistance of self fertilized lines of alfalfa. *Nebr. Agr. Exp. Sta. Res. Bul.* 76. 1934.

PETERSON, M. L., and L. E. MELCHERS. Studies on black stem of alfalfa caused by *Ascochyta imperfecta.* *Phytopath.* **32**:590–597. 1942.

PIERCE, W. H. Viruoses of the bean. *Phytopath.* **24**:87–115. 1934.

PILAND, J. R., and C. F. IRELAND. Application of borax produces seed set in alfalfa. *Jour. Am. Soc. Agron.* **33**:938–939. 1941.

PRICE, W. C. Comparative host ranges of six plant viruses. *Am. Jour. Bot.* **27**: 530–541. 1940.

REITZ, L. P. Reaction of alfalfa varieties, selections and hybrids to *Ascochyta imperfecta*. *Jour. Agr. Research* **76**:307–323. 1948.

REMSBERG, R., and C. W. HUNGERFORD. Black stem of alfalfa in Idaho. *Phytopath.* **26**:1015–1020. 1936.

RICHARDS, B. L. Reaction of alfalfa varieties to stem blight. *Phytopath.* **24**:824–827. 1934. *Utah Acad. Sci., Arts, Letters Proc.* **14**:33–38. 1936–1937.

ROSEMBLIT, A. La Ascochyta de la alfalfa en la Argentina. *Rev. Argent. Agron.* **17**:89–97. 1950.

ROSS, A. F. The concentration of alfalfa-mosaic virus in tobacco plants at different periods of time after inoculation. *Phytopath.* **31**:410–420. 1941.

SACKETT, W. B. A bacterial disease of alfalfa. *Colo. Agr. Exp. Sta. Bul.* 158. 1910.

SAMPSON, K., and J. H. WESTERN. Diseases of British grasses and herbage legumes. Cambridge University Press. London. 1941.

SANFORD, G. B., and M. W. CORMACK. On varietal resistance of Medicago and Melilotus to root rots caused by *Sclerotinia* sp. and *Plenodomus meliloti* D. and S. *Proc. Worlds Grain Exhib. Conf. Can.* **2**:290–293. 1923.

SCHENCK, N. C., and J. W. GERDEMANN. Taxonomy, pathogenicity and host-parasite reactions of *Phoma trifolii* and *Phoma herbarum* var. *medicaginis*. *Phytopath.* **46**:194–200. 1956.

SCOTT, I. T. A study of certain Fusarial diseases of plants. *Mo. Agr. Exp. Sta. Bul.* 244. 1926.

SMITH, F. F., and F. W. POOS. The feeding habits of some leafhoppers of the genus Empoasca. *Jour. Agr. Research* **43**:267–285. 1931.

SMITH, O. F. Rhizoctonia root canker of alfalfa (*Medicago sativa*). *Phytopath.* **33**:1081–1085. 1943.

———. Effect of soil temperature on the development of Rhizoctonia root canker of alfalfa. *Phytopath.* **36**:638–642. 1946.

SNYDER, W. C., and S. RICH. Mosaic of celery caused by the virus of alfalfa mosaic. *Phytopath.* **32**:537–539. 1942.

SPRAGUE, R. Host range and life history studies of some leguminous Ascochytae. *Phytopath.* **19**:917–932. 1929.

STEINMETZ, F. H. Winter hardiness in alfalfa varieties. *Minn. Agr. Exp. Sta. Tech. Bul.* 38. 1926.

STONER, W. N. Leaf hopper transmission of a degeneration of grape in Florida and its relation to Pierce's disease. *Phytopath.* **43**:611–615. 1953.

SWENSON, K. G. Aphid transmission of a strain of alfalfa mosaic virus. *Phytopath.* **42**:261–262. 1952.

TOOVEY, F. W., *et al.* Observations on the black-stem disease of lucerne in Britain. *Ann. Appl. Biol.* **23**:705–717. 1936.

TYSDAL, H. M., and H. L. WESTOVER. Alfalfa improvement. *U.S. Dept. Agr. Yearbook*, pp. 1122–1153. 1937.

WEIMER, J. L. A wilt disease of alfalfa caused by *Fusarium oxysporum* var. *medicaginis* n. var. *Jour. Agr. Research* **37**:419–433. 1928.

———. Some factors involved in the winterkilling of alfalfa. *Jour. Agr. Research* **39**:263–283. 1929.

———. Alfalfa root injuries resulting from freezing. *Jour. Agr. Research* **40**:121–143. 1930.

————. Studies on alfalfa mosaic. *Phytopath.* **24**:239–247. 1934.

————. Alfalfa dwarf, a virus disease transmissible by grafting. *Jour. Agr. Research* **53**:233–347. 1936.

————. Effect of the dwarf disease on the alfalfa plant. *Jour. Agr. Research* **55**:87–104. 1937.

WIANT, J. S., and G. H. STARR. Field studies on the bacterial wilt of alfalfa. *Wyo. Agr. Exp. Sta. Bul.* 214. 1936.

WILSON, O. T. Crown-gall of alfalfa. *Bot. Gaz.* **70**:51–68. 1920.

WIPF, L. Chromosome numbers in root nodules and root tips of certain Leguminosae. *Bot. Gaz.* **101**:51–67. 1939.

ZAUMEYER, W. J. A streak disease of peas and its relation to several strains of alfalfa mosaic virus. *Jour. Agr. Research* **56**:747–772. 1938.

———— and B. L. WADE. The relationship of certain legume mosaics to bean. *Jour. Agr. Research* **51**:715–749. 1935.

———— and ———— Pea mosaic and its relation to other legume mosaic viruses. *Jour. Agr. Research* **53**:161–185. 1936.

CHAPTER 14

CLOVER DISEASES

Varieties of four species of *Trifolium* constitute the major crops of the true clovers, although others are important in local areas. Red, *Trifolium pratense* L., crimson, *T. incarnatum* L., alsike, *T. hybridum* L., and white clover including Ladino, *T. repens* L., are the important species, widely used. There are some 250 described species for the genus. These apparently constitute two polyploid series with basic chromosome numbers of 7 and 8 pairs. *T. pratense* is a diploid with 7 chromosome pairs; *T. incarnatum* and *T. hybridum* are diploids with probably 8 pairs, although 7 pairs are reported for the former species by some investigators; and *T. repens* is a tetraploid with 16 chromosome pairs (Pieters and Hollowell, 1937, Wipf, 1939).

The cultivated clovers are cross-pollinated and highly variable. Local varieties vary greatly in winter hardiness, type of growth, certain morphological characters, and reaction to disease. The plants are self-sterile in the main, and lines are stabilized by sib-pollination and selection. Clonal propagation is useful in comparison of larger populations, especially in disease-reaction studies. Perennial and biennial types are used chiefly in the cultivated varieties. The North American pubescent types of red clover are better adapted to resist damage of insects than are the European types, which have glabrous stems. Many of the diseases are common to more than one species of the genus.

1. Cold and Winter Injury. Varieties of the true clovers vary greatly in their resistance to cold and winter injury. Many of the winter-hardy varieties are the result of survival of the hardy plants during the process of adaptation. The types of injury and relation of these injuries to invasion of the crown and root tissues especially by fungi are similar to those described by alfalfa (Chap. 13).

2. Yellows, Leaf Discoloration, and Dwarfing. The reaction of the clovers to deficiencies of the essential mineral elements is somewhat similar to alfalfa, although as a group these crops apparently are more tolerant than alfalfa (Kline, 1955). Leaf hopper injury in the smooth-stemmed red clovers is severe and prevents their economical production in areas where leaf hopper infestations are common. The symptoms are yellow to red discoloration of leaves and graying of foliage, reduced inter-nodal elongation, and blighted flower heads somewhat similar to the gross symptoms in alfalfa.

3. Mosaics and Yellows, Viruses Transmitted by Aphids and Leaf Hoppers. The virus diseases of the clovers transmitted especially by the pea aphid represent a complex of viruses occurring on legumes and other crop plants. As in alfalfa, the clover virus diseases are perhaps of greater importance on some of the annual leguminous crops than on the clovers themselves. As discussed by Johnson (1933), Johnson (1942), Osborn (1937), and others, two types of symptoms occur on the clovers: (1) the leaf mottling to chlorosis, with or without dwarfing of the plants, and (2) the vein yellowing without appreciable dwarfing of the stems. The literature pertaining to the viruses on clovers has been summarized by Weiss.[1] At least three virus diseases are defined as clover mosaics: red clover vein mosaic, alsike clover mosaic, and subterranean clover mosaic.

Red Clover Vein Mosaic, Virus Transmitted by *Macrosiphum pisi* (Kltb.). The malady is common on a wide range of legumes both peren-nial and annual through temperate North America and probably Europe. This virus is identical with pea stunt virus (Hagedorn and Hanson, 1951). The symptoms on clover are yellow irregular to regular-patterned dis-coloration along the veins of the leaflets of new growth without mottling and with little reduction in plant size or vigor. The symptoms are sim-ilar on the other clovers, except *Trifolium incarnatum* in which the plants are stunted and killed (Osborn, 1937).

The virus is transmitted by expressed juice and the pea aphid *Macrosi-phum pisi* (Kltb). Other viruses occur naturally on the red clover, some of which are undetermined. Red clover is a natural overwintering harbor for the potato yellow dwarf virus as well as a food plant of the vector, the clover leaf hopper, *Aceratagallia sanguinolenta* (Prov.).

Alsike Clover Mosaic, Virus. The mosaic is distributed widely in the United States and Canada, although it is of relatively minor importance on this clover. The alsike clover apparently harbors mosaics of the annual legumes, notably the common pea mosaic (Murphey and Pierce, 1937).

The virus and strains, to date transmitted only by extracted juice, produced stippled interveinal chlorosis with irregular patterns of light and green. The leaflets are smaller but not misshapen. Several viruses occur naturally on alsike and white clover (Johnson, 1942, Zaumeyer and Wade, 1935, 1936), notably those causing pea mottle and pea wilt, which in combination cause a severe mosaic in white clover.

Common pea mosaic, yellow bean mosaic, the various pea streaks, pea mottle, pea wilt, alfalfa mosaic, subterranean clover mosaic, ring spot,

[1] *U.S. Dept. Agr. Plant Dis. Rept. Suppl.* 154. 1945.

broadbean common and mild mosaics, and cucumber mosaic occur on red clover. The symptoms vary greatly and frequently are masked during much of the growing season (Hanson and Kreitlow, 1953).

In addition alfalfa mosaic, red clover vein mosaic, yellow bean mosaic, pea streak, and broadbean mild mosaic occur on white clover (Houston and Oswald, 1953), and red clover vein mosaic, subterranean clover mosaic, common pea mosaic, and New Zealand pea streak occur on alsike clover (Hanson and Kreitlow, 1953).

Subterranean Clover Mosaic, Virus Transmitted by Aphids. This disease is described from Australia where it causes systemic mottling, distortion of the leaflets, dwarfing of the plants, and a reduction in seed set (Aitken and Grieve, 1943). It apparently is transmitted by aphids, and probably it is seed-borne.

The yellows-type viruses transmitted by leaf hoppers occur in limited numbers on the clovers. Witches' broom and alfalfa stunt viruses are transferable to several of the clovers, but apparently do not occur naturally in many (see Chap. 13). The clovers, especially medium red, frequently harbor the potato yellow dwarf virus which is transferred to potatoes nearby by the clover leaf hopper, *Aceratagallia sanguinolenta* (Prov.). Aster yellows occurs occasionally on the clovers. Wound tumer virus, or clover big vein, is transmitted to the clovers.

4. Bacterial Blight, *Pseudomonas syringae* v. Hall, [*Phytomonas trifoliorum* (L. R. Jones *et al.*) Burk.]. The disease occurs sparingly in various parts of North America and Europe and is of little economic importance. The angular leaf spots occur on the leaves, stipules, petioles, and stems of red clover (Jones *et al.*, 1932). The spots are dark brown with exudate on limited portions of the lesions. Infections occur during cool, wet conditions at any time during the growing season.

5. Powdery Mildew, *Erysiphe polygoni* DC. Powdery mildew is distributed widely on red clover in the temperate zones of the world. Prior to 1921, the disease was uncommon in the Central and Eastern United States; yet within approximately three years after this date, the malady was prevalent throughout all North America. Heavy epidemics cause some reduction in yield and quality of the forage and hay (Horsfall, 1930 and Yarwood, 1934).

The light-gray superficial powdery growth of mycelium and conidia is conspicuous on the leaves. Leaves yellow and turn brown when the infection is severe. The mycelium develops on the surface of the epidermal cells of the leaves with haustoria within the cells.

THE FUNGUS

Erysiphe polygoni DC.

Salmon (1900) listed the extensive synonymy for this fungus. Peterson (1938) described the cleistothecia of *Erysiphe polygoni* and *Microsphaera alni* (DC.) Wint. on

Trifolium pratense in the Eastern United States. Cooke (1952) revised the nomenclature of this genus, *Erysiphe polygoni* DC. ex Méret was suggested and *E. communis* Wallr. ex Fries for *E. cichoracearum* DC. in the other species common on field crops.

Cleistothecia are formed commonly in western North America and northern Europe and uncommonly in central and eastern North America. Cleistothecia are scattered with variable simple appendages. The asci are usually ovate with four or eight spores. The ascospores are hyaline, oblong with rounded ends, and are one-celled. The ovate hyaline to gray conidia are borne in chains from an apical generative cell of a short simple conidiophore. Production of conidia is on a diurnal cycle (Yardwood, 1936).

ETIOLOGY. The obligate parasite persists on the foliage in the moderate climates and apparently spreads rapidly during late summer and autumn. Epiphytotics of the disease are favored by relatively dry summer weather. Frequent rains reduce the development and spread of the disease (Yarwood, 1936). The source of inoculum in areas where epiphytotics occur is largely from conidia.

Control of the disease is chiefly by the use of resistant varieties. Jones *et al.* (1941) and Smith (1938) describe the range of resistance and reaction of resistant varieties. Resistant plants are found in most varieties. The European red clover varieties are generally more resistant than the American (Horsfall, 1930, Mains, 1923, 1928).

Specialization of the parasite occurs on the clover species and probably within the species, according to Hammarlund (1925), Neger (1902, 1923), Salmon (1900, 1903), and Yarwood (1936).

6. Spring Black Stem, *Phoma trifolii* E. M. Johnson & Valleau. (See Alfalfa.)

7. Sooty Blotch, *Cymadothea trifolii* (Fr.) Wolf. The leaf blotch is distributed throughout the temperate zones on alsike and white clover and less commonly on the red and crimson, according to Elliott and Stansfield (1924) and Horsfall (1930). The dark-brown or black angular blotches, more abundant on the lower surface, usually cause only minor defoliation. On the lower surface of the leaf the slightly elevated spots are covered with conidiophores bearing conidia of the *Polythrincium* stage. Later in the season the blotches appear black and warty because of the formation of the spermagonia and the perithecial stroma (Fig. 83). The leaves yellow, wither, and brown when the blotches are numerous.

THE FUNGUS

Cymadothea trifolii (Fr.) Wolf
[*Dothidella trifolii* (Fr.) Bayl.-Elliott & Stansf.]
(*Polythrincium trifolii* Schm. & Kunze)
(*Sphaeria trifolii* Pers.)
[*Placosphaeria trifolii* (Pers.) Fcke.]
[*Plowrightia trifolii* (Pers.) Kill.]

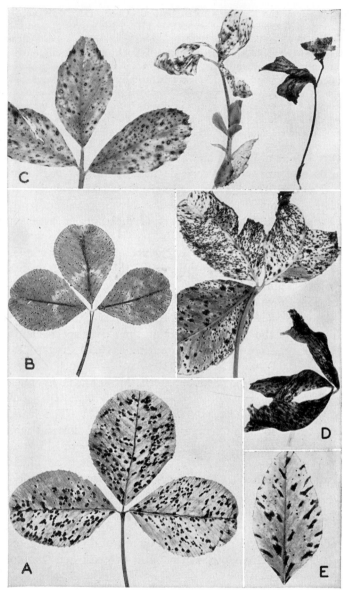

FIG. 83. Leaf spots of alfalfa and clover. (A) Sooty blotch incited by *Cymadothea trifolii* and (B) pepper spot caused by *Pseudoplea trifolii* on white clover and (C) on alfalfa. (D) Later stages of sooty blotch and pepper spot on leaves of white clover. (E) *Cercospora* leaf spot of clover.

Sampson and Western (1941) prefer *Dothidella trifolii;* American usage follows Wolf (1935), *i.e.; Cymadothea trifolii.*

The conidial stage develops during the spring and early summer. Conidiophores are borne on modified cells of the mycelium that form stromata on the undersurface of the leaf. The conidiophores are black, waxy, wavy, simple, and bear conidia followed by apical elongation around the attached conidia. Conidia are pale olivaceous, obovate, 1-septate, and slightly constricted at the septum. The spermagonial stage is formed during the autumn along with the perithecial bearing stroma. Spermagonia are globose to flask-shaped with abundant spermatia extruded in an exudate. Perithecia are embedded in the stroma and are irregular in shape and opening. The asci are clavate with usually eight ascospores. Ascospores are oblong to ovate-oblong, hyaline, 1-septate, and constricted at the septum. Wolfe (1935) reported on the cytology of the ascigerous stage and reviewed the literature on the morphology and etiology of the fungus. The fungus has not been grown on artificial media.

ETIOLOGY. Primary infection occurs from conidia and ascospores. The ascospores develop in early spring, and they are responsible for the early infection. Conidial production is abundant during the summer, and it declines in the autumn as spermagonia and stroma develop. The disease is prevalent on low wet meadowland. Although red and crimson clover are infected in Europe (Killian, 1923), they are generally resistant in the United States.

8. **Pepper Spot or Pseudoplea Leaf Spot,** *Pseudoplea trifolii* (Rostr.) Petr. This leaf spot is distributed generally in the humid, temperate climates on the clovers and alfalfa. Frequently the disease is severe enough on white clover to cause foliage browning. Apparently, this leaf spot occurs only on white clover in Britain (Sampson and Western, 1941).

SYMPTOMS. Numerous small (pin-point), sunken, black spots occur on the leaves and petioles of white clover, including Ladino. These spots rarely enlarge on white clovers. Later brown necrotic areas start from the leaflet margin, progress rapidly, and result finally in curled brown leaflets persisting on the erect petioles. In alfalfa the small sunken spots are brown. On petioles and annual stems they retain the original size. On leaves the spots frequently enlarge to twice the original diameter and appear as brown rings with lighter center. As in clover, general necrosis of leaflets and young stems develops later, resulting in the curled light-brown attached leaflets and, when severe, numerous dead stem apexes (Fig. 83). Perithecia develop later, embedded in the dead leaflet tissues. Quite generally in both clover and alfalfa, *Stemphylium botryosum* Wallr. develops in the tissues during necrosis and results in darker-brown to black areas as conidiophores and conidia develop. Nelson and Kernkamp (1953) and Nelson (1955) associate this latter pathogen with *Pseudoplea* and incorrectly describe it as the conidial stage of *Pseudoplea* on alfalfa.

The Fungus

Pseudoplea trifolii (Rostr.) Petr.
[*Saccothecium trifolii* (Rostr.) Kirsch.]
(*Sphaerulina trifolii* Rostr.)
(*Pleosphaerulina briosiana* Poll.)
(*Pseudoplea briosiana* von Hohn)
(*Pseudoplea medicaginis* Miles)

Sampson and Western (1941) suggested *Saccothecium trifolii* (Rostr.) Kirsch. as the preferable binomial.

The perithecia are flattened to round, embedded in the tissues, with pronounced ostiole opening to the surface. Paraphyses are absent, and asci are few to a perithecium. The asci are broadly ovate with a thickened apex and contain four or eight closely packed spores. The ascospores are cylindrical, tapering at the ends, hyaline to light yellow, triseptate, and muriform late in their development.

ETIOLOGY. The perithecia develop in the late autumn or in the following spring. Ascospores are the source of inoculum during the spring and summer. The disease develops during cool, moist weather. No conidial stage is known (Hopkins, 1923, Horsfall, 1930, Jones, 1916, Miller, 1925).

9. Sclerotinia Root Rot and Crown Rot, *Sclerotinia trifoliorum* Eriks. The disease is distributed extensively, especially in the regions of mild winters or heavy snow cover. In the United States, the disease is of considerable economic importance in the southern clover belt. Damage is severe in the early spring when the plants are susceptible to the invasion of this and similar fungi, according to Loveless (1951), Kreitlow and Sprague (1951), Pape (1937), and Valleau *et al.* (1933). The host range includes a large group of legumes, although the disease is of major importance on the clovers (Kreitlow, 1949, Loveless, 1951, Purdy, 1955).

SYMPTOMS AND EFFECTS. The symptoms of the disease vary with season, weather conditions, and tissues invaded. Brown leaf spots occur in wet periods, especially late autumn. The heavily infected leaflets drop off and are overrun by the mass of white mycelium. Infection spreads downward into the roots and crown. The spring symptoms are largely a soft rot of the crown and roots and less frequently the basal portion of the annual stems. The new growth of the infected plants wilts, and the dead tissues are overgrown by the white mycelium under conditions of high moisture. The sclerotia develop following the mycelium and are numerous in the soil and on the dead tissues.

The Fungi

Sclerotinia trifoliorum Eriks. or
Sclerotinia sclerotiorum (Lib.) DBy.

Purdy (1955) suggests the combination of the several similar species and the use of varieties to designate the sclerotium size. Synonyms for the combined species are given.

The sclerotia are black and vary in size and shape, ranging from smaller than clover seed to several times their diameter. Sclerotia germinate, forming disk-shaped buff to pink apothecia 3 to 8 mm. in diameter, borne on slender stalks. The hymenium is composed of closely packed eight-spored asci interspersed with slender simple paraphyses. The ascospores are usually arranged in one row in the narrow cylindrical ascus. The spores are oblong with rounded ends to elliptical and are one-celled. Spherical microconidia are formed in dense clusters on the aerial mycelium under some conditions.

ETIOLOGY. The apothecia emerge from the soil in autumn, and the ascospores are discharged with force enough to carry them into the foliage above. Leaf and stem infections result under conditions of high moisture, and mycelium also develops on the dead leaves on the soil surface. Infection of the crown and root tissues is initiated in the autumn and spreads during the winter and early spring in mild climates or under heavy snow covering (Eriksson, 1880, Esmarch, 1925, Gussow, 1903, Pape, 1937, Wadham, 1925). In the United States clover belt, the aggressive development of the fungus occurs in the spring as the plants are recovering from winter injury and resuming vegetative activity after the winter dormant period (Kreitlow and Sprague, 1951, Valleau *et al.*, 1933, and Wolf and Cromwell, 1919). The sclerotia develop during the summer and remain viable for long periods. In Europe, considerable attention is given to the presence of sclerotia in seed lots. Doyer (1934) described the sclerotia of three fungi occurring in clover seed lots, *Sclerotinia trifoliorum*, *Typhula trifolii* Rostr., and *Mitrula sclerotiorum* Rostr. Adapted varieties show some resistance to the disease.

Other sclerotium-forming fungi cause disease in the clovers and other legumes. *Sclerotium bataticola* Taub. or *Macrophomina phaseoli* (Maubl.) Ashby is common as a root rot on red clover in the South Central United States, according to Henson and Valleau (1937). *S. rolfsii* Sacc. or *Pellicularia rolfsii* (Curzi) West causes a root and crown rot in the warmer areas of the United States. These fungi are associated with clover survival in pastures and meadows, especially in the Southern United States. Resistant strains of red and white clover are the best means of control (Albrecht, 1942). The clovers are relatively susceptible to the root rot caused by *Phymatotrichum omnivorum* (Shear) Dugg. (Taubenhaus and Ezekiel, 1936).

10. Stemphylium Leaf Spot, *Pleospora herbarum* (Fr.) Rab. or *Stemphylium botryosum* Wallr. and *S. sarcinaeforme* (Cav.) Wiltshire. These leaf, stem, and pod spots are common on especially the clovers and alfalfa in humid climates throughout the world. Frequently in dense

stands, leaf and stem lesions develop aggressively. *S. botryosum* has a
wide host range, although a race appears to be confined to alfalfa chiefly
in the legumes.   *S. sarcinaeforme* apparently has a more restricted host

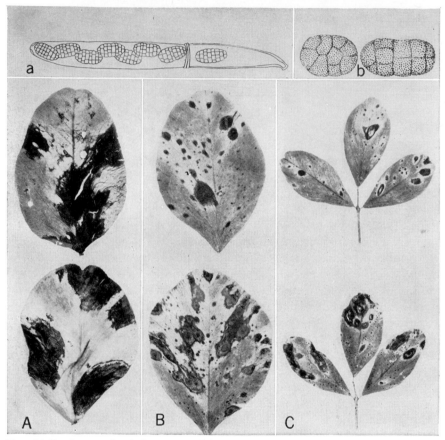

FIG. 84. Leaf spots of red clover caused by *Pleospora herbarum* or *Stemphylium botryosum* (*A*) and *S. sarcinaeforme* (*B*) and of alfalfa caused by *S. botryosum* (*C*).   Drawings of the ascus and ascospores (*a*) and conidia (*b*) of the former species are shown. (*Courtesy of O. F. Smith.*)

range and occurs on the clovers and trefoils (Smith, 1940, Wiltshire,
1938, Graham, 1953).

Graham (1953) described the pathogen on *Lotus corniculata* L. as *S.
loti* Graham.   The morphology of the smooth-spored fungus on clovers
and trefoils varies over a similar range, and more likely they are specialized varieties of the same species.

SYMPTOMS.   The lesions first are small, irregular dark-brown sunken
spots; later development results in irregular concentric zonated light-
and dark-brown lesions (Fig. 84).   The wrinkled dark-brown to sooty

leaves in the final stages of leaf blight remain attached to the plant. Stem, petiole, and pod lesions are restricted sunken brown spots.

THE FUNGI

1. *Pleospora herbarum* (Fr.) Rab. or
Stemphylium botryosum Wallr.
(*Macrosporium medicaginis* Cugini)

The perithecia are globose, membranous, and black. Asci are oblong to clavate with outer and inner walls. Ascospores are elongate to ovate, characteristically 7-septate, and muriform when mature. The conidiophores are short, arise singly or in groups, are occasionally branched, septate, and swollen at the apex. The conidia are ovate to elongate, 3- to 4-septate, somewhat constricted at the septa, muriform, and olivaceous, and echinulations to fine warts are numerous over the surface. Specialized races of this fungus occur on the clovers and alfalfa, according to Smith (1940).

2. *Stemphylium sarcinaeforme* (Cav.) Wiltshire
(*Macrosporium sarcinaeforme* Cav.)

The conidiophores and conidia are similar to the former species, except that the spores are smooth. This species apparently occurs on red clover only. Smith (1940) and Wiltshire (1938) reviewed the taxonomy and morphology of the several species on the legumes.

11. Southern Anthracnose, *Colletotrichum trifolii* Bain & Essary. This anthracnose occurs in the warmer sections of North America, Europe, Kenya, and South Africa on the clovers and alfalfa. The disease is frequently severe on red clover in the southern clover belt of the United States, especially in unadapted susceptible varieties, and it occurs in mild form in mid-summer as far north as Canada, according to Monteith (1928) and Neal (1924).

SYMPTOMS AND EFFECTS. The light- to dark-brown lesions on the stems and petioles are the characteristic symptoms. The girdling of the stems results in the killing and browning of the foliage. The presence of the acervuli and numerous dark-brown setae on the mature sunken lesions differentiates this malady from the northern anthracnose and other stem diseases (Fig. 85).

THE FUNGUS

Colletotrichum trifolii Bain & Essary

Acervuli develop in scattered erumpent form and vary considerably in diameter. Setae are few to numerous, uniseptate or continuous, dark brown to black, and much longer than the conidia. The conidiophores are cylindrical, varying in length with moisture, hyaline, and produce a succession of conidia apically. Conidia are hyaline to pink, straight, rounded at the ends, and without septations.

ETIOLOGY. The fungus persists on the stems, crowns, and roots and spreads rapidly in warm, humid weather. The lesioning and killing of stems occurs largely on the new growth of the second crop. Resistant varieties apparently keep the disease under control in the southern clover

Fig. 85. Red clover attacked by *Kabatiella caulivora;* stem lesions and petiole necrosis followed by death and browning of the leaves. The symptoms are relatively similar for both the northern and southern anthracnose.

belt of the United States, according to Bain and Essary (1906), Monteith (1928), Pieters and Monteith (1926), and Wolf and Kipps (1926). Sampson (1928) studied comparatively the fungi causing the southern and northern anthracnoses.

Two other species, *Colletotrichum destructivum* O'Gara and *C. graminicolum* (Ces.) G. W. Wils., are reported on clovers. The former species occasionally causes some damage. These several species are similar morphologically.

12. Northern Anthracnose, *Kabatiella caulivora* (Kirch.) Karak. The northern anthracnose is restricted largely to clovers, especially red and crimson, in the cooler sections of North America, Europe, and Asia. Damage is severe during periods of cool, wet weather, according to Sampson (1928).

SYMPTOMS AND EFFECTS. The brown lesions develop as sunken linear areas on the stems and petioles. Necrosis and cracking of the stem tissues are pronounced. Girdling of the stems and browning of the foliage symptoms are similar to the southern anthracnose. The white masses of conidiophores and conidia borne in irregular acervuli without setae are conspicuous in the deeper stem depressions and cracks (Fig. 85).

THE FUNGUS

Kabatiella caulivora (Kirch.) Karak.
(*Gloeosporium caulivorum* Kirch.)

The acervuli are small, irregular, hyaline, and without setate. The clusters of short cylindrical, hyaline conidiophores bear one or more hyaline, curved, nonseptate conidia.

The etiology and control of the disease are similar to the former anthracnose. The fungus develops at low temperatures and invades the tissues during cool, humid conditions. Varieties resistant to the southern anthracnose are frequently susceptible to the northern anthracnose.

13. Black Patch Disease, Unidentified Fungus. The disease apparently occurs in scattered locations in the clover areas of the United States. The disease is more prevalent in the southern area where it occurs on red and white clovers, soybean, cowpea, kudzu, and blue lupine. Other legumes are infected by inoculation.

Symptoms frequently appear on leaves, stems, flowers, and seeds. Leaf symptoms are similar to those incited by *Stemphylium* spp. Black mycelial growth occurs on the stems and petioles, frequently invading and girdling the stems. Dark lesions and dark coarse mycelium are evident on the flowers and seeds. Seed transmission and blighted or lesioned seedlings overgrown by the black mycelium are characteristic. The fungus mycelium apparently persists in plants in the field and on as well as in the seed. Sporulation of the fungus has not been observed in nature or induced in culture (Leach and Elliott, 1951).

14. Curvularia Leaf Spot, *Curvularia trifolii* (Kauf.) Boed. This disease occurs as a leaf spot on white clover as the common host. It is world-wide in distribution and generally of minor importance. Angular yellow and brown lesions, water-soaked, and translucent develop on leaflets and frequently advance into the petiole, resulting in dead, brown leaves. The disease develops during warm moist weather, and sporulation of the fungus is especially evident under this environment.

Curvularia trifolii (Kauf.) Boed. (*Brachysporium trifolii* Kauf.).

The fungus produces grayish-brown, simple, septate conidiophores bearing clusters of usually three-celled, fusiform, brown spores. The conidia are typically curved with the central cell enlarged. The classification and morphology of this genus are uncertain (Bonar, 1924, Hanson and Kreitlow, 1953).

15. Rusts, *Uromyces trifolii* (Hedw. f.) Lév. The clover rusts are distributed widely in the humid and semihumid areas. Usually these rusts cause little damage, although susceptible lines are damaged, especially in the late autumn.

The varieties of *Uromyces trifolii* are all autoecious long-cycled rusts. The aecia occur as swollen light-yellow to orange-yellow sori on the stems, petioles, and leaves. The aecial stage is more common on the white and alsike clovers than on the red clover. The uredia appear as small brown pustules on any portion of the green plant. The telia are similar in appearance to the uredia and occur in the old uredia or independently. The uredia and telia are not associated with hypertrophy of the tissues.

THE FUNGI

1. *Uromyces trifolii* (Hedw. f.) Lév.

Arthur (1934) combined the morphologically similar species on the clovers and indicated their specialization by the use of the trinomials *U. trifolii trifolii-repentis* (Liro) Arth. on *Trifolium incarnatum* L. and *T. repens* L, *U. trifolii hybridi* (W. H. Davis) Arth. on *T. hybridum* L., and *U. trifolii fallens* (Desm.) Arth. on *T. incarnatum* L., *T. medium* L., and *T. pratense* L. The synonymy is included by Arthur (1934).

The pycnia are in groups, chiefly on the leaves. Aecia with cup-shaped peridia are in groups on any tissues. The aeciospores are globoid, hyaline, and finely verrucose. Uredia are generally on the undersurface of the leaves and light brown. Urediospores are globoid or broadly ellipsoid, have a light-brown wall, are thick, echinulate, and have two to six pores scattered or equatorial. Telia are in the uredia or are independent, brown, and not covered by the epidermis. Teliospores are one-celled, globoid or broadly ellipsoid, brown, and smooth or have a few scattered warts and short fragile pedicles.

2. *Uromyces nerviphilus* (Grog.) Hotson (*U. flectens* Lagerh.)

This species is similar in morphology to the former and occurs on the veins and petioles of *Trifolium repens* L. and possibly *T. hybridum* L. Frequently it is in association with the former species. The uredial stage of this questionable species is omitted in the cycle.

Other diseases occurring on the clovers are discussed in Chap. 13. Table 14 gives a summary of the more important diseases of alfalfa, sweetclover, and the clovers and the interrelationship of suscepts and parasites.

TABLE 14. DISEASES OF ALFALFA, THE CLOVERS, AND SWEETCLOVER, WITH THE CAUSAL AGENT LISTED UNDER THE MORE COMMON SUSCEPT

Disease	Alfalfa	Medium red clover	White, alsike, and other perennial clovers	Sweetclover
Mosaic	Several viruses	Several viruses	Several viruses	Several viruses
Dwarf	Virus	Virus rare	Virus rare	Virus
Witches' broom	Virus	Virus	Same as alfalfa	Virus
Bacterial wilt	Corynebacterium insidiosum	None	None	None
Bacterial blight	Pseudomonas medicaginis	Pseudomonas syringae	None	None
Crown wart	Urophlyctis alfalfae	Urophlyctis trifolii (Europe)	Urophlyctis trifolii	None
Leaf curl	None	Olpidium trifolii (Europe)	Olpidium trifolii	None
Pythium blight	Pythium debaryanum and other species	Same	Same	Same
Phytophthora root rot	Phytophthora cryptogea	None	None	Phytophthora cactorum
Downy mildew	Peronospora trifoliorum	Rare	Rare	Rare
Powdery mildew	Rare	Erysiphe polygoni	Rare	Rare
Spring black stem	Ascochyta imperfecta	Phoma trifolii	Less common	Mycosphaerella lethalis
Summer black stem	Cercospora zebrina	Same	Same	Mycosphaerella davisii
Stem hypertrophy	None	None	None	Ascochyta caulicola
Leaf spot and root rot	Leptosphaeria pratensis	None	Same	Same
Zonate leaf spot	Pleospora herbarum	Stemphylium sarcinaeforme	Same	Uncommon
Sooty blotch	None	Uncommon	Cymadothea trifolii	None
Pseudoplea leaf spot	Same	Same	Pseudoplea trifolii	Uncommon
Common leaf spot	Pseudopeziza medicaginis	Pseudopeziza trifolii	Same	Uncommon
Yellow leaf spot	Pseudopeziza jonesii	None	None	None
Sclerotinia root rot	Same	Sclerotinia trifoliorum	Same	Same
Fusarium, Cylindrocarpon root rots	Fusarium and Cylindrocarpon	Same	Same	Same
Southern anthracnose	Colletotrichum destructivum	Colletotrichum trifolii	Same	Rare
Northern anthracnose	Rare	Kabatiella caulivora	Uncommon	None
Blackpatch	Rare	Fungus unnamed	Same	Rare
Curvularia leaf spot	Rare	Rare	Curvularia trifolii	Rare
Rhizoctonia root rot	Rhizoctonia spp.	Same	Same	Rare
Smuts	Same	Same	Same	Thecophora deformans and Entyloma meliloti
Rusts	Uromyces striatus	Uromyces trifolii	Same + Uromyces nerviphilus	Rare

REFERENCES

ALBRECHT, H. R.   Effect of diseases upon survival of white clover, *Trifolium repens* L., in Alabama.   *Jour. Am. Soc. Agron.* **34**:725–730.   1942.

ARTHUR, J. C.   Manual of the rusts in United States and Canada.   Purdue Research Foundation. LaFayette, Ind.   1934.

BAIN, S. M., and S. H. ESSARY.   Selection for disease resistant clover: a preliminary report.   *Tenn. Agr. Exp. Sta. Bul.* 75.   1906.

BONAR, L.   Studies on the biology of *Brachysporium trifolii.*   *Am. Jour. Bot.* **11**:123–158.   1924.

CHILTON, S. J. P., *et al.*   Fungi reported on species of Medicago, Melilotus, and Trifolium.   *U.S. Dept. Agr. Misc. Pub.* 499.   1943.   (Good reference list.)

COOKE, W. B.   Nomenclatural notes on the Erysiphaceae.   *Mycologia* **44**:570–574. 1952.

DICKSON, B. T.   Studies concerning mosaic diseases.   *Macdonald Col. Tech. Bul.* 2. 1922.

DOYER, L. C.   De gezondheidstoestand van Klaverzaad, in verband met de keuring van det zaad en de invloed van ontsmetting op dezen toestand.   *Tijdschr. Plantenziekt.* **40**:54–61.   1934.

ELLIOTT, J. S. BAYLIS, and O. P. STANSFIELD.   The life history of *Polythrincium trifolii* Kuntze.   *Trans. Brit. Myc. Soc.* **9**:218–228.   1924.

ERIKSSON, J.   Om klofverrotan med sarskildt afseude pi dess upptrdande i vart landunder aren 1878–79.   *Kgl. Landtbruks-Akad. Handl. Tid.* **19**:28–42.   1880.

ESMARCH, F.   Das Auswintern des Klees durch Kleekrebs.   *Kranke Pflanze* **2**:3–6. 1925.

GRAHAM, J. H.   A disease of birdfoot trefoil caused by a new species of *Stemphylium.* *Phytopath.*   **43**:577–579.   1953.

HAGEDORN, D. J., and E. W. HANSON.   A comparative study of the viruses causing Wisconsin pea stunt and clover vein mosaic.   *Phytopath.* **41**:813–819.   1951.

HAMMARLUND, C.   Zur Genetik, Biologie, und Physiologie einiger Erysiphaceen. *Heriditas* **6**:1–126.   1925.

HANSON, E. W., and K. W. KREITLOW.   The many ailments of clover.   *U.S. Dept. Agr. Yearbook*, pp. 217–228.   1953.

HENSON, L., and W. D. VALLEAU.   *Sclerotium bataticola* Taubenhaus, a common pathogene on red clover roots in Kentucky.   *Phytopath.* **27**:913–918.   1937.

HOPKINS, E. F.   The *Sphaerulina* leaf-spot of clover.   *Phytopath.* **13**:117–126.   1923.

HORSFALL, J. G.   A study of meadow-crop diseases in New York.   *N.Y. (Cornell) Agr. Exp. Sta. Mem.* 130.   1930.   (Good literature list.)

HOUSTON, B. R., and J. W. OSWALD.   The mosaic virus disease complex of ladino clover.   *Phytopath.* **43**:271–276.   1953.

JOHNSON, E. M.   A ringspot-like virus disease of red clover.   *Phytopath.* **23**:746–747. 1933.

JOHNSON, F.   The complex nature of white-clover mosaic.   *Phytopath.* **32**:103–116. 1942.

JONES, F. R.   *Pleosphaerulina* on alfalfa.   *Phytopath.* **6**:299–300.   1916.

—— *et al.*   Evidence of resistance in alfalfa, red clover, and sweetclover to certain fungus parasites.   *Phytopath.* **31**:765–766.   1941.

JONES, L. R., *et al.*   Bacterial leaf spot of clovers.   *Jour. Agr. Research* **25**:471–490. 1923.

KILLIAN, C.   Le *Polythrincium trifolii* Kunze parasite du tréfle.   *Rev. path. veg. et d'ent. agr. France* **10**:202–219.   1923.

KREITLOW, K. W. *Sclerotinia trifoliorum*, a pathogen of ladino clover. *Phytopath.* **39**:158–166. 1949.

―――― and V. G. SPRAGUE. Effect of temperature on growth and pathogenicity of *Sclerotinia trifoliorum*. *Phytopath.* **41**:752–757. 1951.

LEACH, J. G., and E. S. ELLIOTT. The blackpatch disease of red clover and other legumes in West Virginia. *Phytopath.* **41**:1041–1049. 1951.

LOVELESS, A. R. The confirmation of the variety *fabae* Keay of *Sclerotinia trifoliorum* Eriksson. *Ann. Appl. Biol.* **38**:252–275. 1951.

――――. Observations on the clover rot. *Ann. Appl. Biol.* **38**:642–664. 1951.

MAINS, E. B. Differences in the susceptibility of clover to powdery mildew. *Ind. Acad. Sci. Proc.* **1922**:307–313. 1923.

―――― Observations concerning clover diseases. *Ind. Acad. Sci. Proc.* **37**:355–364. 1928.

MILLER, J. H. Preliminary studies on *Ploesphaerulina briosiana*. *Am. Jour. Bot.* **12**:224–237. 1925.

MONTEITH, J. Clover anthracnose caused by *Colletotrichum trifolii*. *U.S. Dept. Agr. Tech. Bul.* 28. 1928.

NEAL, D. C. A pathological and physiological study of the anthracnose fungus (*Colletotrichum trifolii* Bain). *Miss. Agr. Exp. Sta. Rept.* 37. 1924.

NEGER, F. W. Beiträge zur Biologie der Erysipheen. II. Die Keimungserscheinungen der Conidien. *Flora* **90**:221–272. 1902.

――――. Beiträge zur Biologie der Erysipheen. III. Der Parasitismus der Mehltaupilze-eine Art von geduliter Symbisose. *Flora* **116**:331–335. 1923.

OSBORNE, H. T. Vein-mosaic of red clover. *Phytopath.* **27**:1051–1058. 1937.

PAPE, H. Beiträge zur Biologie und Bekampfung des Kleekrebses (*Sclerotinia trifoliorum* Erikss.). *Biol. Reichsanst. Land. Forstw. Arb.* **22**:159–247. 1937.

PETERSON, G. A. Perithecial material of *Erysiphe* and *Microsphaera* on *Trifolium pratense*. *Mycologia* **30**:299–301 1938.

PIETERS, A. J., and E. A. Hollowell. Clover improvement. *U.S. Dept. Agr. Yearbook*, pp. 1190–1214. 1937.

―――― and J. MONTEITH. Anthracnose as a cause of red clover failure in the southern part of the clover belt. *U.S. Dept. Agr. Farmers' Bul.* 1510. 1926.

PURDY, L. H. A broader concept of the species *Sclerotinia sclerotiorum* based on variability. *Phytopath.* **45**:421–427. 1955.

SALMON, E. S. A monograph on the Erysiphaceae. *Mem. Torrey Bot. Club* **9**:1–292. 1900.

――――. On specialization of parasitism in the Erysiphaceae. *Bot. Centbl., Beihefte* **14**:261–315. 1930.

SAMPSON, K. Comparative studies of *Kabatiella caulivora* (Kirchn.) Karak. and *Colletotrichum trifolii* Bain and Essary. *Trans. Brit. Myc. Soc.* **13**:103–142. 1928.

―――― and J. H. WESTERN. Diseases of British grasses and herbage legumes. Cambridge University Press. London. 1941.

SMITH, O. F. Host-parasite relations in red clover plants resistant and susceptible to powdery mildew, *Erysiphe polygoni*. *Jour. Agr. Research* **57**:671–682. 1938.

――――. Stemphylium leaf spot of red clover and alfalfa. *Jour. Agr. Research* **61**:831–846. 1940.

TAUBENHAUS, J. J., and W. N. EZEKIEL. A rating of plants with reference to their relative resistance or susceptibility to Phymatotrichum root rot. *Tex. Agr. Exp. Sta. Bul.* 527. 1936.

VALLEAU, W. D., *et al.*   Resistance of red clovers to *Sclerotinia trifoliorum* Eriks. and infection studies.   *Ky. Agr. Exp. Sta. Bul.* 341.   1933.

WADHAM, S. M.   Observations on clover rot (*Sclerotinia trifoliorum* Eriks.) *New Phytol.* **24**:50–56.   1925.

WILTSHIRE, S. P.   The original and modern conceptions of Stemphylium.   *Trans. Brit. Myc. Soc.* **21**:211–239.   1938.

CHAPTER 15

SOYBEAN DISEASES

The soybean, *Soja max* (L.) Piper [*Glycine max* (L.) Merrill], ranks high as an industrial and feed crop in the United States, especially during recent years.   It is an ancient crop of eastern Asia.   While the plant is largely self-pollinated, the species is represented by diverse types.   Cold tolerance, date of maturity, disease reaction, as well as morphological characters, vary considerably in the collections introduced and in the varieties developed in the United States (Morse and Cartter, 1937, Weiss, 1949).   The cultivated species as well as the wild species of Asia, *G. ussuriensis* Regel & Maack, have 20 chromosome pairs.

Compared with other field crops, damage from diseases at present is comparatively light.   However, the relatively large number of diseases occurring on the crop, the recent intensive breeding program in the United States, and the extensive acreage devoted to the crop, all combine to increase losses from diseases unless special attention is given to the selection of disease-resistant varieties.

1. Leaf Spot and Leaf Discoloration, Nonparasitic.   Marginal leaf yellowing, firing, and browning are associated with mineral deficiencies, especially low potash, and other environmental disturbances (Kornfeld, 1933, Warington, 1954).   Leaf scald occurs under conditions of high temperature and low humidity.   Aphids and leaf hoppers cause damage when infestations are heavy (Gibson, 1922).   According to Johnson and Hollowell (1935), the glabrous varieties are susceptible to injury by the potato leaf hopper, whereas the pubescent types that are grown commercially are relatively resistant.

2. Mosaics, Viruses.   The soybean mosaic is distributed widely on the crop and causes some damage in susceptible varieties.   The leaves are dwarfed, margins are curled downward, the surface is puckered with dark-green areas between the veins and sometimes chlorotic spotting. The petioles and internodes are shortened, especially with early infections.   The pods are stunted, flattened, and curved.   Seed setting is delayed and reduced greatly.   The virus apparently is limited to the soybean.   It is transmitted in seed from infected plants, by mechanical means, and by aphids.   *Myzus persicae* (Sulz.) and *M.* (*convolvuli*) *solani*

383

(Kalt.) are the more common vectors, although six other species are reported as vectors (Gardner and Kendrick, 1921, Heinze and Kohler, 1940, Kendrick and Gardner, 1924). The yellow bean mosaic and several other legume viruses produce symptoms on the soybean as well as the tobacco ring spot virus that produces bud blight.

3. Bud Blight, Tobacco Ring Spot Virus. The disease apparently is increasing in prevalence in the soybean area in the Central United States.

Fig. 86. Bud blight of the soybean caused by the tobacco ring spot virus. (*A*) Twisting and blighting of stem apex (*B*) pod necrosis and browning.

The symptoms vary depending upon the age of the plant at the time infection occurs. Infections prior to the completion of terminal elongation of the stem result in unilateral elongation of the stem apex, a bronzed appearance of the young leaves, and ultimately necrosis and brittleness of the apical growing point (Fig. 86). Internal reddish-brown discoloration is evident, first in the pith near the nodes and later spreading through the stems and accompanied by necrotic areas. The plants do not produce seed, and they remain green until frosted. Infections with the virus near the flowering period, which is generally more common, result in withering and dropping of young pod clusters and dark-brown blotches on the remaining poorly developed pods (Allington, 1946).

The tobacco ring spot virus incites bud blight of soybean. The virus is transferred by contact, it is seed-borne, and insect transmission is indicated by the behavior of the disease in the field (Desjardines *et al.*, 1954). Sweetclover and some other legumes are the common hosts of the virus. Soybeans growing near these legumes frequently are damaged severely. Control of bud blight by avoiding association with these hosts appears advisable. Seed from disease-free fields should be used where possible. Resistant varieties are not reported.

4. Bacterial Blight, *Pseudomonas glycinea* Coerper (*Bacterium glycineum* Coerper), (*B. sojae* Wolf). The bacterial blight is distributed widely with the crop. The disease appears on the leaves, stems, and pods as light- to dark-brown irregular spots without a marginal halo. The lesions become confluent as they enlarge, the necrotic tissues dry out, and portions drop out. A limited amount of gray or brown exudate accumulates on the surface of the lesion (Coerper, 1919). Wolf (1920, 1921) retains the two species *Pseudomonas sojae* Wolf causing the southern blight and *P. glycinea* Coerper, the northern. As in the halo blight of oats and in wildfire of tobacco, varieties of the pathogens incite the halo (Chamberlain, 1952).

The bacteria are seed-borne and persist in crop residue (Coerper, 1919, Kendrick and Gardner, 1921). Lesions on the cotyledons of infected seed and crop residue are a probable source of primary infection. Secondary spread occurs under favorable environmental conditions throughout the growing season. Defoliation and injury to the pods are minor under most conditions. Varieties vary widely in reaction to the disease.

5. Wildfire, *Pseudomonas tabaci* (Wolf & Foster) Stevens [*Phytomonas tobacae* (Wolf & Foster) Bergey *et al.*], (*Bacterium tobacum* Wolf & Foster). The disease occurs on soybeans as well as on tobacco in most of the intensive producing areas of North America. The lesions on the leaves are variable in size, and they are surrounded by a wide yellow halo under most environments. The restricted type of spot with indistinct halo is usually dark brown in contrast to the light-brown, more extensive type of lesion. Exudate does not occur on the surface. The wildfire frequently is associated with bacterial blight and bacterial pustule. Entrance of the pathogen is facilitated, and there is increasing evidence that the production of the halo-producing toxin is not restricted to the one species (Johnson and Chamberlain, 1953). When beating rains are frequent, the lesions spread rapidly and coalesce to involve the entire leaflets. Defoliation occurs around the base of the plant under such conditions. The infection usually occurs in definite areas rather than spread uniformly over the field, and it occurs relatively late in the season. The organism persists on crop refuse and in the soil (Allington, 1945, Graham, 1953).

6. Bacterial Pustule, *Xanthomonas phaseoli* var. *sojense* (Hedges) Starr & Burk.   The disease is more common in the southern soybean area, although it occurs in Northern sections of the United States.   Apparently the disease occurs in the Asiatic soybean areas.   The lesions are common on the leaves as yellow pustular outgrowths on both surfaces. The spots later change to reddish brown with marginal yellowing under some conditions.   The bacteria enter the stomatal cavity where they develop between the cells.   Cell enlargement occurs without water soaking of the tissues.   The slightly elevated pustule, usually without exudate, differentiates this disease from the bacterial blights.   Small reddish-brown spots occur on the pods of susceptible varieties.

THE BACTERIUM

Xanthomonas phaseoli var. *sojense* (Hedges) Starr & Burk.
[*Phytomonas phaseoli* var. *sojense* (Hedges) Burk.]
(*Bacterium phaseoli* var. *sojense* Hedges)
(*Pseudomonas glycines* Nak.)
[*Bacterium glycines* (Nak.) Elliott]
[*Phytomonas glycines* (Nak.) Mag.]

The rod-shaped organism with single polar flagellum, capsules, and no spores appears yellow in culture (Hedges, 1924).

The organism persists in crop residues, especially in the Southern areas. Secondary spread is rapid on susceptible varieties, especially during periods of beating rains.   Resistant varieties are reported (Lehman and Woodside, 1929, Wolf, 1924).   Elrod and Braun (1947) studied the serology of the pathogen.

7. Pythium Root Rot, *Pythium debaryanum* Hesse.   The light-brown soft rot of the roots and basal portions of the young stems is caused by the group of *Pythium* spp. common on legumes.   The disease described by Lehman and Wolf (1926) is general throughout the soybean areas of the world, although relatively less damage occurs on this crop than on the small-seeded legumes (Chap. 13).   Inheritance of resistance is complex (Hartig and Lehman, 1951).

8. Downy Mildew, *Peronospora manshurica* (Naum.) Syd.   The disease occurs rather generally with the culture of the soybean.   Damage is confined to the regions of relatively cooler climates and susceptible varieties, where both foliage and beans are infected.

SYMPTOMS.   The lesions on the leaves appear first as large or small chlorotic spots on the upper surface of the leaves.   The mature lesions are grayish to dark brown with the downy mass of gray conidiophores and gray to violaceous conidia on the lower surface of the lesions.   In susceptible varieties, the lesions spread rapidly over the leaf surface, causing yellowing and finally browning of the leaf.   Oospores are abun-

Fig. 87. The downy mildew of soybean caused by *Peronospora manshurica*. (*A*) Characteristic leaf lesions on susceptible varieties, (*B*) contrast between large and restricted leaf lesions on different varieties. Oospores from infected beans are shown in the insert.

dant in the brown leaf tissue. Pod infections occur at any of the nodes, but they are not evident from the exterior. In infected pods, the interior of the pod and the seed coat of the beans are incrusted by the grayish mass of mycelium and oospores (Fig. 87). The symptoms are similar to those described by Snyder (1934) for the downy mildew on the pea, *Pisum sativum* L.

The Fungus

Peronospora manshurica (Naum.) Syd.
(*P. trifoliorum* var. *manshurica* Naum.)
(*P. sojae* Lehman & Wolf)

The conidiophores are slender, branched, gray to pale violet, 5 to 9 microns wide by 240 to 500 long, and occur in groups from the stomata. The conidia are oblong to nearly round, pale-gray violet, and 18 to 21 microns wide by 24 to 27 long. The oospores are globose, smooth, hyaline to pale yellow, and 20 to 28 microns in diameter (Gäumann, 1923, Lehman and Wolf, 1924, Naoumoff, 1914).

ETIOLOGY. The oospores persist in crop residue and on the seed. The oospores on infected seed apparently produce both a local and a systemic infection in the young seedlings. Probably infection occurs in the hypocotyl from the oospores on the seed coat or in the crop residue. The mycelium develops through the hypocotyl into the first several trifoliate leaves and into some nodal buds (Jones and Torrie, 1946). Resistant varieties are the best means of control. Some of the high-yielding yellow selections adapted to the soybean belt of the United States are very susceptible. However, resistant varieties are available from hybrid selections adapted to the area. Resistance is conditioned by single factor pairs frequently occurring in combination in the hybrids studied (Geeseman, 1950). Four races of the pathogen are differentiated on the basis of the factors for resistance presently known (Geeseman, 1950, Lehman, 1953). Seed treatment is effective in reducing seedling infection (Sherwin *et al.*, 1948).

9. Powdery Mildew, *Erysiphe polygoni* DC. The powdery mildew occurs sparingly on the soybean in the Southern United States and in Europe (Lehman, 1931). Apparently this is a specialized variety of *Erysiphe polygoni* (see Chap. 14). A *Microsphaeria* sp. also occurs on soybean (Lehman, 1947).

10. Anthracnose, *Glomerella glycines* (Hori) Lehman & Wolf and *Colletotrichum truncatum* (Schw.) Andrus & Moore. The disease is prevalent in the warmer areas in Asia, Europe, and North America. In the United States it extends northward into the principal soybean belt. The lesions appear as dark-brown cankers on the cotyledons and hypocotyls of the seedlings. Seedling blight occurs before or after emergence, according to Ling (1940). The stem and pod lesions are indefinite brown areas causing premature death of the plants or plant parts. Seed infection is evident in some instances by the brown lesions. Numerous black setose acervuli develop on the infected parts during wet weather (Lehman and Wolf, 1926, Wolf and Lehman, 1926).

THE FUNGUS

1. *Glomerella glycines* (Hori) Lehman & Wolf
Colletotrichum glycines Hori

The perithecia, submerged in the mycelial tufts, are membraneous, globose, rostrate, and 220 to 340 microns in diameter. The asci are oblong to bluntly clavate, 9.5 to 13.5 microns wide by 70 to 106 long, and interspersed with oblong paraphyses. Ascospores are hyaline, unicellular, bluntly pointed, slightly curved, and 4 to 6 microns wide by 18 to 28 long. Ascervuli are black with numerous brown setae. Conidia are hyaline, unicellular, bluntly tapered, straight, 4 microns wide by 20 to 22 long, and remain viable for relatively short periods.

2. *Colletotrichum truncatum* (Schw.) Andrus & Moore

The acervuli are borne on clearly developed stromata, setae are numerous and large, 3–8 by 60–300 microns. Conidia are hyaline, one-celled, curved and 3–4 by 18–30 microns. This stem pathogen occurs in the warmer areas. Several other species occur on the soybean (Tiffany, 1951, Tiffany and Gillman, 1954).

ETIOLOGY. The mycelium persists for relatively long periods in the crop residue and infected beans. Seedling stands are reduced by seedborne mycelium and soil infestation on crop residue in the milder climates. Secondary stem and pod infections occur during wet weather. Ascosporic inoculum is probably of secondary importance in the areas where the disease is of importance, although it is a factor in the more northern areas. Apparently many of the commercial varieties are relatively resistant. Seed treatment with the organic mercury compounds reduces the loss in seedling stand.

11. Pod and Stem Blight, *Diaporthe phaseolorum* var. *sojae* (Lehm.) Wehm. The disease occurs in North America, Europe, and Asia. Damage usually occurs late as the plants are maturing, although premature killing of the plants and reduced yields of beans are common in the Central United States. The dark-brown lesions with indefinite margins occur on the stems and pods and less commonly on the leaves. The infection usually starts at the junction of stem and branch and girdles the stem or branch. Pycnidia occur, scattered or arranged in rows, in the older dead portions of the lesions (Johnson and Koehler, 1943, Lehman, 1923).

THE FUNGUS

Diaporthe phaseolorum var. *sojae* (Lehm.) Wehmeyer. (*Diaporthe sojae* Lehman)

The perithecia, formed in the black mycelial stroma, are spherical with a slender tapering beak. Asci are sessile, clavate, hyaline, with eight spores borne in one or two rows. The ascospores are hyaline, elongate elliptical, 1-septate, and 2 to 5 microns wide by 10 to 18 long. The *Phomopsis* stage consists of the subglobose osteolate pycnidia with short beaks or frequently without beaks. The conidiophores are slender, tapering, simple, and hyaline. The conidia are oblong to fusiform, straight, continuous, hyaline, and 2 to 3 microns wide by 6 to 7 long.

Diaphorthe phaseolorum var. *batatatis* (Hart. & Field) Wehmeyer, a closely related pathogen, causes a stem canker and premature killing of plants by stem girdling. Apparently pycnidia are absent and the ascospores are shorter, 3–5 by 8–10 microns, than in the former variety of the species (Welch and Gilman, 1948).

The fungi persist in crop residue as mycelium or pycnidia of the imperfect stage and perithecia. The mycelium is seed-borne when conditions are favorable for late pod infection.

12. Fusarium Wilt or Blight, *Fusarium oxysporum* f. *tracheiphilum* (E. F. Sm.) Snyder & Hansen. The *Fusarium* wilt of the cowpea and soybean occurs in the Southern soybean belt of the United States and extends into the Western plains area. The roots and base of the stem show browning as well as the browning or blackening of the vascular bundles. The fungus is a vascular parasite, but true wilting does not occur in the woody plants. The leaves yellow and drop, and the pods are poorly developed. Apparently specialized strains of the parasite occur on the soybean and cowpea (Cromwell, 1919). Resistant varieties are the best means of control, as in the cowpea (Johnson and Koehler, 1943, Orton, 1902). Liu (1940) reported a pod lesioning and seedling blighting caused by a fungus resembling (*Fusarium tracheiphilum* E. F. Sm.) *F. oxysporum* f. *tracheiphilum*. Other *Fusarium* spp. cause seedling blight and root rot. The morphology of the fungus is given in Chap. 18.

13. Cercospora Leaf Spot, *Cercospora sojina* Hara. The frog eyespot occurs in Asia, Europe, and North America. The disease is distributed in the southern section of the soybean belt of the United States, and it is prevalent on late maturing varieties. The typical zonate gray spots with purplish-brown margins are numerous on the leaves. The stem and pod spotting occur as the plants mature, especially on late varieties. The brown to gray spots on the pods are smaller and less zonate than on the leaves. The fungus invades the pods, and mycelium covers the beans, causing a gray to brown discoloration of the seed coat. Zonate spots varying in shades of brown and gray occur on some varieties. Frequently papillate or conical elevation of the testa and mycelium occur on the spots. The color and surface texture of the seed coat is distinct from the purple stain caused by *Cercospora kikuchii* (T. Matsu. & Tomoyasu) Gardner (Lehman, 1928, 1934, Sherwin and Kreitlow, 1952).

THE FUNGI. *Cercospora sojina* Hara (*Cercospora daizu* Miura)

The conidiophores, formed from thin stroma, are pale sooty black, nonseptate to rarely septate. The conidia are hyaline, cylindrical or fusiform, have a rounded apex, somewhat acute base, are nonseptate to as many as 6 septa, not constricted at the septa, and 5 to 7 microns wide by 39 to 70 long.

Another species, *Cercospora kikuchii* (T. Matsu. & Tomoyasu) Gardner first reported from eastern Asia (Matsumoto, 1928, Matsumoto and Tomoyasu, 1925 Murakishi, 1951) causes purple stain of beans.

The parasite persists in leaves and stems as well as in infected seed. Secondary spread from conidia occurs, especially late in the growing season. Early-maturing varieties escape severe injury, although they yield less than the later varieties when free from disease. Seed treatment apparently does not control the disease in the area where damage is severe.

14. Septoria Brown Spot, *Septoria glycines* Hemmi. The disease is distributed widely in North America, Europe, and Asia. The heavy infections are usually early and scattered, and defoliation occurs on the heavily infected leaves. Heavy early infection of the primary leaves is common; subsequent development of the leaf spotting is less common, except during wet seasons. The angular spots are brown turning to reddish brown as they become older. The spots range from small to large, especially where the smaller lesions coalesce. The small spots on the stems and pods occur late in the season. Pycnidia develop in the mature lesions (Hemmi, 1915, 1940). Seed infection is not conspicuous (Wolf and Lehman, 1926).

THE FUNGI

Septoria glycines Hemmi

The pycnidia are submerged, flattened to globose, and open to the surface with a large pore. Conidia are hyaline, filiform, curved, have 1 to 3 indistinct septa, and are mostly 1.4 to 2.1 microns wide by 35 to 40 microns long.

A second species, *Septoria sojina* Thuem, causes light-brown indistinct spots on soybean leaves in Europe and Asia.

A third species, *Septoria sojae* Syd. & Butl., is reported on soybeans. The three species need further comparative study.

15. Sclerotial Root, Stem, and Crown Blights. Several cosmopolitan parasites cause premature killing of the soybean plants. The general symptoms are similar in that the roots, crown, and basal portion of the stem are invaded, resulting in yellowing or wilting and the ultimate weakening or death of the plant. The various fungi are identified by the type of mycelium and the size, shape, and color of the sclerotia.

Sclerotium rolfsii Sacc. or *Pellicularia rolfsii* (Curzi) West occurs in the sandy soils of the Southern area of the United States. The large globose brown sclerotia on the base of the stem and upper portion of the taproot differentiate the Southern blight. This organism attacks a wide range of crop plants in the Southern states.

Macrophomina phaseoli (Maubl.) Ashby, *Sclerotium bataticola* Taub., causes the charcoal rot of crop plants. The disease extends into the corn-belt section of the soybean area and probably causes more total damage to the soybean crop than the former species. Small round to irregular black sclerotia are produced below the epidermis on the dead stems and roots, and the *Macrophomina* pycnidia occur in the same tissues.

Rhizoctonia solani Kuehn and other species occur on the soybean, especially in the northern half of the area in the United States and in Asia. The white mycelial mats in association with the lesioned tissues during wet weather and the characteristic branching of the mycelium are the important distinguishing characters (Atkins and Lewis, 1954, Boosalis, 1950).

Phymatotrichum omnivorum (Shear) Dugg. causes a root rot of the soybean. The dead root and stem base show the characteristic brown discoloration below the bark.

16. Brown Stem Rot, *Cephalosporium gregatum* Allington & Chamberlain. This disease is general in some years in the Middle Western United States. Damage is caused by premature killing in the autumn followed by extensive lodging. Symptoms consist of browning of the pith and xylem of the stem, starting at or below the soil level and progressing slowly upward with only slight external symptoms. Leaf symptoms are occasional early blighting of the lower leaves followed by a rapid interveinal chlorosis of the upper leaves and subsequent necrosis. In advanced stages of the disease, the outside of the stem appears brown and the weakened stems lodge badly. Low temperatures are necessary for disease development. The fungus appears to be soil-borne in crop residue and common in the commercial soybean area of the United States.

THE FUNGUS

Cephalosporium gregatum Allington & Chamberlain

Dense mycelial mat of sterile hyphae is found commonly in the plant tissues, and the fertile hyphae are aerial. Conidiophores are short, usually simple, generally in clusters and bearing conidia terminally in heads. Conidia are ovoid to elliptical, hyaline, one-celled capitate (Allington and Chamberlain, 1948).

Rotation of crops is advisable where the disease is severe (Allington, 1946, Chamberlain and McAlister, 1954).

REFERENCES

ALLINGTON, W. B. Wildfire disease of soybeans. *Phytopath.* **35**:857–869. 1945.
———— and D. W. CHAMBERLAIN. Brown stem rot of soybean. *Phytopath.* **38**:793–802. 1948.
ATKINS, J. G., JR., and W. D. LEWIS. Rhizoctonia aerial blight of soybeans in Louisiana. *Phytopath.* **44**:215–218. 1954.
BOOSALIS, M. G. Studies on the parasitism of *Rhizoctonia solani* Keuhn on soybeans. *Phytopath.* **40**:820–831. 1950.
CHAMBERLAIN, D. W. A halo-producing strain of *Psudomonas glycinea*. *Phytopath.* **42**:299–300. 1952.

—— and D. F. McAlister. Factors affecting the development of brown stem rot of soybean. *Phytopath.* **44**:4–6. 1954.

Coerper, F. M. Bacterial blight of soybean. *Jour. Agr. Research* **18**:179–193. 1919.

Cromwell, R. O. Fusarium blight of the soybean. *Nebr. Agr. Exp. Sta. Res. Bul.* 14. 1919.

Desjardins, P. R., R. L. Latterell, and J. E. Mitchell. Seed transmission of tobacco-ring-spot virus in Lincoln variety soybean. *Phytopath.* **44**:86. 1954.

Elrod, E. P., and A. C. Braun. Serological studies on the genus *Xanthomonas.* III. The *X. vasculorum* and *X. phaseoli* groups: the intermediate position of *X. campestris. Jour. Bact.* **54**:349–357. 1947.

Gardner, M. W., and J. B. Kendrick. Soybean mosaic. *Jour. Agr. Research* **22**:111–114. 1921.

Gäumann, E. Beiträge zu einer Monographie der gattung Peronospora Corda. *Beitr. Kryptogamonoflora Schweiz.* **5**:1–360. 1923.

Geeseman, G. E. Inheritance of resistance of soybeans to *Peronospora manshurica. Agron. Jour.* **42**:608–613. 1950.

——. Physiologic races of *Peronospora manshurica* on soybeans. *Agron. Jour.* **42**:257–258. 1950.

Gibson, F. Sunburn and aphid injury of soybeans and cowpeas. *Ariz. Agr. Exp. Sta. Tech. Bul.* 2. 1922.

Graham, J. H. Overwintering of three bacterial pathogens of soybeans. *Phytopath.* **43**:189–192, 193–194. 1953.

Hartwig, E. E., and S. G. Lehman. Inheritance of resistance to the bacterial pustule disease in soybean. *Agron. Jour.* **43**:226–229. 1951.

Hedges, F. A study of bacterial pustule of soybean and comparison of *Bacterium phaseoli sojense* Hedges with *B. phaseoli* E.F.S. *Jour. Agr. Research* **29**:229–251. 1924.

Heinze, K., and E. Köhler. Die Mosaikkrankheit der Sojabohne und ihre Übertragung durch Insekten. *Phytopath. Zeit.* **13**:207–242. 1940.

Hemmi, T. A new brown-spot disease of the leaf of *Glycine hispida* Maxim, caused by *Septoria glycines* sp. n. *Sapporo Nat. Hist. Soc. Trans.* **6**:12–17. 1915.

——. Studies on Septorioses of plants. VI. *Septoria glycines* Hemmi causing the brown-spot disease of soybean. *Mem. Col. Agr. Kyoto Imp. Univ. Pub.* 47, pp. 1–14. 1940.

Johnson, H. W., and D. W. Chamberlain. Bacteria, fungi and viruses on soybeans. *U.S. Dept. Agr. Yearbook*, pp. 238–247. 1953.

—— and E. A. Hollowell. Pubescent and glabrous characters of soybeans as related to resistance to injury by the potato leafhopper. *Jour. Agr. Research* **51**:371 381. 1935.

—— and B. Koehler. Soybean diseases and their control. *U.S. Dept. Agr. Farmer's Bul.* 1937. 1943.

Jones, F. R., and J. H. Torrie. Systemic infection of downy mildew in soybean and alfalfa. *Phytopath.* **36**:1057–1059. 1946.

Kendrick, J. B., and M. W. Gardner. Seed transmission of soybean bacterial blight. *Phytopath.* **11**:340–342. 1921.

—— and ——. Soybean mosaic: seed transmission and effect on yield. *Jour. Agr. Research* **27**:91–98. 1924.

Kornfeld, A. Die Blättfleckenkrankheit der Soja- eine Kalimangelerscheinung. *Zeit. Pflanzenernah. Dung. Bodenk.* **32**:201–221. 1933.

Lehman, S. G. Pod and stem blight of soybean. *Ann. Mo. Bot. Garden* **10**:119–169. 1923.

————. Frog-eye leaf spot of soybean caused by *Cercospora diazu* Miura. *Jour. Agr. Research* **36**:811–833. 1928.

————. Powdery mildew of soybean. *Jour. Elisha Mitchell Sci. Soc.* **46**:190–195. 1931.

————. Frog-eye (*Cercospora diazu* Miura) on stems, pod, and seeds of soybean, and the relation of these infections to recurrence of the disease. *Jour. Agr. Research* **48**:131–147. 1934.

————. Powdery mildew of soybean. *Phytopath.* **37**:434. 1947.

————. Race 4 of the soybean downy mildew fungus. *Phytopath.* **43**:460–461. 1953.

———— and F. A. WOLF. A new downy mildew on soybeans. *Jour. Elisha Mitchell Sci. Soc.* **39**:164–169. 1924.

———— and ————. Pythium root rot of soybean. *Jour. Agr. Research* **33**:375–380. 1926.

———— and ————. Soybean anthracnose. *Jour. Agr. Research* **33**:381–390. 1926.

———— and J. G. WOODSIDE. Varietal resistance of soybean to the bacterial pustule disease. *Jour. Agr. Research* **39**:795–805. 1929.

LING, L. Seedling stem blight of soybean caused by *Glomerella glycines*. *Phytopath.* **30**:345–347. 1940.

LIU, K. Studies on a Fusarium disease of soybean pods. *Mem. Col. Agr. Kyoto Imp. Univ. Pub.* 47, pp. 15–29. 1940.

MATSUMOTO, T. Beobachtungen über Sporenbildung des Pilzes, *Cercospora kikuchii*. *Ann. Phytopath. Soc. Japan* **2**:65–69. 1928.

———— and R. TOMOYASU. Studies on purple speck of soybean seed. *Ann. Phytopath. Soc. Japan* **1**:1–14. 1925.

MORSE, W. J., and J. L. CARTTER. Improvement in soybeans. *U.S. Dept. Agr. Yearbook*, pp. 1154–1189. 1937.

MURAKISHI, H. H. Purple seed stain of soybean. *Phytopath.* **41**:305–318. 1951.

NAOUMOFF, N. Materiaux pour la flora mycologique de la Russie, Fungi Ussurienses. I. *Bul. Soc. Mycol. France* **30**:64–83. 1914.

ORTON, W. A. The wilt disease of the cowpea and its control. *U.S. Dept. Agr. B.P.I. Bul.* 17. 1902.

SHERWIN, H. S., and K. W. KREITLOW. Discoloration of soybean seeds by the frog-eye fungus, *Cercospora sojina*. *Phytopath.* **42**:568–572. 1952.

————, C. L. LEFEBVRE, and R. W. LEUKEL. The effect of seed treatment on the germination of soybeans. *Phytopath.* **38**:197–204. 1948.

SNYDER, W. C. *Peronospora viciae* and internal proliferation in pea pods. *Phytopath.* **24**:1358–1365. 1934.

TIFFANY, L. H. The anthracnose complex on soybeans. *Iowa State Col. Jour. Sci.* **25**:371–372. 1951.

————. Delayed sporulation of *Colletotrichum* on soybean. *Phytopath.* **41**:975–985. 1951.

———— and J. C. GILMAN. Species of *Colletotrichum* from legumes. *Mycologia* **46**:52–75. 1954.

WARINGTON, K. The influence of iron supply on toxic effects of manganese, molybdenum and vanadium on soybean, pea and flax. *Ann. Appl. Biol.* **41**:1–22. 1954.

WEISS, M. G. Soybeans: diseases and insect pests. *Adv. Agron.* **1**:78–152. 1949.

WELCH, A. W., and J. C. GILMAN. Hetero- and homothallic types of *Diaporthe* on soybeans. *Phytopath.* **38**:628–637. 1948.

WOLF, F. A. Bacterial blight on soybean. *Phytopath.* **10**:119–132. 1920.

————. Bacterial pustule of soybean. *Jour. Agr. Research* **29**:57–68. 1924.

——— and S. G. LEHMAN. Brown-spot disease of soybean. *Jour. Agr. Research* **33**:365–374. 1926.

——— and ———. Diseases of soybean which occur both in North Carolina and the Orient. *Jour. Agr. Research* **33**:391–396. 1926.

——— and I. V. SHUNK. Tolerance to acids of certain bacterial pathogens. *Phytopath.* **11**:244–250. 1921.

CHAPTER 16

PEANUT DISEASES

The peanut, or groundnut, of commerce comprises several types and varieties of *Arachis hypogaea* L. Taxonomically the species is divided into four varieties (Chevalier) or four forms (Burkart) based largely on seed size and color as rearranged by Hermann (1954). Nine wild species indigenous to South America are redescribed by Hermann (1954). These apparently have not been studied extensively (Bukasov, 1930). The chromosome number of *A. hypogaea, A. glabrata* Benth. (*A. prostrata* Benth.), and *A. nambyguarae* Hoene is 20 pairs (Darlington and Janaki Ammal, 1945).

The peanut is a warm-climate legume grown in subtropical and tropical regions of the world. The crop is used primarily as a food crop, the seed comprising an important economic crop of the Southern United States and similar or more tropical areas of the world. It is used less intensively as a forage crop although the hay from seed production represents a large feed supply (Beattie, 1954).

Three commercial types, all annuals, are (1) large-seeded Valencia and Virginia, including both bunch and runner character of plant growth, (2) true runner types, and (3) the smaller-seeded Spanish. Hybrids between the various types result in intermediate forms. Relatively little use has been made of the other species indigenous to South America.

1. Nonparasitic Diseases. A deficiency of the minor elements is manifest by mild chlorosis and necrosis of leaves near the stem apex. The older leaves show dead marginal areas. The general symptoms are similar to those described for alfalfa, Chap. 13. Low available calcium during the period of seed development results in smaller, poorly filled seed, especially in the large-seeded varieties (Beattie, 1954, Harris, 1952).

Heat cankers on the stem near the soil line and heat necrosis of young bud tissues occur on seedlings, especially in fine, dusty soil. The necrotic areas are sunken, dark-colored, and frequently offer avenues for rot-producing organisms, especially when growing conditions are not favorable for rapid wound response of the parenchymatous cells adjacent to the necrotic area. Similar damage occurs in soybean, flax, and other plants (Boyle, 1953).

2. Rosette Disease, Virus Transmitted by *Aphis medicaginis* (Koch) (*A. leguminosae* Theob.) and *A. craccivora* (Koch). This mosaic disease apparently is confined to peanut. The distribution is limited to subtropical and tropical areas of Africa and adjacent India and Java. The disease was first reported on peanuts in native gardens in 1907. Mild mosaic symptoms are chlorotic mosaic patterns with dark-green areas over the veins of the leaves, symptoms that are masked at certain seasons. The symptoms of the severe mosaic, especially following early infection, are (1) first a faint mottling of the young leaflets and chlorosis; both appear transitory; (2) the succeeding leaves show pale yellow mosaic patterns and green over the veins; (3) stem internodal elongation is suppressed resulting in a rosette or witches' broom growth habit; (4) later-formed leaves are progressively smaller, chlorotic, curled, with distorted leaflets. Blossom and pod formation are reduced. The rosette-type symptoms persist, and plants frequently continue slow growth late in the season. Apparently aphids transmit the virus from diseased plants in native gardens and plants left in the fields. The virus is not seed-borne. Resistant varieties are not reported (Evans, 1951, Storey and Bottomley, 1928).

The ring spot of peanuts reported in Brazil is associated with the spotted wild virus of tomato. Other legumes are included in the wide range of hosts of this virus (Costa, 1950).

3. Bacterial Wilt, *Xanthomonas solanacearum* (E. F. Smith) Dowson. Bacterial wilt is distributed widely, but in most areas the losses are rarely serious. Peanuts are relatively resistant to many of the strains of the pathogen on solanaceous hosts.

4. Seed Rot, *Fusarium, Rhizopus, Penicillium, Aspergillus,* etc. Seed rot occurs when poor-quality seed or mechanically damaged seed is retarded in germination by cold, wet, compacted soils or deep planting. Seed-protectant fungicides, especially the organic sulfur compounds, reduce losses greatly. These fungicides also reduce losses from seedling blights incited by *Macrophomina phaseoli* (Maubl.) Ashby or *Sclerotium bataticola* Taub, and *Rhizoctonia* spp., especially during the early seedling stage.

5. Peg and Pod Rots, *Sclerotium, Rhizoctonia, Diplodia, Rhizopus, Penicillium,* etc. The common name peg rot includes rotting of various tissues of gynophore (peg), pod, and kernels from the time these structures reach the soil surface shortly after pollination until the seeds are mature and harvested. These structures frequently are invaded by soil pathogens, are killed by substances produced by pathogens developing on crop residue on the soil surface as in *Sclerotium rolfsii*, and are injured mechanically by nematodes and insects followed by the entrance of various fungi, weakly parasitic or saprophytic. Proper soil drainage and soil type, cultivation, and other management practices help to reduce these losses in both yield and quality (Wilson, 1953).

6. Cercospora Leaf Spots, *Cercospora arachidicola* Hori and *C. personata* (B. & C.) Ell. & Ev.   These leaf spots are distributed widely, especially in the humid areas; as the culture extends into the drier climates damage is less severe.   Extensive defoliation occurs where the disease is not controlled.   The leaves on the soil surface increase the damage from southern blight on plants that are retarded in growth by defoliation from leaf spot.   The severe damage to the crop frequently is the combined effect of the leaf spots and blight.

Early leaf spot generally is incited by *C. arachidicola*.   The somewhat circular spots on the leaves are light tan at first.   As they enlarge, the spots change to reddish brown to black on the lower surface and light brown on the upper.   Usually a yellow halo surrounds each spot.   The late leaf spot incited by *C. personata* appears as dark brown to black; more restricted spots on both surfaces of the leaf.   The halo-like margin is not distinct around the latter spot.   More restricted spots showing less difference in color occur on petioles, stems, gynophores, and pods, especially in the more susceptible Spanish-type varieties.   The two pathogens differ in morphology (Drouillon, 1951, Jenkins, 1938, Vanhoof, 1950, Wilson, 1953).

THE FUNGI

Cercospora arachidicola Hori or
Mycosphaerella arachidicola W. A. Jenkins.

Conidiophores borne from stromatic masses are yellowish brown, geniculate, long with conspicuous scars.   Conidia are hyaline to plane olive green, obclavate, frequently curved, 3 to 12 septations, and 37–108 by 2.7–5.4 microns.   Perithecia are embedded in leaf tissue, ovate to globose, papillate ostioles.   Asci are club-shaped and contain eight spores in uniseriate position.   Ascospores are two-celled, apical cell larger, slightly curved, hyaline, 11 by 3.6 microns.

Cercospora personata (B. & C.) Ell. & Ev. or
Mycosphaerella berkeleyii W. A. Jenkins.

Conidiophores are coarse, short, reddish brown, geniculate, and borne from stromatic basal mat.   Conidia are mostly cylindrical with attenuated tips, straight, olivaceous brown, 18–60 by 5–11 microns.   Mycelium of fungus forms branched haustoria in host cells.   Perithecia are partly embedded in leaf tissues, broadly ovate to globose, papillate ostioles.   Asci cylindrical to club-shaped and contain eight spores. Ascospores are two-celled, constricted at septum, apical cell larger and average longer than in previous species, 15 by 3.4 microns.

The cycle of development is similar in both pathogens.   Primary inoculum is from spores produced on crop residue.   Secondary inoculum from conidia produced on the leaf spots is abundant, and spread of the disease is rapid during favorable weather conditions. Seed-borne inoculum is of minor importance, although pathogens persist on pods and are a factor in spread to new areas.

CONTROL. Fungicidal dusts and sprays are used for leaf spot control. Dusting sulfur alone or with copper (90:10) applied three to four times during the growing season is used generally. Spraying with Bordeaux or carbamates is effective, but more expensive under most conditions. Rotation and sanitation are essential for good disease control as well as good crop management (Beattie, 1954, Wilson, 1953). Suitable resistant varieties are not reported, although differences in susceptibility of varieties to some races of the pathogens are reported.

7. Southern Blight, *Sclerotium rolfsii* Sacc. or *Pellicularia rolfsii* (Curzi) West. This blight, widely distributed in the warm climates of the world on many crop plants, is a serious disease of peanuts, especially when defoliation occurs or vigorous growth is checked by disease or poor crop management. The pathogen develops on fresh crop residue on the soil surface. The mycelium surrounds the plant stems near the soil surface and produces organic acids toxic to living plant tissues. Following necrosis of the plant cells the mycelium invades the stems, gynophores, and pods to incite further rotting and retting of the tissues. The wilted plants and abundant white mycelium and small white to brown sclerotia on the crop residue and rotted plants are the characteristic symptoms (Beattie, 1954, Wilson, 1953).

THE FUNGUS

Sclerotium rolfsii Sacc. or
Pellicularia rolfsii (Curzi) West.

Only the mycelium and sclerotial phases of this fungus were known prior to 1927. The basidial stage was described first as *Corticium* and later as a species in the genus *Pellicularia*. Inconspicuous basidia bearing usually four hyaline sporidia are produced on mycelial mats on the surface of the host. The fungus is identified commonly by the white mass of mycelium and small spherical, white to brown sclerotia (West, 1947).

The control of the disease depends largely upon good crop management. *Rhizoctonia solani* Kühn or *Pellicularia filamentosa* (Pat.) Rogers and other species incite seedling blight, stem cankers, and pod rot on peanuts.

8. Rust, *Puccinia arachidis* Speg. The uredia and telia occur sporadically on peanut in most areas where the crop is grown.

REFERENCES

BEATTIE, J. H. Growing peanuts. *U.S. Dept. Agr. Farmers' Bul.* 2063. 1954.
BOYLE, L. W. Heat canker; a primary phase of collar rot on peanuts. *Phytopath.* **43**:571–576. 1953.

BUKASOV, S. M.   The cultivated plants of Mexico, Guatemala and Columbia with supplementary articles by N. N. Kuleshov, N. E. Zhiteneva and G. M. Popova. *Inst. Rasten. Vaschnil, Leningrad.*   1930.   (In Russian with English summary.)

COSTA, A. S.   Mancha anular do Amendoim cansada pelo virus de vira-cabeca. *Brazantia, S. Paulo* **10**:67–68.   1950.

DARLINGTON, C. D., and E. K. JANAKI AMMAL.   Chromosome atlas of cultivated plants.   George Allen & Unwin, Ltd.   London.   1945.

DROUILLON, R.   La maladie des taches bunes de l'Aracbide.   *Rev. Mycol. Colon* **16**:1–11.   1951.

EVANS, A. C.   Entomological research in Overseas Food Corporation (Tanganyika). *Ann. Appl. Biol.* **38**:526–529.   1951.

HARRIS, H. C.   Effects of minor elements, particularly copper, on peanuts.   *Fla. Agr. Exp. Sta. Bul.* 494.   1952.

HERMANN, F. J.   A synopsis of the genus Arachis.   *U.S. Dept. Agr. Agr. Monogr.* 19, pp. 26.   1954.

JENKINS, W. A.   Two fungi causing leaf-spot of peanut.   *Jour. Agr. Research.* **56**:317–332.   1938.

STOREY, H. H., and A. M. BOTTOMLEY.   The rosette disease of peanuts.   *Ann. Appl. Biol.* **15**:25–45.   1928.

VANHOOF, H. A.   Diseases of groundnut caused by *Cercospora personata* (B. and C.) E. and E. and *Cercospora arachidicola* Hori.   *Contr. Gen. Agr. Res. Sta. Bogor* 114.   1950.

WEST, E.   *Sclerotium rolfsii* Sacc. and its perfect stage on climbing fig.   *Phytopath.* **37**:67–69.   1947.

WILSON, C.   Preventing the diseases of peanut.   *U.S. Dept. Agr. Yearbook*, pp. 448–454.   1953.

CHAPTER 17

DISEASES OF OTHER LEGUME CROPS

Other legume plants are used for forage and soil improvement and seed crops in special areas. The diseases of some of these are similar to those of alfalfa, clovers, and soybean discussed in the previous chapters. Other diseases are distinct in one or more characteristics. A few of the diseases of this nature are included, although frequently information on specific diseases of this group of economic plants is general. The diseases of beans, *Phaseolus vulgaris* L., *P. lunatus* L., and others, and peas, *Pisum sativum* L., *P. arvense* L., and similar species, are discussed by Walker (1952). The genetics of resistance to diseases is summarized by Coons (1953) and Wade (1937) and some sources of resistance by Stevenson and Jones (1953).

A relatively large group of legumes in the genera *Crotalaria, Lathyrus, Lespedeza, Lotus, Lupinus, Pisum, Pueraria, Stizolobium, Vicia,* and *Vigna,* representing both annuals and perennials, are used as forage and legumes for soil improvement (McKee and Pieters, 1937).

Legumes are used for soil improvement, reducing soil erosion, and control of soil-inhabiting plant pathogens. In addition to improved physical state and balance of chemical nutrients, the incorporation of leguminous cover crops into the soil changes the microfloral balance in the rhizosphere, frequently resulting in the control of specific pathogens. The increase in Actinomycetes and other microflora antagonistic to plant pathogens apparently is favored especially by the legumes. Therefore the economical production of leguminous cover crops, especially in the warmer climates where the legume frequently is grown during the winter period, offers another method of disease control (Clark, 1949).

Some of the more important diseases of this group of legumes are discussed briefly. Frequently only general information on the disease is available. Records on disease-resistant varieties are limited. Generally other legumes are substituted when a disease appears on the one being grown.

1. Viruses diseases, Transmitted by Contact or Insect Vectors. *Crotalaria mosaic* apparently is distributed widely on several *Crotalaria* spp., cowpea, and some other legumes (Jensen, 1950). The symptoms range from vein clearing, light-green mosaic pattern, puckering of leaflets to leaves showing wide diversity in shape, light- and dark-green areas, and rosetting of the plant. The virus is transmitted by juice and by *Aphis gossypii* (Glover) and *Myzus persicae* (Sulz.). The virus is not seed-borne.

Lupine Mosaic. A similar mosaic complex occurs on lupines in Europe, Africa, the United States, and other areas. The range in symptoms is similar also. The virus is aphid-transmitted (Klesser, 1953, Weimer, 1952). *Several viruses occur on the cowpea,* apparently a common host for many of the legume viruses. Local lesions are incited on this host by some of these viruses. Many of the pea and bean viruses occur on this group of legumes (Quantz, 1951, Walker, 1952). Tobacco ring spot virus is present in some of these legumes. The specific viruses and their host range are still uncertain. *Witches' broom virus* is transmitted to several legumes in this group by leaf hoppers, but it has not been recorded as occurring naturally (Klostermeyer and Menzies, 1951).

2. Bacterial Diseases. Bacterial blight and wilt are distributed generally on many of these legumes. *Xanthomonas vignicola* Burk. causes damage on cowpea. Stem cankers appear on the stem. Leaf lesions, frequently with halo, are less common on the leaves. *X. lespedezae* (Ayers *et al.*) Starr occurs on both annual and perennial Lespedeza and frequently causes damage. Systemic spread of the bacteria causes water-soaked spots, dead leaves, and rapid wilting of plants; exudate is abundant. *X. phaseoli* (E. F. Smith) Dowson occasionally causes damage on lupines, velvet bean, and cowpea. Angular leaf spots and stem and pod lesions are more common on this group of hosts than severe dwarfing and defoliation as in the bean. *X. solanacearum* (E. F. Smith) Dowson occasionally incites wilt in velvet bean and a few other legumes in addition to peas and beans. *Pseudomonas phaseolicola* (Burk.) Dowson occurs infrequently on kudzu bean and *P. pisi* Sackett on cowpea and vetch. *P. syringae* van Hall incites halo blight on cowpea and vetch as well as the clovers, beans, and soybean. *Corynebacterium flaccumfaciens* (Hedges) Dowson incites dwarfing and wilting of cowpeas.

The bacteria inciting these diseases are seed-borne and persist in crop residue, especially in the warmer climates where this group of legumes is grown. Clean seed and crop rotation and sanitation are the best control measures (Elliott, 1951).

3. Seed Rot and Root Rot, *Pythium* spp., *Phytophthora* spp., and *Aphanomyces euteiches* Drechsl. *Pythium* seed rotting and blighting of seedlings before emergence is a factor in securing stands in many of the legumes (Chap. 14). Poor stands and root rot incited by *Aphanomyces euteiches* Drechsl. and *Phytophthora* spp. occur on some of this group of legumes, especially during periods of cool, wet weather. The symptoms are similar in all of these diseases. The pathogen is differentiated by the morphology of sporangia and zoospores (Weimer, 1952, Weimer and Allison, 1953).

4. Powdery Mildews, *Erysiphe polygoni* DC., *Microsphaera diffusa* C. & P. Specialized races of *E. polygoni* occur on many of these legumes. Usually the disease occurs when plant growth is well advanced and damage is not severe. The powdery mass of mycelium and chains of conidia of the oidial stage on the leaf surface are the characteristic symptoms. Cleistothecia with simple appendages form on many of these legumes (see Chap. 14). The powdery mildew on *Lespedeza, Crotalaria,* and less commonly on some of the other legumes is incited by *Microsphaera diffusa.* The oidial stage is similar to *E. polygoni.* The cleistothecia are differentiated by the rigid appendages with dichotomous apical branches. Resistant plants occur in many populations of these legumes. Rowan Lespedeza is resistant.

5. Black Stem and Leaf Spot, *Ascochyta* spp. and *Mycosphaerella* spp. The blackened stems and dark-colored cankers are common symptoms of this group of pathogens on most legumes. Brown to black spots on leaves and pods occur also. Several species of *Ascochyta,* some of which the *Mycosphaerella* stage is known, are the incitants of this disease. The disease is distributed widely, and usually damage occurs during periods of cool weather. The *Ascochyta* spp. on legumes are somewhat similar in morphology. The pycnidia are globose, ostiolate, embedded in the lesioned or discolored areas, and later frequently develop abundantly in dead stems. The pycnospores are hyaline or light yellow, and two-celled and vary considerably in size and in ratio of width to length. Species in this group are differentiated largely by spore morphology and spore measurements. The morphological characters of *Mycosphaerella* spp. are also somewhat similar (see Chap. 13). Some species, as in the pea, attack different plant parts: *Ascochyta pinodella* L. K. Jones incites basal stem rot, *A. pisi* Lib. a leaf and pod spot, and *Mycosphaerella pinodes* (Berk. & Blox.) Stone restricted spots and discoloration on leaves, stems, and pods. *A. pisi* occurs on Vicia and Lathyrus, and *A. viciae* Lib. similar in morphology is reported on vetch. *A. gossypii* Syd. occurs on lupines in the Southern states (Weimer, 1952). *Phoma* species also incite leaf and stem lesions on these legumes. The pathogens are seed-borne as well as associated with crop residue.

6. Cercospora Spot, *Cercospora* spp. This group of diseases, like summer black stem of sweetclover and clover, are distributed widely and more prevalent during the warmer or later periods of growth. Defoliation and seed damage occur when the diseases are severe. Zonate, brown to gray, round to angular spots, later enlarging and coalescing, occur on the leaves. Small sunken brown spots are characteristic on the stems and pods. The conidiophores are usually short and arranged in fascicles; some few species are characterized by long, less densely packed conidiophores. Mycelium around the base of the conidiophores is abundant or scanty. The conidia are hyaline, usually obclavate or tapering toward the apex, and multiseptate. Many species are differentiated largely on the host basis, and cross-inoculation studies suggest a rather wide host range (Johnson and Valleau, 1949). These authors grouped many of the species occurring on legumes under *Cercospora apii* Fres. Many of the following species listed by host probably could be added: crotalaria-*C. demetrioniana* Wint., lathyrus-*C. viciae* Ell. & Holw.; lespedeza-*C. lespedezae* Ell. & Dearn.; lupine-*C. apii* Fres., velvet bean-*C. stizolobii* Syd.; kudzu-*C. pueraricola* Yam. or *Mycosphaerella pueraricola* Weimer & Luttrell.; vetch-*C. viciae* Ell. & Holw.; cowpea-*C. vignae* Ell. & Ev. or *C. apii*. Some of the pathogens are seed-borne, and all persist in crop residue. Spread from other hosts occurs in some (Chupp, 1937, Solheim, 1929, Weimer and Luttrell, 1948).

7. Anthracnose, *Colletotrichum* spp. and *Glomerella* spp. The anthracnose diseases are distributed widely on the legumes. Damage frequently is severe during the period of seedling development, as well as stem girdling by cankers later. Sunken black lesions on the cotyledons and hypocotyl are the first symptoms. Seedling blight is evident, especially during periods of cool, wet weather. Black stem cankers occur on the stems, and sunken spots, frequently with light-colored centers, appear on the leaves. Later dark pod lesions develop, first as sunken dark spots, later spreading as a pod rot. Badly infected seeds show lesions and mycelium; lightly infected seed is evident by the poorly developed, natural seed coloration.

Several pathogens are associated with anthracnose on legumes. In addition to the species occurring on alfalfa and clovers (Chaps. 13 and 14), several species somewhat general in host range are described. *Colletotrichum lindemuthianum* (Sacc. & Magn.) Briosi & Cov. incites anthracnose on the bean group, including some races on cowpea and kudzu bean; *C. pisi* Pat., possibly synonymous with *C. truncatum* (Schw.) Andrus & Moore, incites stem canker on beans, peas, vetch, lotus, and others; *C. destructivum* O'Gara is general in host range; *C. villosum* Weimer on vetch, and *Glomerella cingulata* (Ston.) Spauld. & Vons. on lupine, kudzu, and possibly lespedeza are some of the additional anthracnose pathogens (Weimer, 1946, 1952). Tiffany and Gilman (1954) discuss the morphology of this group of fungi and suggest species and race relationships.

The pathogens are seed-borne and persist in crop residue. Clean seed is essential to good stands and plant development, especially during cool weather. Resistant species and varieties are recorded mainly from observational data in field plantings.

8. Fusarium Wilt, *Fusarium oxysporum* Schlect. Several forms of this species are distributed widely on the legumes. Many of them are described as separate species in the section *Elegans* which is now reduced to the one species, *F. oxysporum.* The forms most prevalent in inciting wilt, and their respective hosts are *F. oxysporum f. tracheiphilum* (E. F. Sm.) Snyder & Hansen—soybean and cowpea; *f. lupini* Snyder & Hansen —lupine; *f. pisi* (Linford) Snyder & Hansen races 1 and 2—pea; *f. phaseoli* Kend. & Snyder—bean; *f. udum* (Butl.) Snyder & Hansen—chickpea; *f. crotalariae* Carrera—crotalaria. Some of these forms are restricted in host range, others incite disease on a wider group of hosts. Padwick (1940) and Armstrong (1951) transfer the chickpea and crotalaria forms to *Fusarium udum* var. *cajani* Pad. and *F. udum* var. *crotalariae* Pad. (See *Fusarium* wilt under other crop plants.)

9. Southern Blight, *Sclerotium rolfsii* Sacc. or *Pellicularia rolfsii* (Cruzi) West. The disease occurs in the warmer climates on many of these legumes. The damage generally is limited to local areas. The presence of the white mycelium and small white or brown sclerotia over dead leaves on the soil surface under diseased plants is characteristic.

Both foliage blight and stem cankers are incited by *Rhizoctonia solani* Kühn or *Pellicularia filamentosa* (Pat.) Rogers, especially in older, rank growth of the plants (see Chap. 16).

REFERENCES

ARMSTRONG, J. K., and G. M. ARMSTRONG. Physiological races of Crotalaria wilt Fusarium. *Phytopath.* **41**:714–721. 1951.

CHUPP, C. *Cercospora* spp. and their host genera. Department of Plant Pathology, Cornell University. Ithaca, N.Y. 1937. (Processed.)

CLARK, F. E. Soil microorganisms and plant roots. *Adv. Agron.* **1**:241–288. 1949.

COONS, G. H. Breeding for resistance to disease. *U.S. Dept. Agr. Yearbook*, pp. 174–192. 1953.

JENSEN, D. D. A Crotalaria mosaic and its transmission by aphids. *Phytopath.* **40**:512–515. 1950.

JOHNSON, E. M., and W. D. VALLEAU. Synonymy in some common species of *Cercospora*. *Phytopath.* **39**:763–770. 1949.

KLESSER, P. J. Virus diseases of Lupines. *Fmg. So. Africa* **28**:347–350. 1953.

KLOSTERMEYER, E. C., and J. D. MENZIES. Insect transmission of alfalfa witches'-broom virus to other legumes. *Phytopath.* **41**:456–458. 1951.

McKEE, R., and A. J. PIETERS. Miscellaneous forage and cover crop legumes. *U.S. Dept. Agr. Yearbook*, pp. 999–1031. 1937.

PADWICK, G. W. The genus Fusarium. V. *Fusarium udum* Butler, *F. vasinfectum* Atk. and *F. lateritium* var. *uncinatum* Wr. *Indian Jour. Agr. Sci.* **10**:863–878. 1940.

QUANTZ, L. Eine Virose der Erbse und anderer Leguminosen. *Phytopath. Zeit.* **17**:472–477. 1951.

SOLHEIM, W. G. Morphological studies of the genus Cercospora. *Ill. Biol. Monogr.* 12. 1929.

STEVENSON, F. J., and H. A. JONES. Some sources of resistance in crop plants. *U.S. Dept. Agr. Yearbook,* pp. 192–216. 1953.

TIFFANY, L. H., and J. C. GILMAN. Species of *Colletotrichum* from legumes. *Mycologia* **46**:52–75. 1954.

WADE, B. L. Breeding and improvement of peas and beans. *U.S. Dept. Agr. Yearbook,* pp. 251–282. 1937.

WALKER, J. C. Diseases of vegetable crops. McGraw-Hill Book Company, Inc. New York. 1952.

WEIMER, J. L. Lespedeza anthracnose. *Phytopath.* **36**:524–533. 1946.

———. Diseases of cultivated lupines in the Southeast. *U.S. Dept. Agr. Farmer's Bul.* 2053. 1952.

——— and E. S. LUTTRELL. Angular leaf spot of Kudzu caused by a new species of *Mycosphaerella. Phytopath.* **38**:348–358. 1948.

SECTION 4

DISEASES OF FIBER AND OTHER FIELD CROPS

CHAPTER 18

COTTON DISEASES

Cotton is one of the major fiber crops of the world. The several *Gossypium* spp. cultivated for fiber and seeds and the wild species were placed by Longley (1933), Webber (1939), and others in four groups, as follows: (1) the wild Australian species (*Gossypium sturtii* F. Muell.) with 13 chromosome pairs, (2) the wild and cultivated Asiatic species with 13 chromosome pairs, (3) the wild American species with 13 chromosome pairs, and (4) the American semiwild and cultivated species with 26 chromosome pairs. Hybridization between the Asiatic and American cultivated species is difficult because of sterility and incompatibility in chromosome pairing (Harland and Atteck, 1941, Silow, 1941, Stephens, 1942). Crossing within the Asiatic and within the American cultivated species is general. Three types of American cotton are grown: (1) the sea island, (2) the American-Egyptian, and (3) the upland. Commercial varieties of all three types are probably of hybrid origin, and all three are adapted to somewhat different environments and industrial uses.

The cultivated species and varieties are grown over a wide range of environments within the warmer climatic zones of the world. In the tropical regions, many of the species are perennial. Adaptation of the hybrids to a shorter growing season results in annuals with indeterminate growth habit.

Cotton breeding received attention early in the history of the crop in North America. Probably John Griffin was the first to employ what is essentially the backcross method fifty years before the geneticists fully established the method as a sound procedure in plant breeding (Ware, 1936).

The general spread of the boll weevil and *Fusarium* wilt over the cotton belt at the beginning of the present century caused a change in types of cotton grown. The earlier maturing varieties with determinate growth and fruiting habit as well as boll types better adapted to withstanding boll weevil injury replaced many of the older varieties, according to Ware (1936). By 1895, the *Fusarium* wilt disease had spread over the southeastern cotton belt of the United States, and breeding for wilt resistance was started (Orton, 1900, Orton and Gilbert, 1912, Ware, 1936). Orton, in 1899, probably was the first man in the history of agriculture to submit plant populations to an epidemic of a disease in order to obtain resistant selections.   However, selections on naturally infested soil were made by E. F. Smith and E. L. Rivers in 1895.   Insects, diseases, and foreign economic competition all have played an important part in the changing of varietal types and the regions where cotton was grown.

Plant diseases are estimated as causing an average annual loss of 17.1 per cent of the crop, or 1,737 bales, between the years 1930 to 1939 (Plant Disease Survey).   The physiology and diseases of cotton are reviewed by Adams *et al.* (1950).

1. Leaf Discoloration or Rust, Nonparasitic.   Mineral deficiencies cause leaf discoloration and dropping in cotton.   Potash is the most common mineral deficiency in cotton production in humid areas.   The malady is called "rust" because the leaves develop a rusty-brown color. Potash deficiency is manifest by a reduction in plant growth, by the tip and margin of the leaves developing a yellow to brown coloration, and by curling of the leaf margins.   The leaves finally turn reddish brown and shed prematurely, and the bolls are small and fail to open (Cooper, 1939, Neal and Gilbert, 1935).   The potash deficiency also accentuates the development of the wilt disease (Young and Tharp, 1941).

Not only the vegetative development and fruiting of the cotton plant are affected by unfavorable environmental conditions, but also the retention or, conversely, the shedding of the bolls after they have formed. Dunlap (1945), Wadleigh (1944), and others have investigated the influence of environmental factors, such as light intensity, temperature, moisture, and mineral nutrients, on plant development and boll shedding.

2. Virus Diseases, Viruses Transmitted by Grafting, White Fly, and Aphids.

Leaf Curl, Virus Transmitted by White Fly, *Bemisia* spp., and Grafting. Leaf curl, first reported in Nigeria in 1912, appears to be confined to Africa, especially in the Sudan and Nigeria.   The disease was severe during the early period of cotton expansion with susceptible varieties. Symptoms range from no apparent effect upon the plant carrying the virus to twisted, elongated, barren plants with dwarfed and curled leaves. Leaf curling, vein clearing, and necrosis, some mottling, and progressively smaller, curled leaves, bracts, and flowers are characteristic symptoms. The leaf malformations are associated mainly with the lower leaf surface. The range in symptoms is associated with degree of resistance of varieties. Other hosts, *Hibiscus* spp., *Sida* spp., *Althaea rosea* Cav., and *Malvaviscus arboreus* Cav., show similar symptoms.   Meristematic activity is stimulated, and cellular hypertrophy is evident in parenchyma tissues of leaves and stems.   Two strains of the virus, mosaic and leaf curl, are suggested.   The common vectors are *Bemisia gossypiperda* (M. & L.) (*B.*

tabaci Genn) and *B. goldingi* (Corb.). The virus persists in the white fly, but it is not transmitted through eggs. Control is accomplished by elimination of infected plants, especially cotton regrowth (ratoons) and other hosts, during the season cotton is not grown and by the use of resistant varieties. Tarr (1949) gives a full account of the disease and resistant varieties in the different cotton types. A Sakel cotton selection carrying the B_2 and B_3 factor pairs for resistance to angular leaf spot is resistant to leaf curl (Anson and Barlow, 1951, Hutchinson *et al.*, 1950).

Stenosis and *Cyrtosis*, Probably Virus Transmitted by Grafting. The dwarfed leaves and apical bud proliferations described from Haiti, China, India, etc., appear similar to the aster yellows virus on a number of hosts. Graft transmission has been reported.

Cotton Leaf Curl, Probably Virus Transmitted by Aphids. A leaf curling and mottling with blossom shedding is reported from Russia. Symptoms, host range, and vectors are different from the African leaf curl.

Leaf Crumple, Virus Transmitted by White Fly. This virus disease has increased on Acala cotton in the irrigated valleys of southern California, Arizona, and adjacent Mexico. Leaves are thickened, turned under, and show hypertrophy of the interveinal tissue. Flowers show similar modifications. The symptoms and development of the disease as well as host range apparently are different from leaf curl. Two white flies, *Trialeurodes abutilonea* (Haldeman) and *Bemisia inconspicua* (Quaint.), when transferred from diseased plants, incited symptoms (Dickson *et al.*, 1954).

Mosaics. Vein clearing mosaic, yellow mosaic, and ring spot viruses are reported on cotton in Brazil and Argentina (Costa and Forster, 1938, Difronzo, 1941).

3. Angular Leaf Spot, or Bacterial Blight,[1] *Xanthomonas malvacearum* (E. F. Sm.) Dows. The bacterial blight is distributed widely in the cotton areas of the world. In the United States, the disease is more severe in the south central and southwestern sections of the cotton belt, where it occurs on the cultivated cottons and develops from inoculation on *Gossypium thurberi* Tod (*Thurberia thespesioides* A. Gray). The disease is severe in the cotton belt of the Nile and in irrigated cotton areas in Mexico.

SYMPTOMS AND EFFECTS. The small round spots occur on the cotyledons during the early seedling stage of growth, and angular lesions on the older plants. The disease on the leaves appears first as water-soaked spots that enlarge to angular brown to black lesions. The spots occur on the leaves from the seedling to the mature plant stage during periods of

[1] Credit is given the Sub-committee on Cotton Diseases, Southern Experiment Station Committee or Research Council, for the information contained in the report Cotton Diseases, mimeographed reports. Good literature lists are included.

high temperatures.   Black elongate lesions occur on the young stems, sometimes causing girdling and death of the stem.   This type of symptom is referred to as "black arm."   Angular to circular black sunken spots and rotting occur on the bolls (Brown and Streets, 1937, Neal and Gilbert, 1935) (Fig. 88).   Boll rot occurs during hot, humid weather, especially when insect punctures are numerous.   Secondary organisms frequently enter through the bacterial lesions to discolor or rot the boll.   Entrance of the bacteria is through stomata or wounds (Edgerton, 1912, Tennyson, 1936).

THE BACTERIUM

Xanthomonas malvacearum (E. F. Sm.) Dows.
[*Phytomonas malvacearum* (E. F. Sm.) Bergey *et al.*]
(*Bacterium malvacearum* (E. F. Sm.))
(*Pseudomonas malvacearum* E. F. Sm.)

The rods are rounded at the ends, with one polar flagellum, no spores, and form pale-yellow colonies in culture.

ETIOLOGY.   The organism survives in the fuzz on the seeds, probably in the seed, and on crop residue.   The external infestation of the seed and early seedling infection of the cotyledons are common.   Under conditions of high moisture at harvest or overwintering in the fields, seed infection apparently occurs (Archibald, 1927).   Volunteer seedlings from diseased fields frequently show infection of the cotyledons.   Crop residue offers a means of infecting the new crop (Hare and King, 1940).   The spread of the bacteria from infected cotyledons to leaves and other plant parts is associated with high temperatures, high humidity, wind, rain, and irrigation water, according to Brown (1942), Faulwetter (1917), Massey (1927), Stoughton (1928, 1930, 1931, 1932, 1933), and others.   Water soaking of the tissues prior to infection probably aids infection.   Apparently environmental conditions of the Western drier section are more favorable for the disease than those of the Eastern section of the United States.

Control of the disease by seed treatment, crop rotation, and resistant varieties is practical.   The use of sulfuric acid in delinting the seed gives good control in noninfested soil, according to Bain (1939) and Brown and Gibson (1925).   The mercury dusts control seedling blight when they are used after the acid treatment as well as without delinting.   Rotation of crops and deep plowing with irrigation and tillage well in advance of planting are useful in reducing the incidence of the disease from crop residue.

Apparently resistant plants occur in plant populations of most cotton types.   However, complete resistance exists in some of the Old World

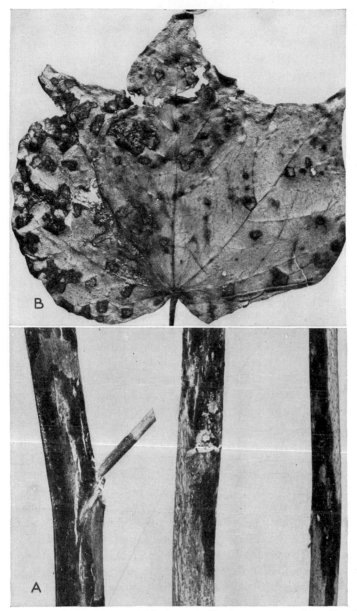

Fig. 88. Angular leaf spot (*A*) and black arm (*B*) of cotton caused by *Xanthomonas malvacearum*.

cottons cultivated in Asia. The selection and use of resistant plants in the several types of cotton has facilitated greatly the combining resistance with adaptation of varieties and lint quality (Brinkerhoff *et al.*, 1952, Knight, 1948, Knight and Hutchinson, 1950). Resistance is conditioned by a number of factor pairs obtained from cottons of various types. Knight and associates have studied the world distribution and the inheritance of five major genes for blight resistance. They vary in dominance; several segregate independently; and genes designated B_1 and B_2 are additive in effect; also B_4 is additive for resistance when in combination with B_2 and B_3. They have hybridized the allopolyploid New World cottons with the diploid Old World varieties (Knight, 1948, Knight and Clouston, 1939, 1941, 1944, Knight and Hutchinson, 1950). Resistant varieties have been selected in the United States although relatively limited investigations on the genetics of resistance have been reported. Resistance in Stoneville 20 is reported to be controlled by a single recessive factor pair (Blank, 1949, Brinkerhoff *et al.*, 1952). Thiers and Blank (1951) studied the nature of resistance to bacterial blight. Resistance appears to be associated with physiological incompatibility between the resistant host studied and the isolates of the pathogen. Seedling reaction to the disease is similar to mature plant reaction (Weindling, 1944).

4. Anthracnose, *Glomerella gossypii* Edg. The disease is distributed widely on cotton. Severe damage to seedling stands and to bolls and seed occurs in the eastern humid section of the cotton belt in the United States. The disease is rare in the western drier area. The malady ranks first of those causing boll and seedling damage in the Southeastern United States, according to Edgerton (1912), Weindling *et al.* (1941), and others.

SYMPTOMS AND EFFECTS. The lesions and rotting occur on the cotyledons, leaves, stems, and bolls. Small reddish to light-colored spots or necrosis of the marginal tissues are common on the cotyledons. Similar oblong brown cankers occur on the hypocotyl and young stem (Harrold, 1943). Girdling of the stem occurs under conditions of high humidity. The lesions frequently are covered with the pink mass of spores in poorly developed acervuli. Limited dead brown spots occur on the leaves and stems, especially associated with injuries or angular leaf spots (Fig. 89). The symptoms on the bolls vary, depending upon the time and manner of infection. Infection on the side of the partly developed boll results in a sunken brown spot with reddish margin. Slimy pink masses of spores are common in the center of the lesion during humid weather. Infection through the dead pistil usually results in an internal rot and drying out of the boll. The general rot of the boll is associated with other organisms, especially *Xanthomonas malvacearum*. Acervuli with dark setae develop on the lesions after necrosis of the tissues.

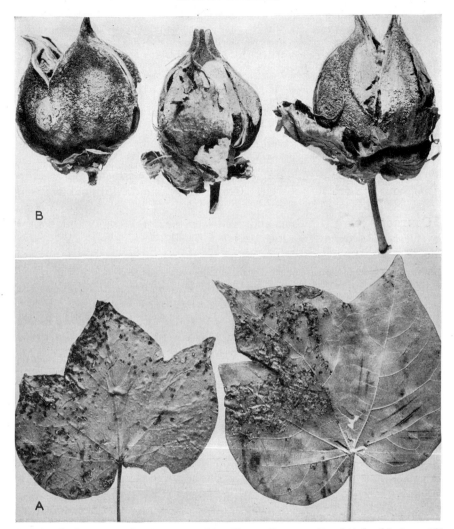

FIG. 89. Anthracnose of cotton caused by *Glomerella gossypii*. (*A*) Leaf lesions, (*B*) boll lesions.

THE FUNGUS

Glomerella gossypii Edg.
Colletotrichum gossypii South.

The perithecia, usually submerged in the dead tissues, are subglobose to pyriform, dark brown, with the beaks protruding, and 80–120 by 100–160 microns in size. The asci are numerous, clavate in shape, and intermixed with slender paraphyses. The ascospores, arranged uniseriately or irregularly biseriately, are elliptical, sometimes slightly curved, hyaline, and measure 5–8 by 12–20 microns. The conidia develop on the mycelium or in acervuli, depending upon moisture.   Acervuli vary in size and shape, and the setae are dark with hyaline tips.   Conidiophores are short, closely packed, and hyaline.   Conidia are cylindrical, generally straight, hyaline, one-celled, 3.5–7 by 12–25 microns in size, and are held in a mucilaginous mass on the acervulus. Ullstrup (1938) reported considerable variability in the fungus.

ETIOLOGY.   External seed infection is general, the internal infection is limited; both result in poor stands and abundant conidial inoculum from the lesions on the cotyledons.   Low temperature and low moisture increase the amount of seedling blight and damping-off of seedlings (Arndt, 1944, 1946, 1953).   Spread of the infection to the plant tissues aboveground from secondary inoculum occurs during humid weather throughout the growing season.   Crop residue is an additional source of inoculum.

CONTROL.   Crop rotation and seed treatment are important control measures.   Panogen, Ceresan, and Improved Ceresan control the seed infection borne externally and thus reduce the seedling infection and inoculum produced on infected seedlings.   Varieties show some differences in susceptibility.   Varieties resistant to *Xanthomonas malvacearum* show less boll rot from anthracnose.

5. Boll Rots and Seedling Blight.   A number of organisms are associated with boll rots and seedling blight in various sections of the world. The fungi associated with these diseases vary in different areas.   The two diseases discussed previously are probably the most important of the group.   Many of the secondary boll rots are due to infection following boll weevil injuries.   *Aspergillus* and *Rhizopus* spp. cause boll rots following injuries and are common in the Southwestern United States. Early infections cause severe damage, and later invasion discoloration of the fibers (Pearson, 1947).   The former produces pink to brown lesions, the latter olive green to brown.   Pink boll rot and root rot caused by *Gibberella fujikuroi* (Saw.) Wr., *Fusarium moniliforme* Sheldon is relatively common   (Harrold,   1943,   Woodruff,   1927).   In addition *Fusarium*, *Diplodia*, *Sclerotium* spp., and other fungi are associated with either boll rots, seedling blights, or both (Gottlieb and Brown, 1941, Ray and McLaughlin, 1942).

Seed treatment is used extensively as a control of the seedling blights. The organic mercury compounds, especially Ceresan and Panogen, are effective in increasing the stands of healthy plants.   Rotation of crops using winter legumes in the rotation, soil preparation, and soil fertility is an important preventative measure.

6. Fusarium Wilt, *Fusarium oxysporum* f. *vasinfectum* (Atk.) Snyder & Hansen.   The wilt is distributed throughout the world in areas of sandy acid soil.   The disease is among the oldest cotton maladies in the United States and causes heavy losses when susceptible varieties are grown on the sandy soils (Orton, 1900). *Fusarium* wilt is interrelated closely with avenues of entrance into the cotton roots produced by various nematodes: *Meloidogyne* spp.-inciting root knot; *Pratylenchus* spp.-meadow nematodes; *Belonolaimus* spp.-sting nematodes; *Trichodorus* spp.-stubby root nematodes; *Xiphinema* spp. and *Tylenchorynchus* spp.-dagger nematodes; *Heliocotylenchus* spp.-spiral nematodes; *Criconemoides* spp.-ring nematodes; *Rotylenchulus* spp.-reniform nematodes and others (Graham and Holdeman, 1953, Neal, 1954, Smith, 1941). The effect of certain of the nematodes on general plant vigor also influences wilt development and the expression of wilt resistance. Unbalanced fertility, especially low potassium, also influences the plant response to wilt and the expression of resistance.

Fusarium wilt is prevalent in the Southeastern United States into eastern Texas on cotton, okra, Burley tobacco, and the coffee weed (*Cassia tora* L.). It is important again in the lighter soils of the Southwestern irrigated sections where Acala Upland and American-Egyptian varieties are grown.

SYMPTOMS AND EFFECTS. The plants show considerable variation in symptoms, depending upon the degree of resistance of the variety and environmental conditions. The diseased plants are small with smaller leaves and bolls. The leaves are yellow to brown and drop from the base upward. Wilting of the leaves and premature death of the plants occur in susceptible varieties. Discoloration of the vascular elements and some plugging of the bundles are evident in sections of the root and stem. The symptoms apparently are caused by products of a toxic nature produced by the association of fungus and cotton plant rather than by the plugging of the xylem vessels, as reviewed by Gottlieb (1944) and Neal (1928) and shown by the recent studies of wilt in tomato (Gothoskar *et al.*, 1955, Waggoner and Dimond, 1955). The variability in pathogenicity of the fungus is discussed by Armstrong *et al.* (1940).

THE FUNGUS

Fusarium oxysporum f. *vasinfectum* (Atk.) Snyder & Hansen
(*Fusarium vasinfectum* Atk.)

The species *Fusarium oxysporum* Schl., as amended by Snyder and Hansen (1940, 1941), gives sufficient latitude morphologically to include the *Fusarium* spp. of section *Elegans* (Wollenweber and Reinking, 1935) or the members of this genus producing the wilt diseases. The physiologic varieties producing wilt on the respective plants or groups of plants are designated by the use of trinomials. Smith (1899), in the early investigations on this group, suggested this when he considered watermelon, cotton, and cowpea wilts as caused by varieties of the same species, and certain others have followed this practice.

The conidiophores are verticillately branched, and they form in sporodochia, reduced pionnotes, or less frequently, on the mycelium direct. Ellipsoidal unicellular microconidia, averaging 2–3.5 by 5–12 microns in size, occur commonly. Sickle-shaped macroconidia, hyaline, mostly 3-septate, but sometimes 4- and 5-septate, typically 3–4.5 by 40–50 microns in size are common. Vegetative resting cells (chlamydospores) form either terminally or intercalary, and sclerotia are common. The ascigerous stage is not known.

ETIOLOGY. Soil infestation is common in light acid or neutral soils. Infection occurs through the root system, with the fungus developing in the xylem. The wilting occurs during the growing season because of excretory toxic substances produced in the vascular elements.

Control is largely by the use of resistant varieties and crop practices. Varieties resistant to both wilt and nematodes give better yields in infested soils (Neal, 1954, Smith, 1941, Young and Humphrey, 1943). Wilt resistance is probably conditioned by a single factor pair (Fahmy, 1931, Kulkarni, 1937). In Egypt, the Asiatic cottons are susceptible as a group, whereas the American and Egyptian cottons contain resistant strains. The incorporation of organic material in the soil and a proper balance of potash fertilizer decrease the losses from wilt (Young and Tharp, 1941). Rotations to reduce nematode infestation decrease wilt damage.

7. Verticillium Wilt, *Verticillium albo-atrum* Reinke & Berth. The disease is distributed generally, and apparently it is associated with alkaline soils (Drummond, 1949, Garrett, 1947). This wilt is prevalent in the Mississippi delta and the Southwestern United States, Mexico, and similar areas. Verticillium wilt is severe in cool soils, especially when the plants are blossoming or later. Mottling of the leaves with pale-yellow irregular areas, later turning brown, are the most characteristic early symptoms. The infected plants ripen prematurely, and boll development is stopped. The interior xylem of the root and stem shows browning as in other plants infected with this wilt fungus.

THE FUNGI. *Verticillium albo-atrum* Reinke & Berth. and *V. dahliae* Kleb.

Conidiophores are branched verticillately, and conidia are borne singly at the apex of the branches. Conidia are globose to ovoid, hyaline to lightly colored. The two species commonly associated with the wilts are differentiated by Berkeley *et al.* (1931) by the difference in type of resting mycelium. *Verticillium albo-atrum* produces dark torulose resting hyphae, whereas *V. dahliae* produces pseudosclerotia and sclerotia. The fungus shows wide variability in morphology (Presley, 1950). Wilting and discoloration of the vascular bundles are associated with proteinaceous materials and polysaccharides, respectively (Green, 1944). Apparently the presence of these compounds is the result of enzymes synthesized by the pathogen in a system somewhat similar to that in the *Fusarium* wilts.

Both species are common in most of the areas where the disease is severe on crop plants. The rapid dissemination of the disease on cotton in the Southwestern United States and Mexico, even on newly broken land, is unexplained (Rudolph and Harrison, 1944). Tolerance to this wilt is associated with late maturity in most varieties. Wilt-tolerant varieties of the American types are being used in the Southwestern United States.   Resistant varieties of the *Gossypium barbadense* L. type are more general than in the other species (Wiles, 1953).

8. Phymatotrichum Root Rot, *Phymatotrichum omnivorum* (Shear) Dugg.   The *Phymatotrichum*, or cotton, root rot is found on more than

FIG. 90. *Phymatotrichum* root rot of cotton, showing plants in various stages of wilting and browning (*A*) and the discoloration of the cortex of the root and stem base (*B*).

2,000 different kinds of cultivated and wild plants.   Losses in cotton are high in Texas to Arizona and Mexico.   In so far as is known, the disease is limited to southwestern North America.

SYMPTOMS AND EFFECTS.   In local areas in the fields the plants wilt slightly and turn brown.   The first symptom visible on the plant is the yellowing and bronzing of the leaves.   Slight wilting of the leaves occurs as the next transitory symptom.   Soon after wilting the leaves become brown and dry but remain attached to the plant (Fig. 90).   Root rot is evident below the soil surface by the browning of the bark and cambium tissues and by the strands of the mycelium on the surface of the rotted roots. Spore mats appear on the soil surface near the dead plants when the soil is moist after summer rains or irrigation. The mats are first cottony and white, later becoming tan in color and powdery in texture. The root rot appears in patches when the plants are partly grown, and the areas of dead plants spread as the season advances.

THE FUNGUS

Phymatotrichum omnivorum (Shear) Dugg.
(*Ozonium omnivorum* Shear)

Streets (1937) gives a good description of the fungus and reviews the literature. The mycelium consists of two types: the less common large-celled mycelium and the fine-celled.   The mycelial strand consists of closely compressed hyphae with slender tapering side branches.   These cruciately branched acicular hyphae are characteristic of the fungus when the strands are actively vegetative.   Old strands are small, dark brown, and with little branching.   The spore mat develops from the strands into a large-celled much-branched hyphal mass with swollen tips (conidiophores) bearing numerous globose to ovate smooth conidia 4.8 to 5.5 microns in diameter or 5–6 by 6–8 microns in the ovate-shaped conidia.   Conidia apparently do not germinate or cause infection.   Sclerotia are round to irregular, small, light to dark brown, produced singly or in chains, and germinate to produce mycelium.

ETIOLOGY.   The mycelial strands and sclerotia persist in the soil for long periods.   Soil infestation in the alkaline dark soils once established is rather permanent (Taubenhaus and Ezekiel, 1936).   Infection and rotting of the roots of the numerous plants affected occur during the growing season, especially from mid-summer into the autumn.   The northern boundary of the general infestation apparently is determined by winter temperatures (Ezekiel, 1945).   The monocotyledonous plants are immune to the root rot under field conditions, although corn is attacked under greenhouse conditions.   Substances within the tissues apparently contribute to the immunity (Ezekiel and Fudge, 1938, Greathouse and Rigler, 1940).   The fungus development in the soil is checked by the competition and antibiotic action of other soil organisms.   Soil conditions favorable for the development of soil microflora apparently offer the best means of holding the root rot parasite in check (Crawford, 1941, Mitchell, 1941).

CONTROL.   Economical control measures frequently are inadequate, especially in the alkaline soils to which the fungus is adapted.   Rotation with oats, corn, sorghum, and grasses reduces the damage.   The use of winter cover crops, especially legumes, reduces the root rot in cotton in most of the areas where the disease is severe.   In the central black-lands of Texas, the disease is controlled by growing a legume cover crop (Chap. 17) during the winter and turning the green legume under before cotton is planted in the spring (Presley, 1954). The use of organic material high in nitrogen or with nitrogen fertilizer, especially when combined with deep tillage and aeration, decreases root rot damage and increases cotton yields (Adams *et al.*, 1939, Blank, 1944, Jordan *et al.*, 1939). Early-maturing varieties tend to escape the damage. Some difference in reaction of cotton varieties and selections is indicated by Goldsmith and Moore (1941), Streets (1937), and Taubenhaus and Ezekial (1936), although the large number of *Gossypium* spp. and varieties tested to date do not indicate the presence of resistance to this disease.

9. Ascochyta Blight, *Ascochyta gossypii* Syd. The disease is distributed widely in humid areas and frequently is a factor in economic production of cotton in cooler humid climates. The disease is important along the northern section of the cotton belt of Eastern and Central United States, although this was not fully recognized prior to 1947. This blight limits somewhat the growing of cotton at high elevations in Mexico and South America and in similar cool wet locations in Asia and Africa (Wallace, 1948). Small brown spots turning gray occur on the leaves. Elongated brown to gray cankers develop on the young stems. The lesions spread and enlarge during wet cool weather and result in defoliation and blighting of young plants. Seedling blight soon after emergence occurs also. Later local lesions develop on leaves, young apical stems, and balls. The disease frequently develops in association with anthracnose (Presley, 1954, Smith, 1953).

THE FUNGUS

Ascochyta gossypii Syd.

Pycnidia, formed in lesions and on old stems, are globose, ostiolate, dark brown. Spores are cylindrical, hyaline, two-celled, with slight constriction at septum. The morphology of many species in this genus is similar, especially when the influence of environment in the expression of morphological characters is considered. The similarity in morphology and host range of this species with those on okra and legumes is significant (Crossan, 1953, Thompson, 1953).

ETIOLOGY. The pathogen is seed-borne and persists on crop refuse. Both sources of inoculum are important in the seedling and young-plant infections. Spores produced on crop residue and leaf and stem lesions provide inoculum for later spread. The disease is severe when cotton follows cotton.

Control is satisfactory when rotation and clean plowing are practiced. Seed treatment with organic mercury compounds controls seed-borne inoculum. Some varieties are more resistant to later stem cankers than others. Dusting with basic copper compounds gives better control than sulfur.

Cercospora gossypina Cke. incites a circular zonate leaf spot on cotton during warm weather. The disease rarely is important.

10. Soreshin, *Rhizoctonia solani* Kühn or *Pellicularia filamentosa* (Pat.) Rogers. The seedling stem canker known as "soreshin" is distributed generally on cotton and is common in the United States. Losses in seedling stands due to the disease are large in some areas and seasons, making necessary the replanting of large acreages.

SYMPTOMS. The characteristic symptom is the stem canker or soreshin near the soil line, although seedling blight is very common earlier in the season. The seedlings blight before or after emergence during cold, wet weather. The seedling rot develops first in the hypocotyl with little discoloration of the cortical tissues. Necrosis of the hypocotyl tissues and blighting of the seedling occur in cold, wet soils. Under conditions less favorable for disease development, the hypocotyl lesions later appear as reddish-brown sunken cankers extending into the base of the stem. These cankers range from linear cortical lesions to completely girdling the stem near the soil line. Partial recovery of the plants with new root growth occurs, especially in older plants, as the soil becomes warm. Angular brown shot-hole-type leaf spots occur less frequently than the stem lesions (Neal, 1942, 1944).

The fungus is soil-borne and attacks a wide range of plants. The disease develops on cotton during periods of cold, wet weather. Crop rotations, good soil preparation, seed treatment with the mercury dusts and Arasan, and planting after the soil is warm are important control measures (Fahmy, 1931). Seed treatment is not effective in controlling the later cankering of the stems, although early hypocotyl invasion and seedling blight are reduced by the residual protective action of the dust fungicide.

11. Rust, *Puccinia stakmanii* Presley. The aecial stage of *Puccinia stakmanii* occurs on cotton in the Southwestern United States and Mexico. Under favorable conditions, infection is heavy and causes reduction in yields. The uredial and telial stage of the rust occur on *Bouteloua* spp., desert grasses common through the area (Presley, 1954, Presley and King, 1943).

SYMPTOMS. The circular slightly elevated orange-yellow to citron-yellow aecia occur chiefly on the undersurface of the cotton leaves (Fig. 91).

THE FUNGUS

Puccinia stakmanii Presley
(*Aecidium gossypii* Ell. & Ev.)
(*Puccinia hibisciata* Kell.)
(*Puccinia schedonnardi* Kell.)

The aecia develop chiefly on the leaves as slightly elevated spots, 2 to 5 mm. in diameter, with the conspicuous peridia first orange and later fading to yellow, and the

FIG. 91. Part of a cotton leaf showing the densely massed aecia of *Puccinia stakmanii* on the leaf surface.

margin of the peridium lacerate or recurved. Aeciospores are globose or broadly oblong, pale yellow to colorless, and finely verrucose. Uredia are formed chiefly in the leaves of the grasses, and they are pale brown with the epidermis of the leaf turned back. Urediospores are globoid or broadly ellipsoid, echinulate, with equatorial germ pores which distinguish this species from the others common on the grasses. Telia are abundant in the uredia, and they are formed also independently as naked sori.

Teliospores are oblong to broadly ellipsoid, only slightly or not constricted at the septum, and rounded at both ends; the pedicel is long and sometimes attached at an angle.

ETIOLOGY. The uredia and telia develop on the grasses during the spring and early summer. Sporidial inoculum from the teliospores in the grasses is abundant during the summer rains and results in the aecial infection on the cotton. The presence of rusted grasses in the fields or in waste areas near the fields and summer rains determines the amount of rust damage on the cotton. Most American cotton varieties are susceptible, whereas the Asiatic cottons are resistant or moderately susceptible.

A second rust, *Cerotelium desmium* (Berk. & Br.) Arth., occurs sparingly on cotton in the Southeastern United States and more generally in South America, West Indies, India, and the South Pacific. Only the uredial and telial stages are known, more recently listed as *Phakopsora desmium* (Berk. & Br.) Cumm.

REFERENCES

ADAMS, J. E., *et al.* Chemistry and growth of cotton in relation to soil fertility and root-rot. *Soil Sci. Soc. Am. Proc.* **4**:329–332. 1939.
———. Cotton. *Adv. Agron.* **2**:2–74. 1950.
ANSON, R. R., and K. C. BARLOW. Progress reports from the Experiment Stations, season 1949–1950. *London Empire Cotton Growing.* 1951.
ARCHIBALD, R. G. Blackarm disease of cotton with special reference to the existence of the causal organism, *Bacterium malvacearum*, within the seed. *Soil Science* **23**:5–10. 1927.
ARMSTRONG, G. M., *et al.* Variation in pathogenicity and cultural characteristics of the cotton-wilt organism, *Fusarium vasinfectum*. *Phytopath.* **30**:515–520. 1940.
ARNDT, C. H. Infection of cotton seedlings by *Colletotrichum gossypii* as affected by temperature. *Phytopath.* **34**:861–869. 1944.
———. Effect of storage conditions on the survival of *Colletotrichum gossypii*. *Phytopath.* **36**:24–29. 1946.
———. Survival of *Colletotrichum gossypii* on cotton seed in storage. *Phytopath.* **43**:220. 1953.
BAIN, D. C. Effect of sulfuric acid treatment on fungi and bacteria present in cotton seed from diseased bolls. *Phytopath.* **29**:879–883. 1939.
BERKELEY, H. G., *et al.* Verticillium wilts in Ontario. *Sci. Agr.* **11**:739–759. 1931.
BLANK, L. M. Effect of nitrogen and phosphorus on the yield and root-rot response of early and late varieties of cotton. *Jour. Am. Soc. Agron.* **36**:875–888. 1944.
BRINKERHOFF, L. A., J. M. GREEN, R. HUNTER, and G. FINK. Frequency of bacterial blight-resistant plants in twenty cotton varieties. *Phytopath.* **42**:98–100. 1952.
BROWN, J. G. Wind dissemination of angular leaf spot of cotton. *Phytopath.* **32**:81–90. 1942.
——— and F. GIBSON. A machine for treating cotton seed with sulfuric acid. *Ariz. Agr. Exp. Sta. Bul.* 105. 1925.
——— and R. B. STREETS. Diseases of field crops in Arizona. *Ariz. Agr. Exp. Sta. Bul.* 148. 1934.

COOPER, H. P.   Nutritional deficiency symptoms in cotton.   *Soil Sci. Soc. Am. Proc.*
4:322–324. 1939.

CRAWFORD, F. R.   Root-rot and its control.   *N.Mex. Agr. Exp. Sta. Bul.* 283. 1941.

CROSSAN, D. F.   Comparative studies on species of *Ascochyta* from okra, bean and
cotton in North Carolina.   *Phytopath.* 43:469. 1953.

DICKSON, R. C., M. McD. JOHNSON, and E. F. LAIRD, JR.   Leaf crumple, a virus
disease of cotton.   *Phytopath.* 44:479–480. 1954.

DIFRONZO, M.   Las enfermedades del Algodonero en la República Argentina.   *Bol.
Junta Algodón, Buenos Aires,* 80:951–978. 1941.

DRUMMOND, O. A.   Notas sôbre a murcha do Algoderio, causada pelo *Verticillium
dahliae.*   *Lilloa Rev. Bot. Tucumán* 21:54–56. 1949.

DUNLAP, A. A.   Fruiting and shedding of cotton in relation to light and other limiting
factors.   *Tex. Agr. Exp. Sta. Bul.* 677. 1945.

EDGERTON, C. W.   The rots of the cotton boll.   *La. Agr. Exp. Sta. Bul.* 137. 1912.

EZEKIEL, W. M.   Effect of low temperatures on survival of *Phymatotrichum omnivorum.*   *Phytopath.* 35:296–301. 1945.

———— and J. F. FUDGE.   Studies on the cause of immunity of monocotyledonous
plants to Phymatotrichum root-rot.   *Jour. Agr. Research* 56:773–786. 1938.

FAHMY, T.   The genetics of resistance to the wilt disease of cotton and its importance
in selection.   *Minn. Agr. Egypt, Tech. Sci. Serv. Bul.* 95. 1931.   *Rev. Appl.
Myc.* 11:178–179. 1932.

————.   The sore-shin disease and its control.   *Minn. Agr. Egypt, Tech. Sci. Serv.
Bul.* 108. 1931.   *Rev. Appl. Myc.* 11:454–455. 1932.

FAULWETTER, R. C.   Dissemination of the angular leaf spot of cotton.   *Jour. Agr.
Research* 8:457–475. 1917.

GARRETT, S. D.   Report on investigations of Verticillium wilt.   *Empire Cotton
Growing Rev.* 24:101–102. 1947.

GOLDSMITH, G. W., and E. J. MOORE.   Field tests of the resistance of cotton to
Phymatotrichum omnivorum.   *Phytopath.* 31:452–463. 1941.

GOTHOSKAR, S. S., R. P. SCHEFFER, M. A. STAHMANN, and J. C. WALKER.   Further
studies on the nature of Fusarium resistance in tomato.   *Phytopath.* 45:303–307.
1955.

GOTTLIEB, M., and J. G. BROWN.   *Sclerotium rolfsii* on cotton in Arizona.   *Phytopath.* 31:944–946. 1941.

GRAHAM, T. W., and Q. L. HOLDEMAN.   The sting nematode *Belonolaimus gracilis*
Steiner: a parasite on cotton and other crops in South Carolina.   *Phytopath.*
43:434–439. 1953.

GREATHOUSE, G. A., and N. E. RIGLER.   The chemistry of resistance of plants to
Phymatotrichum root-rot.   IV. Toxicity of phenolic and related compounds.
Am. Jour. Bot. 27:99–108. 1940.

———— and ————.   The chemistry of resistance of plants to Phymatotrichum root-
rot.   V. Influence of alkaloids on growth of fungi.   *Phytopath.* 30:475–485.
1940.

GREEN, R. J., JR.   A preliminary investigation on toxins produced in vitro by
Verticillium albo-atrum.   *Phytopath.* 44:433–437. 1954.

HARE, J. F., and C. J. KING.   The winter carry-over of angular leaf spot infection in
Arizona fields.   *Phytopath.* 30:679–684. 1940.

HARLAND, S. C., and O. M. ATTECK.   The genetics of cotton.   XVIII.   *Jour. Gen.*
42:1–19. 1941.

HARROLD, T. J.   Histological studies of infections of the cotton hypocotyl by *Glomerella gossypii* and *Fusarium moniliforme.*   *Phytopath.* 33:666–673. 1943.

HUTCHINSON, J. B., R. L. KNIGHT, and E. O. PEARSON. Response of cotton to leaf-curl disease. *Jour. Gen.* **50**:100–111. 1950.

JORDAN, H. V., *et al.* Relation of fertilizers, crop residues, and tillage to yields of cotton and incidence of root-rot. *Soc. Soil Sci. Am. Proc.* **4**:325–328. 1939.

KIRKPATRICK, T. W. Further studies on the leaf curl of cotton in the Sudan. *Bul. Entom. Research* **22**:323–363. 1931.

KNIGHT, R. L. The genetics of black-arm resistance. VI. Transference of resistance from *Gossypium arboreum* to *G. barbadense*. *Jour. Gen.* **48**:359–369. 1948.

———. The role of major genes in the evolution of economic characters. *Jour. Gen.* **48**:370–387. 1948.

——— and T. W. CLOUSTON. The genetics of blackarm resistance. I. Factors B_1 and B_2. *Jour. Gen.* **38**:133–159. 1939. III. Inheritance in the crosses with the *Gossypium hirsutum* group. *Jour. Gen.* **41**:391–409. 1941. IV. *Gossypium punctatum* (Sch. and Thon) crosses. *Jour. Gen.* **46**:1–27. 1944.

KULKARNI, G. S. Studies on the wilt disease of cotton in the Bombay Presidency. *Indian Jour. Agr. Sci.* **4**:976–1048. 1934.

LONGLEY, A. E. Chromosomes in Gossypium and related genera. *Jour. Agr. Research* **46**:217–227. 1933.

MASSEY, R. E. On the relation of soil temperature to angular leaf spot of cotton. *Ann. Bot.* **41**:497–507. 1927.

MITCHELL, R. B., *et al.* Soil bacteriological studies on the control of the Phymato-trichum root rot of cotton. *Jour. Agr. Research* **63**:535–547. 1941.

NEAL, D. C. Cotton wilt: A pathological and physiological investigation. *Miss. Agr. Exp. Sta. Tech. Bul.* 16. 1928.

———. Rhizoctonia infection of cotton and symptoms accompanying the disease in plants beyond the seedling stage. *Phytopath.* **32**:641–642. 1942.

———. Rhizoctonia leaf spot of cotton. *Phytopath.* **34**:599–602. 1944.

———. The reniform nematode and its relationship to the incidence of Fusarium wilt of cotton at Baton Rouge, Louisiana. *Phytopath.* **44**:447–450. 1954.

——— and W. W. GILBERT. Cotton diseases and methods of control. *U.S. Dept. Agr. Farmers' Bul.* 1745. 1935.

ORTON, W. A. The wilt disease of cotton. *U.S. Dept. Agr. Div. Veg. Physiol. Path. Bul.* 27. 1900.

——— and W. W. GILBERT. The control of cotton wilt and root-knot. *U.S. Dept. Agr. B.P.I. Cir.* 92. 1912.

PEARSON, E. O. The development of internal boll diseases of cotton in relation to time of infection. *Ann. Appl. Biol.* **34**:527–545. 1947.

PRESLEY, J. T. Verticillium wilt of cotton with particular emphasis on variation of the causal organism. *Phytopath.* **40**:497–511. 1950.

———. Cotton diseases and methods of control. *U.S. Dept. Agr. Farmers' Bul.* 1745. 1954.

——— and C. J. KING. A description of the fungus causing cotton rust, and a preliminary survey of its hosts. *Phytopath.* **33**:382–389. 1943.

RAY, W. W., and J. H. McLAUGHLIN. Isolation and infection tests with seed- and soil-borne cotton pathogens. *Phytopath.* **32**:233–238. 1942.

RUDOLPH, B. A., and G. J. HARRISON. The unimportance of cotton seed in the dissemination of Verticillium wilt in California. *Phytopath.* **34**:849–860. 1944.

SMITH, A. L. The reaction of cotton varieties to Fusarium wilt and root-knot nematode. *Phytopath.* **31**:1099–1107. 1941.

———. Anthracnose and some blights. *U.S. Dept. Agr. Yearbook*, pp. 303–311. 1953.

Smith, E. F.   Wilt disease of cotton, watermelon, and cowpea (*Neocosmospora* n. gen.).   *U.S. Dept. Agr. Veg. Physiol. Path. Bul.* 17.   1898.

Snyder, W. C., and H. N. Hansen.   The species concept in Fusarium.   *Am. Jour. Bot.* **27**:64–67.   1940.

——— and ———.   The species concept in Fusarium with reference to section Martiella.   *Am. Jour. Bot.* **28**:738–742.   1941.

Stephens, S. G.   Colchicine-produced polyploids in Gossypium.   I.   *Jour. Gen.* **44**: 272–295.   1942.

Stoughton, R. H.   The influence of environmental conditions on the development of angular leaf-spot disease of cotton.   *Ann. Appl. Biol.* **15**:333–341.   1928.   **17**: 493–503.   1930.   **18**:524–534.   1931.   **19**:370–377.   1932.   **20**:590–611.   1933.

Streets, R. B.   Phymatotrichum (cotton or Texas) root rot in Arizona.   *Ariz. Agr. Exp. Sta. Tech. Bul.* 71.   1937.   (Good review of literature.)

Tarr, S. A. J.   Leaf curl disease of cotton.   Commonwealth Mycological Institute. Kew, Surrey.   1951.

Taubenhaus, J. J., and W. M. Ezekiel.   A rating of plants with reference to their relative resistance or susceptibility to Phymatotrichum root rot.   *Tex. Agr. Exp. Sta. Bul.* 527.   1936.

Tennyson, G.   Invasion of cotton seed by *Bacterium malvacearum*.   *Phytopath* **26**: 1083–1085.   1936.

Thiers, H. D., and L. M. Blank.   A histological study of bacterial blight of cotton. *Phytopath.* **41**:499–510.   1951.

Thompson, G. E.   A comparison of *Ascochyta abelmoschi* Harter and *A. gossypii* Sydow in culture and inoculation experiments.   *Phytopath.* **43**:293–294.   1953.

Ullstrup, A. J.   Variability of *Glomerella gossypii*.   *Phytopath.* **28**:787–798.   1938.

Wadleigh, C. H.   Growth status of the cotton plant as influenced by the supply of nitrogen.   *Ark. Agr. Exp. Sta. Bul.* 446.   1944.

Waggoner, P. E., and A. E. Dimond.   Production and role of extracellular pectic enzymes of *Fusarium oxysporum f. lycopersici*.   *Phytopath.* **45**:79–87.   1955.

Wallace, W. M.   Ascochyta blight of cotton.   *E. Africa Agr. Jour.* **14**:10–11. 1948.

Ware, J. O.   Plant breeding and the cotton industry.   *U.S. Dept. Agr. Yearbook*, pp. 647–744.   1936.

Webber, J. M.   Relationships in the genus Gossypium as indicated by cytological data.   *Jour. Agr. Research* **58**:237–261.   1939.

Weindling, R.   A technique for testing resistance of cotton seedlings to the angular leaf-spot bacterium.   *Phytopath.* **34**:235–239.   1944.

——— et al.   Fungi associated with diseases of cotton seedlings and bolls, with special consideration of *Glomerella gossypii*.   *Phytopath.* **31**:158–167.   1941.

Wiles, A. B.   Reaction of cotton varieties to Verticillium wilt.   *Phytopath.* **43**:489. 1953.

Woodroof, N. C.   A disease of cotton roots produced by *Fusarium moniliforme* Sheld.   *Phytopath.* **17**:227–237.   1927.

Wollenweber, H. W., and O. A. Reinking.   Die Fusarien.   Paul Parey.   Berlin. 1935.

Young, V. H., and L. M. Humphrey.   Varietal resistance to the Fusarium wilt disease of cotton.   *Ark. Agr. Exp. Sta. Bul.* 437.   1943.

——— and W. H. Tharp.   Relation of fertilizer balance to potash hunger and the Fusarium wilt of cotton.   *Ark. Agr. Exp. Sta. Bul.* 410.   1941.

CHAPTER 19

FLAX DISEASES

Varieties of common flax, *Linum usitatissimum* L., consist of two types, fiber and seed. These differ considerably in character of plant growth. The fiber types develop tall slender stems, produce a high content of good-quality fiber, and bear seeds of low oil content. Seed-flax varieties develop shorter stems with more tendency to branch and usually bear larger seeds of higher oil content. Flax is adapted to a wide range of environmental conditions but does best in cool climates. It is grown in the warmer regions of the temperate zones as a winter crop and in the cooler regions as a summer crop. Seed-flax varieties are more generally grown in the drier areas, whereas fiber flax is grown in the humid sections.

A wild species, *Linum angustifolium* Huds., apparently is related closely to the cultivated flax. Both are of Asiatic or European origin, have 15 pairs of chromosomes (Kiluchi, 1929, Ray, 1944, Tammes, 1928), and hybridize readily. The other wild *Linum* spp. of the Old and New World that have been investigated have not been compatible in crossing with common flax, although several have the same number of chromosomes (Ray, 1944). Some of these wild *Linum* spp. are susceptible to important diseases of cultivated flax, notably pasmo and rust.

Flax improvement in North America has centered primarily about the successful battle against diseases (Dillman, 1936), although more recently attention has been directed toward combining disease resistance with improved oil content and quality. Flax wilt played an important role in the development of the crop in North America. As early as 1890, Lugger (1896) conducted experiments on the control of wilt by the use of fertilizers and seed treatments. Bolley (1901) described the disease, named the parasite, and reported on wilt-resistant varieties; although unknown to him, Broekema (1893), in the Netherlands, described a similar disease complex and reported resistance to the malady; and in 1896, in Japan, Hiratsuka described the disease as being caused by a species of *Fusarium*. While losses from flax wilt, rust, and pasmo are greatly reduced by the use of resistant varieties, these and other diseases cause heavy losses in this crop in some years.

The diseases of both fiber and seed flax, especially in Europe, are discussed by McKay (1947), Millikan (1951), and Muskett and Colhoun (1947). The recent review of flax diseases, races of the pathogens, and resistant varieties by Flor and Christensen is included in "Seed Flax Improvement" (Culbertson, 1954).

1. Heat Canker, Nonparasitic. Heat injury of the cortical tissues of hypocotyl and stem near the soil line is common on flax (Reddy and Brentzel, 1922) and other succulent herbaceous plants (Hartley, 1918, Harvey, 1923, Tubeuf, 1914). Surface temperatures in the dark-colored soils frequently are high enough to kill the cells of young cortical tissues before the plants are large enough to shade the surface. Cankers resulting from such injury are common in flax in the semihumid plains and at high altitudes throughout the world. The cortical tissues collapse, resulting in the death of the young seedlings or in sunken brown lesions on the stems. The stems usually enlarge above the canker in the plants that survive the initial injury (Fig. 92). Cortical rotting organisms frequently invade the injured tissues to increase the damage (Reddy and Brentzel, 1922). Preventing excessive soil temperatures by early planting, by drilling the rows north and south so as to secure maximum shading, by higher rates of seeding, by use of a nurse crop, by mulching the soil surface, and by irrigation are means of reducing this type of damage.

Deficiency of minor elements, especially zinc, causes necrosis of the growing point and necrotic spots on the leaves. The growth and seed production of flax are reduced materially in soils low in available minor elements (Cass-Smith and Harvey, 1948, Millikan, 1951).

2. Virus Diseases, Transmitted by Leaf Hoppers. Apparently the flax plant is relatively free from virus diseases, although two occur in limited areas. The curly top virus occurs in cultivated and wild flax in western North America (Giddings, 1948). Aster yellows virus occurs on flax in central and western North America (Severin and Houston, 1945). In recent years flax plants with yellow-green apical proliferation of flowers and leaves have been found, especially along the margins of fields.

3. Seed Rot, Seedling Blight, and Root Rot, *Pythium* spp. and Other Fungi. The seedling disease complex known as seedling blight, scorch, fire, and root rot is common in the flax-producing areas of the world. The *Pythium* root rot is probably the most important single factor in this complex. The literature is reviewed by Berkeley (1944).

The symptoms of the disease complex vary considerably. The seedlings blight in the early stages of germination or after emergence in the more severe manifestations of the malady. When the disease develops following the seedling stage, the plants are dwarfed, the lower leaves turn brown, and the root system is reduced by a brown soft rot. The cortical tissues are invaded, followed by the rotting of the vascular tissues under conditions of high soil moisture and high temperature.

The fungus *Pythium debaryanum* Hesse reduces stands and causes root

Fɪɢ. 92. Heat canker of flax caused by high soil temperatures near the soil surface.

rot, especially in the prairie soils of North America. Its morphology is given in Chap. 13.

According to Diddens (1932) and Middleton (1943), *Pythium aphanidermatum* (Edson) Fitzp., *P. splendens* Braun, *P. vexans* DBy., *P. megalacanthum* DBy., *P. mamillatum* Meurs, *P. irregulare* Buis., and *P. intermedium* DBy. are associated less commonly with root rot of flax. *Olpidium brassicae* (Wor.) Dang. (*Asterocystis radicis* DeWild.), *Thielaviopsis basicola* (Berk. & Br.) Ferr., and other fungi are associated with the root rot.

Several other pathogens, *Polyspora lini* Laff., *Colletotrichum linicolum* Pethyb. & Laff., *Fusarium* spp., *Coniothyrium olivaceum* Bonn., *Rhizoctonia* spp., cause seed rotting and seedling blight of flax.

The soil-borne organisms invade the seedling tissues under favorable environmental conditions, and they cause severe damage to stands when mechanically injured seed is used. Crop rotation and seed treatment with the organic mercury compounds or Arasan help protect stands.

4. Powdery Mildew, *Erysiphe* spp. The powdery mildew is probably of no economic importance in North America, Europe, and Asia, although it is reported abundant on flax in Siberia. The powdery superficial gray mycelium and conidia are formed on the leaves, stems, and floral structures.

Frequently only the oidium (conidial) stage develops, and the identification of the pathogen is based on this alone. Two species are reported on flax (Allison, 1934, Homma, 1928, McKay, 1947).

THE FUNGI

Erysiphe polygoni DC.

See Chap. 14 for morphology of this species.

Erysiphe cichoracearum DC. or
Erysiphe communis Wallr. According to Cooke (1952)

Cleistothecia are scattered, reddish brown, and develop simple, slightly colored appendages and contain 10 to 15 asci. The asci are slightly stalked, ovate to broadly ovate, hyaline, and usually contain two oblong to ovate, single-celled ascospores. The ovate-elliptic, hyaline, continuous conidia are borne in chains on short unbranched conidiophores. The morphology of the two species is similar. The latter species is prevalent in the Asiatic flax areas.

5. Pasmo, *Septoria linicola* (Speg.) Gar., *Mycosphaerella linorum* (Wr.) Garcia-Rada. Pasmo occurs in the seed- and fiber-producing areas of North and South America, New Zealand, and Europe. The disease causes defoliation, reduces yield of seed, and damages the fiber (Brentzel, 1930, Sackston, 1950). Newhook (1942) reported the disease as general on wild flax (*Linum marginale* A. Cunn.) in New Zealand.

SYMPTOMS. The symptoms of the disease are striking and easily recognized, especially during the latter part of the growing season. Lesions develop first on the cotyledons and later on the lower leaves of the seedlings. The lesions are generally circular in outline and vary in color from greenish yellow to dark brown, depending upon age. Pycnidia develop abundantly in the older lesions on the cotyledons and on the leaves. Later the stem lesions develop, first as small elongated lesions which then enlarge and coalesce, extending around the stem as well as longitudinally. The infected areas alternate with green tissue until infection becomes severe; then the stems brown as the plants are defoliated in the spots where the disease is severe (Fig. 93). Lesions also occur on the bolls. The pycnidia develop on the stem lesions as they turn brown. Perithecia occur sparingly on the old dead stems.

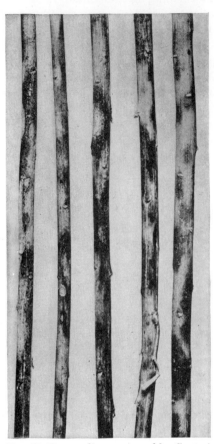

FIG. 93. Pasmo disease caused by *Septoria linicola* or *Mycosphaerella linorum* on flax stems.

THE FUNGUS

Mycosphaerella linorum (Wr.) Garcia-Rada
Septoria linicola (Speg.) Gar. Conidial stage
(*Sphaerella linorum* Wr.)
(*Phlyctaena linicola* Speg.)
(*Septogloeum linicola* Speg.)

The binomial *Sphaerella linorum* Wr. used by many plant pathologists is considered invalid (see W. B. Grove. Sphaerella v. Mycosphaerella. *Jour. Bot.* **50**:89–92, 1912). The British mycologists proposed the acceptance of *Sphaerella* as a *nomen conservandum* at the Stockholm meeting of the International Botanical Congress, but no action was taken.

The perithecia formed on the dead stems are spherical to oval, 70–100 by 60–90 microns in size, and scattered. The asci are oblong to clavate or subcylindrical, sessile, straight or tortuous, hyaline, with eight spores irregularly biseriate or uniseriate. The ascopores are fusiform, mostly curved, two-celled, hyaline, and measure 2.5–4 by 11–17 microns. The submerged pycnidia are subglobose, 62 to 126 microns in diameter, with small ostioles when fully developed. The conidia are subcylindrical, tapering slightly to the ends, straight or curved, usually 3-septate, but some are 7-septate when mature; the 3-septate conidia measure 1.5–3.0 by 12–28 microns and are hyaline. Wollenweber (1938) described the ascigerous stage as *Sphaerella linicola* Wr. but in a note at the time of publication changed the binomial to *S. linorum* Wr., as Naoumoff (1926) had described *Mycosphaerella linicola* Naum. on flax stems in Russia. Wollenweber decided the two were not the same, as the perithecia described by Naoumoff were larger (200 microns). The descriptions are similar otherwise. Naoumoff (1926) described *Ascochyta linicola* Naum. & Wass. causing stem browning and *Phoma linicola* Naum. on the stems late in the season. The latter fungus is probably the same as described on *Linum mucronatum* in 1914 from Mesopotamia as *P. linicola* Bub., from Belgium in 1926 on *L. usitatissimum* as *P. linicola* March. & Verpl., and from Ireland as *Phoma* sp. Pethybridge *et al.*, 1921. Rost (1937) discussed the distribution of *M. linorum* and Garassini's transfer of the conidial stage to the genus *Septoria*. Specialization of the parasite was studied by Rodenhiser (1930).

ETIOLOGY. The fungus persists on old straw, both as mycelium and conidia. The perithecial stage apparently is unimportant in the etiology of the fungus, and it is not common in North America. Primary and secondary inoculum is largely from conidia. Continued wet weather late in the growing season increases the spread and damage.

CONTROL. Rotation and removal of straw or covering the straw in plowing helps reduce the inoculum. Seed treatment reduces the seed-borne conidia. Resistant varieties, such as (C.I. 975), Bison (C.I. 389), Koto (C.I. 842), and Crystal (C.I. 982), reduce the damage materially.

6. Browning and Stem Break, *Polyspora lini* Laff. The disease is distributed generally on flax. It is of little economic importance in North America, but causes damage to fiber flax in Europe, New Zealand, and Asia.

SYMPTOMS. The symptoms of browning are conspicuous on the seedlings and on plants approaching maturity. Browning is largely a seed-borne disease, and the primary light-gray to brown circular lesions with darkened margin develop on the cotyledons. From the primary lesions the infection spreads to the cotyledonary node where a canker is formed. At any subsequent stage the stem may break at the canker. Since most stems are not completely severed, they become semierect, but ripen prematurely, producing little seed. Circular gray to brown lesions develop on the leaves, capsules, and stems late in the season. The development of the fungus is largely in the cortical and parenchymatous tissues. Minute acervuli-like structures bearing abundant conidia develop on the lesions during periods of high moisture.

THE FUNGUS

Polyspora lini Laff.

Small indefinite stromata are formed usually over the stomatal cavities. The hyaline conidiophores form as branches from this mycelial mat and bear masses of conidia from the swollen tips. No setae are formed. The conidia are oval to cylindrical, straight, with bluntly pointed ends, hyaline, and nonseptate. The width of the conidia is constant at about 4 microns; the length is 9 to 20 microns (Henry, 1925, 1938, Lafferty, 1921, Schilling, 1922).

Krenner (1954), studying the pathogen in Egypt, found two types of spores and suggested the transfer to *Microstroma lini* (Laff.) Krenner in the Exobasidiaceae.

ETIOLOGY. The fungus is primarily seed-borne but may persist in the crop residue. Inoculum from the cotyledons and infected seedlings initiate the general infection. Secondary spread occurs during periods of high moisture (Flor, 1936, Henry, 1925, Lafferty, 1921).

CONTROL. The use of disease-free seed, seed treatments, and resistant varieties are important in control of browning. Seed treatment with the organic mercury compounds reduces seed infection (Henry, 1938, Muskett and Calhoun, 1941). Hot-water treatment, 10 minutes at 126°F., is recommended by Baylis (1941). The variety Rio (C.I. 280) is highly resistant to the disease (Flor, 1925, Baylis, 1941).

7. Anthracnose, *Colletotrichum linicolum* Pethyb. & Laff. The anthracnose occurs on both the grain and fiber-flax varieties in humid, cool areas throughout the world. Damage is severe in the California seed-flax area on Punjab (C.I. 20) grown in the winter and in the fiber-flax regions of Europe and Asia. Bolley (1903) first described the disease as "flax canker" and named the organism *Colletotrichum lini* Bolley, but he did not include a description of the fungus.

SYMPTOMS. The symptoms are typical of most of the anthracnose diseases. Cankers on the cotyledons are circular zonated sunken brown spots that spread under cool and moist conditions to involve the cotyledons and apex of the stem. Seedling blight occurs either before or after emergence, in the latter case usually as a stem canker at the soil line. Leaf spots and stem cankers are common during the growing season, especially under conditions of high moisture. Brown spots form on the capsules, and less conspicuous lesions are found on the seed. Acervuli develop on the mature lesions.

THE FUNGUS

Colletotrichum linicolum Pethyb. & Laff.
(*Colletotrichum lini* Bolley)
[*Colletotrichum lini* (West.) Toch.]

The acervuli are formed subepidermally, and they rupture the epidermis. Setae are erect, usually 3-septate, and dark brown. Conidiophores are short, hyaline, and mostly simple. Conidia are cylindrical, tapering toward both ends, straight to slightly curved, hyaline, and one-celled. Westerdijk (1916) reported a somewhat similar disease, suggested the *Colletotrichum* nature of the fungus, and named the fungus *Gloeosporium lini* Westerdijk; however, the fungus was probably *Polyspora lini*.

ETIOLOGY. Infected seed and crop residue are the important sources of infection. Conidia produced on the seedling cankers furnish the primary inoculum. Secondary infection occurs whenever weather conditions are favorable. Stands are reduced, and fiber is damaged by the disease (Hiura, 1924, Pethybridge and Lafferty, 1918, 1920).

CONTROL. The disease is controlled largely by the use of sound seed, seed treatment, rotation of crops, and resistant varieties. Buda (C.I. 326) and Crystal (C.I. 982) are resistant, whereas Punjab (C.I. 20) is highly susceptible (Ray, 1945).

8. Fusarium Wilt, *Fusarium oxysporum* f. *lini* (Bolley) Snyder & Hansen. The wilt disease is distributed generally with flax culture (Baylis, 1940, Bolley, 1906). Wilt has been associated with the culture of flax in North America (Bolley, 1901) to the extent that the flaxseed industry was threatened prior to the development of resistant varieties.

SYMPTOMS. Flax plants are attacked by wilt at any stage in their development, and symptoms vary with varieties and with environmental

FIG. 94. Flax wilt caused by *Fusarium oxysporum f. lini.* The disease is controlled by the use of resistant varieties, right.

conditions. Although primarily a wilt, seedling blight occurs when susceptible seedlings are grown at high temperatures (Millikan, 1951, Schuster and Anderson, 1944, Wilson, 1946). In typical wilt, the leaves turn yellow or grayish yellow, the apical leaves thicken, growth stops, and the plants die and turn light brown (Fig. 94). Frequently the plant is only stunted, in which case the leaves turn yellow and fall prematurely, or the primary stem dies and new, apparently healthy lateral branches develop from the first node. A late infection or a weak attack may be evidenced by premature ripening.

THE FUNGUS

Fusarium oxysporum f. *lini* (Bolley) Snyder & Hansen
(*Fusarium lini* Bolley)

Conidiophores are short and branched and usually form in erumpent sporodochia. Conidia are fusiform to falcate, 3-septate, and hyaline to light pink in mass; microconidia are not abundant (see Chap. 18 for morphology). Specialization of the parasite is reported by Borlaug (1945), Broadfoot (1926), and Millikan (1949).

ETIOLOGY. The fungus is primarily soil-borne; it persists for several years in the soil to invade the young plant through the roots and develops chiefly in the xylem vessels. High temperatures and low moisture are important factors in the development of the disease and the expression of resistance in most flax varieties (Tisdale, 1916, 1917). Fungus mycelium apparently must be present in the plant tissues to produce wilting (Schuster, 1944). Seed infection occurs and accounts for the spread of the parasite to new areas (Bolley, 1924, 1926, Bolley and Manns, 1932).

CONTROL. Wilt-resistant varieties constitute the chief means of control (Bolley, 1901, 1932, Stakman *et al.*, 1919, Tisdale, 1917). Some of the more wilt-resistant varieties are Bison (C.I. 389), and Sheyenne (C.I. 1073). Wilt resistance is conditioned by several factor pairs. Selected strains breed true for different degrees of resistance to given races of the pathogen within the range of environment in which resistance is expressed (Barker, 1923, Burnham, 1932, Christensen, 1954, Houston and Knowles, 1953). The nature of wilt resistance has been studied by Nelson and Dworak (1925), Reynolds (1931), Tisdale (1917), and others.

9. Rust, *Melampsora lini* (Pers.) Lév. Flax rust occurs in the major flax-producing areas of the world. Specialized races of the rust parasite occur on both the cultivated and wild species of *Linum*. The rust causes damage to the fibers and reduces seed production and quality. The loss in the United States in 1951 exceeded 10 million dollars (Flor, 1954).

SYMPTOMS AND EFFECT. The flax rust parasite is an autoecious, long-cycle fungus producing pycnia (spermagonia), aecia, uredia, and telia on the flax plant. Pycnia and aecia usually occur during the early part of the growing season, and they appear as light-yellow to orange-yellow sori on the leaves and stems. The reddish-yellow uredia occur on the leaves, stems, and capsules during the growing season. The brown to black telia, covered by the epidermis, occur chiefly on the stems, but also on the leaves and capsules late in the growing season (Fig. 95).

FIG. 95. Flax rust caused by *Melampsora lini* showing the old aecia, the uredia, and the telia on the leaves and stems.

THE FUNGUS

Melampsora lini (Pers.) Lév.
(*Uredo miniata* f. *lini* Pers.)
(*Xyloma lini* Ehrenb.)
(*Melampsora liniperda* Palm)

The pycnia are subepidermal and are usually formed in the stomatal cavity. The round orange-yellow naked aecia occur on both surfaces of the leaf. Aeciospores are globoid, hyaline, and finely verrucose. The uredia are round to elongate, naked, reddish yellow changing to pale yellow as they mature, with paraphyses intermixed with the spores. Urediospores are elliptical to obovate, walls yellow, contents orange yellow, finely verrucose, and pores usually equatorial (Fig. 95). Telia may be round, but often are elongated and confluent, covered by the epidermis, slightly elevated, and brown to black. Teliospores are formed in a closely packed single layer, prismatic in shape, one-celled, smooth, brown, and germinate in place (Fig. 95).

ETIOLOGY. The autoecious long-cycle rust produces all stages on the flax plant. The teliospores on the crop refuse germinate in the spring to produce the sporidia, which infect the young tissues of the flax plant.

The pycnial stage develops, fusion of compatible haploid cells occurs to initiate the binucleate phase, and the aecial stage forms from the binucleate fusion hyphae. According to Allen (1934) and Flor (1954), the fungus is heterothallic. The primary uredial infection develops from the aeciospores throughout the early part of the growing season, and secondary infections from urediospores account for much of the later spread. Telia are formed around the uredia and from independent uredial infections as the flax plant matures. The telia persist in the flax straw to renew the cycle. The uredial stage continues development in regions where the flax plants are growing in both summer and winter; however, the telial material on the crop refuse is the common source of primary inoculum in most flax-producing areas.

CONTROL. Crop rotation and removal or plowing under of the flax refuse are important in controlling the epidemic development of the disease. The teliospores on small pieces of infected tissue frequently are carried with the seed; therefore careful cleaning of flax seed is important, especially when seeding on new land.

RESISTANT VARIETIES AND SPECIALIZATION OF THE PATHOGEN. While resistant varieties offer the best means of rust control, resistant varieties in culture soon become susceptible because of the increase of a race of the pathogen with high virulence on the variety. The epidemiology of flax rust over the world has shown this shift in races of the pathogen (Cruickshank, 1952, Flor, 1953, 1954, Straib, 1939, Vallega, 1944, Waterhouse and Watson, 1944). Flor (1954) during more than twenty years of intensive study of this disease has contributed much to the basic knowledge of pathogen and host and has laid the foundation for rust control. Physiological specialization has been transferred from the empirical classification to a genetic basis of factors for pathogenicity in the pathogen compared with factors for resistance or susceptibility in the host.

Virulence in the pathogen and susceptibility in the host are expressed as compatible reactions between pathogen and host conditioned by the interplay of factors or genes in both genotypes that permit rust development. The expression of avirulence and resistance, the incompatible reaction of pathogen and host is conditioned by the same or different factor pairs in the fungus and the flax plant. Placed on a gene basis of interpretation, the use of modern genetic techniques is applicable alike to both fungus and flax.

Genetic analysis of the pathogen elucidates sound concepts. Races as they exist in nature generally are heterozygous. Factors determining virulence on specific flax varieties are generally recessive, consequently a pathogenicity test of a uredial culture does not necessarily indicate the virulence of its progeny. Specific factors conditioning virulence can be isolated or combined with a considerable degree of accuracy in predicting their behavior. Race and biotype analyses on this basis give way to the determination and isolation of specific factors for pathogenicity in the pathogen on differential flax varieties carrying single rust-conditioning factors. Flor (1954, p. 158) gives a list of differential varieties and their probable genotype. These include factors conditioning resistance to races of the pathogen from other areas or not yet found in nature.

The factors conditioning resistance in the 32 varieties listed also represent the current breeding stocks for rust resistance. Flor (1947), Kerr (1952), and others found generally greater resistance was dominant to lesser resistance. Nineteen factor pairs were found to condition rust reaction: seven were in the L series of alleles, four were in the M series, five were in the chromosome carrying the Bombay N factor pair, and three were not located. The flax breeder in the North Central United States has available ten or twelve factor pairs that condition resistance to all known North American races of the pathogen. The breeder knows, however, that all of these varieties are susceptible to one or more factors for pathogenicity in *Melampsora lini*, some of which are known to occur in races in South America and elsewhere.

These factors conditioning resistance were each backcrossed to Bison (C.I. 389), using Bison as the maternal parent. From the progeny of the seventh backcross, lines are being selected that are pure for the desired rust-conditioning factor pair and that have in addition the agronomic type, wilt resistance and pasmo tolerance of Bison (Flor, 1954). The periodic shift in the use of one or more of these lines or the concerted shift with the appearance and increase of a specific race of the pathogen offers a practical and systematic means of rust control. In addition the collection of agronomically adapted lines containing the various factors conditioning resistance to specific factors for pathogenicity in the pathogen represents a source of breeding material, a collection of rust-conditioning factors in a common genotype for the study of the nature of resistance and for other investigations.

REFERENCES

ALLEN, R. F. A cytological study of heterothallism in flax rust. *Jour. Agr. Research* **49**:765–791. 1934.

ALLISON, C. C. Powdery mildew of flax in Minnesota. *Phytopath.* **24**:305–307. 1934.

BARKER, H. D. Study of wilt resistance in flax. *Minn. Agr. Exp. Sta. Tech. Bul.* 20. 1923.

BAYLIS, G. T. S. Flax wilt (*Fusarium lini*) in New Zealand. *New Zealand Jour. Sci. Tech.* **22**:157–162A. 1940.

———. Stem-break and browning (*Polyspora lini*) of flax in New Zealand. *New Zealand Jour. Sci. Tech.* **23**:1–8A. 1941.

BERKELEY, G. H. Root-rots of certain non-cereal crops. *Bot. Rev.* **10**:67–123. 1944.

BOLLEY, H. L. Flax wilt and flax-sick soil. *No. Dak. Agr. Exp. Sta. Bul.* 50. 1901. *Bul.* 55. 1903. *Bul.* 71. 1906. *Bul.* 174. 1924. *Bul.* 194. 1926. *Bul.* 256. 1932.

———— and T. F. MANNS. Fungi of flax seed and flax-sick soil. *No. Dak. Agr. Exp. Sta. Bul.* 259. 1932.

BORLAUG, N. E. Variation and variability of *Fusarium lini*. *Minn. Agr. Exp. Sta. Tech. Bul.* 168. 1945.

BRENTZEL, W. E. The pasmo disease of flax. *No. Dak. Agr. Exp. Sta. Bul.* 233. 1930.

BROADFOOT, W. C. Studies on the parasitism of *Fusarium lini* Bolley. *Phytopath.* **16**:951–978. 1926.

BROEKEMA, L. Einige warrnemingen en Denkbeelden over den Vlasbrand. *Landbouwkund. Tijdschr. Nederlandsh Genootschap Landbouwwent.,* pp. 59–71, 105–128. 1893.

BURNHAM, C. R. The inheritance of Fusarium wilt resistance in flax. *Jour. Am. Soc. Agron.* **24**:734–748. 1932.

CASS SMITH, W. P., and H. L. HARVEY. Zinc deficiency of flax. *Jour. Dept. Agr. W. Aust.* (2) **25**:136–142. 1948.

CHRISTENSEN, J. J. The present status of flax disease other than rust. *Adv. Agron.* **6**:161–168. 1954.

COOKE, W. B. Nomenclatural notes on the Erysiphaceae. *Mycologia* **44**:570–574. 1952.

CRUICKSHANK, I. A. M. Resistance of linseed and linen flax varieties to flax rust, (*Melampsora lini*) in New Zealand. I. II. *New Zealand Jour. Sci. Tech.* A **31**:54–57. 1950. **33**:62–64. 1951.

CULBERTSON, J. O. Seed flax improvement. *Adv. Agron.* **6**:144–182. 1954.

DIDDENS, H. A. Untersuchungen über den Flachsbrand. *Phytopath. Zeitschr.* **4**: 291–313. 1932.

DILLMAN, A. C. Improvement in flax. *U.S. Dept. Agr. Yearbook,* pp. 745–784. 1936.

DRECHSLER, C. Three species of Pythium with large oogonial protuberances. *Phytopath.* **29**:1005–1031. 1939.

FLOR, H. H. Physiologic specialization of *Melampsora lini* on *Linum usitatissimum*. *Jour. Agr. Research* **51**:819–837. 1935.

————. Browning disease of flax in the United States. *Phytopath.* **26**:93–94. 1936.

————. New physiologic races of flax rust. *Jour. Agr. Research* **60**:575–591. 1940.

————. Inheritance of rust reaction in a cross between the flax varieties Buda and J.W.S. *Jour. Agr. Research* **63**:369–388. 1941.

————. Relation of rust damage in seed flax to seed size, oil content and iodine value of oil. *Phytopath.* **34**:348–349. 1944.

————. Genes for resistance to rust in flax. *Agron. Jour.* **43**:527–531. 1951.

————. Epidemiology of flax rust in the North Central States. *Phytopath.* **43**: 624–628. 1953.

————. Flax rust. *Adv. Agron.* **6**:152–161. 1954.

GIDDINGS, N. J. Some studies of curly top of flax. *Phytopath.* **38**:999–1002. 1948.

HARTLEY, C. Stem lesions caused by excessive heat. *Jour. Agr. Research* **14**:595–604. 1918.

HARVEY, R. B. Conditions for heat canker and sun scald in plants. *Minn. Hort.* **51**:331–333. 1923.

HENRY, A. W. Browning disease of flax in North America. *Phytopath.* **15**:807–808. 1925.

————. Observations on the variability of *Polyspora lini*. *Can. Jour. Research*
  C 10:409–413. 1934. 16:331–338. 1938.
HIURÁ, M. On the flax anthracnose and its causal fungus, *Colletotrichum lini* (West.)
  Toch. *Jour. Bot. Japan* 2:113–132. 1924.
HOMMA, V. On the powdery mildew of flax. *Bot. Mag. Tokyo* 42:331–334. 1928.
HOUSTON, B. R., and P. F. KNOWLES. Studies on Fusarium wilt of flax. *Phytopath.*
  43:491–495. 1953.
KERR, H. B. Rust resistance in linseed. *Nature, London* 169:159. 1952.
KIKUCHI, M. Cytological studies of the genus Linum. I. *Japan Jour. Gen.* 4:
  202–212. 1929.
KRENNER, J. A. Über die Stengelbruchkrankheit des Flachses. *Agrártud. Egypt*
  6:1–33. 1954.
LAFFERTY, H. A. The browning and stem-break disease of cultivated flax (*Linum
  usitatissimum*) caused by *Polyspora lini* n. gen. n. sp. *Sci. Proc. Roy. Dublin Soc.*
  16:248–274. 1921.
LUGGER, O. A treatise on flax culture. *Minn. Agr. Exp. Sta. Bul.* 13. 1896.
McKAY, R. Flax diseases. *Flax Development Board, Dublin.* 1947.
MIDDLETON, J. T. Taxonomy host range and geographic distribution of the genus
  Pythium. *Mem. Torrey Bot. Club* 20:1–171. 1943.
MILLIKAN, C. R. Studies of strains of *Fusarium lini*. *Proc. Roy. Soc. Vict.* (N.S.)
  61:1–24. 1949.
————. Diseases of flax and linseed. *Dept. Agr. Victoria, Aust., Tech. Bul.* 9. 1951.
MUSKETT, A. E., and J. COLHOUN. Prevention of stem-break, browning, and seedling
  blight in the flax crop. *Nature* 147:176–177. 1941.
———— and ————. The diseases of the flax plant. W. and G. Baird. Belfast.
  1947.
NAOUMOFF, N. A. Novosti mestnoi mikroflora (Novelties of the local microflora).
  Mycology, Leningrad. Vol. 1. 1926.
NELSON, C. I., and M. DWORAK. Studies on the nature of wilt resistance in flax.
  No. Dak. Agr. Exp. Sta. Bul. 202. 1926.
NEWHOOK, F. J. "*Sphaerella linorum*" on flax in New Zealand. *New Zealand Jour.
  Sci. Tech.* 24:102A–106A. 1942.
PETHYBRIDGE, G. H., and N. A. LAFFERTY. A disease of flax seedlings caused by a
  species of *Colletotrichum* and transmitted by infected seed. *Sci. Proc. Roy.
  Dublin Soc.* 15:359–384. 1918.
———— and ————. Investigations on flax diseases. *Jour. Dept. Agr. Tech. Inst.
  Ireland* 20:325–344. 1920. 21:167–187. 1921.
RAY, C., JR. Cytological studies on the flax genus, Linum. *Am. Jour. Bot.* 31:241–
  248. 1944.
————. Anthracnose resistance in flax. *Phytopath.* 35:688–694. 1945.
REDDY, C. S., and W. E. BRENTZEL. Investigations of heat canker of flax. *U.S.
  Dept. Agr. Dept. Bul.* 1120. 1922.
REYNOLDS, E. S. Studies on the physiology of plant diseases. *Ann. Mo. Bot. Gard.*
  18:57–95. 1931.
RODENHISER, H. A. Physiologic specialization and mutation in *Phlyctaena linicola*
  Speg. *Phytopath.* 20:931–942. 1930.
ROST, H. Die Pasmo-Krankheit des Leins in Europa. Erreger: *Septoria linicola*
  (Speg.) Garassini. *Angew. Bot.* 19:163–171. 1937.
SACKSTON, W. E. Effect of Pasmo disease on seed yield and thousand kernel weight
  of flax. *Can. Jour. Research* C 28:493–512. 1950.
SCHILLING, E. Beobachtungen über eine durch *Gloeosporium lini* verusachte Flachs-
  krankheit in Deutschland. *Faserforschung* 2:87–113. 1922.

SCHUSTER, M. L.   The nature of resistance of flax to *Fusarium lini*.   *Phytopath.* **34**:356. 1944.

———— and E. J. ANDERSON.   Seedling blight and root rot of flax in Washington. *Phytopath.* **37**:466–473. 1947.

SEVERIN, H. H. P., and B. R. HOUSTON.   Curly-top and California-aster-yellows disease of flax.   *Phytopath.* **35**:602–606. 1945.

STAKMAN, E. C., *et al.*   Controlling flax wilt by seed selection.   *Jour. Am. Soc. Agron.* **2**:291–298. 1919.

STRAIB, W.   Untersuchungen über den Wirtsbereich und die Aggressivtät physiologischer Rassen von *Melampsora lini* (Pers.) Lév.   *Züchter* **11**:130–136, 162–168. 1939.

TAMMES, T.   The genetics of the genus Linum.   *Bibliog. Genetica* **4**:1–36. 1928.

TISDALE, W. H.   Relation of soil temperature to infection of flax by *Fusarium lini*. *Phytopath.* **6**:412–413. 1916.

————.   Flax wilt: a study of the nature and inheritance of wilt resistance.   *Jour. Agr. Research* **11**:573–606. 1917.

————.   Relation of temperature to the growth and infecting power of *Fusarium lini*. *Phytopath.* **7**:356–360. 1917.

TUBEUF, C. VON.   Hitzetod und Einschnürungs krankheiten der Pflanzen.   *Naturw. Zeit. Forst. Landw. Jarg.* **12**:19–36, 67–88, 161–169. 1914.

VALLEGA, J.   Observaciones sobre la resistencia a la roya de algunos lines ensayados en el Instituto Fitotécnico de Llavallolo.   *Santa Catalina Inst. Fitotéc. Pub.* 1. 1938.

————.   Especializacion fisiologica de *Melampsora lini*, en Argentina.   *Santa Catalina Inst. Fitotéc. Pub.* 39. 1944.

WATERHOUSE, W. L., and I. A. WATSON.   A note on determinations of physiological specialization in flax rust.   *Jour. Roy. Soc. N.S. Wales* **75**:115–117. 1941.

———— and ————.   Further determinations of specialization in flax rust caused by *Melampsora lini* (Pers.) Lév.   *Jour. Roy. Soc. N.S. Wales* **77**:138–144. 1944.

WESTERDIJK, J.   Jaarverslag van de directie, Jaarverslag 1915.   *Phytopath. Lab. Willie Commelin Scholten, Amsterdam*, pp. 6–7. 1916.

WILSON, I. M.   Observations on wilt disease in flax.   *Trans. Brit. Myc. Soc.* **29**: 221–231. 1946.

WOLLENWEBER, H. W.   "*Sphaerella linicola*" n. sp. die ursache der Amerikanischen Leinpest (Pasmo- oder Septoria-Krankheit.).   *Lilloa Revista Bot. Nat. Univ. Tucuman Inst. Miquel Lillo* **2**:483–494. 1938.

CHAPTER 20

TOBACCO DISEASES

The cultivated tobaccos comprise two species, *Nicotiana tabacum* L. and *N. rustica* L.; varieties of the former constitute the major commercial crop. A large number of additional species, however, are indigenous in the Western Hemisphere. The two cultivated species hybridize readily, as each contains 24 chromosome pairs apparently of similar origin. *N. tabacum* has been crossed with some 12 other species, with chromosome pairs ranging from 9 to 24. While in present experimental practice these wide crosses are not utilized extensively in breeding programs, there is, nevertheless, the probability of obtaining disease-resistance factors from a number of the wild species (Clayton, 1953, East, 1928, Garner *et al.*, 1936).

The tobacco plant is adapted over a wide range of climatic and soil conditions of the world. The species are largely self-pollinated, and they are mostly annuals, varying greatly in type of growth and length of growing period.

Both the leaves and stems are used in commerce. The stems are used in the preparation of nicotine extracts. The quality of the leaf is influenced greatly by climatic conditions and soil composition, which tend to localize production of different types of commercial tobacco.

Diseases of the crop are important in both yield of leaf and quality of the product. Plant diseases frequently cause large losses in the crop, and disease investigations and control are important factors in the improvement of the crop. Wolf (1935) has discussed tobacco diseases in detail, and reference should be made to this volume for the more complete information.

1. Leaf Spotting and Yellowing, Nonparasitic. The quality of the tobacco leaf is associated with soil fertility and climate as well as the tobacco variety. "Firing," "rusting," "spotting," and similar descriptive terms have been used to describe various soil nutrient deficiencies and other nonparasitic manifestations. Deficiencies in potash, phosphate, and magnesium influence the quality of the leaf and if acute are manifest by crinkling and yellowing of the leaf tip and margins, brown leaf blotches, and bleaching or chlorosis. The general use of commercial fertilizers in tobacco culture has reduced this type of malady in the more important tobacco areas. Other nonparasitic diseases such as frenching, sunscald, lightning injury, and hail spots are not uncommon (Anderson, 1940, Johnson, 1924, Valleau *et al.*, 1942, Wolf, 1935).

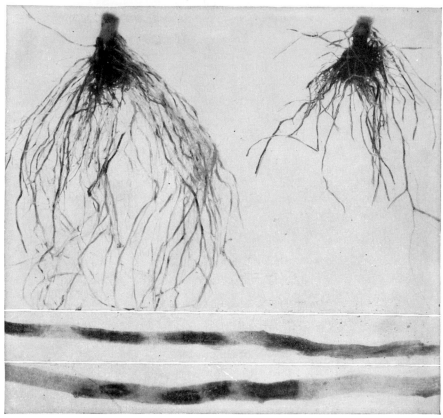

FIG. 96. Tobacco roots showing the characteristic symptoms of nematode injury and brown root rot. Enlarged pieces of roots show necrotic lesions. (*Courtesy J. J. Johnson.*)

2. Brown Root Rot and Nematode Root Rot, Crop Residues and Nematodes. The root rot complex first described in 1915 by Johnson (1915, 1939) on tobacco and other crops has since been shown to be associated closely with soil-inhabiting nematodes. The disease originally was described as occurring in the more northern tobacco areas of North America. In these areas the brown necrosis of the roots and sloughing off of the cortex and the final collapse of the root system (Fig. 96) appear differently from the root rot complex associated with the root knot nematode, *Meloidogyne* spp., generally distributed in the warmer tobacco areas. The recent study of plant nematodes defines the Northern distribution of the meadow nematodes, *Pratylenchus* spp., dagger nematodes, *Tylenchorynchus* spp., and spiral nematodes, *Heliocotylenchus* spp., and their ability to incite wounds and deplete the root systems of plants. These and other nematodes also provide avenues of entrance for soil-infesting pathogenic fungi. The association of these nematodes with brown root rot was demonstrated by Valleau (1947). The complicating relationship of specific crop residues in the soil and brown necrosis of roots and the correction of this condition by the addition of nitrate nitrogen apparently in the absence of nematodes indicate that a complex of factors is involved (Stover, 1951). These root rots reduce the growth of plants. The increase in nematode populations is associated with specific crops. Some pasture and weed grasses and corn, cotton, and peanuts are favorable for the population increase of nematodes and increase the incidence of root rot where tobacco follows these in rotation. Soil fumigation is becoming economically practical in many areas of intensive tobacco production.

3. Mosaics, Viruses Transmitted by Contact, Insects, or Grafting. Tobacco mosaic has been associated closely with the developmental study of virus diseases of plants. Mayer (1886) first described the mosaic on tobacco. Ivanowski (1892) demonstrated the transmissibility and the filterability of the virus principle using tobacco mosaic, and his results were confirmed and interpreted by Beijerinck (1898). The insect transmission of a virus to tobacco, probably the cucumber mosaic virus, was demonstrated by Allard (1914); however, Ball and Takami had demonstrated earlier the insect transmission of the curly top virus of beet and the virus of the dwarf disease of rice, respectively (Chap. 7). The first chemical purification of a virus was accomplished (Stanley, 1935, 1936), using the tobacco mosaic. Cross protection by means of strains of the same virus was demonstrated first with this virus complex (Thung, 1931) and soon confirmed by Salaman (1933) and others. Numerous physical and chemical studies and new techniques have evolved from the study of these relatively stable viruses during the past thirty years. The tobacco plant is used extensively in the assay and identification of the plant viruses.

Gradually information is accumulating on individual diseases of the tobacco plant incited by specific viruses. The viruses and strains occurring in tobacco have been listed and described by Bawden (1939), Holmes (1939), Johnson et al., (1939),[1] Smith (1937), and others.

[1] Manuscript of report of the Committee on Description and Nomenclature of Plant Viruses to the International Botanical Congress, 1939. Sections have been published in various journals.

Common Tobacco Mosaic, Virus Strains Transmitted Largely by Contact.   The common tobacco mosaic is world-wide in distribution, and it is the most common virus disease of this crop plant.   The damage caused by mosaic depends largely upon the stage of plant development when infection occurs and upon the variety grown.   Where mosaic infection is extensive early in the season, losses in yield and quality occur (Clayton and Murtrey, 1950, Johnson and Ogden, 1939, Valleau *et al.*, 1942).

SYMPTOMS.   The characteristic symptom is leaf mottling with yellow-green and dark-green pattern intermingled.   In addition to the mosaic pattern, younger leaves frequently are puckered and malformed (Fig. 97). In later infections plants often show mottling in the tip leaves, some of which develop necrotic areas.

The virus is spread by contact chiefly.   The virus remains infectious in dry mosaic plant tissues for long periods, and it frequently is spread to plants in the seed bed or during transplanting.   Secondary spread by leaf contact, farm implements, and man are common (Johnson and Ogden, 1939, Valleau and Johnson, 1937).   The increase of the virus in the host is associated with conditions favorable for the growth of the host (Chessin, 1951, Chessin and Scott, 1955, Weathers and Pound, 1954).   Strains of the virus with different properties and inciting variations in symptoms occur.   Cross protection either systemically or locally is demonstrable with strains of the virus (Fulton, 1951, Johnson, 1947).   Electron micrograms of the virus particles indicate rods of varying length and association, approximately 15 by 280 millimicrons in size (Sigurgeirsson and Stanley, 1947).   Serological differentiation of the virus and some strains is possible (Moorehead and Price, 1953).   The virus infects a wide range of hosts (Holmes, 1946).

CONTROL.   The common tobacco mosaic is controlled by sanitation and the use of resistant varieties.   Resistant varieties are the more practical means of control, as summarized by Clayton (1953) and Valleau (1952).   The use of the Ambalema type of resistance and the expression of resistance when the two factors are in the recessive condition were defined by Nolla (1938) and expanded to include modifying factors by Clayton *et al.* (1938).   Resistance as expressed by failure to show systemic chlorosis after inoculation with mosaic virus was demonstrated in *Nicotiana glutinosa* L. and the first-generation hybrid between this species and *N. tabacum* L. by Allard (1914).   Clauson and associates continued the study of these hybrids, and the first synthetic, fertile, allopolyploid named *N. digluta* was described (Clausen and Goodspeed, 1925).   Holmes (1938) made available a line of tobacco homozygous for the local lesion reaction to mosaic virus through backcrossing *N. digluta* to tobacco.   The use of this material in practical breeding for mosaic

Fig. 97. Plant and leaves showing the characteristic symptoms of the common tobacco mosaic. (*Courtesy of J. J. Johnson.*)

resistance soon showed mosaic-resistant selections had smaller leaves, and other types developed systemic necrosis rather than local lesions under some field conditions (Clayton and McKinney, 1941). Gerstel (1943) showed that this mosaic-resistant tobacco contained a *glutinosa* chromosome in the *tabacum* genom. Further backcrossing to *N. tabacum* and screening resulted in the transfer of the factor pair for resistance into a normal-appearing tobacco usable in the production of mosaic-resistant hybrids. Holmes (1951) and Valleau (1952), on the basis of the geographic distribution of factors for mosaic resistance, postulate the origin of the virus.

Tobacco Etch, Virus Transmitted by Contact and Aphids. The virus is distributed widely on tobacco and pepper as well as many weeds. The mottling is somewhat similar to the mottling produced by common mosaic; however, the young apical leaves do not show mottling in the etch disease. Severe chlorotic mottle, necrosis, and etch or breaking out of the necrotic areas on the several leaves are characteristic of the disease on especially the light burley types. Plants are stunted frequently. Symptoms on the flue-cured tobacco and cigar, dark types, are mild mottle and etching or frequently no expression of symptoms or no reduction in plant growth. The association of the severe symptoms of etch with the duplicate, recessive allelomorphs g_1g_2, essential for the light burley genotype, is postulated by Stover (1951). The virus particle is probably rod-shaped (Johnson, 1951).

Tobacco Necrosis, Virus in Roots Transmitted in Soil. The necrosis virus first reported in England (Smith and Bald, 1935), later in Australia and Europe, was isolated in Wisconsin in 1950 (Fulton, 1950). The virus in the roots frequently produces no symptoms beyond reduced plant growth. Mechanical inoculation on the leaves produces local lesions on cowpea and tobacco. The virus incites a severe necrosis and death of tulips under natural conditions. The host range of the virus as determined by inoculation with the virus from roots is wide (Fulton, 1950, Price, 1940). Strains of the virus were demonstrated by leaf symptoms and physical properties of the virus. The virus particle is described as round (Johnson, 1951, Markham *et al.*, 1947).

Tobacco Ring Spot, Virus Transmitted by Contact, Seed-borne, and Probably Insects. Ring spot was first described in 1922, although it may be similar to the disease described in Russia in 1890. It is distributed widely on many hosts, especially sweetclover and soybean (Chaps. 13 and 15). Ring spot is essentially a foliage disease of tobacco. The concentric light-green to yellow patterns on the leaves and irregularly parallel lines along the veins appear on all foliage (Fig. 98). The necrotic tissue becomes dry as new rings form around the first. Plants showing symptoms early in the season usually produce leaves later that

Fig. 98. Tobacco leaves showing (A) ring spot and (B) streak symptoms. (*Courtesy of J. J. Johnson.*)

are free of symptoms. Plants grown following sweetclover, lespedeza, or alfalfa or near these crop plants and some common weed hosts show high incidence of ring spot (Berkeley and Koch, 1940, Valleau *et al.*, 1942, Wingard, 1928).

The virus is seed-borne in many hosts and persists in dried tissue. The legume and weed hosts are sources of virus for ready-contact infection in tobacco. The field-distribution pattern of the ring spot on several crop plants suggests insect spread. Strains of the virus occur, differing in time of appearance and damage incited (Valleau, 1932). The ring spot virus particle is approximately spherical and 22 millimicrons in diameter (Desjardins *et al.*, 1953). Apparently all varieties of tobacco and many of the *Nicotiana* spp. are susceptible.

Tobacco Streak, Virus Transmitted by Contact. The virus is distributed widely and probably includes Brazilian streak. Streak frequently appears in severe form on occasional plants in the field. Plants suddenly develop necrosis, puckering, and curling of the young leaves. The areas along the veins darken, and necrotic streaks extend along the midrib (Fig. 98). The several leaves are severely damaged; however, the plant apparently soon recovers, and subsequent leaves show little or no symptoms. The host range of the virus is wide, including legumes, corn, and common weeds. Guar, *Cyamopsis tetragonalobus* (L.) Taub., produces local lesions when infected with this virus (Fulton, 1948).

The virus persists largely in leguminous hosts, especially sweetclover. No insect vector is known, although the sudden appearance in tobacco suggests insect transmission. The virus content in both systemic and local-lesion hosts fluctuates widely. Apparently the distribution of the virus in the tobacco plant varies with tissues and time following inoculation. The plant soon develops a tolerance, and recovery results, based on leaf symptoms and virus content in some tissues (Costa, 1952, Fulton, 1949). The presence of a virus inactivator in both healthy and diseased plants that is removed from action by reducing agents is demonstrable (Fulton, 1949). The role of the inactivator in the development of tolerance is not clear, as the tolerant plants frequently are high in virus content. The virus particle obtained from tobacco and corn is rod-shaped (Johnson, 1951). Apparently resistance to the initial severe symptoms is not evident in tobacco or many related species. *Nicotiana glauca* and *N. gossii* show some resistance (Diachun and Valleau, 1947).

Leaf Curl, Virus Transmitted by White Fly, *Bemisia* spp., and Grafting. The disease is prevalent in Africa and Java, and a similar disease occurs in Venezuela (Storey, 1932, 1935, Tarr, 1951, Thung, 1932, Wolf *et al.*, 1949). The leaf curl virus on tobacco in Africa is not transmissible to cotton, nor cotton to tobacco (Chap. 18), although both viruses occur in some species of *Hibiscus*, *Althaea*, and *Sida* (Tarr, 1951). The symptoms vary from mild curling of leaf edges and thickenings on leaf veins to the severe leaf curling and twisting, leaf puckering and thickening, overgrowths and enations along the veins on the lower surface of the leaf, reduced leaf size, and reduced internodal elongation of the stem.

The physical properties of the virus occurring in Africa are not reported in detail; those of the virus from Venezuela only partly. The virus particle appears to be different from those of other viruses in tobacco in both size and shape; it is soft, approximately spherical, and averages 39 millimicrons (Sharp and Wolf, 1951).

A disease in Maryland broadleaf tobacco suckers with symptoms similar to leaf curl, but without leaf-vein enations, was reported in Maryland (Morgan and McKinney, 1951). Round, disk-like particles with irregular margin were associated with the disease.

Tobacco Rattle, Virus Transmitted by Contact and through Soil. Apparently this virus disease of tobacco and a similar disease of potato are common in Europe. The disease on tobacco was described first in 1906 as canker. It causes large losses in tobacco quality in Continental Europe (Quanjer, 1943). Following inoculation local lesions appear first, later systemic symptoms develop. Field symptoms are gray necrotic spots, leaf margin curling downward, and enlargement and drying out of the necrotic areas. The dry tissues develop tension in the leaves and rattle as the tissues break. Gray streaks develop on the stems. Apparently the virus persists in the soil (Böning, 1931, Rozendaal and Van der Want, 1948). Short, rod-like particles without aggregation are associated with the virus (Thung, 1948). *Nicotiana tomentosa* and *N. tomentosiformis* are resistant. Hybrids with susceptible tobacco varieties indicate that resistance is expressed as a recessive (Thung, 1951).

Cucumber Mosaic on Tobacco, Virus Strains Transmitted by Contact and Aphids. The tobacco is infected readily by the virus causing cucumber mosaic. Occasional fields show heavy infection with this virus when near a source of inoculum and when aphids are abundant and move onto the tobacco plants. Allard's demonstration of aphid transmission of a virus to tobacco was probably cucumber virus (Hoggan, 1929).

The symptoms on tobacco are similar to those of common mosaic. The present varieties resistant to tobacco mosaic are susceptible to cucumber mosaic. The virus is spread readily by aphids, which is not the case with tobacco mosaic virus. The particles of the cucumber mosaic are round (Johnson, 1951).

Tumor Gall, Virus Graft Transmitted. The disease occurs infrequently on tobacco and probably is widely distributed, as is the case in other virus tumors (Black, 1945, Vallcau, 1947). Gall-like tumors occur on roots and stems. The virus is systemic, based on graft transmission from apical stem shoots of tobacco.

Other Viruses. As stated earlier, the tobacco plant is used as an assay plant in the study of many virus diseases and in the reaction of tobacco to specific viruses, especially where the local-lesion reaction is useful (Pound and Weathers, 1953). Some additional viruses occur in nature, but infrequently, and their identity is uncertain.

4. Bacterial or Granville Wilt, *Xanthomonas solanacearum* (E. F. Sm.) Dowson. The bacterial wilt of tobacco and other solanaceous plants is a common disease on tobacco in the Southern United States and the warmer areas of other countries. Losses from the disease are high in local sections in the Southeastern United States.

SYMPTOMS.   The symptoms usually appear several weeks after transplanting.   The leaves wilt suddenly, turn yellow, and finally brown as the plants die.   The stems show yellow discolored areas in the xylem tissues and yellow bacterial masses in the vascular bundles of stems and leaf veins.

THE BACTERIUM

Xanthomonas solanacearum (E. F. Sm.) Dowson
(*Pseudomonas solanacearum* E. F. Sm.)
[*Phytomonas solanaceara* (E. F. Sm.) Bergey *et al.*]
(*Bacillus solanacearum* E. F. Sm.)
(*Bacterium solanacearum* E. F. Sm.)

The rods are motile by means of one polar flagellum, have no spores, no capsules develop, and the colonies are opalescent light yellow to brown.

ETIOLOGY AND CONTROL.   The bacteria persist in the soil in association with crop residue.   The disease occurs on other solanaceous crops, some legumes, and several weeds.   Infection occurs through the roots, especially where injured in transplanting, by nematodes, or by insects. Crop rotation, use of cover crops, balanced fertility, and seed-bed sterilization aid in reducing the disease (Garriss and Ellis, 1941).

Moderately resistant varieties combined with short rotations offer the best present means of control.   Resistant varieties are produced from hybrids between adapted varieties and resistant selections obtained from Colombia, South America.   The new hybrids also are moderately resistant to nematode invasion.   Inheritance is polygenic, and better sources of resistance are needed (Clayton, 1953, Clayton and Smith, 1942).

5. Bacterial Leaf Spot or Wildfire, *Pseudomonas tabaci* (Wolf & Foster) Stevens.   Wildfire occurs in the tobacco districts of North America and Europe (Gigante, 1950).   The disease is found on other plants, notably the soybean.   The occurrence of the disease is sporadic, and damage is associated with weather conditions, although losses on tobacco are probably greater than from any of the other bacterial leaf diseases.

SYMPTOMS.   Spots on the leaves are circular to angular as they coalesce. The spot consists of a small light tan to brown dead infection center surrounded by a halo of chlorotic tissue similar to that of halo blight of oats (Chap. 6).   No exudate is formed on the lesions.   Infection of very young tissues results in a more general chlorosis of the infected area (Fig. 99).   The disease generally appears on the lower leaves first and spreads rapidly during heavy rains or long periods of high moisture when the tissues are water-soaked or the stomatal cavities are filled with moisture, as reviewed by Diachun and Troutman (1954) and Valleau *et al.* (1943).

Fig. 99. (A) Leaves of tobacco seedlings showing the lesions of wildfire produced by *Pseudomonas tabaci.* (B) A mature leaf showing yellow blotches of potash deficiency and the angular black lesions with light centers characteristic of blackfire, or angular leaf spot, caused by *P. angulala.* (*Courtesy of J. J. Johnson.*)

THE BACTERIUM

Pseudomonas tabaci (Wolf & Foster) Stevens
[*Phytomonas tabacae* (Wolf & Foster) Bergey *et al.*]
(*Bacterium tabacum* Wolf & Foster)

The rods are motile by means of polar flagella, have no spores, no capsules are formed, and the colonies are white.

ETIOLOGY AND CONTROL. The bacteria apparently persist in crop refuse to furnish inoculum for the new crop. Conditions favorable for water soaking of the leaf tissue or for filling the stomatal cavities with water are associated with rapid spread and development of the disease. Open stomata are important in the entrance of the inoculum in water into the stomatal cavity (Valleau *et al.*, 1943). Seed-bed sanitation is essential to prevent the damage and spread of the disease on the seedlings. Applications of Bordeaux mixture to the seed bed help control the disease. Rotation of crops and elimination of the infected crop residues are important in the field control of the disease. Some of the commercial varieties are moderately resistant to the disease.

Monogenically inherited, high resistance obtained from *N. longiflora* is used in producing wildfire- and angular leaf spot-resistant varieties (Clayton, 1947, 1953).

6. Angular Leaf Spot, or Blackfire, *Pseudomonas angulata* (Fromme & Murray) Holland. The disease occurs in all tobacco areas of North America, Africa, and the Philippines, where under favorable environmental conditions late in the season the quality of leaf tissue is damaged. The small spots resulting from the late infections are conspicuous in the cured tobacco.

The leaf spots vary considerably in size and shape. In the seed bed the spots are small, angular, and black or dark brown in color. In the field the spots are small and angular at first but frequently develop to larger lesions (Fig. 99). The larger spots are zonate, tan to dark brown, and small amounts of exudate occur on the surface (Fromme and Wingard, 1922). The disease develops rapidly under high moistures and temperatures (Valleau *et al.*, 1943).

THE BACTERIUM

Pseudomonas angulata (Fromme & Murray) Holland
[*Phytomonas angulata* (Fromme & Murray) Bergey *et al.*]
(*Bacterium angulatum* Fromme & Murray)

The only difference between this bacterial pathogen and *P. tabaci* is the capacity of the latter to produce a toxin. Comparative studies of the two, as in the bacterial pathogens on oats (Chap. 6), indicate strains of the same species. The strains producing the toxin show a wider host range than those without the toxin.

The rods are motile by means of polar flagella, have no spores, no capsules are formed, and the colonies are dull white.

The etiology and control are similar to that for wildfire.

7. Pythium Damping-off and Stem Rot, *Pythium* spp. The disease occurs in the seed bed and soon after transplanting into the field. Damage from the disease is general, but especially severe in the tropics. The disease has been studied extensively in India and Sumatra, as summarized by Middleton (1943).

SYMPTOMS. The damping-off occurs in the early to late seedling stage as a seed-bed rot. The characteristic brown soft rot and white surface mycelium occurring in local areas in the bed are indicative of the disease. In the field the dark-brown lesions develop at the base of the stem and frequently extend into the leaves (Anderson, 1940).

THE FUNGI. Several species of *Pythium* cause the malady. The most common species is *Pythium debaryanum* Hesse. (The morphology is given in Chap. 13.) A second species occurs on tobacco in the Eastern United States (Anderson, 1940) and generally through Asia, Africa, and the Pacific area.

Pythium aphanidermatum (Edson) Fitzp.
(*Rheosporangium aphanidermatus* Edson)
(*Pythium butleri* Subrm.)

Sporangia are lobulate, branched, inflated, and freely produced. Oogonia are spherical, usually terminal, and 22 to 27 microns in diameter. Antheridia are usually monoclinous, typically intercalary, and occur one or two per oogonium. Oospores are aplerotic, single, smooth, moderately thick-walled, and 17 to 19 microns in diameter. *Pythium deliense* Meurs occurs on tobacco in Sumatra.

ETIOLOGY AND CONTROL. These fungi are common on certain types of decomposing organic material in the soil. The invasion and rotting of the plant tissues occur under conditions of high moisture and low light intensity. Sterilization of seed beds, drainage, ventilation, and the use of fungicidal sprays are all important as control measures.

8. Black Shank, *Phytophthora parasitica* var. *nicotianae* (Breda de Haan) Tucker. The disease occurs in the Southern United States where it was reported first in 1924. Since then it has spread from the Florida-Georgia tobacco sections. The black shank occurs generally in the tropical sections of South America and the Pacific tobacco areas. The disease causes considerable damage in certain types of tobacco, but resistant varieties are coming into general use (Clayton and McMurtrey, 1950, Gratz and Kincaid, 1938, Tisdale, 1931).

SYMPTOMS. Symptoms are evident by the blackened stalk and the subsequent rapid wilting of the tops. The disease occurs infrequently in the seed bed as a damping-off of the seedlings and the blackened rotted basal portion of the stems. In the field the black rot of the stem starts near the soil line and extends up the stem and down into the roots. Necrosis of the stem tissues is rapid, and a spongy rot develops. The plants wilt rapidly and collapse. Large brown leaf lesions occur in wet weather. The disease spreads from local areas as the season advances (Tisdale, 1931).

THE FUNGUS

Phytophthora parasitica var. *nicotianae* (Breda de Haan) Tucker
(*Phytophthora nicotianae* Breda de Haan)
(*Phytophthora tabaci* Saw.)

Tucker (1931) placed the fungus as a variety of *Phytophthora parasitica* Dast., as it was similar in morphology and varied chiefly in being parasitic on tobacco stems. Mycelial growth is abundant on media at high temperatures. Sporangia are broadly ovate, papillate, and average 25 by 30 microns in size. Vegetative resting spores (chlamydospores) are abundant in culture. Oospores are globose, about 15 to 20 microns in diameter, although the size is variable; the membrane is thick and smooth.

ETIOLOGY. The fungus persists in the soil in association with crop refuse and reinfects the plants near the soil line. The fungus is spread through infested tobacco refuse, surface water, implements, and other agencies.

The development of the disease is associated with nematode damage as in the Granville wilt and in the *Fusarium* wilt. Avenues of entrance for the pathogens are provided early, and severe disease development frequently is associated with high populations of root knot, meadow, dagger, and other nematodes (Sasser *et al.*, 1955). Powers (1954) summarizes the investigations on pathological histology and the cause of the rapid wilting of diseased plants. Rotation of other crops with tobacco are important, therefore, not only in reducing inoculum of the pathogen, but also in lowering the nematode populations. Resistant varieties developed in Florida and later in Kentucky and elsewhere generally are resistant to both *Phytophthora* and nematode injury (Clayton, 1953, Tisdale, 1931, Valleau, 1952). The factors for resistance obtained from *N. tabacum* earlier are not adequate for disease control. The single dominant factor pair obtained from *N. longiflora* conditions a high type of resistance. A factor or factors conditioning high resistance from *N. plumbaginifolia* also are being incorporated into tobacco varieties.

9. Downy Mildew or Blue Mold, *Peronospora tabacina* Adam. The downy mildew was first reported from Australia, where it occurs on the cultivated and some wild species. The first outbreak of the disease in the United States occurred in Florida and Georgia in 1921. Since then downy mildew or blue mold has spread through the central, eastern, and southern tobacco areas (Anderson, 1937, 1940, Clayton and

McMurtrey, 1950, Koch, 1941). Similar diseases earlier were reported occasionally on wild species of *Nicotiana* in North and South America. Similarly, the disease has become prevalent on cultivated tobacco in Argentina and Brazil (Godoy and Caste, 1940, Wolf, 1939). The disease causes loss of plants in the seed bed and delayed planting. The leaf damage from the disease is sporadic, as the development is influenced greatly by temperature and moisture.

SYMPTOMS. The appearance of the infected plants is variable, depending upon weather conditions, age of the plants, and stage of disease development. The disease in the seed bed first appears as pale-green to yellow indefinite lesions on the upper leaf surface and a gray to brown downy mass of conidiophores on the undersurface of the leaf. In the young plants, and especially the leaf, tissues brown and collapse, giving the bed a scalded appearance (Fig. 100). The lesions with conidiophores and conidia are more characteristic on the older plants in the beds and in the field. The older plants generally recover and produce new leaves.

THE FUNGI

Peronospora tabacina Adam
(*Peronospora nicotianae* Speg.)
Peronospora hyoscyami DBy. on *Hyoscyamus*

According to Adams (1933), Angell and Hill (1932), Clayton and Stevenson (1935), and Wolf (1939), three species occur on different suscepts: *Peronospora tabacina* Adam on tobacco and certain wild species of the genus, *P. nicotianae* Speg. on some wild tobacco species in South America and possibly North America, and *P. hyoscyami* DBy. on the black nightshade (*Hyoscyamus niger* L.) Clayton and Stevenson (1943) show that *P. tabacina* Adam is the correct binomial for the pathogen on both wild and cultivated *Nicotiana* spp. and that it is indigenous to North and South America. *P. hyoscyami* is logically a specialized variety of the species on *Hyoscyamus* spp.

One or more conidiophores emerge from the stomata, usually on the lower surface of the leaf. The conidiophores are five to eight times dichotomously branched, the curvature of the branches increasing to the ultimate branches which diverge obtusely, slightly curved, or recurved and end bluntly. The conidia are ovoid to ellipsoid, hyaline to dilute violet, measure 13–19 by 16–29 microns, and germinate to produce germ tubes from the side of the conidium. Oogonia and oospores are produced in the necrotic tissues. The oogonia are terminal, with antheridia forming as a branch from the oogonial stalk. The oospores are globose, dark brown, epispore smooth or slightly roughened, and 35 to 60 microns in diameter. The intercellular mycelium and branched haustoria are described by Henderson (1937).

Oospore germination after a period of dormancy and in association with tobacco seedlings is by the production of a sessile zoosporangium and motile zoospores (Person and Lucas, 1953).

ETIOLOGY. Both conidia and oospores function as primary inoculum under certain conditions. The conidia produced on tobacco plants, second growth, or perennial wild species in the areas of mild winter reinfect the seed beds. The oospores are the more important source of

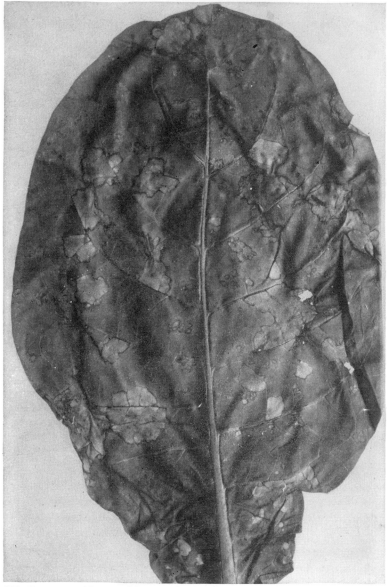

Fig. 100. Downy mildew, or blue mold, of tobacco caused by *Peronospora tabacina.*
(*Courtesy of J. J. Johnson.*)

primary infection for new seed beds in most of the tobacco sections. While the conidia are relatively short-lived, secondary spread and wide distribution of the parasite occur from this source of inoculum (Anderson, 1937, Hill and Angell, 1933, Stover and Koch, 1951, Valleau, 1953). Temperature, sunlight, and moisture are limiting factors in the development of disease (Dixon *et al.*, 1936).

CONTROL. Location of seed beds away from oospore inoculum and use of fungicides offer the best means of control. Steam sterilization as generally practiced, good drainage, and exposure to sunlight are advisable. The use of copper sprays, carbamates, and gases such as benzol and paradichlorbenzene give control in the seed bed (Clayton, 1938). Clayton (1945) and Clayton and McMurtrey (1950) reported the commercial varieties susceptible in the early seedling stage and varying somewhat in susceptibility in the mature plant stage. Angell and Hill (1932) listed all the Australian species as susceptible. Factors for resistance to blue mold and black root are present in *N. debneyi*. These have been transferred to tobacco and, while polygenic, apparently offer no great difficulty to incorporation into commercial tobacco by the backcross method (Clayton, 1953).

10. Brown Leaf Spot, *Alternaria tenuis* Nees. During periods of high temperature and moisture, large circular brown spots develop on the mature leaves and on leaves after harvest. Spots on the leaves from contact infection through dead corolla tissue also occur in shade tobacco. The fungus sporulates on the surface of the lesions.

11. Fusarium Wilt, *Fusarium oxysporum f. nicotianae* (J. Johnson) Snyder & Hansen. The wilt occurs rather generally in the sandy soils of the warmer tobacco sections (Fig. 101). This form of the fungus as well as certain of those parasitizing other crops, as cotton, sweet potato, and tomato, apparently are capable of producing wilt in tobacco especially when nematodes are present (Armstrong and Armstrong, 1948, McClure, 1949). Root knot, nematode root rot, and potassium deficiency apparently increase the damage from wilt. Wilt-resistant varieties offer the best means of control. The morphology of the fungus is given in Chap. 18.

12. Frogeye Leaf Spot, *Cercospora nicotianae* Ell. & Ev. or *C. apii* Fres. This leaf spot is distributed widely throughout the world, and it is of economic importance chiefly in the more tropical, humid areas. The symptoms are zonate white and yellow circular spots with the gray conidiophores present later on the dead bleached portion of the lesion. Infections, just prior to harvest, produce small green spots on the cured leaves. Smaller zonate spots appear on the bracts, calyx, and capsules (Hill, 1936).

FIG. 101. *Fusarium* wilt of tobacco showing the narrow, thickened leaves and brown discolored stem interior.   (*Courtesy of J. J. Johnson.*)

THE FUNGUS

Cercospora nicotianae Ell. & Ev. or *C. apii* Fres.
(*Cercospora raciborskii* Sacc. & Syd.)
(*Cercospora solanicola* Atk.)

The dark-brown septate conidiophores arise in groups from masses of mycelium in the tissue.   The conidia are borne from the terminal cell, and they are displaced by elongation of the conidiophore.   The conidia taper gradually toward the apex, are hyaline, have 0 to 16 septations, and are 90 to 300 microns long.

Morphologically the fungus on tobacco and several other hosts are similar.   Johnson and Valleau (1949) obtained typical lesions on tobacco leaves with *Cercospora* isolates from 16 species in 11 families. They propose combination of the like species, designated originally by host, under the type species of the genus, *C. apii* Fres.

ETIOLOGY. The fungus persists in crop refuse from tobacco, other crops, and some weeds, and it is carried over on seed.

Seed-bed infections occur from plant trash or conidia on the seed. Secondary spread from conidia is general during hot, humid weather. The conidia and mycelium in dry crop refuse remain viable for one or more years. Seed treatment and sanitation apparently control the disease in areas where living plants do not survive from season to season. Fungicidal sprays are necessary in areas where the conidial inoculum is abundant.

13. Black Root Rot, *Thielaviopsis basicola* (Berk. & Br.) Ferr. This fungus produces a root rot on a large number of field crops, including cotton, cowpeas, flax, lupines, peanuts, red clover, soybean, and tobacco, and it is world-wide in distribution, especially in cold, wet alkaline soils (Johnson, 1916). Prior to the use of resistant varieties, the disease caused large losses in yield of tobacco.

SYMPTOMS AND EFFECTS. The disease proper is limited to the root system and the base of the stem. The depletion of the root system and the black rotted root stubs are the characteristic symptoms. The depleted root system results in retarded uneven growth of the plants, a yellow or chlorotic appearance, and temporary wilting (Fig. 102). The rapid recovery of the plants as soon as warm weather prevails is also characteristic of the disease. The fungus invades the root cortex by entrance around the branch root ruptures, mechanical injuries, nematode lesions, or direct penetration and advances through the cortex and central stele, producing the blackened condition simultaneous with the necrosis of the tissues.

THE FUNGUS

Thielaviopsis basicola (Berk. & Br.) Ferr.
[*Thielavia basicola* (Berk. & Br.) Zopf.]
(*Thielavia basicola* Zopf.)

Two types of spores are produced under different environments and ages of the mycelium. The endoconidia are borne on the young mycelium, especially in culture. The endoconidiophores are phialides comprising a bulbous base and an elongated tube 50 to 90 microns long, gradually tapering to 3 to 7 microns in diameter at the end. They arise as a branch from near the center of a hyphal cell. The endoconidia are formed in chains within the conidiophore and are extruded singly or in chains. The conidia vary in size, measuring 8–30 by 3–6 microns, are cylindrical with rounded ends, hyaline, and germinate to produce mycelium (Brierley, 1915). The thick-walled resting spores originate in chains or clusters laterally or terminally from any part of the mycelium; they are hyaline at first, but soon become thick-walled and brown, short cylindric with angular ends, about 12 by 5–8 microns in size, and separate at maturity. The ascigerous stage (*Thielavia*) was formerly associated with this species, but according to McCormick (1925) the relationship is questionable. Rawlings (1940) and Stover (1950) have shown specialization of the parasite.

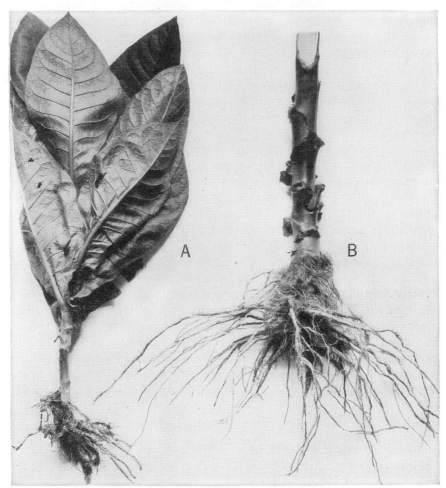

FIG. 102. Black root rot of tobacco caused by *Thielaviopsis basicola*, showing the rotted roots and small plant in a susceptible variety (*A*) compared with the root system of a resistant plant (*B*). (*Courtesy of J. J. Johnson.*)

ETIOLOGY. The fungus infestation persists in alkaline soils for an indefinite period and invades the roots of plants during cool, wet weather. The parasitic activity of the fungus is greatly restricted as the soil temperature rises, probably associated with both the activity of the fungus and the resistance of the plant suscepts (Conant, 1927, Doran, 1929, Jewett, 1938, Johnson and Hartman, 1919).

The control of the disease is largely by means of resistant varieties. Johnson (1916, 1930) demonstrated the practical control by resistant tobacco selections, and he also demonstrated that a complex of several factors was involved in resistance. Breeding for resistance to black root rot started in 1912 by the late J. Johnson represents the beginning of the modern breeding for disease resistance in tobacco. He and associates combined resistance with high yields and good quality and demonstrated the inheritance of resistance on a Mendelian basis (Johnson, 1916, 1930, and Johnson and Hartman, 1919). Valleau and associates soon followed with root rot-resistant varieties (Valleau and Kinney, 1922).

Conant (1927) and Jewett (1938) have studied the nature of resistance. Resistant varieties of the various types of tobacco are used wherever tobacco is grown on soils infested with the parasite. Crop rotations are used in reducing the severity of the infestation.

14. Soreshin, *Rhizoctonia solani* Kuehn and *Sclerotium bataticola* Taub., and Other Fungi. These similar diseases are distributed generally on tobacco and many other field crops. The fungi causing the stem cankering (soreshin) vary in the different tobacco-producing sections. Rhizoctonia is more prevalent in the cooler areas; *Sclerotium bataticola* and other species occur more commonly in the warmer areas. These fungi attack the basal stem tissues and produce a similar type of disease.

SYMPTOMS AND EFFECTS. Dark-colored cankers or rotting at the base of the stem occur in the seed bed or in the field. The lesions usually occur near the soil line and vary from local cortical lesions to cankers developing into the woody tissues and central pith. Under favorable conditions, the lesions extend up the stem and into the lower leaves. The invaded tissues show extensive necrosis and collapse of the tissues. The brown to black rotted lesion is relatively dry, and sclerotia are common on the rotted area. The infected plants are yellow and dwarfed, frequently break over, and in severe rotting they wilt and brown.

THE FUNGI

Rhizoctonia solani Kuehn or *Pellicularia filamentosa* (Pat.) Rogers (*Corticium vagum* Berk. & Curt.)

The morphology of the fungus is given in Chaps. 7 and 11.

Sclerotium bataticola Taub. or
Macrophomina phaseoli (Maubl.) Ashby.

The morphology of the fungus and its several stages are discussed in Chap. 9.

Sclerotium rolfsii Sacc. or *Pellicularia rolfsii* (Curzi) West also occurs on tobacco in the Southern United States.

REFERENCES

ADAMS, D. B. Blue mold of tobacco: on the morphology of the fungus and its nomenclature. *Jour. Dept. Agr. Victoria* **31**:412–416. 1933.

ALLARD, H. A. The mosaic disease of tobacco. *U.S. Dept. Agr. Bul.* 40. 1914.

ANDERSON, P. J. Downy mildew of tobacco. *Conn. Agr. Exp. Sta. Bul.* 405. 1937. (Good literature review and reference list.)

———. Diseases and decays of Connecticut tobacco. *Conn. Agr. Exp. Sta. Bul.* 432. 1940.

ANGELL, H. R., and A. V. HILL. Downy mildew (blue mold) of tobacco in Australia *Coun. Sci. Ind. Res. Australia Bul.* 65. 1932.

ARMSTRONG, G. M. and J. K. ARMSTRONG. Nonsusceptible hosts as carriers of wilt Fusaria. *Phytopath.* **38**:808–826. 1948.

BAWDEN, F. C. Plant viruses and virus diseases. Chronica Botanica. Leiden, Holland. 1939.

BEIJERINCK, M. W. Ueber ein Contagium vivum fluidum als Ursache der Flecken-krankheit der Tabaksblätter. *Verhandel. Konink. Akad. Wetenschap., Amsterdam* (2), **65**:1–22. 1898. *Phytopath. Classics* 7. 1942.

BERKELEY, G. H. Root-rots of certain non-cereal crops. *Bot. Rev.* **10**:67–123. 1944.

——— and L. W. KOCH. Diseases of tobacco in Canada. *Can. Dept. Agr. Pub.* 667. *Farmers' Bul.* 85. 1940.

BLACK, L. M. A virus tumor disease of plants. *Am. Jour. Bot.* **32**:408–415. 1945.

BÖNING, K. Zur Ätiologie der Streifen- und Kräuselkirakheit des Tabaks. *Zeitschr. Parasitenk.* **3**:103–141. 1931.

BRIERLEY, W. B. The "endoconidia" of *Thielavia basicola* Zopf. *Ann. Bot.* **29**:483–493. 1915.

CHESSIN, M. The effects of nitrogen deficiency on the properties of tobacco-mosaic virus. *Phytopath.* **41**:235–237. 1951.

——— and H. A. SCOTT. Calcium deficiency and infection of *Nicotiana glutinosa* by tobacco mosaic virus. *Phytopath.* **45**:288–289. 1955.

CLAUSEN, R. E., and T. H. GOODSPEED. Interspecific hybridization in Nicotiana. II. A tetraploid glutinosa-tabacum hybrid, an experimental verification of Winges' hypothesis. *Genetics* **10**:278–284. 1925.

CLAYTON, E. E. Resistance of tobacco to blue mold (*Peronospora tabacina*). *Jour. Agr. Research* **70**:79–87. 1945.

———. A wildfire resistant tobacco. *Jour. Hered.* **38**:35–40. 1947.

———. Control of tobacco diseases through resistance. *Phytopath.* **43**:239–244. 1953.

———*et al.* Control of the blue mold (downy mildew) disease of tobacco by spray-ing. *U.S. Dept. Agr. Tech. Bul.* 650. 1938.

———. Mosaic resistance in *Nicotiana tabacum* L. *Phytopath.* **28**:286–288. 1938.

——— and H. H. MCKINNEY. Resistance to the common mosaic disease of tobacco. *Phytopath.* **31**:1140–1142. 1941.

——— and J. E. MCMURTREY. Tobacco diseases and their control. *U.S. Dept. Agr. Farmers' Bul.* 2023. 1950.

——— and T. E. SMITH. Resistance of tobacco to bacterial wilt (*Bacterium solana-cearum*). *Jour. Agr. Research* **65**:547–554. 1942.

——— and J. A. STEVENSON. Nomenclature of the tobacco downy mildew fungus. *Phytopath.* **26**:516–521. 1935.

COSTA, A. S. Concentration of Brazilian tobacco streak virus in infected plants. *Phytopath.* **42**:231–236. 1952.

DESJARDINS, P. R., C. A. SENSENEY, and G. E. HESS. Further studies on the electron microscopy of purified tobacco ringspot virus. *Phytopath.* **43**:687–690. 1953.

DIACHUN, S., and J. TROUTMAN. Multiplication of *Pseudomonas tabaci* in leaves of burley tobacco, *Nicotiana longiflora*, and hybrids. *Phytopath.* **44**:186–187. 1954.

———— and W. D. VALLEAU. Reaction of 35 species of *Nicotiana* to tobacco streak virus. *Phytopath.* **37**:7. 1947.

DIXON, L. F., *et al.* Relationship of climatological conditions to the tobacco downy mildew. *Phy opath.* **26**:735–759. 1936.

DORAN, W. L. Effects of soil temperature and reaction on growth of tobacco infected and uninfected with black root rot. *Jour. Agr. Research* **39**:853–872. 1929.

EAST, E. M. The genetics of the genus Nicotiana. *Bibliographia Genetica* **4**:243–320. 1928.

FULTON, R. W. Hosts of the tobacco streak virus. *Phytopath.* **38**:421–428. 1948.

————. Virus concentration in plants acquiring tolerance to tobacco streak. *Phytopath.* **39**:231–243. 1949.

————. Variants of the tobacco necrosis virus in Wisconsin. *Phytopath.* **40**:298–305. 1950.

————. Superinfection by strains of tobacco mosaic virus. *Phytopath.* **41**:579–592. 1951.

GARNER, W. W., *et al.* Superior germ plasm in tobacco. *U.S. Dept. Agr. Yearbook,* pp. 785–830. 1936.

GARRISS, H. R., and D. E. ELLIS. Suggested program for the control of Granville wilt and black shank of tobacco. *No. Car. Agr. Ext. Cir.* 247. 1941.

GERSTEL, D. W. Inheritance in *Nicotiana tabacum.* XVII. Cytogenetical analysis of glutinosa-type resistance to mosaic disease. *Genetics* **28**:533–536. 1943.

GIGANTE, R. Le principali batteriosi del tabacco. *Tabacco* **54**:609–610. 1950.

GODOY, E. F., and A. D. COSTE. Tobacco downy mildew in the tobacco growing region of Salta and Jujuy. *Rev. Argentina Agron.* **7**:221–227. 1940.

GRATZ, L. O., and R. R. KINCAID. Tests of cigar-wrapper tobacco varieties resistant to black shank. *Fla. Agr. Exp. Sta. Bul.* 326. 1938.

HENDERSON, R. G. Studies on tobacco downy mildew in Virginia. *Va. Agr. Exp. Sta. Tech. Bul.* 62. 1937.

HILL, A. V. Cercospora leaf-spot (frogeye) of tobacco in Queensland. *Coun. Sci. Ind. Res. Aust. Bul.* 98. 1936.

———— and H. R. ANGELL. Downy mildew (blue mold) of tobacco. I–III. *Jour. Coun. Sci. Ind. Research Australia* **6**:260–268. 1933.

HOGGAN, I. A. The peach aphid (*Myzus persicae* Sulz.) as an agent in virus transmission. *Phytopath.* **19**:109–123. 1929.

HOLMES, F. O. Inheritance of resistance to tobacco-mosaic disease in tobacco. *Phytopath.* **28**:553–560. 1938.

————. Handbook of phytopathogenic viruses. Burgess Publishing Co. Minneapolis. 1939.

————. A comparison of the experimental host range of tobacco-etch and tobacco-mosaic viruses. *Phytopath.* **36**:640–659. 1946.

————. Indications of a new-world origin of tobacco-mosaic virus. *Phytopath.* **41**:341–349. 1951.

IVANOWSKI, D. Über die Mosaikkrankheit der Tabakspflanze. *Acad. Imp. Sci. St. Petersbourg Bul.* (N.S.) **35**:67–70. 1892. *Phytopath. Classics* 7. 1942.

JEWETT, F. L. Relation of soil temperature and nutrition to the resistance of tobacco to *Thielavia basicola.* *Bot. Gaz.* **100**:276–297. 1938.

JOHNSON, E. M., and W. D. VALLEAU. Synonymy in some common species of *Cercospora*. *Phytopath.* **39**:763–770. 1949.

JOHNSON, J. Resistance in tobacco to the root rot disease. *Phytopath.* **6**:161–181. 1916.

———. Tobacco diseases and their control. *U.S. Dept. Agr. Dept. Bul.* 1256. 1924.

———. Breeding tobacco for resistance to Thielavia root rot. *U.S. Dept. Agr. Tech. Bul.* 175. 1930.

———. Studies on the nature of brown root rot of tobacco and other plants. *Jour. Agr. Research* **58**:843–863. 1939.

———. Virus attenuation and the separation of strains by specific hosts. *Phytopath.* **37**:822–837. 1947.

———. Virus particles in various plant species and tissues. *Phytopath.* **41**:78–93. 1951.

——— and R. E. HARTMAN. Influence of soil environment on the root-rot of tobacco. *Jour. Agr. Research* **17**:41–86. 1919.

——— and W. B. OGDON. Tobacco mosaic and its control. *Wis. Agr. Exp. Sta. Bul.* 445. 1939.

KOCH, L. W. Control of the blue mold disease of tobacco. *Can. Dept. Agr. Pub.* 716 (*Cir.* 171). 1941.

——— and R. J. HASLAM. Varietal susceptibility of tobacco to brown root rot in Canada. *Sci. Agr.* **18**:561–567. 1938.

MARKHAM, R., K. M. SMITH, and R. W. G. WYCKOFF. Electron microscopy of tobacco necrosis virus crystals. *Nature, London* **159**:574. 1947.

MAYER, A. Ueber die Mosaikkrankheit des Tabaks. *Landw. Vers. Sta.* **32**:450–467. 1886. *Phytopath. Classics* 7. 1942.

McCLURE, T. T. Mode of infection of the sweet-potato wilt Fusarium. *Phytopath.* **39**:876–886. 1949.

McCORMICK, F. A. Perithecia of *Thielavia basicola* Zopf. in culture and the stimulation of their production by extracts of other fungi. *Conn. Agr. Exp. Sta. Bul.* 269. 1925.

McKINNEY, H. H. Stability of labile viruses in desiccated tissue. *Phytopath.* **37**:139–142. 1947.

MIDDLETON, J. T. The taxonomy, host range and geographic distribution of the genus Pythium. *Mem. Torrey. Bot. Club* **20**:1–171. 1943.

MOOREHEAD, E. L., and W. C. PRICE. A new serological test for tobacco mosaic virus. *Phytopath.* **43**:73–77. 1953.

MORGAN, O. D., and H. H. McKINNEY. A disease of Maryland broad leaf tobacco with symptoms similar to leaf-curl symptoms. *Phytopath.* **41**:564. 1951.

NOLLA, J. A. B. Inheritance in Nicotiana. *Jour. Hered.* **29**:43–48. 1938.

PERSON, L. H., and G. B. LUCAS. Oospore germination in *Peronospora tabacina*. *Phytopath.* **43**:701–702. 1953.

POUND, G. S., and L. G. WEATHERS. The effect of air and soil temperatures on the multiplication of turnip virus 1 in certain *Nicotiana* spp. *Phytopath.* **43**:550–554. 1953.

——— and ———. The relation of host nutrition to multiplication of turnip virus 1 in *Nicotiana glutinosa* and *N. multivalvis*. *Phytopath.* **43**:669–674. 1953.

POWERS, H. R., JR. The mechanism of wilting in tobacco plants affected with black shank. *Phytopath.* **44**:513–521. 1954.

PRICE, W. C. Comparative host range of six plant viruses. *Am. Jour. Bot.* **28**:530–541. 1940.

QUANJER, H. M.   Bijdrage tot de kennis van de in Nederland voorkomende ziekten van Tabak en van de Tabaksteelt on Kleigrond.   *Tijdschr. Plantenziekt.* **49**: 37–51. 1943.

RAWLINGS, R. E.   Observations on the cultural and pathogenic habits of *Thielaviopsis basicola* (Berk. and Br.) Ferraris.   *Ann. Mo. Bot. Garden* **27**:561–593.   1940.

ROZENDAAL, A., and J. P. H. VAN DER WANT.   Over de Identiteit van het Ratelvirus van de Tabak en het Stengelbontvirus van de Aardappel.   *Tijdschr. Plantenziekt.* **54**:113–133.   1948.

SASSER, J. N., G. B. LUCAS, and H. R. POWERS, JR.   The relationship of root-knot nematodes to black-shank resistance in tobacco.   *Phytopath.* **45**:459–461.   1955.

SHARP, D. G., and F. A. WOLF.   The virus of tobacco leaf curl. II.   *Phytopath.* **41**:94–98.   1951.

SIGURGEIRSSON, T., and W. M. STANLEY.   Electron microscope studies on tobacco-mosaic virus.   *Phytopath.* **37**:26–38.   1947.

SMITH, K. M.   A textbook of plant virus diseases.   The Blakiston Division, McGraw-Hill Book Company, Inc.   New York.   1937.

———— and J. G. BALD.   A description of a necrotic virus disease affecting tobacco and other plants.   *Parasitology* **29**:86–95.   1935.

STANLEY, W. M.   Isolation of a crystalline protein possessing the properties of tobacco-mosaic virus.   *Science* (N.S.) **81**:644–645.   1935.

————.   Chemical studies on the virus of tobacco mosaic. VI. The isolation from diseased Turkish tobacco plants of a crystalline protein possessing the properties of tobacco-mosaic virus.   *Phytopath.* **26**:305–320.   1936.

STOREY, H. H.   Leaf curl of tobacco in Southern Rhodesia.   *Rhod. Agr. Jour.* **29**: 186–192.   1932.

————.   Virus diseases of East African plants. II. Leaf-curl disease of tobacco.   *E. Africa Agr. Jour.* **1**:148–153.   1935.

STOVER, R. H.   The black root rot disease of tobacco. II. Physiologic specialization of *Thielaviopsis basicola* on *Nicotiana tabacum*.   *Can. Jour. Research* C **28**:726–738.   1950.

————.   Association in tobacco of the severe symptom response to etch virus and the white burley character.   *Phytopath.* **41**:1125–1126.   1951.

————.   Some methods and problems in the study of nematode root rot of tobacco in Ontario.   *Phytopath.* **41**:34.   1951.

———— and L. W. KOCH.   The epidemiology of blue mold of tobacco and its relation to the incidence of the disease in Ontario.   *Sci. Agr.* **31**:225–252.   1951.

TARR, S. A. J.   Leaf curl disease of cotton.   Commonwealth Mycological Institute.   Kew, Surrey.   1951.

THUNG, T. H.   De Krulen Kroepoek-ziekten van Tabak en de Oorzaaken van hare Verbreidung.   *Proefstat. Vorstenl. Tabak. Meded.* 72.   1932.

————.   Waarnemingen over enkele Plantenviren met het Electronenmicroscop.   *Chronica Naturae* **104**:342–348.   1948.

TISDALE, W. B.   Development of strains of cigar-wrapper tobacco resistant to black shank (*Phytophthora nicotianae* Breda de Haan).   *Fla. Agr. Exp. Sta. Bul.* 226.   1931.

TUCKER, C. M.   Taxonomy of the genus Phytophthora de Bary.   *Mo. Agr. Exp. Sta. Res. Bul.* 153.   1931.

VALLEAU, W. D.   Seed transmission and sterility studies of two strains of tobacco ring spot.   *Ky. Agr. Exp. Sta. Bul.* 327, pp. 48–80.   1932.

————.   Clubroot of tobacco.   A wound-tumorlike graft transmitted disease.   *Phytopath.* **37**:580–582.   1947.

————. Breeding tobacco for immunity to black shank. *Phytopath.* **42**:288. 1952.

————. Breeding tobaccos for disease resistance. *Econ. Bot.* **6**:69–102. 1952. (Good reference list.)

————. The evolution of susceptibility to tobacco mosaic virus in Nicotiana and the origin of the tobacco mosaic virus. *Phytopath.* **42**:40–42. 1952.

———— *et al.* Tobacco diseases. *Ky. Agr. Exp. Sta. Bul.* 437. 1942.

————. Angular leaf spot and wildfire of tobacco. *Ky. Agr. Exp. Sta. Bul.* 454. 1943. (Good literature list.)

———— and E. M. JOHNSON. Tobacco mosaic sources of infection and control. *Ky. Agr. Exp. Sta. Bul.* 376. 1937.

———— and ————. The relation of meadow nematodes to brown root rot of tobacco. *Phytopath.* **37**:838–841. 1947.

———— and E. J. KINNEY. Strains of standup white burley tobacco resistant to root rot. *Ky. Agr. Exp. Sta. Cir.* 28. 1922.

WEATHERS, L. G., and G. S. POUND. Host nutrition in relation to multiplication of tobacco mosaic virus in tobacco. *Phytopath.* **44**:74–80. 1954.

WINGARD, S. A. Hosts and symptoms of ring spot, a virus disease of plants. *Jour. Agr. Research* **37**:127–153. 1928.

WOLF, F. A. Downy mildew of tobacco in Brazil. *Phytopath.* **29**:291. 1939.

————, W. H. WHITCOMB, and W. C. MOONEY. Leaf-curl of tobacco in Venezuela. *Jour. Elisha Mitchell Sci. Soc.* **65**:38–47. 1949.

WOODS, W. M. Intracellular inclusions in tobacco mosaic-infected *Nicotiana glutinosa* and its hybrids. *Phytopath.* **34**:694–696. 1944.

DISEASES OF FIELD CROPS ARRANGED
BY CASUAL FACTOR

(Suggested for a three-hour course)

1. Nonparasitic Diseases
   Gray speck disease of cereals—manganese deficiency
   Alfalfa yellows—boron deficiency
   Leaf spotting "rust" of cotton—potash deficiency
   Frost injury and winterkilling in wheat
   Cold and winter injury in alfalfa
   Heat canker in flax
   Brown root rot of tobacco
2. Virus Diseases
   Vectors largely aphids
      Common tobacco mosaic
      Tobacco ring spot
      Alfalfa and clover mosaics
      Corn and sugarcane mosaics
   Vectors largely leaf hoppers
      Alfalfa dwarf
      Corn and sugarcane streak diseases
      Chlorotic streak of sugarcane
      Dwarf disease of rice
   Vectors unknown but symptoms similar to streaks
      Wheat mosaics
3. Bacterial Diseases
   Bacterial blights
      Bacterial blight of barley, wheat, rye, grasses—*Xanthomonas translucens*
      Bacterial streak of sorghums—*Xanthomonas holcicola*
      Bacterial stripe of sorghums—*Pseudomonas andropogoni*
      Angular leaf spot of tobacco—*Pseudomonas angulata*
      Angular leaf spot of cotton—*Xanthomonas malvacearum*
      Bacterial pustule of soybean—*Xanthomonas phaseoli* var. *sojense*
   Halo blights
      Halo blight of oats and grasses—*Pseudomonas coronafaciens*

Wildfire of tobacco and soybean—*Pseudomonas tabaci*
Bacterial wilts
    Alfalfa wilt—*Corynebacterium insidiosum*
    Stewart's wilt of corn—*Bacterium stewartii*
    Granville wilt of tobacco—*Xanthomonas solanacearum*
4. Chytridiales
    Crown wart of alfalfa—*Urophlyctis alfalfae* or *Physoderma alfalfae*
    Physoderma disease of corn—*Physoderma maydis*
5. Peronosporales
  Pythium diseases
    Root rot of corn and sugarcane—*Pythium arrhenomanes*
    Root rot of cereals and grasses—*Pythium graminicola*, etc.
    Root rot and damping-off of legumes, etc.—*Pythium debaryanum*
  Downy mildews
    Downy mildew of alfalfa—*Peronospora trifoliorum*
    Blue mold or downy mildew of tobacco—*Peronospora tabacina*
    Downy mildew of the cereals and grasses—*Sclerospora macrospora*
    Downy mildew of millets—*Sclerospora graminicola*
6. Perisporiales (Powdery Mildews)
    Powdery mildew of cereals and grasses—*Erysiphe graminis*
    Powdery mildew of legumes—*Erysiphe polygoni*
7. Hypocreales
    Ergot of cereals and grasses—*Claviceps purpurea* and *C. paspali*
    Gibberella head blight, root rot, and seedling blight of cereals—*Gibberella zeae, Fusarium culmorum*, etc.
    Corn ear and stalk rot and seedling blight—*Gibberella zeae* and *G. fujikuroi*, etc.
    Sooty blotch of clovers—*Cymadothira trifolii*
8. Sphaeriales
    Pasmo disease of flax—*Mycosphaerella linorum*
    Spring black stem of legumes—*Mycosphaerella lethalis, Ascochyta imperfecta, Phoma trifolii*
    Summer black stem and leaf spot of legumes—*Mycosphaerella davissi* and *Cercospora zebrina*
    Culm rot of rice—*Leptosphaeria salvinii*
    Take-all of wheat and grasses—*Ophiobolus graminis*
    Corn leaf blights—*Cochliobolus heterostropus, Helminthosporium turcicum*, and *H. carbonum*
    Spot blotch of barley and wheat—*Helminthosporium sativum* [see also *Cochliobolus miyabeanus* (*H. oryzae*)]
    Net blotch of barley—*Pyrenophora teres* (see also *P. bromi* and *P. avenae*)
    Stripe disease of barley—*Helminthosporium gramineum*
9. Pezizales
    Sclerotinia root rot and crown rot on red clover—*Sclerotinia trifoliorum*
    Pseudopeziza leaf spot of alfalfa—*Pseudopeziza medicaginis, P. trifolii, P. meliloti*
    Yellow leaf spot of alfalfa—*Pseudopeziza jonesii* or *Pyrenopeziza medicaginis*
10. Sphaeropsidales
    Diplodia ear rot and stalk rot of corn—*Diplodia zeae, D. macrospora* (see also *Physalospora zeae* and *P. zeicola*)
    Septoria leaf blotches of cereals and grasses—*Septoria* spp.

11. Melanconiales (Anthracnoses)

    Northern anthracnose of clovers—*Kabatiella caulivora*

    Southern anthracnose of clovers—*Colletotrichum trifolii*

    Anthracnose of cereals and grasses—*Colletotrichum graminicolum*

    Anthracnose of cotton—*Glomerella gossypii*

    Red rot of sugarcane—*Physalospora tucumanensis*

12. Moniliales

    Phymatotrichum root rot—*Phymatotrichum omnivorum*

    Scald of cereals and grasses—*Rhynchosporium secalis* and *R. orthosporium*

    Cercospora leaf spot of rice—*Cercospora oryzae*

    Fusarium wilt of crops—*Fusarium oxysporum*

        (See also *Verticillium* wilt of cotton—*Verticillium albo-atrum*)

13. Autobasidiomycetes

    Rhizoctonia root rot and stem blight of crops—*Pellicularia filamentosa*

    Typhula snow mold or scald of cereals and grasses—*Typhula itoana, T. idahoensis,* etc.

14. Ustilaginales (Smuts)

    Ustilago spp.

    Covered smuts of cereals and grasses—*Ustilago hordei, U. kolleri, U. bullata,* etc.

    Intermediate or black loose smuts of cereals and grasses—*Ustilago avenae, U. nigra*

    Loose smuts of cereals and grasses—*Ustilago nuda, U. tritici*

    Stalk smuts of grasses—*Ustilago spegazzinii,* etc.

    Stripe leaf smut of grasses—*Ustilago striiformis*

    Corn smut—*Ustilago maydis*

    Sphacelotheca spp.

    Kernel smuts of sorghum—*Sphacelotheca cruenta, S. sorghi*

    Tilletia spp.

    Kernel smuts (bunt) of wheat and grasses—*Tilletia caries, T. foetida,* etc.

    Neovossia spp.

    Kernel smut of rice—*Tilletia horrida* or *Neovossia horrida*

    Entyloma spp.

    Leaf spot smut of rice and grasses—*Entyloma dactylidis, E. meloiti,* etc.

    Urocystis spp.

    Flag smut of wheat and grasses—*Urocystis tritici, U. agropyri,* etc.

15. Uredinales (Rusts)

    Melampsora spp.

    Flax rust—*Melampsora lini*

    Uromyces spp.

    Rust of legumes—*Uromyces striatus, U. trifolii*

    Puccinia spp.

    Stem rust of cereals and grasses—*Puccinia graminis*

    Stripe rust of cereals and grasses—*Puccinia glumarum*

    Leaf rust of cereals and grasses—*Puccinia coronata, P. rubigo-vera, P. poae-sudeticae,* etc.

    Cotton rust—*Puccinia stakmanii*

APPENDIX B

BACTERIA AND FUNGI PARASITIC ON FIELD CROPS

SCHIZOMYCETES	More common host	Page
Eubacteriales		
Erwinia flavida (Fawc.) Magrou............	Sugarcane	216
Bacterium albilineans Ashby...............	Sugarcane	212, 213
B. stewartii E. F. Sm.....................	Corn	77, **78**
Xanthomonas axonoperis Starr & Garces.....	Grasses	298
X. holcicola (Elliott) Starr & Burk.........	Sorghums	190, 191
X. itoana (Toch.) Dows...................	Rice	155
X. lespedezae (Ayers *et al.*) Starr............	Legumes	402
X. malvacearum (E. F. Sm.) Dows..........	Cotton	411–413, 414, 416
X. oryzae (Uyeda & Ishiyama) Dows.......	Rice	155
X. panici (Elliott) Savul..................	Millets	115
X. phaseoli (E. F. Sm.) Dows..............	Legumes	402
X. phaseoli var. *sojense* (Hedges) Starr & Burk..............................	Legumes	386
X. rubrilineans (Lee *et al.*) Starr & Burk....	Sugarcane	213, 214, 298
X. rubrisubalbicans (Chris. & Edg.) Savul...	Sugarcane	214
X. solanacearum (E. F. Sm.) Dows.........	Tobacco	397, 402, 451, 452
X. translucens (L. R. Jones, A. G. Johnson, & Reddy) Dows	Cereals	27, **28**, 128, 238, 296, 297
X. translucens f. sp. cerealis Hagb...........	Grasses, cereals	27, 296, 297
X. translucens f. sp. phleipratensis Wallin & Reddy...............................	Grasses	296
X. translucens f. sp. secalis (Redd, Godkin & Johnson) Hagb........................	Rye, grasses	173, 296
X. translucens f. sp. undulosa (E. F. Sm., Jones, & Reddy) Hagb.................	Wheat	231
X. vasculorum (Cobb) Dows...............	Sugarcane	79, 80, 211, 212, 213, 298
X. vignicola Burk........................	Legumes	402
Pseudomonas alboprecipitans Rosen.........	Grasses, corn	115
P. andropogoni (E. F. Sm.) Stapp..........	Sorghums	190,191
P. angulata (Fromme & Murray) Holland...	Tobacco	453, 454
P. atrofaciens (McCull.) Stevens...........	Wheat	231, 232

SCHIZOMYCETES	More common host	Page
Eubacteriales, *P. coronafaciens* (Elliott) Stevens	Oats	126, 127, **128**
P. coronafaciens var. *atropurpurea* (Reddy & Godkin) Stapp	Grasses	296, 297
P. glycinea Coerper.	Soybeans	385
P. medicaginis Sackett.	Alfalfa	344
P. phaseolicola (Burk.) Dows.	Legumes	402
P. pisi Sackett.	Legumes	402
(*P. sojae* Wolf) = *P. glycinea* Coerper.	Soybeans	385
P striafaciens (Elliott) Starr & Burk.	Oats	127, 128, **129**, 296
P. syringae Van Hall.	Sorghums	191, 368, 402
P. tabaci (Wolf & Foster) Stevens.	Tobacco	385, 452–**454**
[*P. trifoliora* (L. R. Jones *et al.*) Burk] = *P. syringae* Van Hall.	Clover	368
Corynebacterium agropyri (O'Gara) Burk. . . .	Grasses	298
C. flaccumfaciens (Hedges) Dows.	Legumes	402
C. insidiosum (McCull.) Jensen.	Alfalfa	342, **343**
C. rathayi (E. F. Sm.) Dows.	Grasses	298
C. tritici (Hutch.) Burk.	Wheat	232
PHYCOMYCETES		
Uniflagellate		
Chytridiales		
Olpidiaceae		
Olpidium brassicae (Wor.) Dang. = *Asterocystis radicis* DeWild.	Flax	430
O. trifolii (Pass.) Schroet.	Clover	346
Physodermataceae		
Urophlyctis alfalfae (Lageh.) Magn. or *Physoderma alfalfae* (Lageh.) Karling . . .	Alfalfa	345
U. trifolii (Pass.) Magn. or *Olpidium trifolii* (Pass.) Schroet.	Clover	346
Physoderma alfalfae (Lageh.) Karling	Alfalfa	345, 346
P. maydis Miyabe = *P. zea-maydis* Shaw.	Corn	80
Biflagellate		
Saprolegniales		
Saprolegniaceae		
Aphanomyces euteiches Dreschl.	Legumes	402
Peronosporales		
Pythiaceae		
Pythium aphanidermatum (Edson) Fitzp. . .	Tobacco, flax	83, 430, 455
P. aristosporum Vanterpool.	Wheat	233
P. arrhenomanes Drechsl.	Cereals, grasses	81, 82, 192, 214, 233, 299, 346, 386
P. debaryanum Hesse.	General	192, 299, 346, 386, 430, 455

PHYCOMYCETES Biflagellate	More common host	Page
Peronosporales, Pythiaceae, *P. deliense* Meurs	Tobacco	455
P. *graminicola* Subrm..................	Cereals	30, 81, 192, 214, 233, 299, 346
P. *intermedium* DBy..................	Flax	430
P. *irregulare* Buis.....................	Flax	430
P. *mamillatum* Meurs..................	Flax	430
P. *megalacanthum* DBy..................	Flax	430
P. *splendens* Braun....................	Alfalfa, flax	346, 430
P. *tardicrescens* Vanterpool.............	Wheat	233
P. *ultimum* Trow......................	Alfalfa	346
P. *vexans* DBy........................	Alfalfa, flax	346, 430
P. *volutum* Vanterpool & Truscott........	Wheat	233
Phytophthora cactorum (Leb. & Cohn) Schroet............................	Sweetclover	347
P. *cryptogea* Pethy. & Laff..............	Flax	347
P. *erythroseptica* Pethy.................	Sugarcane	215, 347
P. *megasperma* Drechsl..................	Sweetclover	347
P. *parasitica* Dast.....................	General	456
P. *parasitica* var. *nicotianae* (Breda de Haan) Tucker......................	Tobacco	455, 456
Peronosporaceae		
Sclerophthora macrospora (Sacc.) Thir., Shaw & Naras. or *Sclerospora macrospora* Sacc.	Grasses	30, 83, **84**, 86, 117, 118, 129, 233, 234, 299
Sclerospora butleri Weston..............	Grasses	86
S. *farlowii* Griff.......................	*Chloris* sp.	86
S. *graminicola* (Sacc.) Schroet...........	Grasses, millets	83, **84**, 86, 115, **116**, 117, 192, 234, 299
S. *indica* Butl.........................	Corn	83, **84**, 86
(S. *javanica* Palm) = S. *maydis* (Rac.) Butl..............................	Corn	84
S. *macrospora* Sacc. or *Sclerophthora macrospora* (Sacc.) Thir., Shaw & Naras.	Cereals, grasses	30, 83, **84**, 86, 117, 118, 129, 215, 233, 234, 299
S. *magnusiana* Sorok....................	Equisetum	86
S. *maydis* (Rac.) Butl..................	Corn	83, **84**, 86
S. *miscanthi* Miy......................	*Miscanthus* sp.	86
S. *noblei* Weston.......................	Sorghum, grasses	86
S. *northi* Weston.......................	*Erianthus* sp.	86
S. *oryzae* Brizi........................	Rice	86
S. *philippinensis* Weston................	Corn, sorghum, sugarcane	83, **84**, 86
S. *sacchari* Miy.......................	Sugarcane, corn	83, **84**, 86, 215
S. *sorghi* (Kulk.) Weston & Uppal........	Sorghum	83, **84**, 86, **192**
S. *spontanea* Weston...................	Sugarcane, corn	**84**, 86

PHYCOMYCETES Biflagellate	More common host	Page
Peronosporales, Peronosporaceae, *Peronospora*		
hyoscyami DBy	Nightshade	457
P. manshurica (Naum.) Syd	Soybeans	386–**388**
P. nicotianae Speg	Tobacco	457
P. tabacina Adam	Tobacco	456, **457**, 458
P. trifoliorum DBy	Alfalfa, clover	347, 348
ASCOMYCETES		
Pyrenomycetes		
Erysiphales		
Erysiphaceae		
Erysiphe cichoracearum DC	Flax	369, 431
E. communis Wallr	Flax	369, 431
E. graminis DC	Cereals, grasses	30, **31**, 33, 52, 129, 299
E. graminis avenae E. Marchal	Oats, grasses	129
E. graminis hordei E. Marchal	Barley, grasses	31, 32, 33, 34
E. graminis secalis E. Marchal	Rye, Grasses	173
E. graminis tritici E. Marchal	Wheat, grasses	235
E. polygoni DC	Legumes	368, 369, 388, 403, 431
Microsphaera alni (DC.) Wint	Clover	368
M. diffusa C. & P	Legumes	403
Hypocreales		
Clavicipitaceae		
Claviceps cinerea Griff	Grasses	300
C. maximensis Theis	Grasses	300
C. microcephala (Wallr.) Tul	Grasses	300
C. paspali F. L. Stevens & Hall	Grasses	300, 301
C. purpurea (Fr.) Tul	Cereals, grasses	37, 156, 174, 175, 176, 178, 179, 235, 300
C. pusilla Ces	Grasses	300
[1]*Ustilaginoidea oryzae* (Pat.) Bref. or *U. virens* (Che.) Tak	Rice	156
U. setariae Bref	Grasses	156
U. virens (Cke.) Tak	Grasses	156
Epichloë typhina (Fr.) Tul	Grasses	301, 302
Hypocreaceae		
Calonectria graminicola (Berk. & Br.) Wr	Cereals, grasses	174, 241, 302
C. graminicola var. *neglecta* Krampe	Cereals	241, 242
C. nivale (Fr.) Snyder & Hansen	General	241, 302
C. nivale f. *graminicola* (Berk. & Br.) Snyder & Hansen	Cereals	241, 242

[1] Ainsworth, G. C., and G. R. Bisby. A dictionary of the fungi. Imperial Mycological Institute, 1943, places this genus only in the Moniliales.

ASCOMYCETES Pyrenomycetes	More common host	Page
Hypocreales, Hypocreaceae, *Gibberella fujikuroi* (Saw.) Wr.	Cereals, grasses	85, 88, 89, 90, **91,** 155, 193, 215, 416
G. fujikuroi var. *subglutinans* Edwards....	Cereals, grasses	85, 88, 89, 90, 91, 215
G. moniliforme (Sheld.) Snyder & Hansen.	Corn	88, 89, 91, 155, 215
G. roseum (Lk.) Snyder & Hansen........	General	238
G. roseum f. *cerealis* (Cke.) Snyder & Hansen.............................	Cereals, grasses	34, 85, 89, **90,** 155, 238, 240, 303
G. zeae (Schw.) Petch..................	Cereals	34, 85, 87, 88, 89, **90,** 155, 193, 238, 240, 303
Sphaeriales		
Phyllachoraceae		
Phyllachora graminis (Fr.) Fckl..........	Grasses	302
Gnomoniaceae		
Glomerella cingulata (Ston.) Spauld. & Vons	Legumes	404
G. glycines Lehman & Wolf.............	Soybeans	388, **389**
G. gossypii Edg......................	Cotton	414, **415**
Diaporthaceae		
Diaporthe phaseolorum var. *batatis* (Hart. & Field) Wehm....................	Soybeans	390
var. *sojae* (Lehm.) Wehm..............	Soybeans	390
D. sojae Lehman = var. *sojae* above......	Soybeans	389
Pseudosphaeriales		
Pleosporaceae		
Physalospora tucumanensis Speg..........	Sugarcane	193, **216, 217**
P. zeae Stout.........................	Corn	95
P. zeicola Ell. & Ev....................	Corn	94, **95**
Leptosphaeria avenaria G. F. Weber.......	Oats, grasses	134
L. herpotrichoides DeNot...............	Cereals	245
L. nodorum Müller.....................	Wheat	249
L. pratensis Sacc. & Briard.............	Alfalfa, clover	352, 353, **354**
L. salvinii Catt.......................	Rice, grasses	156, **157,** 158
L. taiwanensis Yen & Chi...............	Sugarcane	219
Ophiobolus graminis Sacc...............	Cereals, grasses	242, **243,** 244, 245, 303
O. graminis var. *avenae* Turner..........	Oats, grasses	243
O. miyabeanus Ito & Kuribay. See *Cochliobolus miyabeanus* (Ito & Kuribay.) Dickson...........................	Rice	158, 159
O. oryzae Miyake.....................	Rice	158
O. oryzinus Sacc......................	Rice	158
O. sativus Ito & Kuribay. See *Cochliobolus sativus* (Ito & Kuribay.) Drechsl..	Cereals, grasses	47

ASCOMYCETES Pyrenomycetes	More common host	Page
Pseudosphaeriales, Pleosporaceae, *O. setariae* Ito & Kuribay. See *Cochliobolus setariae* (Ito & Kuribay.) Dickson.............	Millets	118
Cochliobolus heterostrophus Drechsl.......	Corn	97, 98, **99**
C. miyabeanus (Ito & Kuribay.) Dickson..	Rice	158, **159**, 160, 161
C. sativus (Ito & Kuribay.) Drechsl. See also *Ophiobolus sativus* Ito & Kuribay...	Cereals, grasses	**47**, 48, 246
C. setariae (Ito & Kuribay.) Dickson. See also *Ophiobolus setariae* Ito & Kuribay..	Millets	118
C. stenospilus (Carpenter) Matsu. & Yamam...........................	Sugarcane	218
C. tritici Dast. See *C. sativus* (Ito & Kuribay.) Dast....................	Wheat	**47**, 48, 246
Pseudoplea trifolii (Rostr.) Petr. or *Sac- cothecium trifolii* (Rostr.) Kirsch........	Clover	370, 371, **372**
Pyrenophora avenae Ito & Kuribay........	Oats	**129**, 130
P. bromi Died.........................	*Bromus* spp.	305, 306
P. graminea Ito & Kuribay. (?) See *Hel- minthosporium gramineum* Rab.........	Barley	42
P. teres (Died.) Drechsl................	Barley	37, **38**, 39, 40, 42, 131
P. tritici-repentis Died..................	Grasses	247, 306
P. tritici-vulgaris Dickson...............	Wheat	247
Pleospora herbarum (Fr.) Rab...........	Clover, alfalfa	350, 373, 374, **375**
P. rehmiana Staritz....................	Legumes	350
P. trichostoma (Fr.) Ces. & deNot........	General	40, 42, 247
Dothideales Dothidiaceae *Cymadothea trifolii* (Fr.) Wolf or [*Dothi- della trifolii* (Fr.) Bayl.-Elliott]........	Clover	**369**, 370, 371
Mycosphaerellaceae *Mycosphaerella arachidicola* Jenkins.......	Peanut	398
M. berkeleyii Jenkins...................	Peanut	398
M. davisii F. R. Jones.................	Alfalfa, clover	351, **352**
M. lethalis Stone.....................	Sweetclover	349, 350
M. linicola Naum.....................	Flax	432
M. linorum (Wr.) Garcia-Rada...........	Flax	431, **432**, 433
M. pinodes (Berk. & Blox.) Stone........	Legumes	403
M. pueraricola Weimer & Luttr..........	Legumes	404
DISCOMYCETES		
Inoperculates		
Helotiales Geoglossaceae *Mitrula sclerotiorum* Rostr..............	Clovers	373

DISCOMYCETES Inoperculates	Most common host	Page
Helotiales, Sclerotiniaceae		
Sclerotinia sclerotiorum (Lib.) DBy. or *S.*		
trifoliorum Eriks....................	Clovers, etc.	**372,** 373
Mollisiaceae		
Phialea temulenta Prill. & Del...........	Grasses	304
Pseudopeziza jonesii Nannf. or *Pyrenopeziza*		
medicaginis Fckl....................	Alfalfa	355, 356, **357**
P. medicaginis (Lib.) Sacc..............	Alfalfa	355, **356**
P. meliloti Syd.......................	Sweetclover	355, **356**
P. trifolii (Biv.-Bern.) Fckl.............	Clover	355, 356
Pyrenopeziza medicaginis Fckl. or *Pseudo-*		
peziza jonesii Nannf.................	Alfalfa	357
BASIDIOMYCETES		
Heterobasidiomycetes		
Ustilaginales		
Ustilaginaceae		
[*Ustilago aculeata* (Ule) Liro] = *U. macro-*	*Agropyron* and	316
spora Desm.	*Elymus* spp.	
U. avenae (Pers.) Rostr................	Oats, barley	56, 58, 59, 60, 134,
		135, **136,** 137,
		138, 140
[*U. bromivora* (Tul.) Fisch. von Waldh.] =		
U. bullata Berk.....................	Grasses	320
U. bullata Berk.......................	Grasses	320, 321
U. crameri Korn......................	Millet, grasses	**118,** 119, 120
U. crusgalli Tr. & Earle...............	Millets	**120**
(*U. echinata Schröt.*) = *U. macrospora*	Reed canary	316
Desm.	grass	
U. halophila Speg.....................	*Distichlis* spp.	318
U. hordei (Pers.) Lagerh...............	barley	56, 59, 60, 61, **62,**
		135, 138, 139,
		140
U. hypodytes (Schlecht.) Fries...........	Grasses	318, 319
U. kolleri Wille. or *U. hordei* (Pers.)	Oats	61, 135, 138, **139,**
Lagerh.		140
U. longissima (Schlecht.) Meyen.........	*Glyceria* spp.	316
(*U. lorentziana* Thüm) = *U. bullata* Berk.	Grasses	320
U. macrospora Desm....................	Grasses	316
U. maydis (DC.) Cda..................	Corn	100, **101,** 102
(*U. medians* Bied.) = *U. nigra* Tapke or		
U. avenae (Pers.) Rostr.............	Barley	55, 58, 134
U. mulfordiana Ell. & Ev..............	*Fescue* spp.	320
U. neglecta Niessl.....................	*Setaria* spp.	118, **119**
U. nigra Tapke or *U. avenae* (Pers.) Rostr.	Barley	55, 58, 61, 134
U. nuda (Jens.) Rostr..................	Barley, wheat	52, **53,** 54, 55, 184,
		252, 253

BASIDIOMYCETES Heterobasidiomycetes	More common host	Page
Ustilaginales, Ustilaginaceae, (*U. nummularia* Speg.) = *U. hypodytes* (Schlecht.) Fries.	Grasses	318, 319
(*U. perennans* Rostr.) = *U. avenae* (Pers.) Rostr.	Grasses	134
U. scitaminea Syd.	Sugarcane	220
(*U. sitanii* G. W. Fisch.) = *U. trebouxii* H. & P. Syd.	Grasses	320
U. spegazzinii Hirschh.	Grasses	318, 319
U. spegazzinii var. *agrestis* (Syd.) G. W. Fisch. & Hirschh.	Grasses	318
U. striiformis (Westend.) Niessl.	Grasses	315, 317
U. trebouxii H. & P. Syd.	Grasses	320
U. tritici (Pers.) Rostr. or *U. nuda* (Jens.) Rostr.	Wheat	184, 252, 253
U. vavilovii Jacz.	Rye	184
U. williamsii (Griff.) Lavr.	*Oryzopsis* and *Stipa* spp.	318
[*U. zeae* (Schw.) Ung.] = *U. maydis* (DC.) Cda.	Corn	101
Sphacelotheca andropogonis (Opiz.) Bubak.	Andropogon	322
S. cruenta (Kuehn) A. A. Potter.	Sorghum	198, 199, 200, 220, 322
S. destruens (Schlecht.) Stev. & A. G. Johnson.	Panicum	**119, 120,** 322
S. holci Jacks.	Johnson grass	200
S. reilina (Kuehn) Clint.	Sorghum, corn	**103,** 104, 201, 322
S. sorghi (Lk.) Clint.	Sorghum	201, 322
Sorosporium cenchri Henn.	Grasses	321
S. confusum Jacks.	Grasses	321
S. ellisii Wint.	Grasses	321
S. everhartii Ell. & Gall.	Grasses	321
[*S. syntherismae* (Pk.) Farl.] = *C. cenchri* Henn		
Tolyposporium bullatum (Schroet.) Schroet.	Millet	118
T. ehrenbergii (Kuehn) Pat.	Sorghum	202
T. penicillariae Bref.	Millet	118
Thecaphora deformans Dur. & Mont.	Legumes	359
Tilletiaceae		
Tilletia asperifolia Ell. & Ev.	Muhlenbergia	322
T. brevifaciens Fisch. = *T. controversa* Kühn.	Wheat, grasses	253, 257
T. caries (DC.) Tul.	Wheat	253, 254, **255,** 256, 259, 260, 323
T. cerebrina Ell. & Ev.	*Deschampsia* spp.	322
T. controversa Kühn.	Wheat, grasses	253, 254, 257, 259, 323
T. corona Scribn.	Southern grasses	169

BASIDIOMYCETES Heterobasidiomycetes	More common host	Page
Ustilaginales, Tilletiaceae, *T. elymi* Diet. & Holw.......................................	*Elymus* sp.	322
T. foetida (Wallr.) Liro..................	Wheat	253, 254, 256, **257,** 259, 260
T. fusca Ell. & Ev......................	*Festuca* spp.	322
T. guyotiana Hariot.....................	*Bromus* spp.	322
T. holci (Westend.) de Toni..............	*Holcus* sp.	322
T. horrida Tak. or *Neovossia horrida* (Tak.) Padw. & Kahn......................	Rice	167, 168, **169**
T. indica Mitra. or *Neovossia indica* (Mitra) Mundk.....................	Wheat	257, 258
(*T. laevis* Kühn) = *T. foetida* (Wallr.) Liro................................	Wheat spp.	257
[*T. tritici* (Bjerk.) Wint.] = *T. caries* (DC.) Tul................................	Wheat	255
Neovossia horrida (Tak.) Padw. & Kahn or *T. horrida* Tak.....................	Rice	167, 168, 169
N. indica (Mitra) Mundk. or *T. indica* Mitra............................	Wheat	257, 258
Entyloma crastophilum Sacc. or *E. dactyl-* *idis* (Pass.) Cif......................	Grasses	318
E. dactylidis (Pass.) Cif................	Grasses	169, 318
E. irregulare Johans. or *E. dactylidis* (Pass.) Cif................................	*Poa* spp.	318
E. lineatum (Cke.) Davis...............	Wild rice, rice	169, 318
E. meliloti McAlp.....................	Legumes	359
E. orzyae Syd. or *E. dactylidis* (Pass.) Cif..	Rice	169
Urocystis agropyri (Preuss) Schröt........	Grasses, wheat	184, 259, **261,** 262, 316, 317
U. occulta (Wallr.) Rabenh..............	Rye	**183,** 184, 261
U. tritici Körn. or *U. agropyri* (Preus) Schröt.	Wheat	184, 259, 261, 262, 263, 316
Uredinales		
Melampsoraceae		
Melampsora lini (Pers.) Lév.............	Flax	436, **437,** 439
Cerotelium desmium (Berk. & Br.) Arth...	Cotton	424
Phakopsora desmium (Berk. & Br.) Cumm.	Cotton	424
Pucciniaceae		
Uromyces nerviphilus (Grog.) Hotson.....	Clover	378
U. striatus Schroet.....................	Legumes	359
U. striatus medicaginis (Pass.) Arth.......	Alfalfa	359
U. trifolii (Hedw. f) Lév...............	Clover	378
U. trifolii fallens (Desm.) Arth...........	Red clover	378
U. trifolii hybridi (W. H. Davis) Arth....	Alsike	378
U. trifolii trifolii-repentis (Liro) Arth......	White clover	378

BASIDIOMYCETES Uredinales	More common host	Page
Pucciniaceae, (*Puccinia anomala* Rostr.) = *P.*		
hordei Otth..........................	Barley	64, 65, 66
P. arachidis Speg.....................	Peanut	399
P. coronata Cda.......................	Oats, grasses	142, **143**, 145, 323–325
var. *avenae* Fraser & Led..............	Oats, grasses	143, 146
(*P. dispersa* Eriks. & Henn.) = *P. rubigo-*		
vera (DC.) Wint......................	Rye, grasses	186
P. glumarum (Schm.) Eriks. & Henn......	Cereals, grasses	64, 185, 273, **274,** 275, 323–325
P. graminis Pers......................	Cereals, grasses	52, 64, 106, 265, 267, 271, 274, 278, 323–325
P. graminis agrostidis Eriks..............	Grasses	323–325
P. graminis avenae Eriks. & Henn........	Oats, grasses	140, 141, 142, 324, 325
P. graminis lolii Waterh................	Grasses	323
P. graminis phlei-pratensis (Eriks. & Henn.) Stakman & Pieme.............	Timothy, grasses	323–325
P. graminis poae Eriks. & Henn..........	*Poa* spp., grasses	323–325
P. graminis secalis Eriks. & Henn........	Rye, grasses	64, 185, 324, 325
P. graminis tritici Eriks. & Henn........	Wheat, grasses	64, 263, **264,** 269, 270, 273, 324, 325
P. hordei Otth.......................	Barley	64, **65**
P. kuehnii (Krueger) Butl..............	Sugarcane	220
P. montanensis Ell.....................	Grasses	326
P. poae-sudeticae (West.) Jørst..........	*Poa* spp.	326
P. polysora Underw....................	Corn, grasses	105, 106, 326
P. purpurea Cke......................	Sorghum	106, 202, 203
P. recondita Rob......................	Wheat	65, 186, 274, 276–278
P. rubigo-vera (DC.) Wint. or *P. recondita* Rob.	Cereals, grasses	52, 65, 185, 186, 274, 278, 323–325, 326
P. rubigo-vera secalis (Eriks.) Carleton or *P. recondita* Rob.....................	Rye, grasses	185, 186
P. rubigo-vera tritici (Eriks. & Henn.) Carleton or *P. recondita* Rob...........	Wheat, grasses	65, 276, 277, 278
P. sacchari Patel, Kamat & Padhye.......	Sugarcane	221
P. sorghi Schw.......................	Corn	**105**, 106
P. stakmanii Presley...................	Cotton, grass	**422**, 423
(*P. triticina* Eriks.) = *P. recondita* Rob...	Wheat, grasses	65, 277
Angiopsora zeae Mains..................	Corn	106
Auriculariales		
Auriculariaceae		
Heliocobasidium purpureum (Tul.) Pat....	Clover	358

HOMOBASIDIOMYCETES Uredinales	More common host	Page
Polyporales		
Thelephoraceae		
Corticium sasakii (Shirai) T. Matsum.		
(*Hypochnus sasakii* Shirai).............	Rice	167
C. solani (Prill. & Del.) Bourd. & Galz....	General	167, 251
(*C. vagum* Berk. & Curt.) = *C. solani*		
(Prill. & Del.) Bourd. & Galz..........	General	167
Pellicularia filamentosa (Pat.) Rogers. See	General	**167**, 251, 399, 405,
also *Rhizoctonia solani* Kuehn		422, 463
P. rolfsii (Curzi) West. See *Sclerotium*	General	373, 391, **399**, 405,
rolfsii Sacc.		422, 463
Clavariaceae		
Typhula graminum Karst................	Grasses	252
T. idahoensis Remsberg.................	Cereals, grasses	251, **252**
T. itoana Imai.........................	Cereals, grasses	251, 252, 315
T. trifolii Rostr.......................	Legumes	373
FUNGI IMPERFECTI		
Sphaeropsidales		
Sphaerioidaceae		
Phoma herbarum var. *medicaginis* West. or		
Ascochyta imperfecta Pk...............	Alfalfa	350
P. linicola Bub........................	Flax	432
(*P. linicola* Naum.) = *P. linicola* Bub....	Flax	432
(*P. linicola* March. & Verpl.) = *P. linicola*		
Bub.............................	Flax	432
P. medicaginis Malbr. & Roum. = *Aso-*		
chyta imperfecta Pk...................	Alfalfa	350
P. meliloti Allesch.....................	Legumes	354
P. trifolii E. M. Johnson & Valleau.......	Clover	349, **350**, 369
Macrophoma zeae Tehon & Daniels or		
Physalospora zeae Stout...............	Corn	**95**
Macrophomina phaseoli (Maubl.) Ashby.	General	197, 373, 391, 397,
or *M. phaseolina* (Tassi) G. Goid. See		463
also *Sclerotium bataticola* Taub.		
M. phaseolina (Tassi) G. Goid. or *Botryo-*		
diplodia phaseoli (Maubl.) Thir........	General	197
Plenodomus meliloti Mark.-Let...........	Sweetclover	358
Ascochyta agropyrina (Fairm.) Trott......	Grasses	311
A. caulicola Laub.....................	Sweetclover	351
A. gossypi Syd........................	Cotton	403, 421
(*A. graminicola* Sacc.) = *A. sorghi* Sacc....	Grasses	311
A. imperfecta Pk......................	Alfalfa	349, 350
A. linicola Naum. & Wass..............	Flax	432
A. maydis Stout.......................	Corn	198

FUNGI IMPERFECTI	More common host	Page
Sphaeropsidales, Sphaerioidaceae, *A. meliloti* (Trel.) J. J. Davis or *Mycosphaerella lethalis* Stone	Sweetclover	349, 350
A. meliloti Trusova has priority over above		
A. pinodella L. K. Jones	Legumes	403
A. pisi Lib	Legumes	403
A. sorghi Sacc	Grasses	311
A. sorghina Sacc	Sorghum	198
A. viciae Lib	Legumes	403
A. zeae Stout	Corn	198
A. zeicola Ell. & Ev	Corn	198
A. zeina Sacc	Corn	198
Coniothyrium olivaceum Bon	Flax	431
Diplodia frumenti Ell. & Ev. or *Physalospora zeicola* Ell. & Ev	Corn, general	94, **95**
D. macrospora Earle	Corn	91, 92, **94**
[*D. maydis* (Berk.) Sacc.] = *D. zeae* (Schw.) Lév	Corn	92
D. zeae (Schw.) Lév	Corn	91, **92**, 93, 94
Stagonospora arenaria Sacc	Grasses	312
S. meliloti (Lasch.) Petr. or *Leptosphaeria pratensis* Sacc. & Briard.	Sweetclover, clover	352–**354**
S. sacchari Lo & Ling	Sugarcane	219
Wojnowicia graminis (McAlp.) Sacc. & D. Sacc	Cereals	243
Selenophoma bromigena (Sacc.) Sprague & A. G. Johnson	*Bromus* spp.	313, 314
S. donocis (Pass.) Sprague & A. G. Johnson	Grasses	314
Septoria avenae Frank or *Leptosphaeria avenaria* G. F. Weber	Oats, grasses	132, **134**
S. avenae f. triticea T. Johnson or *L. avenaria f. triticea* T. Johnson	Grasses	134
S. bromi Sacc	*Bromus* spp.	312, 313
S. elymi Ell. & Ev	*Elymus* and *Agropyron* spp.	314
S. glycines Hemmi	Soybeans	391
S. infuscans (Ell. & Ev.) Sprague	*Elymus* spp.	314
S. jaculella Sprague	*Bromus* spp.	313
S. linicola (Speg.) Gar. *Mycosphaerella linorum* (Wr.) Garcia-Rada	Flax	431, **432**
S. macropoda Pass	*Poa* spp.	313
S. nodorum (Berk.) Berk. or *Leptosphaeria nodorum* Müller	Wheat, grasses	52, 248, **249**, 250
S. oudemansii Sacc	*Poa* spp.	313
S. pacifica Sprague	*Elymus mollis*	314
S. passerinii Sacc	Barley, grasses	**51**, 312
S. secalis Prill. & Del	Rye, grasses	182, 183, **312**

FUNGI IMPERFECTI	More common host	Page
Sphaeropsidales, Sphaerioidaceae, *S. secalis* var. *stipae* Sprague	*Agrostis* and *Stipa* spp.	312
S. *sojae* Syd. & Butl.	Soybeans	391
S. *sojina* Thuem	Soybeans	391
S. *tritici* Rob.	Wheat, oats	**248,** 249, 250, 312
S. *tritici* f. *avenae* (Desm.) Sprague	Grasses	312
S. *tritici* f. *holci* Sprague	*Holcus lanatus* L.	312
S. *tritici* var. *lolicola* Sprague & A. G. Johnson	*Lolium* spp.	312
Melanconiales		
Melanconiaceae		
Gloeosporium bolleyi Sprague	Grasses	247, 308
G. *lini* Westerdijk	Flax	434
Microstroma lini (Laff.) Krenner	Flax	433
Kabatiella caulivora (Kirch.) Karak	Clover	376, **377**
Colletotrichum destructivum O'Gara	Clover	376, 404
C. *falcatum* Went. or *Physalospora tucumanensis* Speg	Sugarcane	193, 216, 217
C. *glycines* Hori	Soybean	388, 389
C. *gossypii* South	Cotton	415
C. *graminicolum* (Ces.) G. W. Wils	Cereals, grasses	51, 132, 180, 181, 193, 194, 216, 217, 247, 308, 376
C. *lindemuthianum* (Sacc. & Mayn.) Briosi & Cov	Legumes	404
(*C. lineola* Cda.) = *C. graminicolum* (Ces.) G. W. Wils	Sorghum	193
C. *linicolum* Pethyb. & Laff	Flax	431, **434**
C. *pisi* Pat	Legumes	404
C. *trifolii* Bain & Essary	Clover	**375,** 376
C. *truncatum* (Schw.) Andrus & Moore	Legumes	388, 389, 404
C. *villosum* Weimer	Legumes	404
Moniliales		
Cryptococcaceae		
Polyspora lini Laff	Flax	431, 433, 434
Moniliaceae		
Penicillium oxalicum Currie & Thom	Sorghum	192
Phymatotrichum omnivorum (Shear) Dugg.	General	358, 373, 392, 419, **420**
Verticillium albo-atrum Reinke & Berth	Cotton, general	418
V. *dahliae* Kleb	Cotton, general	418
Cephalosporium gregatum Allington & Chamberlain	Soybeans	392
Spicaria elegans (Cda.) Harz	Sorghum	198
Rhynchosporium orthosporum Caldwell	Grasses	49, 309, 310
R. *secalis* (Oud.) J. J. Davis	Barley, rye	**49,** 50, 51, 183, 309

FUNGI IMPERFECTI	More common host	Page
Moniliales, Moniliaceae, *Mastigosporium album*		
Riess..............................	Grasses	308, 399
var. *calvum* Ell. & Davis..............	Grasses	308
var. *muticum* Sacc....................	Grasses	309
M. calvum (Ell. & Davis) Sprague........	Grasses	309
M. cylindricum Sprague................	Grasses	309
M. rubricosum (Dearn. & Barth.) Nannf..	Grasses	308
P. grisea (Cke.) Sacc...................	Grasses	163, 310
Piricularia oryzae Cav.................	Rice	162, **163**
Cercosporella herpotrichoides Fron.........	Cereals	245
Dematiaceae		
Thielaviopsis basicola (Berk. & Br.) Ferr..	Tobacco, flax	430, 461, **462**
Nigrospora oryzae (Berk. & Br.) Petch....	Corn, rice	95, **96**
N. sacchari (Speg.) Mason..............	Corn, sugarcane	95
N. sphaerica (Sacc.) Mason............	Corn, barley	95, 96
Periconia circinata (Mangin.) Sacc........	Sorghum	194, **195**
Scolecotrichum graminis Fckl.............	Grasses	310, 311
S. graminis avenae Eriks................	Oats, grasses	132
Helminthosporium avenae Eidam or *Pyreno-*		
phora avenae Ito & Kuribay..........	Oats	40, **129, 130**
H. bicolor Mitra.......................	Wheat	247
H. bromi Drechsl. or *Pyrenophora bromi*		
(Died.) Drechsl.....................	*Bromus* spp.	306
H. californicum Mackie & Paxton........	Barley	49
H. carbonum Ullstrup..................	Corn	97, 98, **99**, 100
H. cynodontis Marig...................	Grasses	307
H. dictyoides Drechsl...................	*Festuca* sp.	305, 307
H. erythrospilum Drechsl................	*Agrostis* spp.	306
H. giganteum Heald & Wolf.............	Grasses	118, 307
H. gramineum Rab....................	Barley	40, 41, **42**
H. hadrotrichoides Ell. & Ev.............	Millets, grasses	118
H. halodes Drechsl.....................	Grasses	118, 247
II. halodes var. *tritici* Mitra.............	Wheat	247
H. leucostylum Drechsl. or *H. hadrotri-*		
choides Ell. & Ev...................	Millet, grasses	118
H. maydis Nisik. & Miy. or *Cochliobolus*		
heterostrophus Drechsl................	Corn	97, 98, **99, 100**
H. miyakei Nisik.....................	*Eragrostis* sp.	308
H. monoceras Drechsl..................	*Echinochloa* sp.	118
H. nodulosum Berk. & Curt.............	Millet, grasses	118
H. oryzae Breda de Haan. or *Cochliobolus*	Rice	158, **159,** 160, 161,
miyabeanus (Ito & Kuribay.) Dickson		162
H. panici-miliacei Nisik................	Millet	118
H. poae Baudys.......................	*Poa* sp.	306
H. ravenelii Curt. & Berk..............	Grasses	307
H. rostratum Drechsl...................	Grasses	99
H. sacchari (Breda de Haan) Butl........	Sugarcane	118, 218, 307

FUNGI IMPERFECTI	More common host	Page
Moniliales, Dematiaceae, *H. sativum* Pam., King, & Bakke. or *Cochliobolus sativus* (Ito & Kuribay.) Drechsl	Cereals, grasses	43, 44, 45, 46, **47**, 48, 118, 131, 159, 227, 232, 246, 304
H. setariae Saw. or *Cochliobolus setariae* (Ito & Kuribay.) Dickson	*Setaria* spp.	118
H. siccans Drechsl	*Lolium* spp.	307
H. sigmoideum Cav. or *Leptosphaeria salvinii* Catt	Rice	**157**, 158
H. sigmoideum var. *irregulare* Cralley & Tullis	Rice	157, 158
H. sorokinianum Sacc. see *H. sativum* Pam., King & Bakke	Grasses	47, 48
H. stenacrum Drechsl	*Agrostis* spp.	306
H. stenospilum Drechsl. = *Cochliobolus stenospilus* (Carpenter) Matsu. & Yamam	Sugarcane	218
H. teres Sacc. or *Pyrenophora teres* (Died.) Drechsl	Barley	37, **38**, 39
H. tetramera McK. *Curvularia specifera* (Bainier) Boed	Wheat	247
H. triseptatum Drechsl	*Agrostis* spp.	306
H. tritici-repentis Died. or *Pyrenophora tritici-repentis* (Died.) Drechsl.	*Agropyron* and *Elymus* spp.	247, 306
H. tritici-vulgaris Nisik. or *Pyrenophora tritici-vulgaris* Dickson	Wheat	247
H. turcicum Pass	Corn, Sudan grass	97, 98, **99**, 100, 193, 194, 307
H. vagans Drechsl	*Poa* sp.	304, 305
H. victoriae Meehan & Murphy	Oats, grasses	**131**
H. yamadai Nisik	Millet	118
H. zeicola Stout	Corn	99
Curvularia geniculata (Tracy & Earle) Boed	Grasses	247, 308
C. lunata (Wakk.) Boed	Wheat, Rice	162
C. ramosa (Bainier) Boed	Wheat	247
C. specifera (Bainier) Boed	Wheat	247
C. trifolii (Kauf.) Boed	Clover	377, 378
Stemphylium botryosum Wallr. or *Pleospora herbarum* (Fr.) Rab	Alfalfa, clover	373, 374, **375**
S. loti Graham	Lotus	374
S. sarcinaeforme (Cav.) Wiltshire	Red clover	373, 374, **375**
Alternaria tenuis Nees	Tobacco	459
Cercospora apii Fres	General	404, 459, 460, 461
C. arachidicola Hori or *Mycosphaerella arachidicola* Jenkins	Peanut	398
C. davisii Ell. & Ev. or *Mycosphaerella davisii* F. R. Jones	Legumes	351, **352**

FUNGI IMPERFECTI	More common host	Page
Moniliales, Dematiaceae, *C. demetrioniana*		
Wint.	Legumes	404
C. gossypina Cke.	Cotton	421
C. graminicola Tr. & Er.	Grasses	310
C. imperate Syd.	Sugarcane	219
C. kikuchii (T. Matsu. & Tomoy.) Gardn.	Soybean	390
C. kopkei Krüger.	Sugarcane	219
C. lespedezae Ell. & Dearn.	Legumes	404
C. longipes Butl.	Sugarcane	219
C. meliloti Oud.	Sweetclover	352
C. nicotianae Ell. & Ev.	Tobacco	459, 460
C. oryzae Miyake.	Rice	164, **165**
C. personata (B. & C.) Ell. & Ev. or *Myco-*		
sphaerella berkeleyii Jenkins.	Peanut	398
C. pueraricola Yamam. or *Mycosphaerella*		
puearicola Weimer & Luttr.	Legumes	404
C. sojina Hara.	Soybean	390
C. sorghi Ell. & Ev.	Corn, sorghum	197
C. stizolobii Syd.	Legumes	404
C. taiwanensis Mat. & Yamam. or *Lepto-*		
sphaeria taiwanensis Yen & Chi.	Sugarcane	219
C. vaginae Krüger.	Sugarcane	219
C. viciae Ell. & Ev.	Legumes	403, 404
C. vignae Ell. & Ev.	Legumes	404
C. zebrina Pass.	Alfalfa, clover	351, **352**
Tuberculariaceae		
Cerebella andropogonis Ces.	*Claviceps*	301
Cylindrocarpon ehrenbergi Wr.	Alfalfa, clover	358
C. obtusisporum (Ckc. & Hark.) Wr.	Alfalfa, clover	358
Fusarium arthrosporoides Sherb.	Alfalfa, clover	358
F. avenaceum (Fr.) Sacc.	Cereals, general	34, 36, 240, 358
F. culmorum (W. G. Sm.) Sacc.	Cereals, general	34, 36, 239, 242,
		246, 303, 358
F. culmorum var. *cereale* (Cke.) Wr.	Cereals, general	239
F. graminearum Schw. or *Gibberella zeae*		
(Schw.) Petch. or.	Cereals, grasses	36, **90**, 238
F. roseum (Lk.) Snyder & Hansen.	All inc.	238
F. roseum f. cerealis (Cke.) Snyder &	Cereals	34, 35, 36, 37, 87,
Hansen		**90**, 193, 238
var. "Avenaceum".	Cereals	35, 36, 238
var. "Culmorum".	Cereals	35, 36, 238
var. "Graminearum" etc.	Cereals	35, 36, 87, 90, 238
F. moniliforme Sheld. or *Gibberella fujikuroi*	Corn	88, **91**, 155, 193,
(Saw.) Wr.		198, 215, 416
F. moniliforme var. *subglutinans* Wr. &		
Reink. or.	General	91, 215
F. moniliforme (Sheld.) Snyder & Hansen		
or *Gibberella moniliforme* (Sheld.) S. & H.	General	91

FUNGI IMPERFECTI	More common host	Page
Moniliales, Tuberculariaceae, *F. nivale* (Fr.) Ces. or *Calonectria graminicola* (Berk. & Br.) Wr..........................	Cereals, grasses	302
F. nivale var. *majus* Wr. or		
F. nivale (Fr.) Snyder & Hansen..........	General	174, 241
F. nivale f. *graminicola* (Berk. & Br.) Snyder & Hansen or *Calonectria nivale* f. *graminicola* (Berk. & Br.) Snyder & Hansen..........................	Cereals	174, 241
var. "Major" etc...................	Grasses	241
F. oxysporum Schlecht.................	Plant wilts	405, 417
F. oxysporum f. *crotalariae* Carnera.......	Crotalaria	405
F. oxysporum f. *lini* (Bolley) Snyder & Hansen.............................	Flax	434–**436**
F. oxysporum f. *lupini* Snyder & Hansen..	Lupine	405
F. oxysporum f. *medicaginis* (Weimer) Snyder & Hansen...................	Alfalfa	358
F. oxysporum f. *nicotianae* (J. Johnson) Snyder & Hansen...................	Tobacco	459, 460
F. oxysporum f. *phaseoli* Kend. & Snyder..	Bean	405
F. oxysporum f. *pisi* (Linford) Snyder & Hansen.............................	Pea	405
F. oxysporum f. *tracheiphilum* (E. F. Sm.) Snyder & Hansen...................	Soybean	390, 405
F. oxysporum f. *udum* (Butl.) Snyder & Hansen.............................	Chickpea	405
F. oxysporum f. *vasinfectum* (Atk.) Snyder & Hansen..........................	Cotton	416, **417**
F. poae (Pk.) Wr......................	Cereals, grasses	358
F. roseum (Lk.) Snyder & Hansen and f. *cerealis* (Cke.) Snyder & Hansen. See under *F. graminearum*	Cereals	34, 36, 87, **90**, 155, 193, 238
F. scirpi var. *acuminatum* (Ell. & Ev.) Wr.	Alfalfa, clover	358
F. udum var. *cajani* Pad...............	Chickpea	405
F. udum var. *crotalariae* Pad............	Crotalaria	405
Gloeocercospora sorghi D. Bain & Edg.....	Sorghum, corn	196
Ramulispora sorghi (Ell. & Ev.) Olive & Lefebvre...........................	Sorghum	196
Mycelia-sterilia		
R. crocorum Fr. See also *Helicobasidium purpureum* (Tul.) Pat.................	Alfalfa, clover	358
R. oryzae Ryker & Gooch...............	Rice	165, 166
R. solani Kuehn or *Pellicularia filamentosa* (Pat.) Rogers (*Corticium vagum* Berk. & Curt.)	General	166, 167, 250, 251, 314, 358, 392, 399, 405, 422, 463

FUNGI IMPERFECTI	More common host	Page
Mycelia-sterilia, *R. violacea* Tul. See also		
Helicobasidium purpureum (Tul.) Pat.....	Alfalfa, clover	358
R. zeae Voorhees.......................	Corn, rice	100,167
Sclerotium bataticola Taub. or *Macrophomina phaseoli* (Maubl.) Ashby	General	197, 373, 391, 463
S. oryzae Catt. or *Leptosphaeria salvinii* Catt...............................	Rice	157, 158
S. rolfsii Sacc. or *Pellicularia rolfsii* (Curzi) West	General	252, 373, 391, 397, **399**, 405, 463

INDEX

Page numbers in **boldface** type refer to principal discussions

A

Aceratagallia sanguinolenta, 367, 368
Aceria tulipae, 230, 295
Aconitum, aecial host, 323
Actaea, aecial host, 323
Aegilops spp., 226, 324
Agallia constricta, 342
  quadripunctata, 342
Agalliopsis novella, 342
Agropyron cristatum, 315, 318, 319
  elongatum, 294, 319
  inerme, 294
  intermedium, 294
  pauciflorum, 321
  repens, 245, 294, 296, 304, 306
  spp. 126, 174, 180, 226, 230, 295, 296,
    298–300, 306, 310, 314, 316, 318,
    324, 326
  trachycaulum, 295
  trichophorum, 319
Agropyron mosaic, 294
Agrostis alba, 303, 307
  var. *gigantea*, 316
  palustris, 316
  spp., 196, 299, 300, 306, 308, 312, 316,
    318, 324
  tenuis, 316
Alfalfa, chromosomes, 335
  disease resistance, 335, 344, 360
  (*See also Medicago*)
Alfalfa diseases, 335, 379
  anthracnose, 375–377, 404
  bacterial stem blight, 344
  bacterial wilt, 342–344
  black stem, 349–352
    spring, 349–351
    summer, 351, 352
  boron deficiency, 339, 340
  cold injury, 336
  common leaf spot, 355, 356
  crown rot, 352
  crown wart, 345, 346

Alfalfa diseases, damping-off, 346
  downy mildew, 347, 348
  dwarf, 340, 341
  dwarfing in, 339
  Fusarium wilt, 357
  heat and drought injury, 340
  insect injury, 339, 366
  leaf discoloration, 339
  leaf hopper injury, 339, 366
  leaf spots, 339, 352–354
  mineral deficiencies, 339, 340
  mosaics, 340
  Pseudopeziza leaf spots, 355–357
  Pseudoplea leaf spot, 370–372
  Pythium root rot, 346
  Rhizoctonia root rot, 358
  root rots, 352, 357
  rootlet rots, 346
  rust, 359, 360
  Sclerotium blight, 358
  smut, 359
  Stemphylium leaf spot, 373–375
  table of, 379
  viruses, 340, 367
  winter injury, 336, 367
  witches' broom, 340, 341
  wound tumor virus, 342
  yellow leaf blotch, 355–357
  yellowing, 339
Alfalfa varieties, Caliverdi, 356
  DuPuits, 356
  Ladak, 360
  Turkestan, 344, 360
Alopecurus fulvus, 155, 296
  spp., 86, 310, 324
Alsike clover (*see* Clover diseases)
Alternaria spp., 189, 227, 246
  tenuis, 459
Althaea rosea, 410
  spp., 450
Aluminum, excesses in, 75
Amastigosporium graminicola, 309
Ammophila arenaria, 324

Anatomy, of Gramineae, 9, 10, 23, 74
    caryopsis, 8
    crown, 9, 74
    kernel, 8
    relation of, to disease, 9, 13
    root, 9
    seedling, 8, 9, 74
    of Leguminosae, 10, 11, 335
    crown, 335
    relation of, to disease, 11, 13
    root, 11, 12, 335, 339
    seed, 11
    seedling, 12
    of other crop plants, 13
Anchusa, aecial host, 185, 323
Andropogon australis, 86
    barbinodis, 295
    spp., 86, 294, 322, 324
Anemone, aecial host, 323
Angiospora zeae, 106
Anguillulina tritici, 232
Angular leaf spot, of cotton, 411
    of tobacco, 453, 454
Anthoxanthum spp., 324
Anthracnose, of alfalfa, 375, 404
    of barley, 51, 180
    of beans, 404
    of cereals, 132, 180, 217, 247
    of clover, 375, 377, 404
    of cotton, 404, 414–416
    of cowpea, 404
    of crotalaria, 404
    of flax, 434
    of grasses, 180, 217, 247, 308
    of Johnson grass, 193, 216
    of kudzu bean, 404
    of lespedeza, 404
    of lotus, 404
    of lupine, 404
    of oats, 132, 180
    of pea, 404
    of rye, **180,** 181
    of sorghum, 193, 216
    of soybean, 388, 389
    of sudan grass, 181, 193, 194, 216, 308
    of sugarcane, 216, 217
    of velvet bean, 404
    of vetch, 404
    of wheat, 180, 247
Aphanomyces euteiches, 402
Aphanomyces root rot, 402
Aphelenchoides oryzae, 154
Aphid injury of plants, 27, 33, 366
Aphid vectors, *Aphis craccivora*, 397
    fabae, 340
    gossypii, 340, 402
    medicaginis, 340, 397

Aphid vectors, *Aphsis, setariae*, 207
    spp., 411
    Carolinaia cyperi, 207
    Macrosiphum dirhodum, 27
    granarium, 27
    pisi, 340, 367
    solanifolii, 340
    Myzus persicae, 76, 341, 402
    solani, 383
    Peregrinus maidis, 76
    Rhopalosiphum fitchii, 27, 76
    maidis, 27, 76
    prunifolia, 27, 76
    (*See also* Leaf hopper vectors)
Aplanobacter (*see Corynebacterium*)
Aquilegia, aecial host, 323
Arachis glabrata, 396
    hypogaea, 396
    nambyguarae, 396
    prostrata, 396
Arasan seed treatment, 83, 192
Arctagrostis latifolia, 324
Aristilda spp., 321
Arrhenatherum elatius, 324
    spp., 129, 180
Ascochyta agropyrina, 311
    caulicola, 351
    gossypii, 403, **421**
    graminicola, 311
    imperfecta, 349, **350**
    linicola, 432
    maydis, 198
    meliloti, 349, **350**
    pinodella, 403
    pisi, 403
    sorghi, 311
    sorgina, 198
    spp., 310, 312, 403
    viciae, 403
    zeae, 198
    zeicola, 198
    zeina, 198
Ascochyta leaf spot, cotton, 421
    grasses, 310
    legumes, 349, 403
Aspergillus spp., 85, 416
Aster yellow virus, 342, 368, 411, 429
Asterocystis radicis, 430
Astragalus sinicus, 155
    spp., 341, 359
Avena abyssinica, 122
    barbata, 122
    brevis, 122, 137
    byzantina, 122
    fatua, 122
    nuda, 122

Avena nudibrevis, 122
  orientalis, 122
  sativa, 122
  spp., 23, 86, 126, 129, 134, 139, 140, 142, 295, 299, 324
  sterilis, 122
  strigosa, 122, 147
Axonopus spp., 298

B

Bacillus flavidus, 216
  sorghi, 190
Bacteria, classification of, 28
  nomenclature of, 28
Bacterial blight, of *Agropyron*, 126, 296, 298
  of alfalfa, 344
  of barley, **27,** 128, 231
  of bean, 402
  of *Bromus*, 126, 296, 297
  of broom corn, 190
  of cereals, 27, 126, 231
  of clover, 368, 402
  of corn, 80, 190, 298
  of cotton, 411–414
  of cowpea, 402
  of grasses, 27, 231, **296**–298
  of Johnson grass, 298
  of kudzu bean, 402
  of lespedeza, 402
  of lupine, 402
  of millets, 115
  of oats, 126, 128
  of pea, 402
  of rice, 155
  of rye, 173, 231
  of sorghum, 190, 191, 298
  of soybean, 385, 402
  of sudan grass, 190, 191, 298
  of sugarcane, 190, 298
  of timothy, 298
  of tobacco, 452–455
  of velvet bean, 402
  of vetch, 402
  of wheat, 27, 231
Bacterial disease types, 27, 126, 190, 296
Bacterial halo blight, 126, 191, 296
Bacterial pustule of soybean, 386
Bacterial stem blight of alfalfa, 344
Bacterial streak, of sorghum, 190, 214, 298
  of sugar cane, 190, 214, 298
Bacterial stripe, of *Agropyron*, 27, 296
  of *Bromus*, 296
  of grasses, 27, 296
  of millets, 115

Bacterial stripe, of oats, 127, 128
  of sorghum, 190, 191, 212, 298
  of sugarcane, 190, 213, 298
Bacterial wilt, of alfalfa, 342–344
  of corn, **77,** 79
  of grasses, 298
  of legumes, 342, 402
  of peanut, 397
  of sugarcane, 79, 211, 298
  of tobacco, 451, 452
Bacterium agropyri, 298
  albilineans, 212, **213**
  andropogoni, 190
  stewartii, 77, **78**
  (*See also Corynebacterium; Pseudomonas; Xanthomonas*)
Bakanae disease of rice, 155
Barley, anatomy of plant, 7
  blighted, Federal grades, 35
  chromosomes, 23
  disease losses in U.S., 23
  disease resistance, 33, 36, 40, 49, **51,** 57, 61, 63, 64, 66
Barley diseases, **23**
  anthracnose, 51
  bacterial blight, **27,** 128, 173, 231
  black semiloose smut, 58, 134
  boron deficiency, 24
  brown spot, 24
  copper deficiency, 24, 125
  covered smut, **61,** 139
  downy mildew, 30, 83, 86
  ergot, 37, 174
  false stripe, 24
  false stripe mosaic, 24
  Fusarium blight, **34,** 85, 237
  head blight, 34, 46, 85, 236
  Helminthosporium blight, 35, 159
  intermediate loose smut, 58
  leaf rust, **64,** 276
  leaf scald, 49
  loose smut, **52,** 54, 252
  mineral deficiency, 24, 123
  net blotch, **37**
  powdery mildew, 30
  Pythium root rot, **30**
  Rhynchosporium scald, **49,** 50
  rusty blotch, 49
  scab, **34,** 236
  seedling blight, 34, 85, 237
  Septoria leaf blotch, 51
  spot blotch, **43**
  stem rust, 64, 263
  stripe, 25, **40**
  stripe mosaic, **24,** 295
  stripe rust, 64, 274
  yellow dwarf, **25,** 126, 295

Barley varieties, Abate, 27
  Abyssinia, 57
  Anoidium, 40, 58
  Arlington, 33
  Black hull-less, 34, 63, 295
  Bolivia, 30, 66
  Brachitic, 63
  Callas, 66
  Chevalier, 49
  Chevron, 24, 30, 33, 34, 36, 43, 48, 53, 64, 295
  Cross, 36
  Dorsett, 58
  Duplex, 33
  Excelsior, 61, 63
  Glabron, 43, 61
  Glacier, 24
  Goldfoil, 34
  Hanna, 51
  Hannchen, 43, 49, 61, 63
  Heils Hanna, 34
  Hillsa, 63
  Himalaya, 61, 63
  Jet, 58
  Juliaca, 66
  Kindred, 40, 48, 64
  Korsbyg, 36
  Lion, 43, 61, 63
  Lompoc, 61
  Lyallpur, 61, 63
  Manchu, 40
  Manchurian, 30, 48, 61
  Mars, 43, 64
  Mecknos-Moroc, 66
  Ming, 40
  Nepal, 34, 61, 63
  Newal, 43, 61
  Oderbrucker, 30, 31
  Odessa, 56, 61, 63
  Orge, 66
  Otrada Beardless, 51
  Pannier, 61, 63
  Peatland, 37, 43, 48, 64
  Persicum, 61, 63
  Peruvian, 34, 66
  Plush, 24
  Quinn, 66
  Rabat, 40
  Regal, 43
  Spartan, 43
  Svansota, 36, 49
  Tifang, 40
  Trebi, 43, 57, 58, 63
  Valki, 58
  Vance, 43
  Velvon, 43
  Wisconsin Barbless, 43, 61

Basal glume rot of wheat, 231
Basidiomycete, low-temperature, 252, 315, 358
Bean diseases, 367, 368, 377, 401, 404
Beckmannia spp., 316, 324
Beet curly top virus, 154, 429, 445
Belonolaimus spp., 417
Bemisia goldingi, 411
  gossypiperda, 410
  inconspicua, 411
  spp., 410, 450
Berberis fendleri, 326
  spp., 146, 264, 267
  vulgaris, 265, 266
Berchemia scandens, 326
Black arm of cotton, 411, 413
Black chaff of wheat, 231
Black loose smut, of barley, 58, 134
  of grasses, 58, 134
  of oats, 58, 134
Black nightshade, downy mildew of, 457
Black patch of legumes, 377
Black point of grains, 46, 232, 246
Black root rot, 461–463
Black shank of tobacco, 455
Black sheath rot of rice, 158
Black stem of legumes, 349, 351, 369, 403
Black stem rust (see Stem rust)
Blackfire of tobacco, 454
Blast, of oats, 123
  of rice, 162
Blighted barley, Federal grades, 35
Blind seed disease of grasses, 304
Blue mold of tobacco, 456–459
Boll rots of cotton, 409, 414, 416
Boron deficiency diseases, 24, 125, 154
  of alfalfa, 339
  of barley, 24
  of oats, 125
Botryodiplodia phaseoli, 197
Bouteloua spp., 120, 324, 422
Brachycladium spp., 247
Brachypodium distachyon, 324
Briza maxima, 296
  spp., 296, 324
Brome grass diseases, bacterial blights, 296
  bends, 294
  brown leaf spot, 304, 305
  downy mildew, 86
  ergot, 174, 299, 301
  head smut, 320, 321
  heritable leaf spots, 294
  kernel smut, 322
  leaf fleck, 308
  leaf scald, 49, 309

Brome grass diseases, mosaic, 126, 294
  Selenophoma blotch, 313, 314
  Septoria blotch, 312, 313
Bromus carinatus, 295, 321
  catharticus, 296, 320, 321
  commutatus, 86
  hordeaceus, 296, 321
  inermis, 49, 178, 294–297, 301, 303,
    305, 306, 309, 310, 313, 323
  polyanthus, 321
  rubens, 296
  spp., 126, 174, 180, 294–296, 300, 306,
    310, 312, 313, 322, 324
  unioloides, 320
  vulgaris, 309
Broom corn (*see* Sorghum)
Brown necrosis, Hope wheat, 231
Brown root rot of tobacco, 444
Brown stem rot of soybeam, 392
Brown stripe of sugarcane, 218
Browning disease of flax, 433
Browning root rot of grasses and wheat,
  232
Buchloë dactyloides, 324
Bud blight of soybean, 384
Bunt, dwarf, 254
  of grasses, 254
  of rye, 254
  of wheat, 253

C

Calamagrostis spp., 300, 301, 308, 310,
  324
Calamovilfa longifolia, 324
Calcium ratio, 125, 154, 396
Calonectria graminicola, 174, **241**, 302
  var. *neglecta*, 241, 242
  nivale, 241, 302
    f. *graminicola*, 241, 242
Cambium development in relation to
  disease, 12, 15, 339
Carbamate fungicides, 399
Carneocephala fulgida, 342
Carolinaia cyperi, 207
Caryopsis, anatomy and disease, 8
Cassia tora, cotton wilt, 417
Catabrosa aquatica, 324
Cell composition and physiology, 15, 16
  relation of, to disease, 15
    to resistance, 15
Cell wall composition and maturity, 15
  relation of, to disease, 15
    to resistance, 15
Cenchrus spp., 300, 321
Cephalosporium gregatum, 392
Cercospora apii, 404, 459, 460, 461
  arachidicola, 398

Cercospora davisii, 351, **352**
  demetrioniana, 404
  gossypina, 421
  graminicola, 310
  imperatae, 219
  kikuchii, 390
  kopkei, 219
  lespedezae, 404
  longipes, 219
  meliloti, 352
  nicotianae, 459, 460
  oryzae, 164, **165**
  personata, 398
  puericola, 404
  sojina, 390
  sorghi, 197
  spp., 403
  stizolobia, 404
  taiwanensis, 219
  vaginae, 219
  viciae, 403, 404
  vignae, 404
  zebrina, 351, **352**
Cercosporella herpotrichoides, **245**
  spp., 245
Cerebella andropogonis, 301
Ceresan seed treatment (*see* Mercury
  seed treatment)
Cerotelium desmium, 424
Chaetochloa italica (*see Setaria italica*)
Charcoal rot, of corn, 197
  of sorghum, 197
Chloris elegans, 86
  gayana, 295
  spp., 324
Chlorosis, nonparasitic, 75
  relation to rust reaction, 269, 270
Chlorotic streak of sugarcane, 76, 209
Choke of grasses, 301
Chromosomes, of *Arachis*, 396
  of *Avena* spp., 122
  of *Gossypium* spp., 409
  of *Hordeum* spp., 23
  of *Linum* spp., 428
  of *Medicago* spp., 335
  of *Melilotus* spp., 335
  of *Nicotiana* spp., 443
  of *Oryza* spp., 153
  of *Panicum* spp., 115
  of *Saccharum* spp., 206
  of *Secale* spp., 173
  of *Setaria* spp., 115
  of *Soja* spp., 383
  of *Sorghum* spp., 188
  of *Trifolium* spp., 366
  of *Triticum* spp., 225
  of *Zea mays*, 74

Cicadula bimaculata, 77
Cicadulina mbila, 76, 209
  spp., 296
  storeyi, 76, 209
  zeae, 76, 209
Cinna spp., 324
Cladosporium spp., 301
Claviceps cinerea, 300
  maximensis, 300
  microcephala, 300
  paspali, 300, 301
  purpurea, 37, 156, 174–179, 235, 300
  pusilla, 300
  spp., 156, 299
Clematis, aecial host, 323
Clover disease resistance, 369, 373, 377
Clover diseases, 366, 379
  anthracnose, 373, 377
  bacterial blight, 368
  black patch, 377
  Cercospora leaf spot, 370, 403, 404
  cold injury, 366
  common leaf spot, 355
  crown rot, 372
  crown wart, 346
  Curvularia spot, 377, 378
  downy mildew, 347
  dwarfing, 366, 368
  Fusarium wilt, 357
  insect injury, 366
  leaf spot, 356, 373, 374
  mineral deficiency, 366
  mosaics, 342, 367
  northern anthracnose, 377
  powdery mildew, 368, 369
  Pseudopeziza leaf spot, 355–356
  Pseudoplea leaf spot, 370–372
  Pythium rot, 346
  Rhizoctonia root rot, 358, 373
  root rots, 357, 372
  rusts, 378
  Sclerotinia root rot, 372, 373
  smut, 359
  sooty blotch, 369–371
  southern anthracnose, 373–375
  spring black stem, 349–351, 369
  Stagonospora leaf spot, 352–354
  Stemphylium leaf spot, 373–375
  summer black stem, 351, 352
  virus, 342, 367–368
  winter injury, 366
  witches' broom, 341, 368
  wound tumor virus, 368
  yellows, 366, 367
Clover leaf hopper, 366
Cob rots of corn, 85, 95
Cobb's bacterial wilt of sugarcane, 79, 211

Cochliobolus heterostrophus, 97–99
  miyabeanus, 158–161
  sativus, **47**, 48, 246
  setariae, 118
  spp., 308
  stenospilus, 218
  tritici, 48, 246
Cold injury, of alfalfa, 336, 338
  of clover, 366
  of corn, 75
  of wheat, 227
Cold-water seed treatment, 57
Coleoptile, 9
  penetration of, by fungi, 9, 59, 62
  relation of, to disease, 9
Collenchyma tissues, 11, 15
  relation of, to disease, 11, 15
Colletotrichum destructivum, 376, 404
  falcatum, 193, 216, 217
  glycines, 388, 389
  gossypii, 415
  graminicolum, 51, 132, **180**, 181, 193, 194, 216, 217, 247, 308, 376
  lindemuthianum, 404
  lineola, 193
  lini, 434
  linicolum, 431, **434**
  pisi, 404
  spp., 193
  trifolii, **375**, 376
  truncatum, 388, **389**, 404
  villosum, 404
Common leaf spot of alfalfa, 355, 356
Common tobacco mosaic, 445, 446
Coniothyrina olivaceum, 431
Contact infection, 237
Copper deficiency, 24, 125, 154, 340
Copper dusts, 200, 258, 399
Corn, botanical varieties of, 74
  chromosome number of, 74
  disease losses, 74, 75
  disease resistance, 74, 77, 100, 106
Corn diseases, **74**
  Ascochyta spot, 198
  bacterial blights, 74, 80, 190, 298
  bacterial *Holcus* spot, 191
  bacterial wilt, **77**, 79, 212
  brome mosaic, 295
  charcoal rot, 197
  cob rot, 95, 96
  Cobb's disease, **79**
  cold injury, 75
  crazy top, 83
  cucumber mosaic, 76
  Diplodia rots, 85, **91**, 93
  downy mildews, **83**, 215
  drought, 75

Corn diseases, ear rots, 85, 87, 97, 100
    frost injury, 75
    Fusarium rots, **85,** 87, 88, 215
    Gibberella rots, **85,** 87, 88, 215
    Gloeocercospora leaf spot, 196
    head smut, **103,** 104, 202
    Helminthosporium leaf blight, 74, **97**
    Helminthosporium leaf spots, **97, 98**
    heritable leaf spot, 75
    high temperature, 75
    kernel mold, 85
    kernel rots, 85, 88, 89, 97
    leaf rust, 105, 203
    mineral deficiency, 75
    mosaics, 75, 207
    mottle mosaic, 76
    Nigrospora rot, **95,** 96
    nonparasitic, **75**
    Physoderma brown spot, **80**
    Pythium root rot, **81**
    Rhizoctonia rot, **100,** 167
    rusts, 11, **105,** 221
    Scutellum rot, 85
    seed rot, 81
    seedling blights, 81
    seedling injury, 85
    smut, 10, 100, 102–104
    stalk rots, **85,** 87, 88, 91, 93, 100
    Stewart's disease, 77
    streak, 75, 76, 209, 214, 296
    stripe, 75, 76, 213
    stunt, 77
    tobacco streak virus, 450
    virus, 75, 77, 450
    wallaby-ear, 77
    yellow dwarf virus, 25
Corn flea beetle, bacterial wilt, 78
Corticium sasakii, 167
    solani, 167, 251
    vagum, 167
Corynebacterium agropyri, 298
    flaccumfaciens, 402
    insidiosum, 342, 343
    rathayi, 298
    tritici, 232
Corynephorus canescens, 324
Cotton, American, 409
    Asiatic, 409
    chromosome number, 409
    disease losses, 410
    disease resistance, 409, 411, 414
    Egyptian, 409
    wild, 409
Cotton diseases, 409
    angular leaf spot, 411–413
    anthracnose, 414–416
    Ascochyta blight, 421

Cotton diseases, bacterial blight, 411–413
    black arm, 411–413
    boll drop, 410
    boll rots, 412, 416
    Cyrtosis virus, 411
    Fusarium wilt, 409, 416–418
    leaf crumple, 411
    leaf curl, virus, 410
    leaf discoloration or rust, 410
    mineral deficiency, 410
    mosaic, 410
    Phymatotrichum root rot, 419–421
    Rhizoctonia canker, 422
    rusts, 422–424
    seedling blight, 416, 422
    soreshin, 422
    stenosis virus, 411
    Verticillium wilt, 418
    virus, 410, 411
    white fly, 410, 411
    wilts, 416–419
Cotyledons, relation of, to disease, 12
Covered kernel smuts, 61, 322
Covered smuts, of barley, 58, 60, **61**
    of grass, 61, 139, 320
    of oats, 61, 139
Cowpea diseases, 377, 401, 402, 404
Criconemoides spp., 417
Crotalaria spp., 401, 403
Crotalaria diseases, 401, 404
Crown, of Gramineae, 9, 74
    relation of, to disease, 7
    root development from, 7
    winter injury in, 227
    of Leguminosae, 11, 335
    structure and disease of, 11
    winter injury in, 336
Crown rots, of alfalfa, 357, 358
    of cereals, 35, 90, 302
    of clover, 357, 358
    of grasses, 34, 302
    of sorghums, 193, 197
    of sweet cover, 347, 357, 358
    of wheat, 34, 302
Crown rust, of grasses, 323
    of oats, 131
Crown wart, of alfalfa, 345, 346
    of clover, 346
Culm rot, of grasses, 156, 245
    of rice, 156
    of wheat, 245
Culm smut, of cereals, 183, 184
    of grasses, 318, 319
Curvularia geniculata, 217, 308
    lunata, 162
    ramosa, 247
    specifera, 247

Curvularia trifolii, 377, 378
Cuticle of plants, formation of, 14
  penetration of, by fungi, 15, 50
  relation of, to disease, 14, 49
  to resistance, 15
Cyamopsis tetragonalobus, 450
Cylindrocarpon ehrenbergi, 358
  obtusisporium, 358
  spp., 357
Cylindrosporium infuscans, 314
Cymadothea trifolii, **369**–371
Cymbopogon citratus, 218, 307
Cynodon dactylon, 295, 324
Cynosurus cristatus, 324

D

Dactylis glomerata, 295, 298, 307, 311,
  316, 324
  spp., 294, 300, 308, 310
Dactyloctenium aegyptiacum, 324
Damaged grain, Federal grades, 35, 85,
  174, 237
Damping-off, influence of anatomy on,
  12, 81
Danthonia spp., 296, 324
Defense mechanisms, plant, 13
Deficiency diseases, 24, 75, 294
  (*See also* Mineral-deficiency maladies)
Delphacodes striatellus, 155
Delphinium, aecial host, 323
Deltocephalus dorsalis, 155
  striatus, 230
Deschampsia spp., 322, 324
Desmodium spp., 359
Diaporthe phaseolorum, var. *batatatis*, 390
  var. *sojae*, 389
Digitaria spp., 296, 300
Diplodia frumenti, 94, 95
  macrospora, 91, 92, **94**
  maydis, 92, 94
  spp., 85, 95, 397, 416
  zeae, 91–94
Diplodia boll rot of cotton, 416
Diplodia ear rot of corn, **91**, 93
Diplodia seedling blight of corn, 91, 94
Diplodia stalk rot of corn, 91, 93
Disease losses (see specific crops)
Disease resistance (*see* specific crops)
Distichlis spicata, 324
  spp., 318
Ditylenchus angustus, 154
Dothidella trifolii, 371
Dothiorella phaseoli, 197
Downy mildew, of alfalfa, 347
  of barley, 30, **84**, 86

Downy mildew, of black nightshade, 457
  of clovers, 347
  of corn, **83**, 86, 192
  of grasses, 83, 86, 233
  of millets, 83, 86, **115**, 117
  of oats, 83, 86, 129
  of pea, 387
  of rice, 86
  of *Setaria viridis*, 84, 86
  of sorghum, 83, 86, 192
  of soybean, 386–388
  of sugarcane, **84**, 86, 192, 206, 215
  of tobacco, 456–459
  of wheat, 84, 86, **233**
Draeculacephala minerva, 342
  portola, 209
Drought injury, 75, 410
Dulbulus elimatus, 77
  maidis, 77
Dwarf bunt of wheat, 254, 257, 263
Dwarf disease, of alfalfa, 340
  of rice, 154, 445

E

Ear rots of corn, 85
  Diplodia, **91**, 93
  Fusarium, **85**, 88
  Gibberella, **85**, 87
  Helminthosporium, 97
  Nigrospora, 95, 96
  Rhizoctonia, 100
Echinochloa colonum, 115
  crusgalli, 118, 120, 296, 324
    var. *edulis*, 155
    var. *frumentacea*, 115
  spp. 295
Elaeagnus commutata, 326
Eleusine coracana, 115, 118
  indica, 295
  spp., 296
Elymus canadensis, 321, 322
  glaucus, 322
  mollis, 314
  spp., 296, 300, 306, 310, 314, 316, 318,
    320, 324, 326
Emetic substances from scabbed grain,
  35, 90
Endodermis, composition of, 15
  development of, 15
  relation of, to disease, 15
Endria inimica, 230
Entyloma crastophilum, 318
  dactylidis, **169**, 318
  irregulare, 318
  lineatum, 169, 318
  meliloti, 359

Entyloma oryzae, 169
  spp., 316
Epichloe typhina, 301, 302
Epicotyl development and disease, 7, 11, 13
Epidermal cells and disease, 13
Epigeal and hypogeal seedling development, 10, 11
Equisetum, 86
Eragrostis aspera, 86
  cilianensis, 296
  pilosa, 308
  spp., 295, 307, 324
Ergot, 11
  of barley, 37, 174
  of durum wheat, 37
  of grasses, 37, 299
  of rye, 37, 156, 174
  sclerotia parasitized, 175, 301
  of wheats, 37, 174, 235
Ergot compounds, ergoclarin, 174
  ergosterol, 174
  ergostetrine, 174
  ergotamin, 174
  ergotoxin, 174
Ergot poisoning, 174, 301
Ergoty grain, Federal grades, 174
Erianthus divaricatus, 105
  maximum, 86
  spp., 326
Erwinia flavida and pokkah-bong, 216
Erysiphe cichoracearum, 369, **431**
  communis, 369, 431
  graminis, 30, 31, 33, 52, 299
    var. *avenae,* 129
    var. *hordei,* 31–34
    var. *secalis,* 173
    var. *tritici,* 235
  polygoni, **368**, 369, 388, 403, 431
  spp., 431
Etch virus of tobacco, 448
Euchlaena luxurians, 86
  mexicana, 86
  spp., 86
Euphorbia, cyparissias, aecial host, 359, 360
Eyespot, *Helminthosporium,* 218, 307
  of grasses, 218, 307
  of sugarcane, 218

F

False smut of rice, 156
False stripe of barley, 24
Festuca elatior, 307, 311
  var. *arundinacea,* 295, 303
  rubra, 294, 295, 302

Festuca, spp., 295, 299, 300, 316, 320, 322, 324
Fiji disease of sugarcane, 210
Flag smut, of grasses, 315–317
  of wheat, 259, 262
Flax, chromosome number, 428
  disease resistance, 428, 436, 438
Flax diseases, 428
  anthracnose, 434
  browning, 433
  curly top, 429
  Fusarium wilt, 434–436
  heat canker, 429, 430
  mineral deficiencies, 429
  pasmo, 428, 431–433
  powdery mildew, 431
  Pythium seed rot and root rot, 430
  root rot, 429
  rust, 428, 436–439
  seedling blight, 429, 431, 435
  stem break, 433
  virus, 429
  wilt, 428
  yellows, 429
Flax varieties, Bison, 433, 436, 439
  Buda, 434
  Crystal, 433, 434
  Koto, 433
  Punjab, 434
  Rio, 434
  Sheyenne, 436
Floral bracts, relation of, to disease, 46
  to fungicides, 8
Floral infection, ergot, 178, 300
  scab, 35, 237
  smuts, 53, 55, 104, 169, 253
  stripe, 42
Foot rot, of grasses, 45, 245, 304
  of rye, 174
  of wheat, 45, 245, 304
Formaldehyde seed treatments, 59, 61, 63, 258
Frogeye leaf spot, of soybean, 390
  of tobacco, 459
Frost injury, of alfalfa, 336, 337
  of cereals, 24, 227
  of clover, 336, 366
  of corn, 75
  of wheat, 227
Fungicides, 6, 33, 83, 242, 302, 314, 399, 459
Fusarium, 238
  arthrosporoides, 358
  avenaceum, 34, 36, 240, 358
  culmorum, 34, 36, **239**, 242, 246, 303, 358
    var. *cereale,* 239

Fusarium graminearum, 36, **90**, 236, **238**
　moniliforme, 88, **91**, 155, 193, **215,**
　　416
　　var. *subglutinans*, **91**, 215
　nivale, 174, 241, 302
　　f. *graminicola*, 174, **241**
　　var. *major*, 241
　　var. *majus*, 242
　oxysporum, 405, 417
　　f. *crotalariae*, 405
　　lini, 434–436
　　lupini, 405
　　medicaginis, 358
　　nicotianae, 459, 460
　　phaseoli, 405
　　pisi, 405
　　tracheiphilum, 390, 405
　　udum, 405
　　vasinfectum, 416, **417**
　poae, 358
　roseum, 238
　　f. *cerealis*, 34, 36, 87, **90**, 155, 193,
　　　238
　　　var. Avenaceum, 35, 36, 238
　　　var. Culmorum, 35, 36, 238
　　　var. Graminearum, 35, 36, 87,
　　　　90, 238
　scirpi var. *acuminatum*, 358
　spp., 85, 129, 155, 156, 174, 175, 189,
　　236, 238, 240, 242, 246, 251, 301,
　　303, 357, 390, 417, 431
　udum var. *cajani*, 405
　　var. *crotalariae*, 405
Fusarium blight, of barley, **34,** 156
　of cereals, 34, 129, 236
　of corn, **85,** 156, 215
　of cotton, 416
　of flax, 431, 435
　of grasses, 34, 236, 303, 315
　of rice, 155
　of rye, 34, 174
　of sorghum, 193
　of soybean, 390
　of sugarcane, 215
　of wheat, 34, 156, 236
Fusarium boll rot of cotton, 416
Fusarium root rot, of alfalfa, 357
　of clover, 357, 390
Fusarium seedling blight, 35, 85, 237
Fusarium wilt, of alfalfa, 357
　of chickpea, 405
　of clover, 357
　of cotton, 409, 416, 417
　of cowpea, 405
　of crotalaria, 405
　of flax, 434–436
　of lupine, 405

Fusarium wilt, of pea, 405
　of soybean, 390
　of tobacco, 456, 459, 460
Fusoma rubicosa, 309

G

Gibberella fujikurio, 85, 88, **91**, 155, 193,
　　215, 416
　　var. *subglutinans*, 85, 88–91, 215
　moniliforme, 88, 89, **91**, 155, 215
　roseum, 238
　　f. *cerealis*, 34, 35, 85, 89, **90**, 155,
　　　238, 241, 303
　saubinetii, 90, 238, 240
　spp., 85, 129, 174, 236, 303, 416
　zeae, **34**, 85, 87–**90**, 155, 193, 236, **238,**
　　240, 303
Gibberella ear rot, 85, 87, 88
Gibberella seedling blight, 34, 85, 155,
　　236, 237, 303
Gloeocercospora sorghi, 196
Gloeocercospora leaf spot of corn and
　　sugarcane, 196
Gloeosporium bolleyi, 247, 308
　lini, 434
Glomerella, cingulata, 404
　glycines, 388, **389**
　gossypii, 414, **415**
Glume rust (*see* Stripe rust)
Glyceria spp., 86, 180, 300, 302, 316, 324
Glycine max, 383
　ussuriensis, 383
Gossypium barbadense, 419
　spp., 409, 421
　sturtii, 409
　thurberi, 411
Grass diseases, 293
　Agropyron mosaic, 294
　anthracnose, 51, 180, 217, 308
　Ascochyta leaf spot, 310, 311
　bacterial blights, 27, 126, 128
　bacterial stripe, 296
　barley stripe mosaic, 24, 295
　barley yellow dwarf, 25, 295
　bends, 294
　black loose smut, 58, 134, 320
　black stem rust (*see* Stem rust)
　blind seed disease, 304
　brome mosaic, 294
　brown leaf blight, 310
　brown stripe rust, 326
　brown stripe smut, 316
　bunt, 254, 322
　chlorotic streak virus, 295
　choke, 301
　corn streak virus, 296

Grass diseases, covered smut, 62, 139, 320
  crown rot, 304
  crown rust, 142, 323
  culm smuts, 318, 319
  downy mildews, 83, 86, 233, 299
  Epichloë head blight, 301
  ergot, 37, 174, 299–301
  flag smut, 315–317
  foot rot, 45, 245, 246, 302, 304
  Fusarium blight, 34, 174, 236, 238, 303
  gray speck, 123, 294
  head smuts, 9, 320, 321
  Helminthosporium blights, 43, 246, 304
  Helminthosporium eyespot, 307
  heritable leaf spot, 294
  kernel smuts, 254, 322
  leaf blights, 304
  leaf rusts, 65, 323
  leaf scald, 309
  leaf smuts, 315, 316, 317
  loose smut, 52, 56
  low-temperature Basidiomycete, 315
  Mastigosporium leaf fleck, 308
  mineral deficiency, 123, 294
  mosaics, 24, 25, 75, 76, 126, 190, 230, 294–296
  nonparasitic maladies, 123, 294
  Piricularia leaf spot, 163, 310
  powdery mildew, 30, 235, 299
  Pythium root rot, 81, 232, 298, 304
  Rhizoctonia blight, 165, 250, 314
  Rhizoctonia crown-root rots, 165, 250, 314
  Rhynchosporium scald, 49, 309
  rusts, 323–325
  Sclerotium blight, 302
  Scolecotrichum spot, 132, 310
  seedling blight, 43, 232
  Selenophoma leaf blotch, 313, 314
  Septoria blotches, 51, 134, 312, 313
  snow mold, 174, 240, 302, 303
  snow scald, 251, 302
  sooty spike, 307
  spot blotch, 43
  Stagonospora leaf spot, 312
  stem rust, 64, 264, 266, 323
  stinking smut, 254, 322
  streak, 76, 229, 295
  stripe mosaic, 24, 76, 230, 295
  stripe rust, 64, 274, 326
  stripe smuts, 315
  sugarcane mosaic, 295
  take-all, 242, 303
  tar spot, 302
  Typhula blight, 251, 314
  Typhyla snow mold, 251, 314

Grass diseases, virus, 24, 25, 75, 207, 229, 294–296
  wheat streak mosaic, 295
  yellow dwarf, 25, 295
  zonate eyespot, 307
Greenbug vector, 27
  Toxoptera graminum, 27
Groundnut (*see* Peanut)

H

Halo blight, of grasses, 126, 296, 297
  of oats, 126
Haynaldia spp., 226
  villosa, 279
Head smuts, of corn, **103**, 104, 202
  of grasses, 320
  of millets, 119, **120**, 320
  of rye, 184
  of sorghum, **103**, 104, 201, 202, 322
Heat canker, 396, 429
Heat scald, 75, 429
Hedysarum spp., 341
Helicobasidium purpureum, 358
Heliocotylenchus spp., 417, 445
Helminthosporium avenae, 40, **129**, 130
  bicolor, 247
  bromi, 305, 306
  californicum, 49
  carbonum, 97–100
  cynodontis, 307
  dictyoides, 305–307
  erythrospilum, 306
  giganteum, 118, 307
  gramineum, 40, **42**, 47
  halodes, 118, 247
    var. *tritici*, 247
  leucostylum, 118
  maydis, 97–100
  miyakei, 308
  monoceras, 118
  nodulosum, 118
  oryzae, 158–162
  panici-miliacei, 118
  poae, 306
  ravenelii, 307
  rostratum, 99
  sacchari, 118, 218, 307
  sativum, 43, 45–49, 118, 131, 159, 227, 232, 246, 304
  setariae, 118
  siccans, 307
  sigmoideum, 157, 158
    var. *irregulare*, **158**
  sorokinianum, 47
  stenacrum, 306

Helminthosporium stenospilum, 218
  spp., 38, 162, 189, 246, 308
  teres, 37–39
  tetramera, 247
  triseptatum, 306
  tritici-repentis, 247, 306
  tritici-vulgaris, **247**
  turcicum, 97–100, 193, 194, 307
  vagans, 304, 305
  victoriae, **131**
  yamadai, 118
  zeicola, 99
Helminthosporium blight, of barley, 43,
    46, 159, 246
  of cereals, 43, 46, 159, 246, 304
  of corn, **97**, 98, 105, 193
  of grasses, 43, 97, 246, 304
  of millets, 118
  of oats, 129, 131, 146, 147, 196
  of rice, 158–161, 163
  of sorghum, 97, 193
  of sudan grass, 97, 193, 194, 307
  of sugarcane, 218, 307
  of wheat, 43, 159. 246, 308
Hepatica, aecial host, 323
Heritable leaf blotches, in corn, 75
  in grass, 294
  in oats, 125
  in sorghum, 189
  in wheat, 228
Hibiscus spp., 410, 450
Hierochloë spp., 325
Holcus lanatus, 306, 312, 322
  spp., 316, 325
Holcus spot, bacterial, of corn, 80
    of sorghum, 191
    of sudan grass, 191
Hordeum distichon, 23
  spontaneum, 52
  spp., 51, 64, 86, 295, 296, 312, 320,
    325
  vulgare (*see* Barley)
Hordnia circellata, 342
Host (*see* botanical names of specific
    hosts)
Hot-water treatment, 25, 56
  of anthracnose, sugarcane, 218
  of browning flax, 434
  of chlorotic streak, 210
  of loose smuts, cereals, 25, 56, 253
  of ratoon stunt, 211
  of sereh, sugarcane, 211
Hydrophyllum, aecial host, 323
Hyoscyamus niger, downy mildew, 457
  spp., 457
Hypochnus filamentosa, 167
  sasakii, 167

Hypocotyl invasion by fungi, 12
Hypogeal type of seedling, 12
Hystrix spp., 325

I

Impatiens, aecial host, 323
Inheritance of factors for resistance to,
    angular leaf spot of cotton, 414
  brown rust of bromes, 4
  bunt of wheat, 258
  corn *Helminthosporium,* 100
  corn rust, 106
  covered smut of cereals, 63
  flag smut, 263
  flax rust, 436–439
  Fusarium wilt, 4, 409, 418, 436
  general, 4, 409
  leaf rusts, 4, 278
  loose smuts, 57, 253
  milo disease of sorghum, 194
  powdery mildew, 33, 235
  Rhynchosporium scald, 51
  rice *Cercospora* spot, 165
  rice *Helminthosporium* spot, 162
  semiloose smut, 61
  soybean downy mildew, 388
  spot blotch of barley, 40, 49
  stem rust, 58, 141, 272
  stripe disease of barley, 43
  stripe rust, 276
  tobacco black rot, 456
  tobacco blue mold, 459
  tobacco mosaic, 446, 448
  tobacco root rot, 463
Insect damage to plants, 16, 27, 214,
    339, 383, 409, 410
  in relation to diseases, 27, 78, 154, 214
Insect dissemination, of bacteria, 78
  of fungi, 175, 178
Insect vectors, aphids, 25, 27, 75, 76,
    126, 207, 230, 340, 341, 445
  leaf hoppers, 75–77, 154, 209–211, 230
  white fly, 410, 411
  (*See also* Aphid vectors; Leaf hopper
    vectors)
Introcellular inclusions, corn mosaic, 76
  rice dwarf, 154
  sugarcane, chlorotic streak, 209
    fiji disease, 211
  wheat mosaic, 229
Iron-excess injury, 75
Isopyrum spp., 278

J

Johnson grass diseases, 188
  anthracnose, 193, 216, 217
  bacterial blight, 190, 213, 298

Johnson grass diseases, *Helminthosporium* blight, 97, 193
  loose kernel smut, 198
  rust, 105, 202

K

Kabatiella caulivora, 376, **377**
Kentucky bluegrass, 295
Kernel mold of corn, 85
Kernel rots, of corn, **85**, 88–91, 95
  of rice, 155
Kernel smuts, of grasses, 254, 322
  of millets, **118**, 119
  of rice, 167, 168
  of rye, 184
  of sorghums, 198, 199, 201
  of wheat, 253
Koeleria cristata, 325
  spp., 301
Kudzu bean diseases, 377, 401, 404

L

Lagurus ovatus, 325
Lamarckia aurea, 325
Lanternfly, 76
Lathyrus spp., 341, 359, 401
Lathyrus diseases, 401, 403, 404
Leaf curl, of cotton, 410
  of tobacco, 450, 451
Leaf hopper injury, 27, 339
Leaf hopper vectors, *Aceratagillia sanguinolenta*, 367, 368
  Agallia constricta, 342
    quadripunctata, 342
  Agalliopsis novella, 342
  Carneocephala fulgida, 342
  Cicadula bimaculata, 77
  Cicadulina mbila, 76, 209
    spp., 230
    storeyi, 76, 209
    zeae, 76, 209
  Delphacodes striatellus, 155
  Deltocephalus dorsalis, 155
    striatus, 230
  Draeculacephala minerva, 342
    portola, 209
  Dulbulus elimatus, 77
    maidis, 77
  Endria inimica, 230
  Hordnia circellata, 342
  Nephotettix apicalis var. *cincticeps*, 155

Leaf hopper vectors, *Orosius argentatus*, 341
  Peregrinus maidis, 76
  Perkinsiella saccharicida, 211
    vastatrix, 211
  Scaphytopius acutus, 341
    dubius, 341
Leaf rust, of cereals, 64, 131, 185, 323
  of grasses, 65, 185, 323
Leaf scald disease, of barley, 49, 309
  of grasses, 49, 309
  of rye, 49, 309
  of sugarcane, 212
Leaf scorch of sugarcane, 219
Leaf smut, of grasses, 169, 315
  of rice, 169
Leaf spot, angular, in cotton, 411
  in tobacco, 454
Legume diseases, 335, 401–405
Legumes, anatomy of, 11
  relation of, to disease, 11, 183
  root of, 11
  seed of, 11
  seedling of, 11
Leptochloa filiformis, 216
Leptosphaeria avenaria, **134**
    var. *triticea*, 134
  herpotrichoides, 245
  nodorum, 249
  pratensis, 352–**354**
  salvanii, 156–**158**
  spp., 308
  taiwanensis, 219
Lespedeza spp., 401, 403
Lespedeza diseases, 401, 404
Lignin formation in walls, 11
  relation of, to disease, 11
Limnodea arkansana, 325
Linum angustifolium, 428
  marginale, 431
  mucronatum, 432
  spp., 428, 436
  usitatissimum, 428, 432
Lithospermum, aecial host, 323
Lolium spp., 86, 180, 296, 300, 304, 307, 312, 316, 325
Long smut, of millet, **118**
  of sorghum, 202
Loose kernel smut of sorghums, 198
Loose smut, of barley, **52**, 54, 58, 252
  flower development in, 53
  hot-water treatment of, 56
  infection of, 53, 55, 57
  linkage of, in resistance factors, 58
  resistance to, 57
  of grasses, 52, 320
  of wheat, 52, 54, 252

Lotus corniculata, 374
    spp., 341, 401
Lotus diseases, 401
Lucerne (*see* under Alfalfa; *Medicago*)
Lupine diseases, 377, 401, 404
Lupinus spp., 359, 401

M

Macrocalyx, aecial host, 323
Macrophoma zeae, 95
Macrophomina phaseoli, 197, 373, 391,
    397, 463
Macrosiphum dirhodum, 27
    granarium, 27
    persicae, 340
    pisi, 340, 367
    solanifolii, 340
Magnesium deficiency, 75, 124, 125, 443
Mahonia spp., aecial hosts, 264, 266
Maize (*see* Corn)
Malvaviscus arboreus, 410
Manganese deficiency, in cereals, 24,
    123, 154
    in grasses, 24, 123
    in oats, 123
Marasmius spp. on sugarcane, 214
Mastigosporium album, 308, 309
        var. *calvum*, 309
        var *muticum*, 309
    calvum, 309
    cylindricum, 309
    rubricosum, 308
Mastigosporium leaf fleck of grasses, 309
Medicago falcata, 335, 358, 360
    hispide, 335
    lupulina, 360
    ruthenica, 360
    sativa, 335, 360
    spp., 335, 341, 359, 360
Melampsora lini, 436, **437**, 439
Melanism in Hope wheat, 231
Melica imperfecta, 326
    spp., 319, 325
Meloidogyne spp., 417, 445
Melolotus alba, 335
    indica, 335
    officinalis, 335
    spp., 335, 341, 342, 352, 359
Membranes, plant, 7
        development of, 7
        relation of, to disease, 7
        to resistance, 7
Mercury, seed-treatment compounds, 8,
        30, 35, 40, 43, 48, 61, 63, 94, 104, 193,
        200, 240, 242, 250, 258, 381, 412, 416,
        417, 433
    soil treatment, 250–252, 302

Microsphaera alni, 368
    diffusa, 403
    spp., 388
Microstroma lini, 433
Middle lamella, relation of, to disease, 15
Mildew, downy (*see* Downy mildew)
    powdery (*see* Powdery mildew)
Milium effusum, 325
Millet chromosome number, 115
Millet diseases, 115
    bacterial blights, 115
    downy mildew, 83, 84, 86, 115, **117**
    head smut, 119, **120**
    Helminthosporium spots, 118
    kernel smut, **118**, 119
    long smut, 118
    mosaics, 207, 295, 296
    Pythium root rot, 81, 115
    virus, 207, 295
Milo disease of sorghums, 194, 195
Mineral deficiencies, and disease, 123,
        418, 459
    and wilt, 418, 459
Mineral-deficiency maladies, of alfalfa,
        339
    of barley, 24, 123
    of clover, 366
    of corn, 75
    of cotton, 410, 418
    of flax, 429
    of grasses, 123, 294
    of legumes, 125
    of oats, 123
    of peanut, 396
    of rice, 154
    of soybean, 383
    of tobacco, 443, 459
    of wheat, 123, 124, 228
Mineral excesses, 24, 75, 125, 154
Miscanthus japonicus, 86
    sinensis, 295
    spp., 86, 219, 295, 298
Mite, grass vector, 230
Mitrula sclerotiorum, **372**, 373
Molinia caerulea, 325
Molybdenum deficiency, 125
Mosaic, of *Agropyron*, 294
    of alfalfa, 340
    of barley, 24, 126, 230, 295
    of brome, 126, 294
    of clovers, 367
    of corn, 75, 76, 207, 230, 296
    of cotton, 410, 411
    of cowpea, 401, 402
    of *Crotalaria*, 401, 402
    of grasses, 24, 75, 207, 230, 294
    of lupine, 402

Mosaic, of oats, **125,** 230, 295
of peanut, 397
of rice, 154, 296
of rye, 173
of sorghum, 75, 190, 207, 295
of soybean, 383, 384
of sugarcane, 75, 76, 207, 208, 295
of sweetclover, 340
of tobacco, 445–448
of wheat, 24, 126, 228, 229, 295, 296
Muhlenbergia spp., 322, 325
Mycosphaerella arachidicola, 398
berkeleyii, 398
davisii, 351, **352**
lethalis, 349, **350**
linicola, 432
linorum, 431–433
pinodes, 403
puraricola, 404
spp., 403
Myosotis, aecial host, 323
Myzus convolvuli, 383
persicae, 76, 341, 383, 402
solani, 383
spp., 384

N

Nematodes, 214, 417, 444–446, 459
dagger, 417, 445, 456
meadow, 417, 445, 456
reniform, 417
rice gall, 154
rice stem, 154
ring, 417
root knot, 417, 444–446, 459
spiral, 417, 445
sting, 417
stubby root, 417
wheat gall, 232
Neovossia horrida, 167–169
indica, 257, 258
Nephotettic apicalis var. *striatellus,* 155
Net blotch of barley, 37, 39
Nicotiana debneyi, 459
digluta, 446
glauca, 450
glutinosa, 446, 448
gossii, 450
longiflora, 454, 456
plumbaginifolia, 456
rustica, 443
spp., 450, 457
tabacum, 443, 446, 448, 456

Nicotiana tomentosa, 451
tomentosiformis, 451
Nigrospora cob and stalk rot, 95, 96
Nigrospora oryzae, 95, **96**
sacchari, 95
sphaerica, 95, 96
Nonparasitic diseases, 24, 75, 189
bends of grasses, 294
blast of cereals, 123
brown leaf spot of barley, 24
brown root rot of tobacco, 444
chlorophyll deficiency, 75, 189, 339
chlorosis, 75, 189, 339, 443
chlorotic leaf spot, 189
cold injury, 74, 227, 336
dry leaf spot, of cereals, 123, 124
of grasses, 124
false stripe of barley, 24
firing of tobacco, 443
frenching of tobacco, 444
frost injury, 24, 75, 227
gray speck of cereals, 123, 124, 228, 294
hail spots, 444
heat canker, 396, 429
heat scald, 75, 444
heritable blotches, 125, 294, 339
leaf discoloration, 75, 189, 294, 339
leaf necrosis, 123, 294, 339
leaf scald, 444
lightning injury, 444
mineral deficiencies, 24, 75, 123, 228, 294, 339, 383, 410, 443
boron, 24, 125, 339
copper, 24, 125, 339
magnesium, 75, 444
manganese, 123
molybdenum, 125, 339
phosphate, 339, 444
potash, 75, 339, 383, 410, 418, 444
proliferations, 294
straight head of rice, 153
sun scald, 444
weak neck of sorghum, 189
white tip of rice, 154
winter injury, 226, 336
Northern anthracnose of clover, **376, 377**

O

Oat, chromosome number, **122**
disease losses, 123
disease resistance, 122, 137, 140, **141,** 146

Oat diseases, **122**
  anthracnose, 132
  bacterial halo blight, 126, **127**
  bacterial stripe blight, 128
  black loose smut, 58, **134**
  blast, **123**
  blue dwarf, 126
  copper deficiency, 125
  covered smut, 61, 139
  crown rust, 131, **142,** 144
  downy mildew, 83, 129
  dry leaf spot, 124
  Fusarium blight, 34, 129
  gray speck, **123,** 124, 228
  halo blight, **126,** 127
  Helminthosporium, blight, **131,** 146,
    147, 196
    leaf blotch, **129,** 130
  heritable leaf spot, 125
  manganese deficiency, 123
  mineral deficiency, 125
  molybdenum deficiency, 125
  mosaic, 27, **125,** 126, 229
  nonparasitic, **123**
  powdery mildew, 30, 129
  red leaf virus, 27, 126
  Rhizoctonia blight, 250
  rice dwarf, 155
  Scolecotrichum leaf blotch, 132
  Septoria leaf blotch, **132,** 133, 312
  stem rust, 131, **140**
  stripe rust, 324
  take-all, 243
  virus, 27, 125, 295
  yellow dwarf, 26, 126, 295
  zinc deficiency, 125
Oat varieties, Andrew, 142
  Anthony, 137, 140
  Ascencao, 147
  Bannock, 140
  Benton, 138
  Black Diamond, 137, 140
  Black Mesdag, 137, 140
  Bond, 128, 134, 144, 147
  Bond × Anthony, 138
  Boone, 138
  Camas, 137, 140
  Canuck, 142
  Clinton, 138
  D69 × Bond, 128, 138, 140
  Fulghum, 128, 137, 138, 140
  Garry, 142
  Gothland, 137, 140
  Green Russian, 142
  Hajira, 141, 142
  Huron, 138
  Idamine, 140

Oat varieties, Iogold, 142
  Joanette, 141
  Jostrain, 142
  Landhafer, 141, 147
  Lee × Victoria, 137
  Lelina, 137, 140
  Marida, 140
  Marion, 138, 142
  Markton, 138, 140
  Marvic, 138
  Mindo, 141
  Minland, 142
  Monarch, 137, 140
  Morota, 138
  Neosho, 138
  Nicol, 137, 140
  Nortex, 138
  Rainbow, 138, 142
  Rangler, 138
  Ransom, 142
  Red Rustproof, 138
  Richland, 131, 138, 141, 142
  Rodney, 142
  Roxton, 142
  Sac, 140
  Santa Fe, 144, 147
  Sevnothree, 142
  Tama, 142
  Trispernia, 147
  Ukraine, 147
  Vicland, 142
  Victoria, 131, 132, 137, 138, 144,
    146
  Victoria × Richland, 128, 132, 137,
    138, 140
  Victory, 137, 138, 140
  White Tartar, 141, 142
Okra, 417
Olpidium brassicae, 430
  trifolii, 346
Onosmodium, aecial host, 323
Ophiobolus graminis, 242–245, 303
  var. *avenae,* 243
  miyabeanus, 158, 159
  oryzae, 158
  oryzinus, 158
  sativus, 47
  setariae, 118
Ornithogalum spp., 65
  umbellatum, aecial host, 65
Orosius argentatus, 341
Orthocide, seed protectant, 83, 192
Oryza sativa, 86, 153
  spp., 153
Oryzopsis spp., 318, 319
Oxalis corniculata, aecial host, **105**
  cymbosa, 105

Oxalis europaea, 105
  spp., 105, 106, 203
  stricta, 105

P

Panicum capillare, 296
  maximum, 300
  miliaceum, 115, 155, 295, 296
  spp. 86, 115, 118, 120, 294, 295, 307, 321
  virgatum, 325
Panogen (*see* Mercury, seed treatment)
Pasmo disease of flax, 431–433
Paspalum dilatatum, 300, 301
  maliacophyllum, 301
  setaceum, 325
Pea, diseases of, 342, 367, 368, 377, 401
Peanut, chromosome number, 396
Peanut diseases, 396
  bacterial wilt, 397
  Cercospora spot, 398
  deficiency, 396
  heat canker, 396
  leaf spots, 398
  peg rots, 397
  pod rots, 397
  ring spot, virus, 397
  rosette, virus, 397
  rust, 399
  seed rot, 397
  seedling blight, 397
  southern blight, 399
  virus, 397
  wilt, 397
Pellicularia filamentosa, **167**, 251, 399, 405, 422, 463
  rolfsii, 373, 391, 399, 405, 463
Pennicillium oxalicum, 192
  spp., 85, 227, 397
Pennisetum glaucum, 115, 295
  purpureum, 218, 307
  spp., 86, 115, 116, 299, 300
Peregrinus maidis, 76
Perennial rust infection, 360
Perennial smut infection of grasses, 9, 315, 318
Periconia circinata, 194, **195**
Periderm function in disease, 15
Perkinsiella saccharicida, 211
  vastratrix, 211
Peronospora hyoscyami, 457
  manshurica, 386–**388**
  nicotianae, 457
  tabacina, 456–458
  trifoliorum, 347, 348

Peruvian flint corn resistant, to rust, 106
  to streak, 77
Phacelia, aecial host, 323
Phaeoseptoria, genus, 312
Phakopsora desmium, 424
Phalaris arundinacea, 295, 316
  paradoxa, 295
  spp., 86, 234, 296, 325
  tuberosa, 234, 295
Phaseolus lunatus, 401
  vulgaris, 401
Phellogen function in disease, 15
Phialea temulenta, 304
Phleum pratense, 295, 307, 317
  spp., 180, 296, 300, 318, 325
Phomaher barum var. *medicaginis*, 350
  linicola, 432
  medicaginis, 350
  meliloti, 354
  spp. 350, 354
  trifolii, 349, **350**, 369
Phosphate deficiency, 339, 443
Phragmites communis, 325
  spp., 86
Phyllachora graminis, 302
  spp., 302
Phymatotrichum omnivorum, 358, 373, 392, 419, **420**
Phymatotrichum root rot, 358, 373, 392, 419–421
Physalospora, spp., 91
  tucumanensis, 193, **216**, 217
  zeae, 95
  zeicola, 94, **95**
Physiologic races of parasites, general, 4, 33, 56, 164, 165, 201, 235, 259, 273, 388
  tables of, 34, 61, 136, 140, 142, 202, 260
Physiological anatomy, of Gramineae, 7
  of Leguminosae, 11, 336–339
Physiological specialization, 4, 56, 164, 165, 235, 273
Physoderma alfalfae, 345, 346
  maydis, 80
  zea-maydis, 80
Physoderma brown spot of corn, **80**
Phytomonas (*see Bacterium; Corynebacterium; Pseudomonas; Xanthomonas*)
Phytophthora cactorum, 347
  cryptogea, 347
  erythroseptica, 215, 347
  megasperma, 347
  parasitica, 456
    var. *nicotianae*, 455, **456**
  spp., 347, 402

Piricularia grisea, 163, 310
  oryzae, 162, **163**
Pisum arvense, 401
  sativum, 387, 401
  spp., 401
Plant defense mechanisms, 13
Plenodomus meliloti, 358
Pleospora graminea, 42
  herbarum, 350, 373–**375**
  rehmiana, 350
  trichostoma, 40, 42, 247
Poa ampla, 141
  pratensis, 155, 295, 296, 304, 305, 317, 326
  secunda, 306
  spp., 174, 180, 299, 300, 302, 304, 313, 316, 318, 320, 325
  trivialis, 306
Pod and stem blight of soybean, 389
Pokkah-bong of sugarcane, 215
Polypogon monspeliensis, 325
Polyspora lini, 431, **433**, 434
Polythrincium trifolii, 369
Potash deficiency, 75, 340, 383, 410, 418, 443, 459
Powdery mildew, of cereals, 30, 129, 173, 235
  of clovers, 368
  control of, 33
  effect of, on composition, 31, 235
  of flax, 431
  of grasses, 30, 129, 235, 299
  of legumes, 368, 403
  resistance to, 33, 173, 235
  of soybean, 388
Pratylenchus pratensis and wilt, 417
  spp. 417, 445
Primordia of stems, development of, 9, 10
  relation of, to disease, 9
  systemic infection of, 9, 59, 62
Pseudo-black chaff of wheat, 231
Pseudomonas alboprecipitans, 115
  andropogoni, 190, 191
  angulata, 453, **454**
  atrofaciens, 231, 232
  coronafaciens, 126–128
    var. *atropurpurea*, 296, 297
  glycinea, 385
  medicaginis, 344
  phaseolicola, 402
  pisi, 402
  sojae, 385
  striafaciens, 127–129, 296
  syringae, 191, 368, 402
  tabaci, 385, 452–454
  trifoliorum, 368

Pseudopeziza jonesii, 355–357
  medicaginis, 355, **356**
  meliloti, 355, **356**
  trifolii, 355, 356
Pseudoplea leaf spot, 370, 371
Pseudoplea trifolii, 370–372
Puccinellia spp., 325
Puccinia anomala, 64–66
  arachidis, 399
  coronata, 142, **143**, 145, 146, 323–325
  dispersa, 186
  glumarum, 64, 185, 273–275, 323–326
  graminis, 52, 64, 106, 265, 267, 271, 274, 278, 323–325
    var. *agrostidis*, 323–325
    var. *avenae*, 140–142, 324, 325
    var. *lolii*, 323
    var. *phlei-pratensis*, 323–325
    var. *poae*, 323–325
    var. *secalis*, 64, 185, 324, 325
    var. *tritici*, 64, 263, **264**, 269, 270, 273, 324, 325
  hordei, 64, **65**
  kuehnii, 220
  montanensis, 326
  poae-sudeticae, 326
  polysora, 105, 106, 326
  purpurea, 106, 202, 203
  recondita, 65, 185, 186, 274, 276, 278
  rubigo-vera 52, 65, 185, **186**, 274, 277, 278, 323–326
    var. *secalis*, 185, 186
    var. *tritici*, 65, 276–278
  sacchari, 221
  sorghi, **105**, 106
  spp., 326
  stakmanii, **422**, 423
  triticina, 65, 277
Pueraria, diseases of, 401
Pueraria spp., 401
Pulmonaria, aecial host, 323
Pyrenopeziza medicaginis, 357
Pyrenophora avenae, **129**, 130
  bromi, 305, 306
  graminea, 42
  spp., 308
  teres, 37–40, 42, 131
  tritici-repentis, 247, 306
  tritici-vulgaris, 247
Pythium aphanidermatum, 83, 430, **455**
  aristosporum, 233
  arrhenomanes, **81**, 82, 192, 214, 233, 299, 346, 386
  debaryanum, 192, 299, 346, 386, 430, 455
  deliense, 455

Pythium graminicola, 30, 81, 192, 214, 233, 299, 346
  intermedium, 430
  irregulare, 430
  mamillatum, 430
  megalacanthum, 430
  splendens, 346, 430
  spp., 30, 81, 115, 192, 214, 232, 299, 304, 346, 386, 429, 455
  tardicrescens, 233
  ultimum, 346
  vexans, 346, 430
  volutum, 233
Pythium damping-off, 12, 298, 346
Pythium root rot, 30, 81, 115, 192, 196, 214, 232, 298, 346, 386, 402
Pythium seed rot, 81, 192, 298, 346
Pythium seedling blight, 81, 192, 232, 298, 346

R

Ramulispora sorghi, 196
Ranunculus, aecial host, 323
Red clover (*see* Clover diseases)
Red rot of sugarcane, 216, 217
Rhamnus cathartica, aecial host, 142, 145
  dahurica, 142
  lanceolata, 142
  spp., 143, 146, 326
Rhizoctonia bataticola, 197
  crocorum, 358
  oryzae, **165**, 166
  solani, 166, **167,** 250, 314, 358, 392, 399, 405, 422, 463
  spp., 100, 165, 214, 302, 358, 397, 431
  violacea, 358
  zeae, 100, 167
Rhizoctonia blight, of cereals, 100, 165, 250
  of corn, 100, 166
  of flax, 431
  of grasses, 165, 250, 314, 315
  of rice, 165, 166
  of soybean, 392
  of wheat, 250
Rhizoctonia canker, of cotton, 422
  of tobacco, 463
Rhizoctonia lesions, on rice, 166
  on wheat, 250
Rhizoctonia rot, of alfalfa, 358
  of clover, 358
  of corn, 100, 167
  of cotton, 422
  of legumes, 405
  of sweetclover, 358
  of tobacco, 463

Rhizopus spp., 85, 397
Rhopalosiphum fitchii, 27, 76
  maidis, 27, 76, 207
  prunifoliae, 27, 76
Rhynchosporium orthosporum, **49,** 309, 310
  secalis, **49,** 50, 51, 183, 309
Rhynchosporium scald, of barley, 49, 50, 309
  of grasses, 49, 309, 310
  of rye, 49, 50, 309
Rice, chromosome number, 153
  disease losses, 153
  disease resistance, 158, 162, 164, 165
Rice diseases, 153
  bacterial blight, 155
  bacterial kernel rot, 155
  bakanae disease, 155
  black sheath rot, **158**
  blast, 162
  Cercospora spot, 164
  culm rot, 153, 156–158
  downy mildew, 86
  dwarf or stunt virus, 154, 296, 445
  false smut, 156
  Fusarium blight, 155
  Helminthosporium blight, 159–161, 163
  kernel rot, 155
  kernel smut, **167,** 168
  leaf smut, 169
  mineral deficiency, 154
  nematodes, 154
  nonparasitic maladies, 153
  Rhizoctonia blight, 165
  Rhizoctonia lesions, 166
  rotten neck, 162
  Sclerotium blight, 156
  seedling blight, 153, 159, 160
  sheath and culm blight, 165
  straighthead, 153
  stripe virus, 155
  white tip, 154
Rice varieties, red rice, 165
  Supreme Blue Rose, 165
Ring spot virus, of sweetclover, 342
  of tobacco, 448, 449
Roots, anatomy in relation to disease, 7, 10, 12
  cambial and noncambial, in legumes, 11
  recovery of, from injury, 7, 12
  types of, 7, 11
  winter injury in, 13
Rosette disease, of peanut, 397
  of wheat, 229
Rotylenchulus spp., 417
Rusts, of alfalfa, 369
  of barley, **64**

Rusts, of cereals, 64, 140, 142, 185, 263, 273, 276, 323
  of clover, 378
  of corn, 105
  of cotton, 422–424
  of flax, 436–439
  of grasses, 64, 105, 323–326
  of legumes, 369, 378
  of oats, 140, 142, 323
  of peanut, 399
  of rye, 185
  of sorghum, 106, 202, 221
  of sugarcane, 220, 221
  of wheat, 64, 65, 263, 273, 276
Rusty blotch of barley, 49
Rye, chromosome number, 173
  disease losses, 173
Rye diseases, 173
  anthracnose, **180**, 181
  bacterial blight, 27, 173, 231
  bunt, 184, 254
  downy mildew, 86
  ergot, 37, **174,** 176, 177
  foot rot, 174
  Fusarium blight, 34, 85, 174, 236
  head smuts, 184
  kernel smuts, 184
  leaf rust, **185,** 186, 276
  leaf scald, 183
  mosaic, 173
  powdery mildew, 31, 173
  Rhynchosporium scald, 49, 183
  rice dwarf, 155
  scab, 34, 174, 237
  Septoria leaf blotch, 182, **183**
  smut, **184**
  snow mold, 174, 240
  stalk smut, **183,** 184
  stem rust, 185, 264
  stinking smut, 184
  stripe rust, 185, 274
  virus, 173

S

Saccarum barberi, 206
  officinarum, 86, 206
  robustum, 206
  sinense, 206, 220
  spontaneum, 84, 86, 206, 220
  spp., 86, 295, 298
Saccothecium trifolii, 372
Scaphytopius acutus, 341
  dubius, 341
Schedonnardus paniculatus, 325
Sclerenchyma tissue in relation to disease, 15

Sclerophthora, genus, 83
  macrospora, 30, **83,** 84, 86, 117, 118, 129, 233, 234, 299
Sclerospora butleri, 86
  farlowii, 86
  graminicola, 83, **84,** 86, 115–117, 192, 234, 299
  indica, 83, **84,** 86
  javanica, 84
  macrospora, 30, 83, **84,** 86, 117, 118, 129, 215, 233, 299
  magnusiana, 86
  maydis, 83, **84,** 86
  miscanthi, 86
  noblei, 86
  northi, 86
  oryzae, 86
  philippinensis, 83, **84,** 86
  sacchari, 83, **84,** 86, 215
  sorghi, 83, **84,** 86, **192**
  spontanea, **84,** 86
  spp., 83, 85, 192, 215
Sclerotinia sclerotiorum, 372
  trifoliorum, 372, 373
Sclerotinia blight of clover, 372
Sclerotium bataticola, 197, 373, 391, 463
  oryzae, 157, 158
  rolfsii, 252, 373, 391, 397, **399,** 405, 463
  spp., 302, 358, 397, 416
Sclerotium blight, of alfalfa, 358
  of cereals, 197, 252
  of clovers, 358, 373
  of cotton, 197, 416
  of grasses, 158, 197, 252, 302
  of legumes, 197, 405
  of peanuts, 397, 399
  of rice, **156**
  of sorghum, 197
  of soybean, 197, 391
  of tobacco, 463
  of wheat, 252
Scolecotrichum graminis, 310, 311
  var. *avenae,* 132
Scutellum rot of corn, 85
Secale cereale, 173
  montanum, 325
  spp., 86, 185, 226, 296
Seed-coat penetration, 11, 12
Seed infection, general, 24, 53
Seed infestation, general, 8, 59
Seed rot, of clover, 346
  of corn, 85, 88, 91, 95, 97
  of flax, 429, 434, 435
  of legumes, 402
  of peanut, 397
Seed treatments (*see* specific compounds, e.g., Mercury)

Seedling anatomy, 8, 11, 13
Seedling blight, of barley, 44
  of corn, 81, **85**, 90, 100
  of cotton, 416
  of flax, 429
  of grasses, 298, 303
  of rice, 155, 159
  of sorghum, 192, 193
  of soybean, 388–390
  of wheat, 43, 236, 237
Selenophoma bromigena, 313, 314
  donocis, 314
Semipermeable membranes, 8, 12
Septoria avenae, 132, 134
    var. *tritici*, 134
  bromi, 312, 313
  elymi, 314
  glycines, 391
  infuscans, 314
  jaculella, 313
  linicola, 431, **432**
  macropoda, 313
  nodorum, 52, 248–250
  oudemansii, 313
  pacifica, 314
  passerinii, **51**, 312
  secalis, 182, **183**, 312
    var. *stipae*, 312
  sojae, 391
  sojina, 391
  spp., 51, 433
  tritici, 248–250, 312
    var. *avenae*, 312
    var. *holci*, 312
    var. *lolicola*, 312
Septoria leaf blotch, of cereals, 51, 183, 248
  of grasses, 51, 183, 248, 312, 313
  of soybean, 391
Sereh disease of sugarcane, 206, 211
Setaria glauca, 86, 118
  italica, 86, 115, 118, 295
  lutescens, 118, 120
  magna, 86
  spp., 118, 295, 296
  verticillata, 86
  viridis, 86, 116, 118, 120
Sheath and culm blight of rice, 156, 158
Shepherdia canadensis, 326
Sida spp., 450
Sitanion hystrix, 312
  jubatum, 298
  spp., 316, 320, 325
Smuts, of barley, 52, 54, 58, 60–62
  of cereals, 52, 62
  of corn, 100, 102–104
  of grasses, 52, 56, 58, 62, 315–323

Smuts, of legumes, 359
  of millets, 118, 321
  of oats, 58, 61, 131
  of rice, 167–169
  of rye, 183, 184
  of sorghum, 103, 104, 322
  of sugarcane, 220
  of wheat, 56, 57, 252, 253, 259
"Smutty," Federal grain grade, 62
Snow mold, of cereals, 174, 240, 302
  of grasses, 174, 240, 302, 303
Snow scald of grasses, 302
Soja max, 383
Sooty blotch of clover, 369, 370
Sooty spike of grasses, 307
Sooty stripe of sorghum, 307
Soreshin, of cotton, 422
  of tobacco, 463
Sorghastrum spp., 325
Sorghum arundinaceum, 86, 296
  halepense, 188, 298
  plumosum, 86
  spp., 86, 106, 193, 294, 322
  vulgare, 86, 188, 295
    var. *sudanense*, 188, 295, 298
Sorghum, chromosome number, 188
  disease losses, 189
  disease resistance, 105, 198, 200–203
Sorghum diseases, 188
  anthracnose, 193, 194, 216
  Ascochyta spot, 198
  bacterial blights, 80, 190, 191, 212, 213, 298
  Cercospora spot, 197
  charcoal rot, **197**, 198
  covered kernel smut, 199, **201**, 322
  crown rot, 193, 197
  downy mildews, 83, 84, 86, **192**
  Fusarium seedling blight, 193
  Gloeocercospora leaf spot, 196
  head smut, 103, 104, **201**
  Helminthosporium blight, 97, 98, 193, 194, 196
  heritable leaf spots, 189, 294
  Holcus spot, 191
  long smut, 118, **202**
  loose kernel smut, **198**, 199, 322
  milo root rot, **194**, 195
  mosaics, 75, 190, 207, 295
  nonparasitic maladies, 189, 294
    leaf spots, 189
    weak neck, 189
  Pythium root and crown rot, 81, 192, 195, 196
  rust, 106, **202**, 221
  Sclerotium blight, 197
  seedling blight, 81, 192

Sorghum diseases, sooty stripe, 196
  stalk rot, 193, 198
  streak and stripe, 76, 190, 213, 214, 298
  virus, 75, 190
  weak neck, 189
Sorghum types, 105, 188
Sorosporium cenchri, 321
  confusum, 321
  ellisii, 321
  everhartii, 321
  spp., 320
  syntherismae, 321
Southern anthracnose of clover, 375
Southern blight, 252, 373, 399, 416
  of cotton, 416
  of legumes, 373
  of peanut, 399
Soybean, chromosome number, 383
  disease losses, 383
  disease resistance, 388
Soybean diseases, 383
  anthracnose, 388, 389
  bacterial blights, 385
  bacterial pustule, 386
  black patch, 377
  brown stem rot, 392
  bud blight, 384
  Cercospora leaf spot, 390
  downy mildew, 386–388
  frogeye leaf spot, 390
  Fusarium wilt, 390
  mineral deficiency, 383
  mosaics, 383, 384
  pod and stem blight, 389
  powdery mildew, 388
  Pythium root rot, 386
  Rhizoctonia blight, 392
  Sclerotium blight, 391
  seedling blight, 388–390
  Septoria spot, 391
  virus, 383, 384
  wildfire, 385
  yellow bean mosaic, 384
Sphacelotheca andropogonis, 322
  cruenta, 198–200, 220, 322
  destruens, 119, **120,** 322
  holci, 200
  panici-miliacei, 120
  reiliana, **103,** 104, 201, 322
  sorghi, 199, **201,** 322
Sphaerella linicola, 432
  linorum, 432
Sphenopholis spp., 325
Spicaria elegans, 198
Sporobolus airoides, 298
  spp., 307, 325

Spot blotch, of barley, 43, 45
  of grasses, 43
  of wheat, 43
Spring black stem of legumes, 349, 369,
  403
Stagonospora arearia, 312
  meliloti, 352–354
  sacchari, 219
  spp., 354
Stalk rots, of corn, *Diplodia*, 91–93
  Gibberella, 85, 90
  Nigrospora, 95
  of sorghum, 193, 197
Stalk smut, of grasses, 318
  of rye, 183
Stem break of flax, 433
Stem rust, of cereals, 64, 131, 185, 263
  of grasses, 64, 185, 264
  linkage for resistance, 58, 272
Stemphylium botryosum, 371, 373–**375**
  loti, 374
  sarcinaeforme, 373–**375**
Stemphylium leaf spot of legumes, 373, 374
Stewart's wilt of corn, 77
Stinking smut, of grasses, 254
  of rye, 184, 254
  of wheat, 253
Stipa spp., 312, 318, 319, 325
Stizalobium spp., 401
Stizalobium, diseases of, 404
Stomata, relation of, to disease, 14
Straighthead of rice, 153
Streak disease, of corn, 75, 76, 209
  of sugarcane, 76, 209
  of sweetclover, 342, 450
Stripe disease, of barley, 40
  of corn, 76
  of rice, 155
  of sugarcane, 76
Stripe rust, of cereals, 64, 185
  of grasses, 64
Stripe smuts of grasses, 315
Stunt disease of corn, 77
Suberin, relation of, to disease, 14, 15
Subterranean clover mosaic, 342, 367, 368
Sudan grass diseases, 188, 216, 217
  anthracnose, 193, 308
  bacterial blights, 190, 191, 213, 298
  covered kernel smut, 199, 201, 322
  Helminthosporium blight, 97, 98, 193
  heritable leaf spots, 189, 294
  Holcus spot, bacterial, 191
  loose kernel smut, 198, 199, 322
  rust, 202
Sugarcane, chromosome number, 206
  disease resistance, 206, 207, 209, 211–
    213, 217, 219

Sugarcane, grass mosaics on, 75, 207, 295
  green type of, 207
  yellow types of, 207
Sugarcane diseases, 206
  anthracnose, 216, 217
  bacterial streak, 214, 298
  bacterial stripe, 190, 213, 298
  brown spot, 219
  brown stripe, 218
  chlorotic streak, 76, 209, 210, 295
  Cobb's bacterial wilt, 79, **211**, 298
  downy mildew, 83, 84, 86, 206, 215
  eyespot, 218
  fiji disease, 77, 210
  Fusarium blight, 215
  Gloeocercospora leaf spot, 196
  gumming disease, 211
  Helminthosporium leaf blights, 218, 307
  kernel smut, 220
  leaf scald, 212
  leaf scorch, 219
  mosaics, 76, 206, 207, 208, 295
  Phytophthora rot, 215
  pokkah-bong, 215
  Pythium root rot, 81, **214**
  ratoon stunt virus, 211
  red rot, 216, 217
  ring mosaic virus, 211
  rust, 220
  sereh, 206, 211
  smut, 206, **220**
  streak virus, 76, 209
  stripe virus, 76
  target blotch, 218
  virus, 76, 207
  wilts, 211, 298
Sulfur fungicides, 33, 83, 399
Sweet potato wilt pathogen, 459
Sweetclover diseases, 335, 379
  common leaf spot, 355, 356
  downy mildew, 347
  Fusarium wilt, 358
  leaf spots, 352, 353
  mosaics, 340–342
  Phytophthora root rot, 347
  Pythium rot, 346
  Rhizoctonia root rot, 358
  ring spot virus, 342
  root rot, 352, 357
  smut, 359
  spring black stem, 349–351
  stem canker, 351
  Stemphylium leaf spot, 373
  streak virus, 342, 450
  summer black stem, 351, 352
  virus, 340–342
  wilt, 357

Sweetclover diseases, winter injury, 336
  witches' broom, 341
  wound tumor virus, 342
Symphytum, aecial host, 323
Systemic infection, by *Helminthosporium gramineum*, 42
  relation of, to primordial culm development, 9
  by smut fungi, 9, 55, 62, 136, 315

T

Take-all, of cereals, 242, 303
  of grasses, 242, 303
  of wheat, 242, 303
Tar spot of grasses, 302
Temperature damage, canker, 396, 429
  high, 75
  low, 75
  scald, 75, 429
Teosinte-corn hybrids, 86
Teosinte diseases (*see* Corn diseases)
Thalictrum polygamum, 278
  speciosissimum, 278
  spp., 278, 323
Thecapora deformans, 359
Thielavia basicola, 461, 462
Thielaviopsis basicola, 430, 461, 462
Thurberia thespesioides, 411
Tilletia asperifolia, 322
  brevifaciens, 257
  caries, 253–257, 259, 260, 323
  cerebrina, 322
  controversa, 253, 254, **257**, 259, 323
  corona, 169
  elymi, 322
  foetens, 257
  foetida, 253, 254, 256, 257, 259, 260
  fusca, 322
  guyotiana, 322
  holci, 322
  horrida, 167–169
  indica, **257**, 258
  laevis, 257
  spp., 169, 184, 254, 322
  tritici, 255
Timothy diseases, anthracnose, 308
  bacterial blight, 296
  brown leaf blight, 310, 311
  ergot, 299
  flag smut, 316
  Helminthosporium leaf spot, 307
  leaf smuts, 315–317
  stem rust, 323–325
  streak or brown leaf spot, 310, 311
  stripe smut, 315, 317
  yellow dwarf virus, 295

Tobacco, chromosome number, 443
  disease losses, 443
  disease resistance, 446, 448, 451
  types of, 443
Tobacco diseases, 443
  angular leaf spot, 453, 454
  bacterial leaf spot, 452–454
  bacterial wilt, 451, 452
  black root rot, 461–463
  black shank, 455, 456
  blackfire, 454
  blue mold, 456–459
  brome mosaic, 294
  brown leaf spot, 459
  brown root rot, 444, 445
  common mosaic, 446, 447
  cucumber mosaic, 451
  downy mildew, 456–459
  etch virus, 448
  frenching, 444
  frogeye leaf spot, 459–461
  Fusarium wilt, 459, 460
  Granville wilt, 451, 452
  hail injury, 444
  leaf curl virus, 450, 451
  leaf spotting, 443
  lightning injury, 444
  mineral deficiency, 443, 453
  mosaics, 445–451
  necrosis virus, 448, 449
  nematode injury, 444, 445
  nonparasitic maladies, 443
  Pythium damping-off, 455
  Pythium stem rot, 455
  rattle virus, 451
  Rhizoctonia rot, 463
  ring spot virus, 342, 384, 448, 449
  soreshin, 463
  streak virus, 342, 450
  sunscald, 444
  tumor gall, 451
  virus, 445–451
  wildfire, 452–454
  yellowing, 443
Tobacco etch virus, 448
Tobacco ring spot virus, on soybean, bud
    blight, 384
  on tobacco, 448, 449
Tobacco streak virus, on sweetclover, 342
  on tobacco, 450
Tolyposporium bullatum, 118
  ehrengergii, 202
  penicillariae, 118
Tomato wilt pathogen, 417, 459
Toxin, bacterial, 27
  fungal, 417
Toxoptera graminum, 27

Trialeurodes abutilonea, 411
Trichodorus spp., 417
Trifolium hybridum, 366, 378
  incarnatum, 366, 367, 378
  medium, 378
  pratense, 359, 366, 368, 378
  repens, 366, 378
  spp., 342, 359, 366
Trigonella spp., 342, 359
  smut on, 359
Triodia flava, 325
Tripsacum laxum, 295
  spp., 105, 326
Trisetum spp., 325
Triticum aegilopoides, 225
  aestivum, 225
  compactum, 225
  dicoccoides, 225
  dicoccum, 225, 226, 253
  durum, 225, 235
  macha, 225
  monococcum, 225, 226, 235, 272, 279
  orientale, 225
  persicum, 225
  pyramidale, 225
  spelta, 225
  sphaerococcum, 225
  spp., 23, 64, 86, 225, 226, 230, 270, 273,
      312
  thaoudar, 225
  timococcum, 226, 272
  timopheevi, 225, 226, 235, 253, 272,
      279
  turgidum, 225
  vavilovi, 225
  vulgare, 225, 226, 235, 258, 279
Tylenchorynchus spp., 417, 445
Typha latifolia, 159
Typhula graminum, 252
  idahoensis, 251, **252**
  itoana, 251, **252**, 315
  spp., 302, 314
  trifolii, 373
Typhula blight, of clover, 373
  of grasses, 251, 302, 314
  of wheat, 251, 302

U

Urocystis agropyri, 184, 259, **261, 262,**
    316, 317
  occulta, **183**, 184, 261
  spp., 315
  tritici, 184, 259, 261–263, 316
Uromyces nerviphilus, 378
  spp., 326
  striatus, 359

Uromyces striatus medicaginis, 359
  trifolii, 378
    var. *fallens*, 378
    hybridi, 378
  trifolii-repentis, 378
Urophlyctis alfalfae, 345
  trifolii, 346
Ustilaginoidea oryzae, 156
  setariae, 156
  virens, **156**
Ustilago aculeata, 316
  avenae, 56, **58**, 59, 60, 134–138, 140
  bromivora, 320
  bullata, 320, 321
  crameri, 118–120
  crusgalli, **120**
  echinata, 316
  halophida, 318
  hordei, 56, 59–**62,** 135, 138–140
  hypodytes, 318, 319
  kolleri, 61, 135, 138–140
  levis, 61
  longissima, 316
  lorentziana, 320
  macrospora, 316
  maydis, 100–102
  medians, 55, 58, 134
  mulfordiana, 320
  neglecta, 118, **119**
  nigra, 55, 56, 58, 61, 134
  nuda, 52–55, 184, 252, 253
  nummularia, 318, 319
  perennans, 134
  scitaminea, 220
  spegazzinii, 318, 319
    var. *ayrestis*, 318
  spp., 52, 184, 315
  striiformis, 315, 317
  trebouxii, 320
  tritici, 184, 252, 253
  vavilovii, 184
  williamsii, 318
  zeae, 101

V

Vectors (*see* specific vectors, e.g., Aphid
  vectors)
Velvet bean diseases, 402, 404
Verticillium albo-atrum, **418**
  dahliae, 418
Verticillium wilt of cotton, 418, 419
Vicia spp., 359, 401, 403, 404
Vicia, diseases of, 359, 401
Vigna spp., 401
Vigna, diseases of, 401

Virus, characteristics of, 5, **445**
  history of diseases, 5, 207, 229, 445
  intracellular bodies associated with, 76,
    154, 201, 209, 211
  transmission of, 24, 229, 383, 384, 410,
    445, 451
Virus diseases, of *Agropyron*, 294
  of alfalfa, 340–342
  of barley, 24, 76, 126, 295
  of bean, 367
  of beets, 154, 294
  of brome, 294, 295
  of clover, 367
  of corn, 75–77, 207, 209, 228, 295, 296
  of cotton, 410, 411
  of cowpea, 401, 402
  of *Crotalaria*, 401
  of flax, 429
  of grasses, 24, 25, 75, 76, 155, 190, 207,
    209, 228, 295–296
  of lupine, 402
  of millets, 207, 295
  of oats, 26, 125, 126, 155, 295
  of pea, 367
  of peanut, 397
  of rice, 154, 207, 296
  of rye, 24, 155, 173
  of sorghum, 75, 190, 207
  of soybean, 383, 384
  of sugarcane, 75–77, 207–211, 295
  of sweetclover, 340, 341
  of teosinte, 77
  of tobacco, 207, 294, 402, 445–451
  of wheat, 24, 75, 126, 155, 173, 228
Viruses, *Agropyron* mosaic, 294
  alfalfa, mosaic, 340, 342, 367, 368
    stunt, or dwarf, 341, 368
    witches' broom, 341, 368, 401
    yellows, 342
  alsike clover mosaic, 342, 367
  aster yellows, 342, 368, 411, 429
  barley, stripe mosaic, **24**, 230, 296
    yellow dwarf, **25,** 126, 230, 296
  bean mosaics, 402
    yellow, 342, 367, 368
  beet curly top, 154, 429, 445
  broadbean, common mosaic, 368
    mild mosaic, 342, 368
  brome mosaic, 126, 294
  clover mosaic, 367
  corn, mosaic, 76
    mottle, 76, 77
    streak, 76, 209, 230, 296
    stripe, 76
    stunt, 76
    wallaby-ear, 77
  cotton, crumple, 411

Viruses, cotton, leaf curl, 410, 411
  mosaic, 410, 411
  stenosis, 411
Crotalaria mosaic, 401
cucumber mosaic, 76, 368, 445, 451
  celery strain, 76
  lily strain, 76
fiji, corn, sugarcane, 77, 210
lupine mosaic, 402
New Zealand pea streak stunt, 368
oat mosaic, 125
pea, common mosaic, 342, 367, 368
  mottle, 342, 367
  streak, 342, 367, 368
  stunt, 367
  wilt, 342, 367
peanut rosette, 397
potato yellow dwarf, 367, 368
red clover vein mosaic, 342, 367, 368
rice, stripe, 154, 296
  stunt, 154, 296
subterranean clover mosaic, 342, 367, 368
sugarcane, chlorotic streak, 76, 209, 295
  fiji, 210
  mosaic, 76, 207, 295
  ratoon stunt, 211
  ring mosaic, 211
  ring spot, 211
  sereh, 211
  streak, 76, 209
tobacco, common mosaic, 446, 447
  etch, 448
  leaf curl, 450
  mosaic, 445
  necrosis, 448
  rattle, 451
  ring spot, 342, 367, 402, 448, 449
  streak, 342, 449, 450
tomato spotted wilt, 397
tumor gall, 342, 368, 451
wheat, mosaic, 173, 228
  rosette mosaic, 173, 229
  streak mosaic, 126, 229, 295
  striate yellows, 229, 296
witches' broom, 341, 368
wound tumor, 342, 368
yellow bean mosaic, 368
yellow dwarf, alfalfa, 341, 368
  barley, 25, 126, 230, 295
  potato, 367

W

Weak neck of sorghum, 189
Wheat, chromosome number, 225
  disease losses, 227

Wheat, disease resistance, 226, 229, 230, 235, 253, 258, 263, 275, 278
Wheat diseases, anthracnose, 247
  bacterial blights, 27, 231, 232
  basal glume rot, 231
  black chaff, 231
  black point, 246
  brown necrosis, 231
  browning root rot, 232
  bunt, 225, **253**, 256
  corn streak virus, 230
  crown rot, **246**
  culm rot, 245
  downy mildew, 30, 83, 84, 86, **233**
  dwarf bunt, 254, 257, 263
  ergot, 37, **235**
  flag smut, **259**, 262
  foot rot, 227, 240, 242, 245
  frost injury, 227
  Fusarium blight, 34, 85, **236**
  Gibberella blight, 34, 85, **236**
  glume blotch, 248, 249
  glume rust, 64, 273
  gray speck, 123, 124, 228
  head blight, 34, 85, **236**, 240
  Helminthosporium, blight, 43, 159, 227, 232, **246**
    leaf spot, **247**
  heritable leaf spot, 228
  kernel smut, 253
  leaf necrosis, 228
  leaf rust, 225, **276**
  loose smut, 52, 54, 56, 57, **252**
  manganese deficiency, 228
  melanism, 231
  mineral deficiency, 228
  mosaics, 173, 228, **229**, 295
  nematode, 229, 232
  Neovossia smut, 257
  nonparasitic maladies, 228
  powdery mildew, 31, 225, 235
  pseudo-black chaff, 231
  Pythium root rot, 30, 232
  Rhizoctonia blight, 165, **250**
  rice dwarf, 155
  root rot, 232, 246, 247
  rust, 64, 225, 263, 273, **276**
  scab, 34, **236**
  Sclerotium blight, 252
  seedling blight, 85, 236
  Septoria, glume blotch, 52, **248**, 249
    leaf blotch, **248**, 249
  smuts, general, 225, 252, 253
  snow mold, **240**, 251
  stem rust, 64, 225, 227, **263**
  stinking smut, 253, 256
  streak virus, 76, 126, 228, 229, 295

Wheat diseases, striate virus, 229, **230,** 296
  stripe mosaic, 24, 230, 295
  stripe rust, 64, **273,** 275
  take-all, **242**
  Typhula blight, 251
  virus, 228, 229, 295
  winter killing, 227
  yellow dwarf, 27, 228, 230, 295
Wheat varieties, Albit, 260
  Apex, 246
  Axminister, 235
  Canus, 260
  Ceres, 271
  Chino, 166, 276
  Churl, 235
  Currell, 253
  Dixon, 235
  Eureka, 271
  Florence, 258
  Golden, 263
  Grüne Dame, 253
  H-44, 226, 231, 272
  Henry, 235
  Hohenheimer, 258, 260
  Hope, 226, 231, 235, 253, 258, 271, 272, 279
  Huron, 235
  Hussar, 258, 260
  Jubilee, 253
  Kanred, 253, 272
  Kawvale, 253
  Kenya, 272
  Kenya Farmer, 272
  Kolben, 228
  Leap, 253
  McMurachy, 272
  Marquillo, 272
  Marquis, 48, 230, 260
  Martin, 258, 260
  Mida, 235
  Mindum, 260
  Norka, 235
  Oro, 258, 260, 263
  Oro × Federation, 263
  Preston, 253
  Redman, 24
  Regent, 230
  Renown, 230
  Reward, 230
  Rex, 263
  Ridit, 260
  Rio, 258, 263
  Rival, 235
  Sonora, 235
  Thatcher, 235, 246
  Thorne, 253

Wheat varieties, Timstein, 272
  Trumbull, 253
  Turkey, 229, 237, 253, 258
  Ulka, 260
  White Odessa, 260
White fly, of cotton, 410, 411
  of tobacco, 450
White tip of rice, 154
Wildfire, of soybean, 385
  of tobacco, 452
Wilt (*see* Bacterial wilt; *Fusarium* wilt; *Verticillium* wilt)
Winter injury, of alfalfa, 13, 336
  of clover, 366
  of wheat, 227
Witches' broom of alfalfa, 340
Wojinowicia graminis, 243
Wound response in roots, 337

X

Xanthomonas axonoperis, 298
  holcicola, 190, 191
  itoana, 155
  lespedezae, 402
  malvacearum, 411–414
  oryzae, 155
  panici, 115
  phaseoli var. *sojense,* 386, 402
  rubrilineans, 213, 214, 298
  rubrisubalbicans, 214
  solanacearum, 397, 402, 451, **452**
  translucens, 27, **28,** 128, 231, 296, 297
    f. sp. cerealis, 27, 296, 297
    f. sp. phleipratensis, 296
    f. sp. secalis, 173, 296
    f. sp. undulosa, 231
  vasculorum, 79, 80, 211, **212,** 213, 298
  vignicola, 402
Xiphinema spp., 417

Y

Yellow leaf blotch of alfalfa, 355
Yellows, of alfalfa, 339
  of clover, 366, 367

Z

Zea mays, 74, 86
  var. *amylacea,* 74
  everta, 74
  indentata, 74
  indurata, 74
  saccharata, 74
  tunicata, 74
Zinc deficiency, 125, 340, 429
Zizania aquatica, 169, 318